The Cell Division Cycle

Temporal Organization and Control of Cellular Growth and Reproduction

The Cell Division Cycle

Temporal Organization and Control of Cellular Growth and Reproduction

David Lloyd
Professor of Microbiology
University College, Cardiff, Wales

Robert K. Poole
Lecturer in Microbiology and Nuffield Foundation Science Research Fellow
Queen Elizabeth College, University of London

Steven W. Edwards
University of Wales Fellow
University College, Cardiff, Wales

1982

Academic Press
A Subsidiary of Harcourt Brace Jovanovich, Publishers
London New York
Paris San Diego San Francisco
São Paolo Sydney Tokyo Toronto

ACADEMIC PRESS INC. (LONDON) LTD.
24/28 Oval Road
London NW1

United States Edition published by
ACADEMIC PRESS INC.
111 Fifth Avenue
New York, New York 10003

British Library Cataloguing in Publication Data

Lloyd, D.
The cell division cycle.
1. Cell division
I. Title II. Poole, R.K. III. Edwards, S.W.
574.87′62 QH605
ISBN 0-12-453760-X
LCCCN 81-67918

Filmset in Plantin and printed by Page Bros (Norwich) Ltd.

Preface

Whenever information is obtained from a population of organisms, it represents the result of many individual contributions. Averaging occurs if the population is heterogeneous. If the cells are all of one type, or if the organism has been grown in pure culture, the resulting population is often mistakenly referred to as homogeneous. Such a population is the usual starting material for the classical biochemical methodology of analysis. But because cellular growth and division are to a large extent discontinuous processes, the philosophy underlying this approach is oversimplistic. Data obtained in this way are time-averaged over an interval equal to the cell division time, the fine details of sequenced events that comprise the cell division cycle being hidden. Oversimplification is also implicit in the microbiologists' usual reference to "growth", which does not distinguish between growth of individuals or population growth, and which usually means a combination of both. A preoccupation with "biomass" is symptomatic of this attitude.

However, the past decade has seen an increasing awareness of the rich complexity of sequentially-ordered processes that characterize the cell division cycle, and the elaborate network of controls that ensures faultless cyclic progression, generation after generation. Continuing growth of the horizons of our awareness of these processes relies increasingly on cooperation between cell biologists, geneticists, biochemists and biophysicists. This book harvests some notable recent successes.

We begin by orientating the time domain occupied by cell cycle-dependent events within the larger vistas of time that characterize biological reactions and events. As a rule, open systems often exhibit oscillatory behaviour; the cell cycle is itself a type of oscillation and insights into its mechanisms come from considering it as such. Later chapters detail the special methods used to probe time structure of cells, and a description of DNA replication, transcription and synthesis of proteins and other cellular components, leads to a discussion of the time course of the assembly of multicomponent substructures, membranes, organelles and cell walls. The organization of the energy supply for these feats of biosynthesis is also described, as is the invaluable information obtained from that special class of mutants, the cell division cycle mutants, which under nonpermissive conditions cannot complete the natural cycle of events. Finally we highlight the far-reaching possibilities for application of these fundamental researches to problems of cytodifferentiation and cancer therapy.

The literature search for information presented in this volume was completed on 31.12.1980.

Acknowledgements

The authors wish to thank Miss R. H. Matthews, Miss B. Parker, Mrs M. Adams, Mrs W. Stott, Mrs J. Williams, Mrs J. Phillips and Miss D. Brunskill for secretarial help. Thanks are also due to Mrs M. Lloyd and Dr J. C. Poole for checking references and index, to Drs B. W. Bainbridge, H. Degn, L. F. Olsen, J. W. Pollard, C. F. Thurston, A. P. J. Trinci, W. A. Venables and A. T. Winfree for reading parts of the manuscript and providing helpful suggestions, and to all those who provided reprints and unpublished information.

Contents

When a thing was new, people said 'It is not true'.
Later when its truth became obvious, people said
'Anyway it is not important', and when its importance
could not be denied people said, 'Anyway it is not new'.

William James (1842–1910)

I. The Time Domains of Living Systems

"What are called structures are slow processes of long duration, functions are quick processes of short duration"

von Bertalanffy
Problems of Life

I. Introduction: Spatial and Temporal Hierarchies

In order to study the cell division cycle (Edwards, 1981), it is necessary to consider all those processes that occur while a single cell grows and eventually divides. Any integrated view of the multiplicity of biochemical reactions involved in these processes is not possible without consideration of the various levels of organization which obtain in both space and time (von Bertalanffy, 1952).

Investigation of the organization of the spatial hierarchy has been a major preoccupation of biologists for more than 50 years, and has resulted in a detailed appreciation of ultrastructure from the submolecular to the cellular level. Based primarily on the complementary approaches of subcellular fractionation and separation of individual macromolecular species, membranes and organelles, and the increasingly powerful ultrastructural analyses made possible by a battery of physical techniques, a clear picture of subcellular architecture has emerged. Progressing upwards through the structural hierarchy we encounter small molecules diffusing freely and rapidly through intracellular aqueous micro-environments but constrained by specific permeability barriers into discrete subcellular compartments. Macromolecules (mol. wt $> 10\,000$) occur both in solution and associated to a varying degree of intimacy with the phospholipid–protein hydrophobic membranes; these macromolecules are often constructed from subunits in multimeric or oligomeric structures. The membranes often enclose distinct organelles of which more than a dozen different types have been characterized both structurally and functionally. The cell is bounded by the cytoplasmic membrane, which is the real functional interface with the environment, although it may itself be enclosed within other structures (pellicle or cell wall).

This static picture is divorced from the dynamic perspective which is especially relevant in the study of a growing cell. The temporal hierarchy stretches through from fast reactions proceeding independently of diffusion, through "metabolic time" and "epigenetic time" to the cell division cycle (genetic) time domain, which can itself overlap

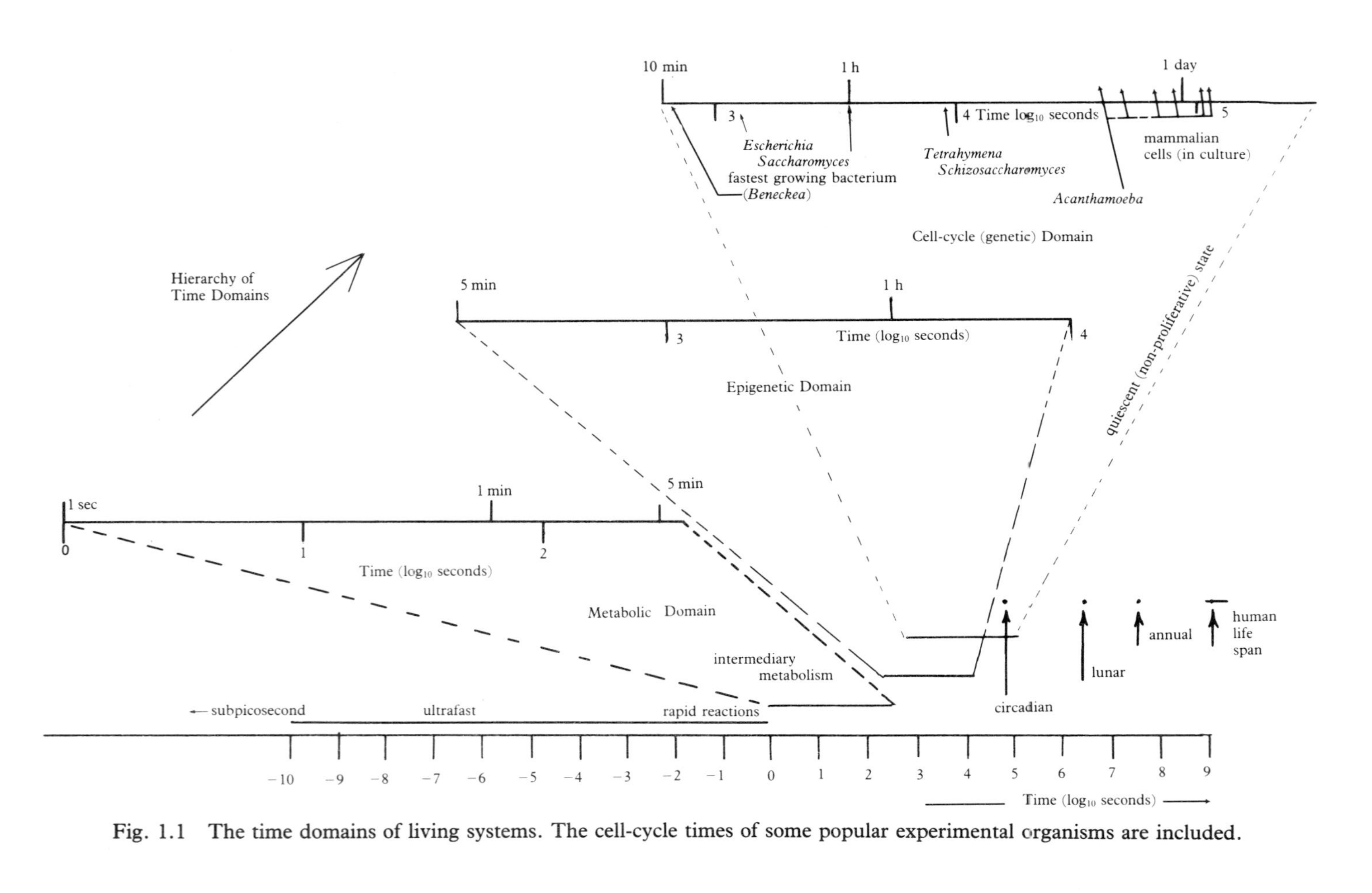

Fig. 1.1 The time domains of living systems. The cell-cycle times of some popular experimental organisms are included.

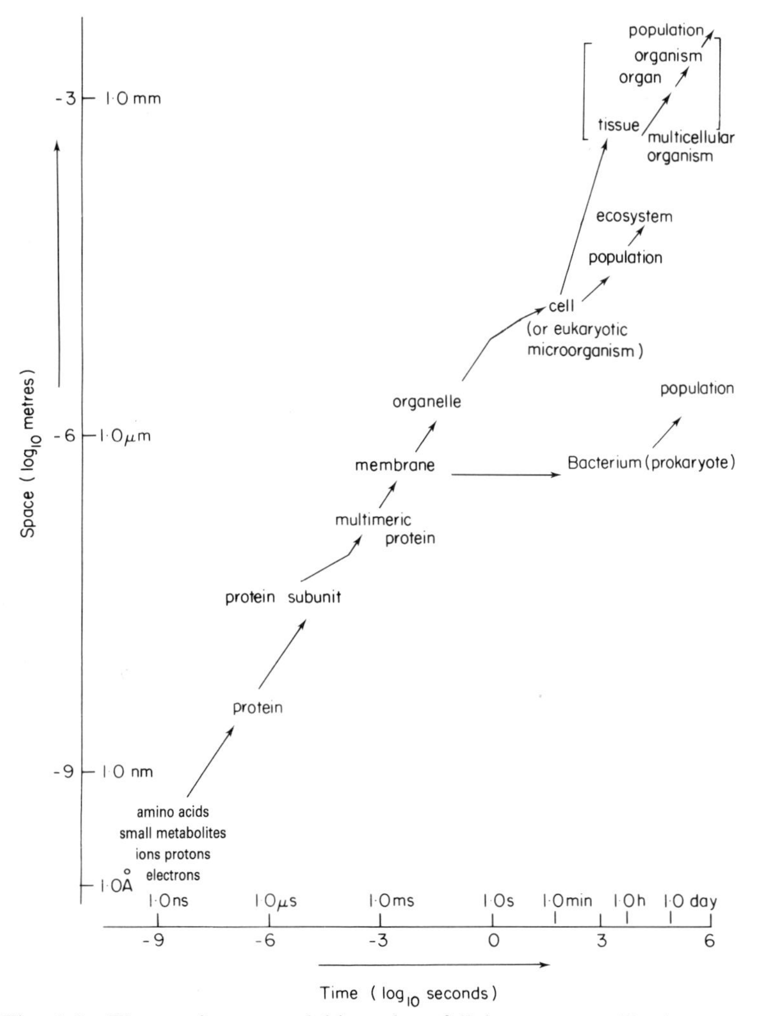

Fig. 1.2 The spatio-temporal hierarchy of living systems. During traverse of a diagonal, we ascend the hierarchies of space and time, from rapidly-occurring processes involving small components (function), through successively more highly organized structures which may be studied on short or long time-scales (structure).

with the circadian time domain (Fig. 1.1). Spatial and temporal hierarchies may also be represented on Cartesian coordinates (Fig. 1.2); this presentation emphasizes the continuum more usually dissected into "function" (fast reactions) and "structure" (slow reactions).

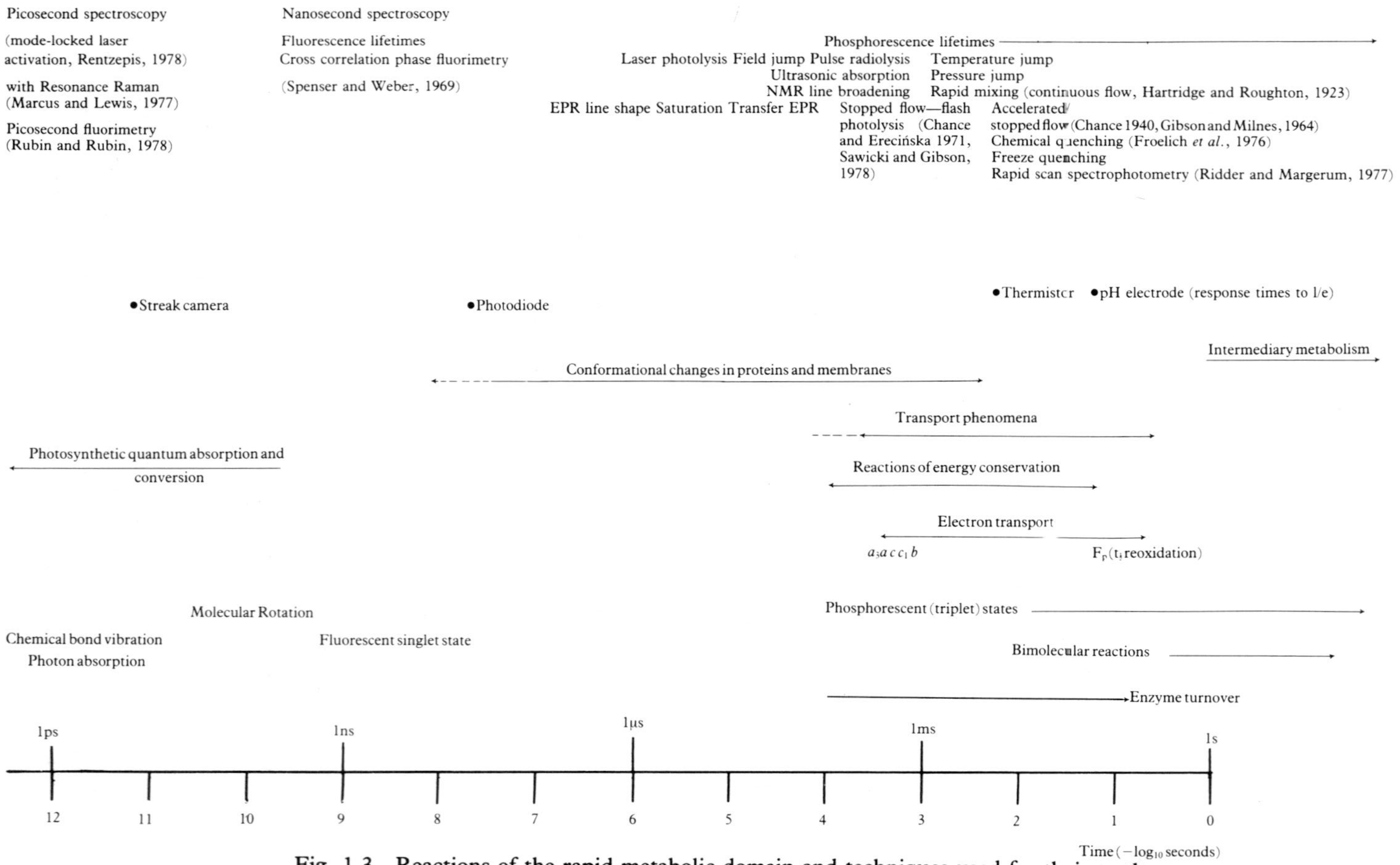

Picosecond spectroscopy

(mode-locked laser activation, Rentzepis, 1978)

with Resonance Raman (Marcus and Lewis, 1977)

Picosecond fluorimetry (Rubin and Rubin, 1978)

Nanosecond spectroscopy

Fluorescence lifetimes

Cross correlation phase fluorimetry

(Spenser and Weber, 1969)

Phosphorescence lifetimes

Laser photolysis Field jump Pulse radiolysis

Ultrasonic absorption

NMR line broadening

EPR line shape Saturation Transfer EPR

Stopped flow—flash photolysis (Chance and Erecińska 1971, Sawicki and Gibson, 1978)

Temperature jump

Pressure jump

Rapid mixing (continuous flow, Hartridge and Roughton, 1923)

Accelerated stopped flow (Chance 1940, Gibson and Milnes, 1964)

Chemical quenching (Froelich *et al.*, 1976)

Freeze quenching

Rapid scan spectrophotometry (Ridder and Margerum, 1977)

•Streak camera

•Photodiode

•Thermistor

•pH electrode (response times to 1/e)

Intermediary metabolism

Conformational changes in proteins and membranes

Transport phenomena

Photosynthetic quantum absorption and conversion

Reactions of energy conservation

Electron transport

$a_3 a c c_1 b$

F_p ($t_{\frac{1}{2}}$ reoxidation)

Molecular Rotation

Chemical bond vibration

Photon absorption

Fluorescent singlet state

Phosphorescent (triplet) states

Bimolecular reactions

Enzyme turnover

1ps 1ns 1μs 1ms 1s

12 11 10 9 8 7 6 5 4 3 2 1 0

Time ($-\log_{10}$ seconds)

Fig. 1.3 Reactions of the rapid metabolic domain and techniques used for their study.

II. Rapid Reactions in the Metabolic Time Domain

The most rapidly-occurring biological reactions that have been measured (Fig. 1.3) occur on a picosecond (10^{-12} s) time scale, e.g. mode-locked laser activation of the primary events of photosynthesis and vision (Busch and Rentzepis, 1977). Fluorescence lifetime measurements provide information in the nanosecond (10^{-9} s) range. Weber (1976) has pointed out why we should be interested in these extremely short time ranges:

> "The diffusion of a small metabolite such as glucose over a distance of 1 nm, takes place in about 1 ns. Thus the environment of a macromolecule changes appreciably, through the diffusion of the small molecules in its vicinity, in this length of time. In 100 ps the small metabolite would diffuse over a distance smaller than its own molecular radius, so that in periods of time shorter than this the macromolecular environment may be considered constant. We must conclude that to understand the elementary changes in molecular disposition that must precede any physiologically significant change, we need to know the molecular dynamics over periods of a fraction of a nanosecond but no shorter than that."

Measurements of the rotational and lateral mobilities of proteins and phospholipids in membranes by means of techniques that use pulsed laser excitation of fluorochromes (Cone, 1972), or electron paramagnetic resonance probes (Scandella *et al.*, 1972), give information in the μs (10^{-6} s) range, and suggest that a phospholipid can migrate laterally along the length of the bacterial cytoplasmic membrane (1 μm) in less than 1 ms (10^{-3} s). Electron transfer from the primary acceptor system of bacterial chromatophores to quinone pools within the membrane and concomitant proton uptake occurs in 200 μs (Petty and Dutton, 1976).

Much more information is available over the ms time scale (10^{-3} s), as many different rapid-reaction techniques for the study of biological systems in this range have been developed (Chance *et al.*, 1964b). Thus the half-time for the oxidation of cytochrome *c* oxidase, the terminal member of the mammalian mitochondrial electron transport chain, at body temperature is about 0·3 ms. The most slowly re-oxidized components of the reduced chain are the flavoproteins with half-times of the order of 200 ms. The associated changes of membrane potential and transmembrane pH gradients which drive ATP synthesis must also occur in this range of time scales. Decreases in reaction rates by a hundredfold accompany a reduction in temperature to -30°C; this approach to the study of rapid reactions has been perfected by

Douzou (1977) and has been especially valuable for prolonging the lifetimes of transient enzyme–substrate intermediates.

How relevant are the studies of rapid reactions in the context of cell cycle-related phenomena? Discontinuous expression of membrane-associated components implies a changing stoichiometry of these components within a membrane and consequently the reaction kinetics of a multi-enzyme complex might be quite different in cells at different cell-cycle stages. Changes in the fluidity of the surface membrane of mammalian cells during the cell cycle have already been observed (Fox *et al.*, 1971; de Laat *et al.*, 1980). Displacement instabilities in membranes indicate a time scale of 0·1 s for vesicle formation (Crum *et al.*, 1979).

III. Slower Reactions in the Metabolic Time Domain

The metabolic system consists of the reaction sequences and cycles of intermediary metabolism; the major processes that determine rates of change are diffusion, interaction and enzyme-catalysed transformation of small molecules. Activation and inhibition by effectors play a major role in the control systems operating in this system. The complete metabolic map comprises a network of as many as 10^4 or 10^5 kinds of interactions. As yet, only single enzymes and rather small pathways have been extensively characterized by kinetic methods, and the more integrated studies necessary for an understanding of the complete system are not yet possible. An approach to this problem by structural simplification of the mathematical model of the system has been outlined by Reich and Sel'kov (1974).

The time course of establishment of steady states in the metabolic system is determined by the turnover rate of substrate molecules (10^2–10^5 molecules/enzyme molecule/s). Using an average enzyme turnover number of $10^3\ s^{-1}$ and an estimated intracellular enzyme concentration of 10^{-8} M, it is evident that a period of a second can give a measurable change in substrate concentration. It can also be shown that the relaxation time for a typical reaction of the metabolic system (the time taken for the effect of a perturbation to have decreased to e^{-1} of its initial value) is of the order of 2 min (Goodwin, 1976). Over this time-interval it can be reasonably assumed that enzyme concentrations will have remained constant.

Studies of the molecular basis of metabolic control have provided detailed insights into the phenomena of feedback inhibition (Umbarger, 1961), the allosteric behaviour of enzymes (Monod and Jacob,

1961), nucleotide coupling of pathways, multi-site control, and cooperative enzyme kinetics (Monod *et al.*, 1965). These principles allow an understanding of the control circuits operating in the metabolic network. Perhaps the most studied example of the molecular dynamics of a metabolic pathway is that of glycolysis (Fig. 1.4). The first

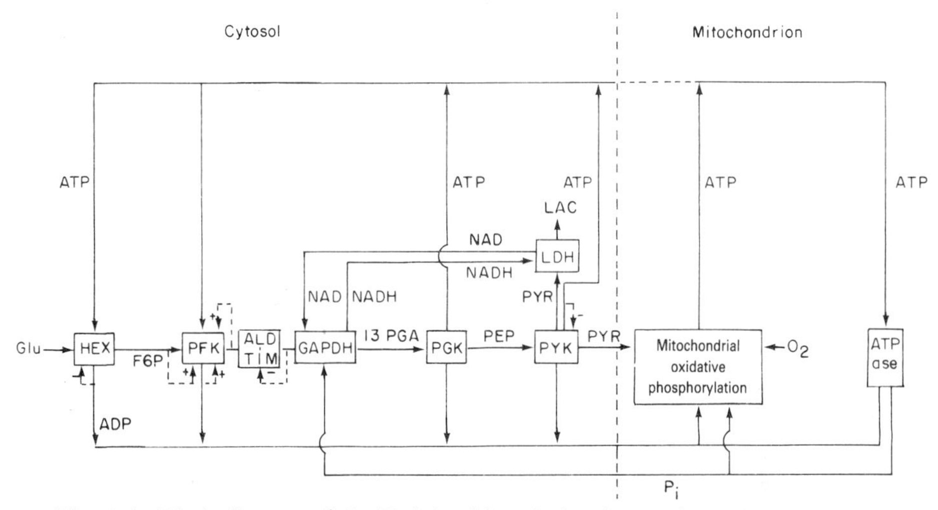

Fig. 1.4 Block diagram of the Embden-Meyerhof pathway of glycolysis and interactions with the mitochondrial system of ATP generation. (Reproduced with permission from Hess, 1968.)

suggestion of the inherent instability of metabolite levels *in vivo* came from the observations of an overshoot following glucose addition to a starved yeast suspension (Chance, 1955); and of oscillatory transients of the photosynthetic carbon cycle (Wilson and Calvin, 1955). Since that time, a variety of waveforms have been observed, including damped and persistent sinusoids (Fig. 1.5), both in populations of cells and in individual organisms (Chance *et al.*, 1973a). The pool sizes of all the intermediates of glycolysis have been shown to oscillate in the range 10^{-5} to 10^{-3} M and with periods of the order of minutes.

The oscillating state is a consequence of (a) the nonlinear kinetics of the reactions involved which is provided by a multiplicity of cyclic interactions including positive and negative feed-backs (Franck, 1978) and (b) the system being maintained far from equilibrium (Glansdorff and Prigogine, 1971; Prigogine and Nicolis, 1971; Nicolis and Prigogine, 1977; Prigogine, 1980) by continuous supply of the reactants and removal of the products (Degn, 1972; Nicolis and Portnow, 1973).

Furthermore, the oscillatory state where the variables (i.e. metabolite concentrations) orbit on a stable-limit cycle surrounding an unstable steady state is only one of several states predicted from theory (Andronov *et al.*, 1966; Minorsky, 1962) and observed experimentally: e.g. multiple steady states, limit cycles or a combination of both as well as non-periodic oscillatory behaviour, usually referred to as chaotic behaviour (Gurel and Rössler, 1979; Degn *et al.*, 1979).

The dynamics and mechanism of glycolytic oscillations have been analysed in detail in cell-free extracts and has led to the identification

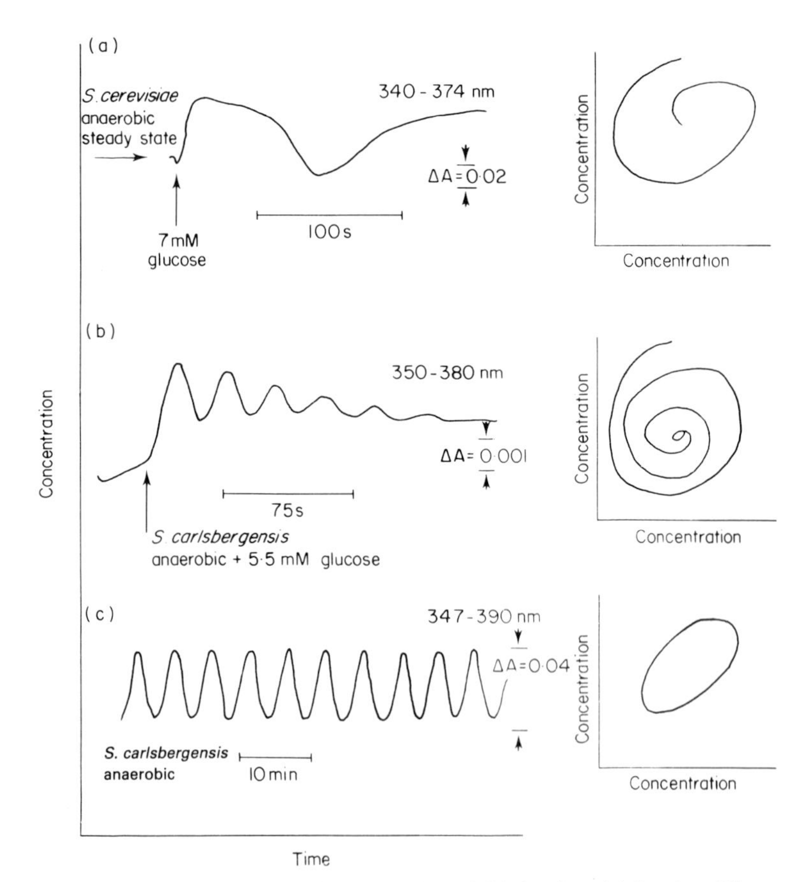

Fig. 1.5 Waveforms generated in yeast (a) and (b) *in vivo*, (c) *in vitro*. The traces are continuous readout of NADH concentration; right hand panels show associated phase–plane behaviour: (a) Overshoot–undershoot; (b) Damped oscillatory; (c) Sustained sinusoid (limit cycle). (Redrawn from Chance, 1955; Chance *et al.*, 1964a; Chance *et al.*, 1967.)

of the "primary oscillophore" of the system as the enzyme phosphofructokinase (Boiteux *et al.*, 1975); this conclusion is supported by computer studies of models of glycolytic oscillations (Higgins, 1967; Sel'kov, 1968; Chance *et al.*, 1973b; Goldbeter and Nicolis, 1976). The temperature dependency of the oscillation of NADH in glycolyzing yeast extracts between 19 and 40°C has been investigated and has given a Q_{10} of between 2·2 and 4·0 (Hess *et al.*, 1966).

Oscillations in mitochondrial functions have also been observed; thus periodicities in ion transport, respiration and redox states of NADH, flavoprotein and cytochrome *b* accompany those in mitochondrial volume (Fig. 1.6) and ATP pools (Gooch and Packer, 1974).

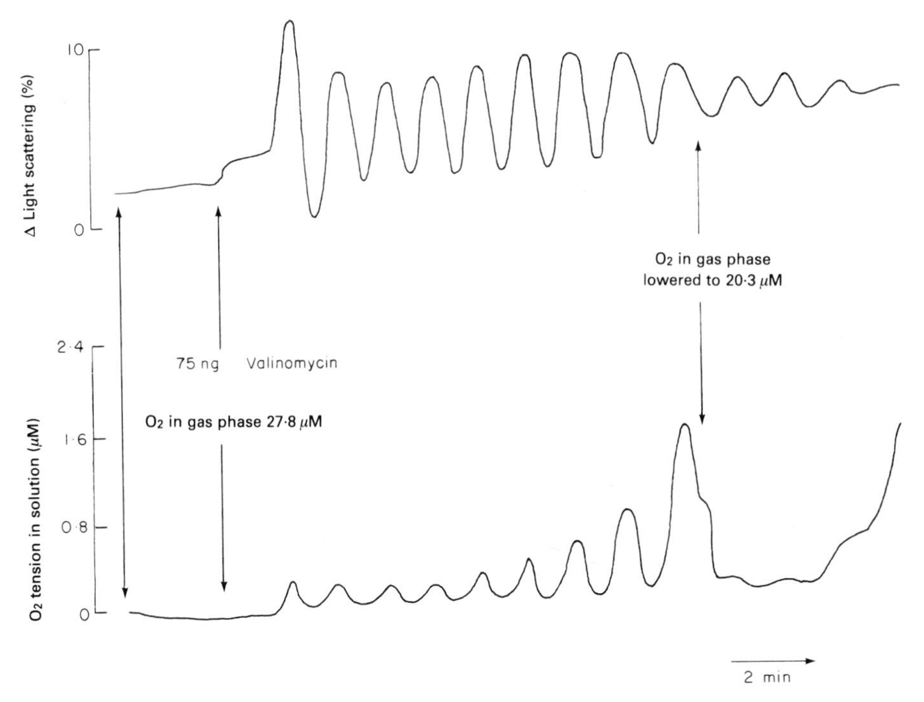

Fig. 1.6 Oscillations in mitochondrial functions. Record of mitochondrial volume changes (monitored by light scattering) and oxygen tension in a suspension of pigeon heart mitochondria. (Reproduced with permission of A. Boiteux and H. Degn.)

While these phenomena provide valuable data on the dynamic organization of metabolic processes, no clear physiological function for oscillatory behaviour in the metabolic time domain has been established. A possible function may result from the necessity to separate incompatible reactions catalysed by enzymes found in a single subcellular compartment; the only means by which this can be achieved is to have the reactions occurring at different times (Sel'kov, 1972). For instance, the so called "futile cycles" of intermediary metabolism in which two metabolites are interconverted by two different irreversible reactions may not operate *in vivo*.

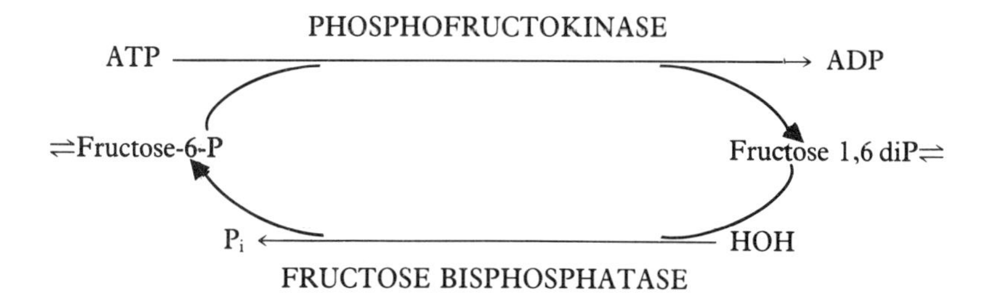

The apparently useless hydrolysis of ATP predicted as a net consequence of these reactions is suppressed by allosteric regulation which switches the antagonist enzymes on and off reciprocally and leads to self-oscillatory changes in enzyme activities (Sel'kov, 1980).

IV. The Epigenetic Time Domain

In the epigenetic time domain we are concerned with the biosynthesis, processing, transport, and interaction of macromolecules, i.e. with the consequences of gene expression and altered activities of genes. The classical studies of Monod on the process of adaptation in bacteria led eventually to the model of the *lac* operon (Jacob and Monod, 1961) which explains the control of transcription by a protein repressor, itself a product of a regulator gene (Fig. 1.7). It has been estimated that the time taken for the synthesis of a single protein molecule is about 5 s in bacteria (McQuillan *et al.*, 1959), and the lag between the addition of inducer and the appearance of measurable enzyme activity is about 5 min (Pardee, 1962). The molecular basis of the catabolite repression exerted by glucose via lowered levels of 3′–5′ cyclic AMP also involves the binding of a protein (cyclic AMP receptor protein) to a gene, in this case the promoter, so that the feedback loop again involves transcriptional control (Riggs *et al.*, 1971). Several operon control circuits in bacteria are now extensively documented (Lewin, 1974); the best characterized are those involved in the synthesis of

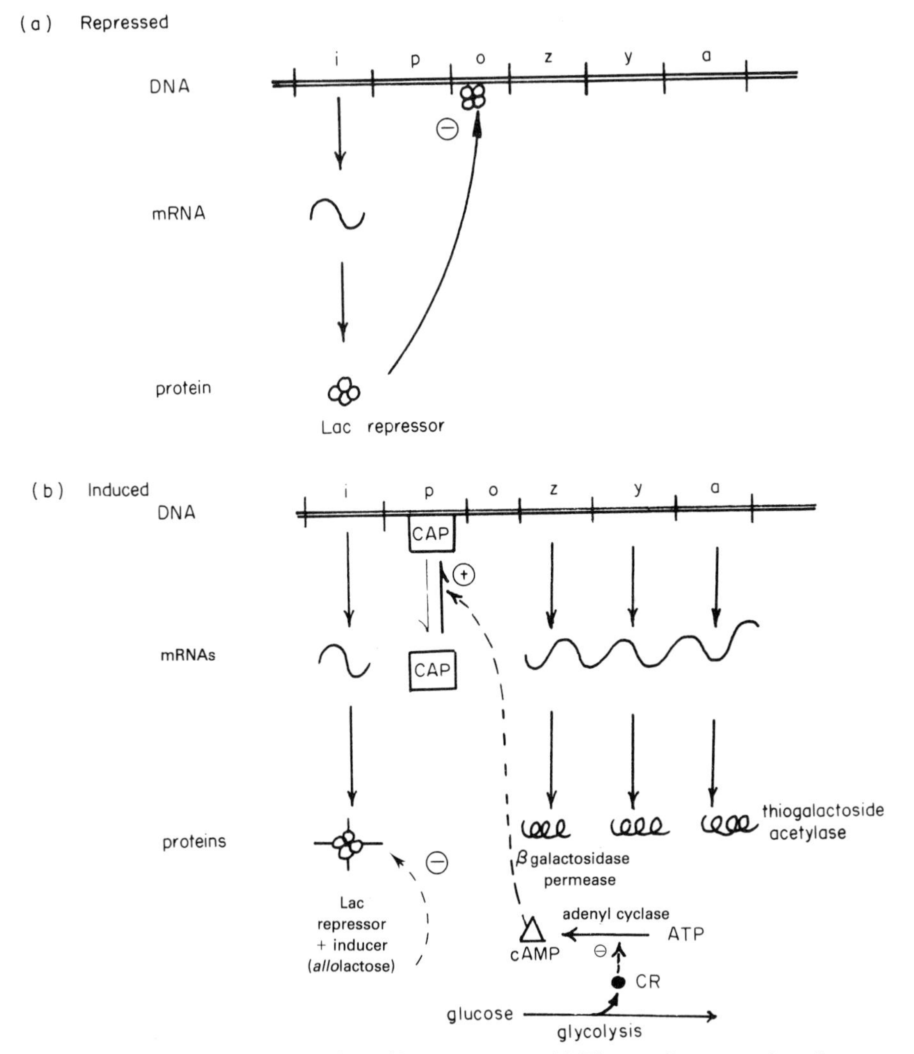

Fig. 1.7 Model for the regulation of lactose operon; (a) The regulator gene *i* produces a repressor which combines with the operator to prevent transcription; (b) an inducer (e.g. *allo*lactose) combines with the repressor removing it from the operator allowing transcription of the structural genes *z*, *y* and *a*. The promotor p is essential for the initiation of transcription, as is bound catabolite activator protein (CAP); this binding of CAP requires 3′,5′-cyclic AMP. Under conditions of high glycolytic activity catabolite repressor (CR) inactivates adenyl cyclase.

histidine and arginine, and the metabolism of arabinose. The control of the synthesis of ribosomal RNA in bacteria by the stringent factor protein and guanosine tetraphosphate (Pedersen *et al.*, 1973) also involves well-defined feedback loops.

The negative and positive control circuits operating in bacterial transcription, form the basis for models of the control of gene expression in eukaryotes (e.g. Britten and Davidson, 1969) but the higher levels of organization of the DNA in the latter, the lack of functional gene-clustering, the presence of split genes and the association with histone proteins in chromatin, are all factors which suggest substantial differences. Transcriptional and translational delays are longer in eukaryotic systems, and messenger RNA species are often stable. Nevertheless, the epigenetic system may still be regarded as having relaxation times within the range 10^2–10^4 s (1·5 min to 3 h), the time varying according to the type of cell under study (Goodwin, 1963).

V. The Cell Division Cycle (Genetic) Domain

The cell cycle is characterized by a recurrent sequence of discrete events, Edwards (1981): DNA replicates and growth leads to an overall approximate doubling in amount of each cellular constituent prior to segregation at cell division (Fig. 1.8). The time domain of the cell cycle is thus a higher order of the temporal hierarchy, and nesting within it are the ultrafast, the metabolic, and the epigenetic domains (Fig. 1.9).

The time taken for the completion of a cell cycle can be less than 10 min for some prokaryotes growing under ideal conditions, e.g. *Benekea natriegens* (Eagon, 1962). The shortest generation times reported for lower eukaryotes are about one hour, e.g. 57 min for the water mould *Achlya bisexualis* (Griffin *et al.*, 1974), 66 min for the fungus *Geotrichum candidum* (Trinci, 1972), and 1·7 h for the amoeboflagellate *Naegleria gruberi* (Fulton, 1977). The cell cycles in amphibian embryos can be as short as 15 min (e.g. in *Xenopus*, Graham and Morgan, 1966) but in mammalian embryos they last much longer (e.g. 10–20 h in mice, Graham, 1961); these cycles are, however, atypical in that cell division in early embryos occurs without net growth. Generation times for cultured mammalian cells lie mostly within the range 7·7 to 30 h (Klevecz, 1976). Cell cycle times also fall in this range in those tissues of adult mammals which are continually being renewed (Prescott, 1976a).

Cell cycle times depend on nutritional sufficiency and on temperature, and are thus evidently subject to control by external environ-

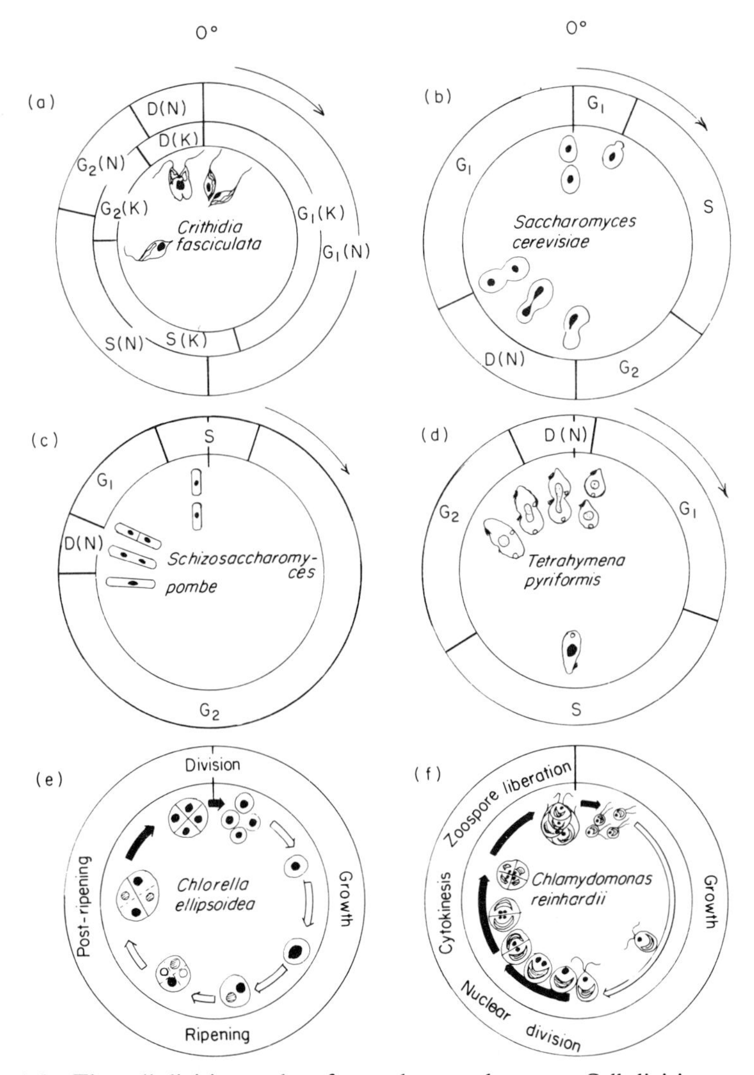

Fig. 1.8 The cell-division cycles of some lower eukaryotes. Cell division occurs at zero degrees and traverse is clockwise in all cases: (a) *Crithidia fasciculata* showing events occurring in nucleus (N) and kinetoplast (K). (b) *Saccharomyces cerevisiae*; (c) *Schizosaccharomyces pombe*; (d) *Tetrahymena pyriformis* showing nuclear events, division of contractile vacuole and stomatogenesis; (e) *Chlorella vulgaris*; (f) *Chlamydomonas reinhardii*. In (e) and (f) white arrows indicate processes in the light period, and black ones those in dark period. D (N) = nuclear division; D (K) = kinetoplast division.

mental influences. However, the individual cells within a culture require widely different time intervals to complete their cell cycle traverse under constant conditions. This is so for bacteria (Koch and Schaechter, 1962; Kubitschek, 1971), yeasts (Nurse and Thuriaux, 1977; Fantes and Nurse, 1977), algae (Cook and Cook, 1962) and for

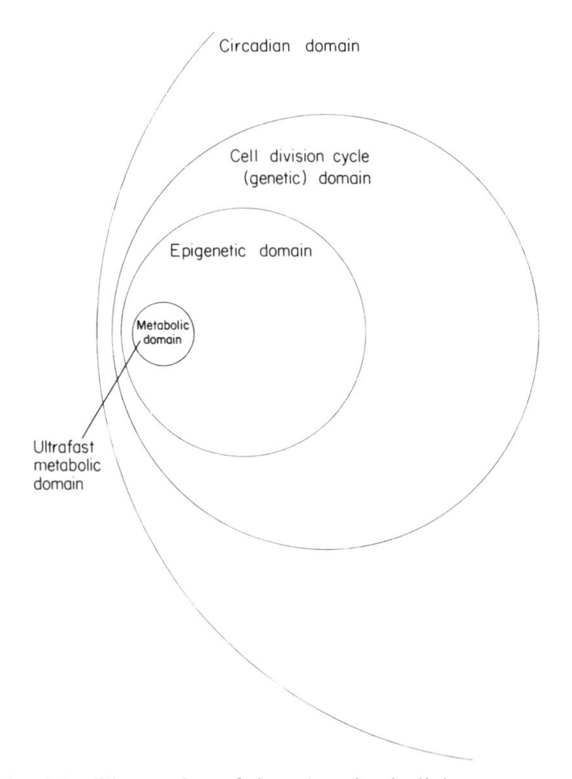

Fig. 1.9 The nesting of time domains in living systems.

animal cells in culture (Peterson and Anderson, 1964; Sisken and Morasca, 1965; Fox and Pardee, 1971; Smith and Martin, 1973). This variability of cycle time within a genetically homogeneous population places a limit on the degree of synchrony which can be experimentally imposed and maintained within a culture, and has important implications in mechanisms of control of the timing of cell division. Because the cell cycle is the product of interaction of a large number of regulated sub-systems which are loosely coupled to one another, variability in its duration is inevitable, and it has been suggested that it is not necessary to invoke probabilistic factors (Engelberg, 1968).

However, observations of the wide distribution of cell generation times of mammalian cells in cloned cultures (as much as 20%) leads to the alternative view, that at least one critical step in cell cycle traverse, perhaps dependent on the action of a small number of molecules, provides a transition probability (Smith and Martin, 1973).

A. CONTROL OF CELLULAR GROWTH AND DIVISION

The many models which have been proposed to account for the control of the timing of cellular growth and division are discussed more fully in Chapter 8, but can be conveniently considered to fall into three basic classes:

1. *The Transition Probability Model*

This model (Fig. 1.10a) proposes that the cell cycle has two parts, an indeterminate part (A state) and a determinate part (B state) (Smith and Martin, 1973). The spread of cell generation times (which is observed even between sibling cells, Minor and Smith, 1974) is accounted for by a random step from the A state to the B state which occurs at some time in G_1. This transition may depend on a critical amount of a single compound regulated by a number of coupled feedback loops; the random fluctuations in achieving the transition might be accounted for by the suggestion that the active species is rare and diffuses slowly. After the transition, the remainder of the cell-cycle (B state) proceeds by a programmed sequence of DNA duplication and cell division. Thus the probabilistic event is the key regulatory step that controls cell proliferation rates. This proposal is in accord with the widely-observed invariance of the duration of ($S + G_2 + M$) for given organisms and cell lines (Prescott, 1976a,b): the variability of cell cycle times is confined to variability in the duration of G_1.

Another observation that relates to this model is due to Pardee (1974a) who points out that the two alternative modes of existence of animal cells *in vivo*, proliferative or quiescent, also apply to cells in culture (Fig. 1.10). Thus, suboptimal nutritional conditions (high cell density, nutrient or serum insufficiency, high cAMP) can bring about quiescence: this state is associated with the production of phosphorylated proteins (Kletzien *et al.*, 1977). Several diverse blocks to proliferation (deprivation for isoleucine, glutamine, serum or phosphate, elevation of cAMP levels by several stategies, and inhibition by cytochalasin B) put cells into the *same* quiescent state, in that cells escape at the same point in G_1 when nutrition is restored (see, however, Moses *et al.*, 1980; Rossow *et al.*, 1979). The specific time in the cell cycle when this critical release event occurs is the "restriction point". Malignant cells are proposed to have lost their restriction point control. The analogy between this restriction point, the probabilistic event of Smith and Martin (1973) and the "start" point of the yeast cell cycle (Hartwell *et al.*, 1974) has been stressed by Shilo *et al.* (1976, 1977).

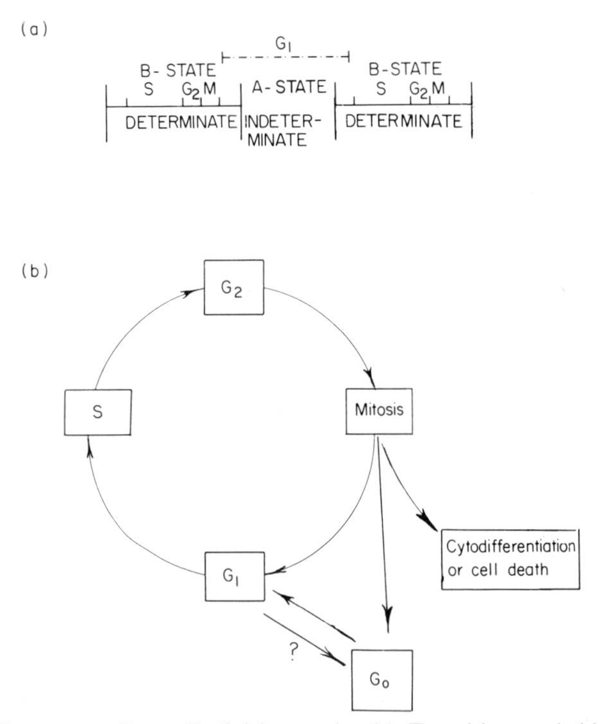

Fig. 1.10 The mammalian cell division cycle. (a) Transition probability model (reproduced with permission from Smith and Martin, 1973). (b) G_0 is the state into which cells are postulated to move when the cycle is arrested in G_1 by various environmental conditions. Irreversible transitions to terminal cytodifferentiation or cell death are also shown (redrawn from Lajtha, 1963 and Prescott, 1976a.)

However, the identification of "start" as the rate-limiting step of the cell cycle has objections (Nurse and Fantes, 1977), and Wheals (1977) has shown that the data on bud-emergence is open to several interpretations and does not strictly obey the first-order kinetics demanded by the transition probability model. The possible presence of subpopulations of cells with long doubling times also complicates the interpretation of some data that have been used as evidence for this model (Koch, 1980).

Further evidence for the validity of the transition probability model has come from the observation of the exponential nature of the distribution of sister cell times which is a unique feature of cell cycle models containing an exponentially distributed phase (Shields, 1977, 1978, 1979; Shields *et al.*, 1978). That the G_1 phase has a much lower Q_{10} (= 1·3 between 22 and 34°C) than the other cell cycle phases in suspensions of higher plant cells also points to a regulatory step involving diffusion of a relatively rare molecular species (Gould, 1977). Direct evidence for the applicability of the transition probability model *in vivo* has come from studies with hepatocytes (Domingo *et al.*, 1978).

A kinetic model of the cell cycle postulating a probabilistic event as being *solely* responsible for entry into the S phase may be too simple (Yen and Pardee, 1979). An approximately twofold variation in nuclear volume (with uniform DNA content) was observed in Swiss 3T3 cells arrested in G_0 (the quiescent state, Lajtha, 1963; Burns and Tannock, 1970) after lowering the serum content of the medium. The earliest cells to embark on DNA synthesis were those with the largest nuclei, whereas cells with the smallest nuclei were among the latest. As all G_0 cells did not have equal probabilities of entry into S at the given moment, the regulation of the transition from G_0 to S was partly deterministic. It was also shown that all cells having the same nuclear volume did not initiate DNA synthesis at the same moment, so that factors other than nuclear volume must also be involved. An extension of the original transition probability model by the addition of a second random transition and a cell size component (Brooks, 1979; Brooks *et al.*, 1980) is necessary to account for the observations that subsequent generations of cells do not appear to be influenced by their ancestral history. Another convolution of the original model takes into account that the rate-constant for the transition from the indeterministic A-state of the cell cycle depends on the time which a particular cell has spent in that state (Svetina, 1977; Svetina and Žecš, 1978; Nedelman and Rubinow, 1980). Finally, it has recently been claimed that all the properties exhibited by a system attributed to transition-probability can be accounted for by a "G_1 rate model" (Castor, 1980).

2. *The Division Protein or Mitogen Model (Mitotic Oscillator Model)*

Many unicellular eukaryotes respond to a variety of chemical and physical agents in a similar way. Thus when synchronized cultures of *Tetrahymena pyriformis* or *Schizosaccharomyces pombe* are given non-lethal heat pulses, they show a pattern of increasing division delay as pulses are given later and later in the cycle (Swann, 1957; Zeuthen and Rasmussen, 1972; Miyamoto *et al.*, 1973; Zeuthen, 1971, 1974). Similar set-backs can also be produced with pulses of cycloheximide (Polanshek, 1977). A transition point occurs late in the cell cycle beyond which no further delay (or a constant delay) in cell division is effected. The pattern of delay before the transition point has received much attention because it provides the basis for the idea that the continuous accumulation of a compound needed for cell division spans a major portion of the cell cycle. Much evidence has been provided, especially by the *Tetrahymena* system (Zeuthen, 1974) that the accumulating substance is protein ("division proteins") perhaps part of a structure (subpellicular oral fibres(?); Williams and Zeuthen, 1966) which is unstable until completely assembled at the transition point (Fig. 1.11). Thus inhibitors of protein synthesis, or amino acid analogues give set-backs and transition points similar to those produced by heat pulses (Hamburger, 1962; Frankel, 1962, 1967a,b). Polanshek (1977) has however shown that in *Schiz. pombe* several kinds and periods of sensitivity to heat shocks and cycloheximide pulses can be distinguished. In Chinese hamster cells, initiation of DNA synthesis is set-back by cycloheximide or puromycin (Schneiderman *et al.*, 1971; Highfield and Dewey, 1972) and this suggests that the accumulation of different kinds of initiators may occur to threshold levels at different stages of the cell cycle. Initiators of DNA synthesis have also been implicated in bacteria (Donachie *et al.*, 1973), and the evidence for an initiator of mitosis in *Physarum polycephalum* (Fig. 1.11b) is extensive (Rusch *et al.*, 1966; Sachsenmaier *et al.*, 1972; Bradbury *et al.*, 1974a, b; Sudbery and Grant, 1976; Tyson *et al.*, 1979). In this slime mould, plasmodia contain up to 10^8 nuclei in a single cytoplasm in virtually complete synchrony, dividing every 10–12 h. Plasmodia at different phases may be fused, and the fused pair undergoes mitosis at an intermediate phase. It is suggested that mitogen accumulates during the cycle, triggers mitosis at a critical level and is then destroyed. If however any two neighbouring phases, A and B, separated by less than 80 min are fused, the pair ultimately synchronizes to a phase on the short 80 min phase arc between A and B (Kauffman, 1974; Kauffman and Wille, 1975); this observation suggests the presence of a continuous oscillator—the mitotic oscillator (Sel'kov, 1970). Further modelling of the results of phase synchronization and heat shock experiments encourages this interpretation (Tyson and Kauffman, 1975; Kauffman and Wille, 1977). The limit cycle biochemical oscillator model for the control of mitotic timing in *Physarum* postulates

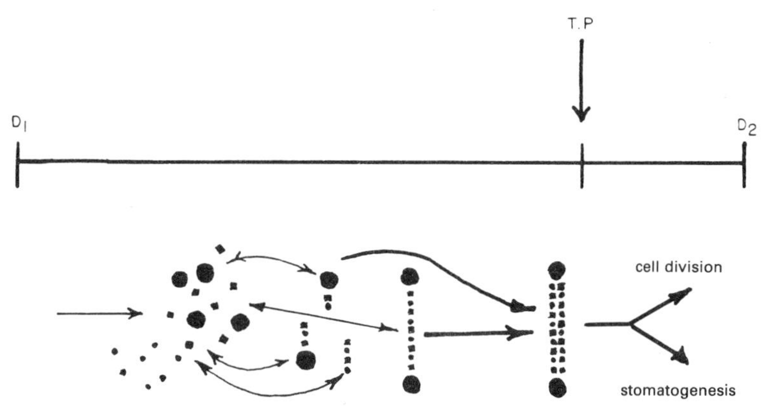

(b) "Hourglass" (Relaxation oscillator) model for *Physarum polycephalum*

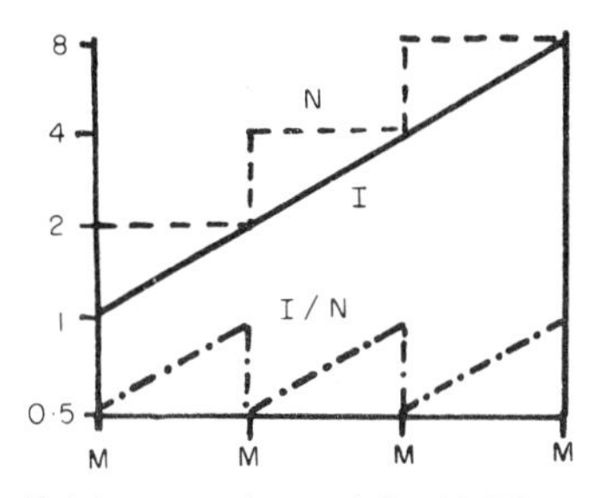

Fig. 1.11 Mitogen or division protein models; (a) The assembly of a division protein structure in *Tetrahymena pyriformis* is completed by the transition point (T.P.). Before this time, heat shocks lead to complete disassembly of this structure (and have "set-back"). (Reproduced with permission from Zeuthen and Williams, 1969.) (b) Kinetics of total mass (——) and number of nuclei (– – –) of exponentially growing synchronous plasmodia of *Physarum polycephalum*. (N) = nuclear receptor sites. (I) = initiator. (M) = synchronous nuclear mitoses.

the synthesis of an inactive mitogen precursor, X at a constant rate through the cycle, which is enzymatically converted to an active form Y, at a constant rate. The active form of the mitogen autocatalyses the production of more of itself from the inactive form at another rate, and decays with first order kinetics. This system leads to a sustained oscillation in the concentrations of precursor and active forms of mitogen, and mitosis is triggered when the level of the active form exceeds a critical concentration, Y_c (Fig. 1.12, Wille *et al.*, 1977).

That biochemical oscillations continue, even when mitosis is suppressed, is implicit in this model and confirmed in experiments. However the data obtained do not reveal the existence of a non-oscillating state as predicted by the limit cycle model (Wille, 1979) which is the central distinction between this and alternative models

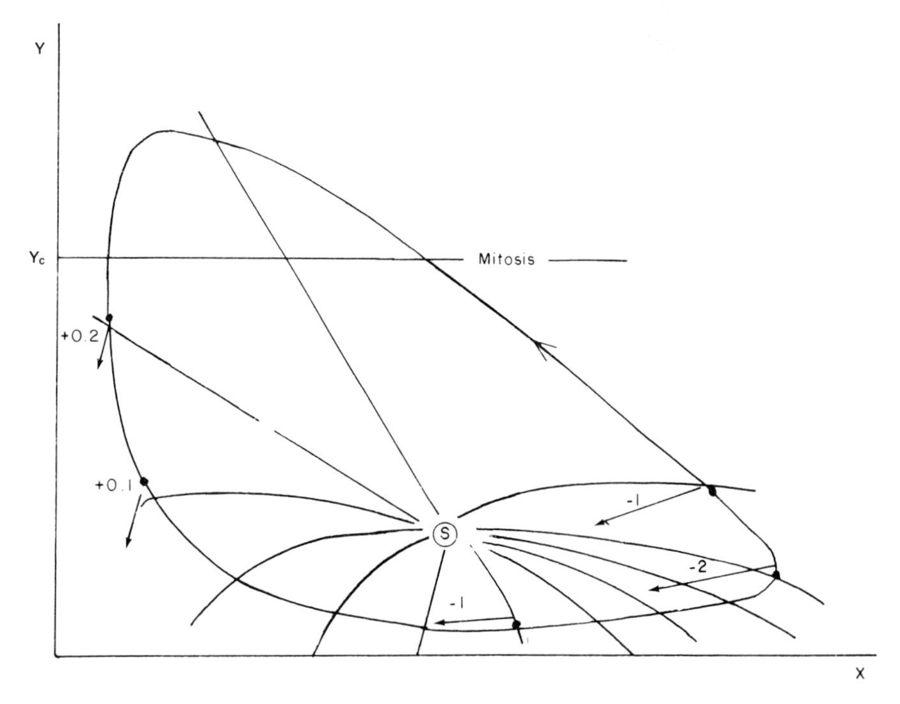

Fig. 1.12 The limit cycle in the concentrations of X and Y. A critical level of Y triggers mitosis. The nearly radial lines emanating from a point inside the cycle are "isochrons" separating equal intervals of time along trajectories, and along the limit cycle path, and indicates hours before mitosis. The effect of 20% destruction of X and Y from five phases is shown as destruction vectors (arrows) which indicate the direction and magnitude in delay (−) or advance (+) of the next mitosis. All isochrons meet at the steady state singularity, S, inside the limit cycle. (Reproduced with permission from Wille, 1979.)

of the *Physarum* system. For instance, Sachsenmaier (1976) suggests that the initiation of mitosis is a genuine phase discontinuity occurring once in each mitosis; this constitutes an extreme relaxation oscillator form of the mitogen accumulation model (Fig. 1.11). Objections to a central timer hypothesis, based on the unverified assumption of a *direct* effect of heat shocks on control variables, have been put forward by Tyson and Sachsenmaier (1978) who suggest that the "phase shifts" of Kauffman and Wille (1975) may be only transient effects. Further support for the concept of timekeeping by a limit cycle oscillator comes from the work of Klevecz *et al.* (1978; 1980a,b) with cultured mammalian cells which show characteristic patterns of phase response behaviour after mild perturbation (see p. 39). Both Klevecz *et al.*

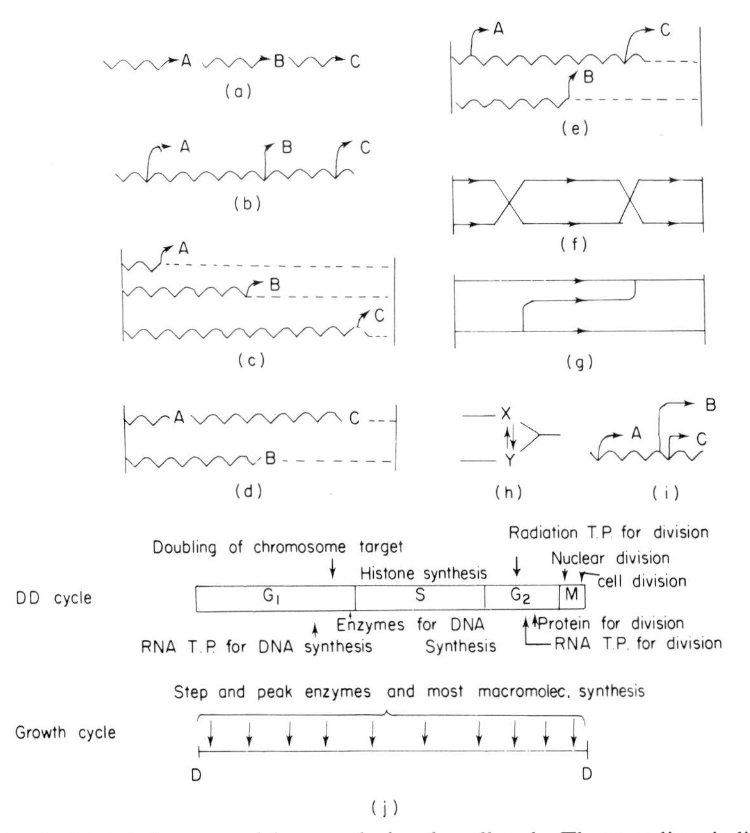

Fig. 1.13 Models for sequential events during the cell cycle. The wavy lines indicate the progress of timers. (a) Dependent sequence; (b) Independent, single timer (IST) sequence; (c) Independent, multiple timer (IMT) sequence; (d) Two parallel dependent sequences; (e) Two parallel IST sequences; (f) Two sequences with two check points. The events of each sequence before a check point have to be completed if there is to be progress beyond the check point. (g) Two sequences in which an early event in the lower sequence has to have been completed before a late event in the upper sequence can occur. (h) "Interdependency" of two events. (i) Delay in the full expression of an event B in an IST sequence may alter its order. (j) The "DNA-division cycle" and the "growth cycle". T.P. = transition point; D = division. (Reproduced with permission from Mitchison, 1971.)

(1978) and Gilbert (1978a,b,c) conclude that the oscillator concept also explains the data on which the transition probability model is based.

3. *The Dependent Pathways Model*

In this model, it is proposed that events of the cell cycle may constitute one or more causal sequences which eventually join to give a closed loop of states. Thus each event of a single sequence is necessary before all later events can occur. Mitchison (1974) has summarized the evidence for the existence of causal sequences which comes largely from the mapping of temperature-sensitive cell-cycle mutants of *Saccharomyces cerevisiae* (Hartwell *et al.*, 1974) and *Chlamydomonas reinhardii* (Howell, 1974) and events of the bacterial cycle (Donachie *et al.*, 1973). Parallel sequences of events in the cell cycle of *Schizosaccharomyces pombe* (Fig. 1.13j) lead to the suggestion that there is a "DNA division cycle", which has as its main events, DNA synthesis, nuclear division and cell division (and perhaps also histone synthesis and various transition points). A second sequence includes the main processes of protein and RNA synthesis which constitute the "growth cycle" and includes some periodic events which may be used as markers of cell cycle traverse (Mitchison, 1971; Mitchison and Creanor, 1971; Sissons *et al.*, 1973). These two sequences are to some extent independent, or at least only loosely coupled, in that each can proceed under circumstances where the other is blocked. Mitchison (1974) goes on to distinguish alternative models in which the events are either themselves *part* of a timing mechanism or are merely *underlain* by a timing mechanism (Fig. 1.13a–i).

VI. Circadian Time

Many daily rhythms of biological activity persist for long periods even when organisms are placed in conditions of constant temperature and continuous darkness. These circadian rhythms are driven by an endogenous oscillation in cellular controls and play the key role of time-measuring for the organism. All eukaryotic cells (but probably not bacteria) possess a light-entrainable cellular pacemaker (circadian clock) characterized by the following properties: (1) a persistent free-running period of about 24 h; (2) entrainability and phase-responsiveness to single or multiple externally-applied signals (Zeitgebers) which include light, temperature, nutrients, oxygen, D_2O, Li^+, K^+, ethanol,valinomycin, theobromine, theophylline, and cycloheximide; (3) temperature compensation of the period (Vanden Driessche, 1975; Palmer *et al.*, 1976; Hastings and Schweiger, 1976; Saunders, 1977; Brady, 1978). The lower eukaryotes have provided some of the most fruitful experimental systems for the study of circadian phenomena, particularly in attempts to identify the molecular basis of the biological clock. The wide variety of processes showing circadian periodicities in these organisms include mating behaviour in *Paramecium*, phototactic response in *Euglena* and *Chlamydomonas*, cell division in *Gonyaulax*, *Euglena*, *Paramecium*, *Tetrahymena* and *Chlamydomonas*, photosynthetic capacity in *Gonyaulax*, *Acetabularia* and *Phaeodactylum*, changes in chloroplast shape and ultrastructure, bioluminescence, enzyme activities and concentrations of metabolic intermediates (Wille, 1979). It is widely believed that one fundamental oscillator may drive all these rhythmic functions, although its identification has resisted strenuous attempts; alternatively the cell may resemble a "clockshop" in so far as it may have a high order of redundancy of timekeepers (Winfree, 1975, 1976, 1980).

VII. Interactions between Time Domains

Analysis of the time structure of living organisms in terms of a hierarchy is convenient but incomplete. Evidently interactions occur not solely within a single hierarchical level but also between levels.

A. THE CELL DIVISION CYCLE AND THE CIRCADIAN CLOCK

Daily light–dark and temperature cycles can be used to synchronize mitotic and cell-division rhythms, and these rhythms in some cases persist for a few cycles in the absence of any entraining cycle. Furthermore under "normal physiological conditions" the mitotic rhythms of many higher animal cells in tissues (Rensing and Goedeke, 1976) and the cell cycle times of lower eukaryotes in natural environments, often have periods of about 24 h (Bruce, 1965). The obvious conclusion is that there is a matching of the "natural" growth rate of some cells with the environmental periodicity, but that in most cases only a weak coupling occurs between the circadian clock and mitotic and cell-division rhythms. These early observations of circadian periodicities of cell division e.g. in *Euglena gracilis* (Edmunds, 1965) and *Gonyaulax polyedra* (Sweeney and Hastings, 1958), together with the discovery that the non-phototrophic protozoon, *Tetrahymena pyriformis*, displays a light synchronizable endogenous circadian rhythm of cell division (Wille and Ehret, 1968, Fig. 1.14) have led to several interesting developments, including the chronon theory of circadian timekeeping

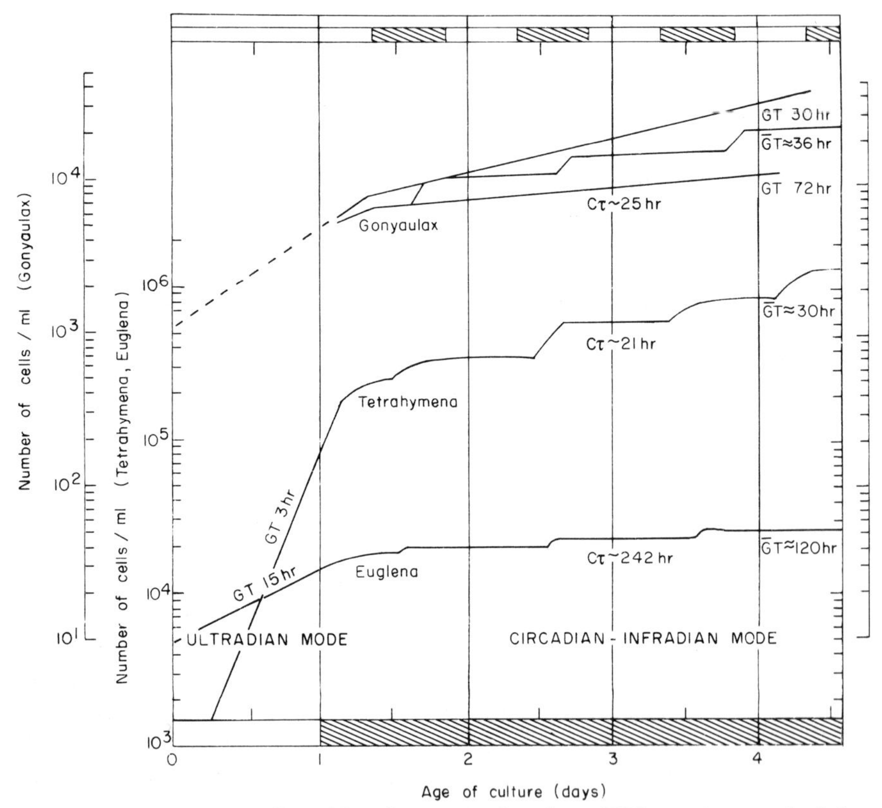

Fig. 1.14 The G-E-T effect. Ultradian generation times (GT) are given on the left, and infradian generation times on the right. It is during the infradian mode that cells are light-synchronizable; although infradian generation times are long and variable, in each case the consequence is a circadian output; the circadian period ($C\tau$) is ≈ 1 day (by definition). *Gonyaulax* is shown in continuous light (top, open bar) as an asynchronous population with average infradian generation times in the range 30–70 h and in light–dark cycles of 12 h of each as a synchronized population with an integral generation time of ≈ 36 h. The curves for *Euglena* and *Tetrahymena* might also have included asynchronous infradian slopes, but these have been omitted here for clarity; the synchronized populations represented in each of these cases show free-running endogenous circadian rhythms following photoinduction by a switch-down from light (bottom, open bar) to darkness (bottom, hatched bar) at a critical transition point between the two modes of growth. (After Ehret and Wille, 1970, reproduced by permission.)

(Ehret and Trucco, 1967). All three organisms show mitotic entrainment by a Zeitgeber (e.g. in Fig. 1.14, a switch-down in light intensity) at a special phase of population growth between the ultradian (average generation time < 24 h) and the infradian (average generation time > 24 h) growth modes (Ehret and Wille, 1970). In the ultradian mode, cell division cycle controls can override the regulatory circadian oscillator (Edmunds, 1975). Cyclic temperature changes are also effective in entraining a circadian rhythm of cell division in continuous cultures of *T. pyriformis* growing under infradian conditions (Meinert *et al.*, 1975).

It would seem plausible to ascribe light-induced division synchrony to the same mechanism that leads to synchrony by heat shocks, inhibitor blocks or any repetitive shift between different environmental conditions to which specific stages of the cell division cycle are differentially sensitive. Compelling evidence that this model is not sufficient to explain all the observations with the *Euglena* system has been presented by Edmunds (1978). When a light–dark cycle (LD: 10, 14) was employed, a doubling of cell numbers occurred every 24 h (Edmunds, 1965; Fig. 1.15), but diurnal LD cycles with $L < 10$ h gave less than a doubling in cell numbers at each step (Edmunds and Funch, 1969a; Fig. 1.16). In the experiment shown, the doubling time of the culture (i.e. the average cell cycle time) was about 36 h, and in each burst of cell division only about two-thirds of the cells divided. But the cells that *did* divide during any given burst did so every 24 h during a relatively narrow time interval ("gate") occurring in each dark period. It is suggested that the endogenous, light-sensitive, self-sustaining circadian oscillation (which underlies the overt diurnal population rhythm) has been entrained to a 24 h period by the 24 h LD cycle; this clock then permits those cells that are capable of cell division to divide when the gate is open. Those cells that are not prepared for division are somehow prevented from dividing until the next gate opens 24 h later. So the circadian clock blocks (or differentially promotes) cell cycle progress. This "gating" hypothesis was initially based on data obtained with autotrophically cultured wild-type *E. gracilis*; in these experiments light played a dual role (energy source and time cueing). Simpler systems for study are provided by light-entrainable, non-photosynthetic mutants of *E. gracilis* and a number of other organisms (Table 1.1; Edmunds, 1975). In all of the studies with the five different photosynthetic mutants, cell division is confined primarily to the dark intervals of the LD cycle (or to subjective night in DD or LL regimes), and this phasing agrees with other work on the organisms shown in Table 1.1, all of which show

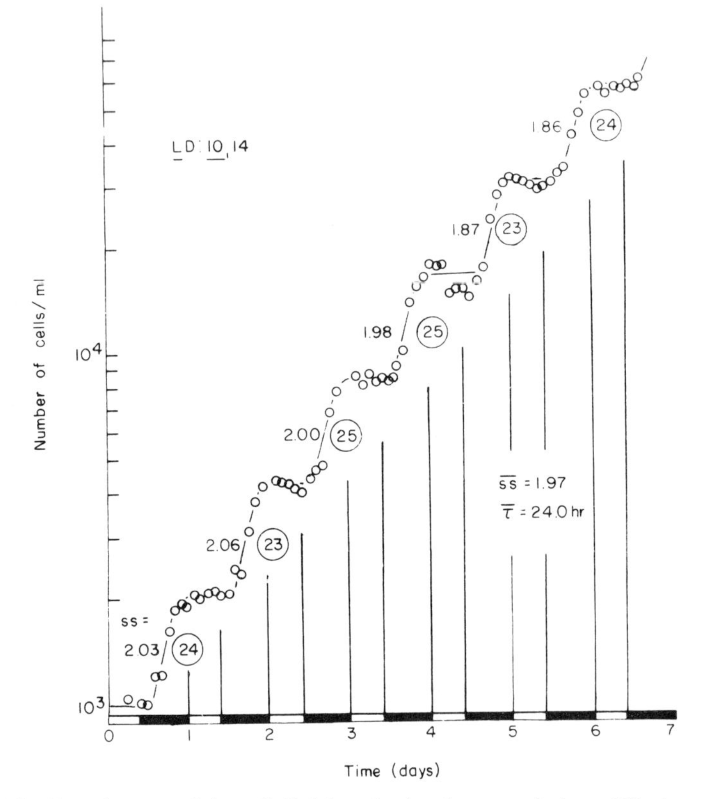

Fig. 1.15 Entrainment of the cell division rhythm in a population of *Euglena* grown autotrophically at 25°C in LD: *10*,14. Ordinate: cell concentration (cells per millilitre); abscissa: elapsed time (days). Step-sizes ($\overline{ss}$., ratio of number of cells per millilitre following a division-burst to that just before the onset of divisions) are indicated for the successive division-bursts. The period of the rhythm is also given in hours (encircled just to the right of each burst). The average ($\bar{\tau}$) of the rhythm in the culture is essentially identical to that of the synchronizing cycle, and a doubling of cell number usually occurs every 24 h. (Reproduced with permission from Edmunds and Funch, 1969a.)

persistent circadian rhythms of cell division. The entrainment of one nonphototrophic organism (*S. cerevisiae*) appears to involve mitochondrial cytochromes as photoreceptors (Ułaszewskłi *et al.*, 1979).

Circadian oscillations in photosynthetic capacity (Laval–Martin *et al.*, 1979), cell-settling rates, phototaxis, dark motility, amino acid incorporation, cell volume and shape changes, and enzymatic activities (alanine, lactate, and glucose-6-P dehydrogenases, L-serine and L-threonine deaminases) persist in *E. gracilis* in infradian cultures where

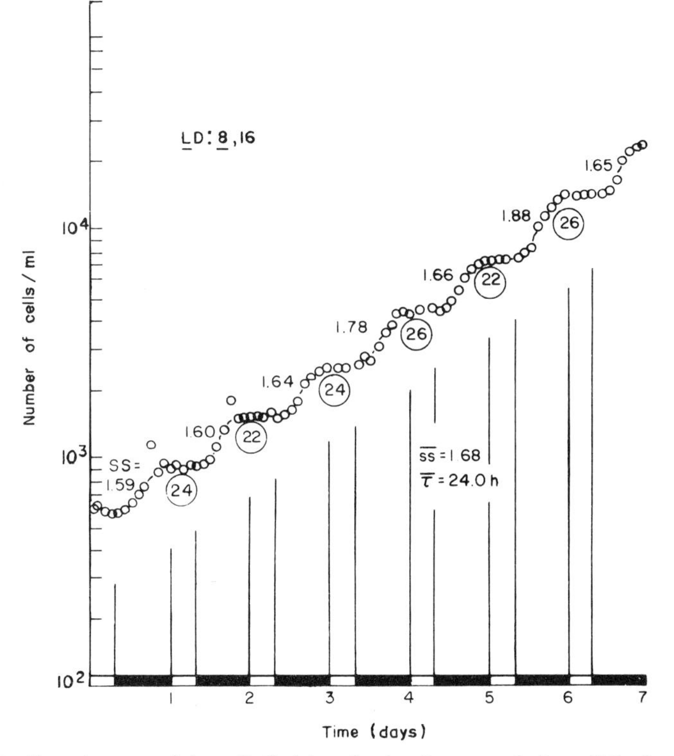

Fig. 1.16 Entrainment of the cell division rhythm in a population of *Euglena* grown photoautotrophically at 25°C in LD: *8*,16. Other labels as for Fig. 1.15. Although the average period of the rhythm is precisely that of the synchronizing LD cycle, the average stepsize of the successive fission bursts is substantially less than 2·0, indicating that not all cells divide during any one cycle. (Reproduced with permission from Edmunds and Funch, 1969a.)

the doubling time greatly exceeds 24 h (Edmunds, 1975), and even in cells not proceeding through their normal cell division cycle (Edmunds and Cirillo, 1974). The distinction between the generation time of individuals of the population, doubling time of the culture, and the circadian cycle in slowly-dividing cell populations is shown in Fig. 1.17 (Ehret and Dobra, 1977). Cell division cycles can take on any period, between a lower genetically-determined value (under optimal conditions) and 24 h (depending on environmental factors like nutritional status etc., but see p. 36); in this ultradian mode the cell-cycle times are temperature dependent. In the infradian mode, the doubling time is longer than 24 h, the cell cycle time approximates

Table 1.1 *Some Lower Eukaryotes in which Persistent Circadian Rhythms of Cell Division have been observed* (after Edmunds, 1975)

Organism	Reference
ALGAE	
Chlamydomonas reinhardii	Bruce (1970)
Chlorella pyrenoidosa	Pirson and Lorenzen (1958)
Chick strain 211-8b	Hesse (1972)
Euglena gracilis Klebs Strain Z	Edmunds (1966, 1971) Edmunds and Funch (1969a,b)
Euglena gracilis (photosynthetic-deficient mutants	
P_4 ZUL mutant	Jarrett and Edmunds (1970) Edmunds (1971) Edmunds *et al.* (1971)
P_7ZNgL mutant	Mitchell (1971)
W_6 ZHL mutant	Edmunds *et al.* (1976)
W_nZUL mutant	Mitchell (1971)
Y_9 NaIL mutant	Edmunds *et al.* (1976)
Gonyaulax polyedra	Sweeney and Hastings (1958)
Gymnodinium splendens	Hastings and Sweeney (1964)
FUNGI	
Candida utilis	Wille (1974)
Saccharomyces cerevisiae	Edmunds *et al.* (1979)
PROTOZOA	
Paramecium bursaria	Volm (1964)
Paramecium multimicronucleatum	Barnett (1969)
Tetrahymena pyriformis	
Strain W	Wille and Ehret (1968) Edmunds (1974)
Strain GL	Edmunds (1974)

a circadian interval of 24 h and becomes almost temperature-independent. The "gating hypothesis" implies that no real values for cell-cycle times can fall in the domain $24 < g < 48$ h, i.e. that under infradian conditions cell cycle time is multimodally variable, and only shows values which are multiples of 24 h.

B. ULTRADIAN–CIRCADIAN INTERACTIONS

There is no unequivocal demonstration of any circadian rhythm in rapidly proliferating (ultradian) populations, and it is difficult to

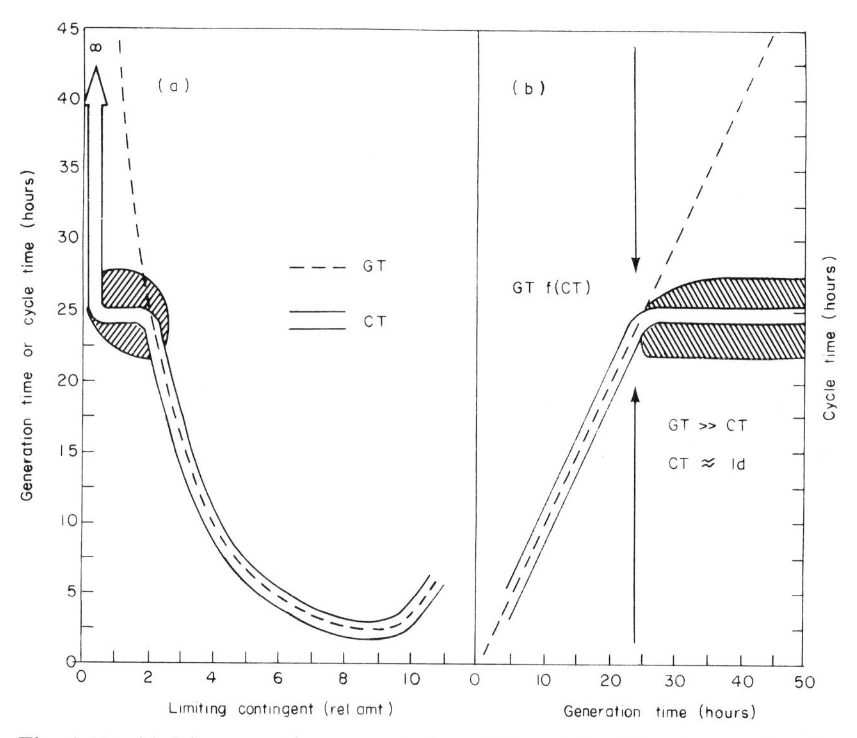

Fig. 1.17 (a) Diagrammatic representation of the relationship of generation time (broken line) and "cell cycle time" (open path) to an environmental factor (limiting contingent). Only in lethal environments does the cycle time (CT) exceed circadian values (approaching infinity, arrow). Hatched area represents the range for CT. (b) Generation time is a function of cycle time only during ultradian growth. As GT approaches infinity, CT remains approximately constant (circadian). (Reproduced with permission from Ehret and Dobra, 1977.)

envisage how 24 h timekeeping could take place when all the individuals of the population are dividing every few hours. One way around this problem is to assume that the circadian period is generated from some higher frequency oscillation(s), and a number of possible mechanisms for such a frequency reduction have been proposed (Fig. 1.18).

1. *Subharmonic Resonance or Frequency Demultiplication*

A pair of coupled conservative oscillators may exhibit a low frequency collective mode of oscillation induced by a resonance phenomenon (Goodwin, 1963). No systematic analyses of this proposed subharmonic response system have been made. If an endogenous oscillation is driven by an external oscillation having a similar frequency, then the free-running frequency of the endogenous oscillator will lock onto the driving frequency. Subharmonic resonance to give a low frequency output occurs when the driving frequency is close to an integral multiple of the free running frequency. It is perhaps, however, unreasonable to expect any significant coupling between oscillators having

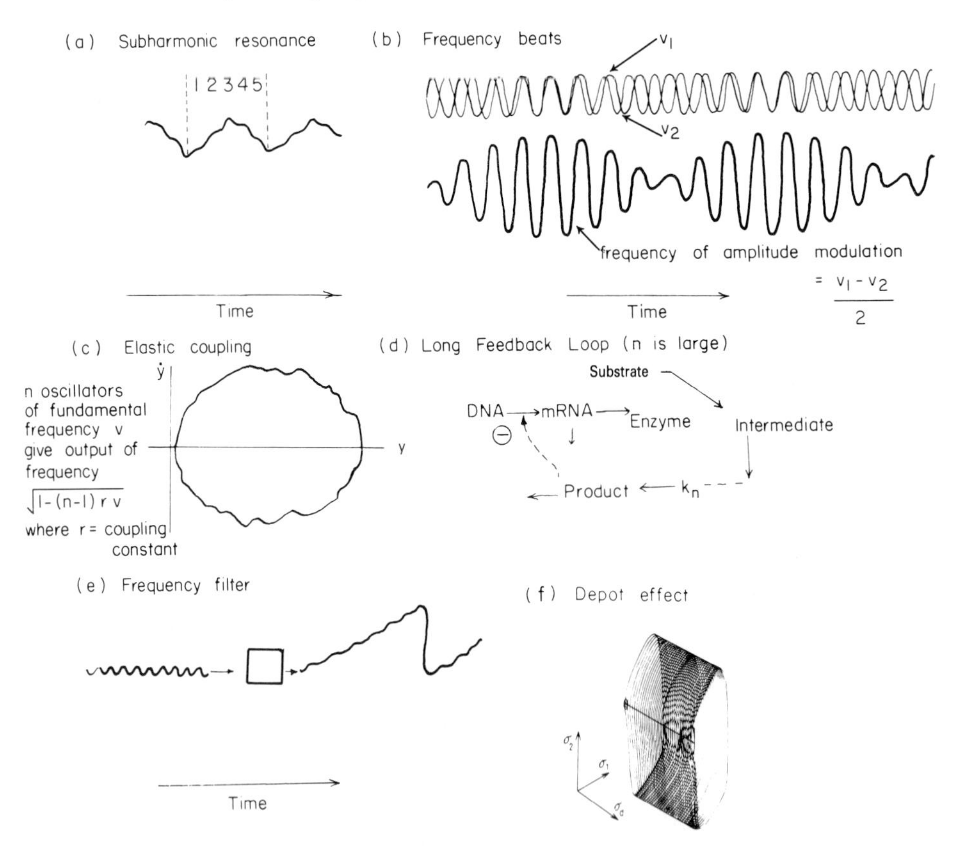

Fig. 1.18 Frequency reduction mechanisms: (a) Frequency demultiplication or subharmonic resonance (Goodwin, 1963); (b) Frequency beats by superposition of two sinusoids of differing periods; (c) Elastic coupling; summed output of several coupled oscillators gives limit cycle with long period (Pavlidis, 1969); (d) Long feedback loop. The period of oscillation in a negative feedback loop is much longer than the half-life of the most stable element of the loop (Tyson, 1979); (e) Frequency filter or counter; (f) Depot effect. Simulated phase plane plot of fructose 6-phosphate concentration (σ_1) *vs.* fructose 1,6 bisphosphate concentration (σ_2) *vs.* glycogen concentration (σ_d) gives highly spiralized stable limit cycles due to nonlinear interaction between two self-oscillatory mechanisms one of short and one of long period (Sel'kov, 1980).

grossly different frequencies of the kind required by this model, which is derived from an experimental demonstration of forced vibrations in a mechanical system having a non-linear restoring force (Ludeke, 1946). Frequency entrainment in a set of coupled relaxation oscillators is analysed mathematically by Grasman and Jansen (1979); this treatment yields an unexpected number of stable synchronized wave solutions.

2. *Frequency Beats*

Long period oscillations could be generated as a consequence of the phenomenon of frequency beats (Chance *et al.*, 1967). However, Winfree (1967) has pointed out that if two oscillators are coupled because of the proximity of their individual frequencies (in order to produce low beat frequencies) then mutual entrainment will occur and the beats will disappear. Another objection is that minor variations in the frequencies of the individual oscillators result in major changes (proportionally) in the resulting frequency.

3. *Elastic Coupling*

Strong coupling in a population of oscillators can give synchronization at a new frequency, much lower than the one at which they would have oscillated individually (Pavlidis, 1969). Even for weaker coupling, if one oscillator continues at a high frequency, the observed sum of their outputs will present a low frequency oscillation. An objection to this proposal is that it is difficult to synchronize the oscillators from arbitrary initial conditions (Tyson, 1976).

4. *Low Frequency Generation in Long Feedback Loops*

The period of oscillation of a system is much longer than the half-life of the longest lived intermediate in a feedback loop, unless the feedback inhibition is highly non-linear (Gilbert, 1978a; Tyson, 1979).

5. *Frequency Filtering*

A low frequency could be produced from a high frequency by a frequency filter which "counts" the high frequency.

6. *Deposition Effect*

The reversible exchange of a metabolic intermediate (e.g. hexosemonophosphate), with a huge amount of a reserve energy store (e.g.

glycogen) which acts as a buffer, can dramatically increase the oscillation period. This model (Sel'kov, 1971, 1973; Schulmeister and Sel'kov, 1978) displays complex behaviour including highly spiralized stable limit cycles.

Ultradian oscillations undergoing frequency demultiplication to generate circadian periods must possess the following properties:

(a) They must show a temperature-compensated frequency.

(b) Phase resetting by external perturbating influences (light, heat, chemical) must give a response curve of the correct shape (see Fig. 1.24).

Many models for possible temperature-compensation mechanisms have been proposed e.g:

(i) Temperature-compensated product inhibition (Pittendrigh and Bruce, 1957) might resemble the temperature independence of a resistance-capacitance time-constant circuit (Chance *et al.*, 1967);

(ii) A relatively temperature-independent diffusion-controlled rate-limiting step (Ehret and Trucco, 1967);

(iii) Adaptation to lowered enzyme content compensates for higher enzyme activities at high temperatures (Pavlidis and Kauzmann, 1969);

(iv) The direct effect of temperature on membrane fluidity is opposed by compensatory adaptation which alters unsaturated fatty acid composition (Njus *et al.*, 1974);

(v) Compensation for the positive temperature coefficient of protein synthesis by a membrane assembly step with a negative Q_{10} (Schweiger and Schweiger, 1977).

Although it is easy to devise mechanisms suitable for temperature compensation of circadian timekeeping, most of the ultradian oscillating systems which have been studied have temperature-dependent periods *in vivo* as well as in cell-free extracts. All the oscillators in the metabolic time domain, including those best understood examples, the glycolytic oscillator (Chance *et al.*, 1973) and the morphogenetic intercellular communication system in *Dictyostelium discoideum* (Wurster, 1976; Nanjundiah *et al.*, 1976), have temperature-dependent frequencies (Fig. 1.19). Metabolic oscillations are thus unlikely to represent the high frequency "ticks" of the circadian clock.

More likely candidates in this role are to be found in the epigenetic time domain; it is attractive to suggest that in this case problems of the accuracy of frequency demultiplication would be minimized by using hourly periodicities rather than those with periods of the order of minutes. Of the epigenetic oscillators studied, only those in one

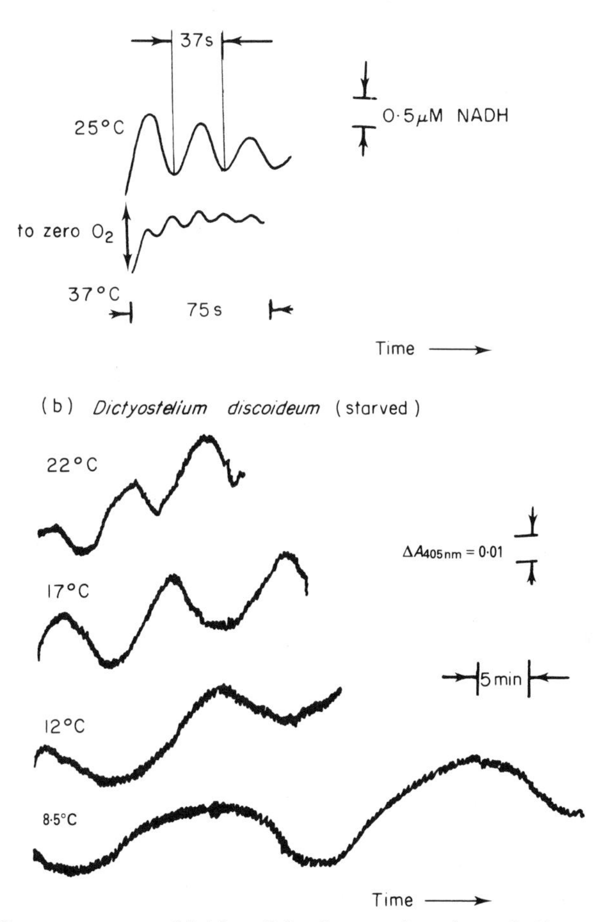

Fig. 1.19 Temperature sensitivities of the frequencies of metabolic oscillations; (a) Oscillations in intracellular NADH in a 6% (w/v) suspension of yeast induced by a transition to anaerobiosis. (Reproduced with permission from Chance *et al.*, 1964a.) (b) Oscillations of light-scattering by an aerated suspension of amoebae (5×10^7 organisms per ml after 4 h starvation at 23°C. (Reproduced with permission from Wurster, 1976.)

system have been shown to have a period which is independent of temperature (Lloyd and Edwards, 1982). The respiration rate of *Acanthamoeba castellanii* oscillates during the growth of synchronous cultures with a period of about 1 h irrespective of temperature over the range 25–30°C (Fig. 1.20a–c). Measurements of adenine nucleotide pools, redox states of mitochondrial flavoproteins, and experiments

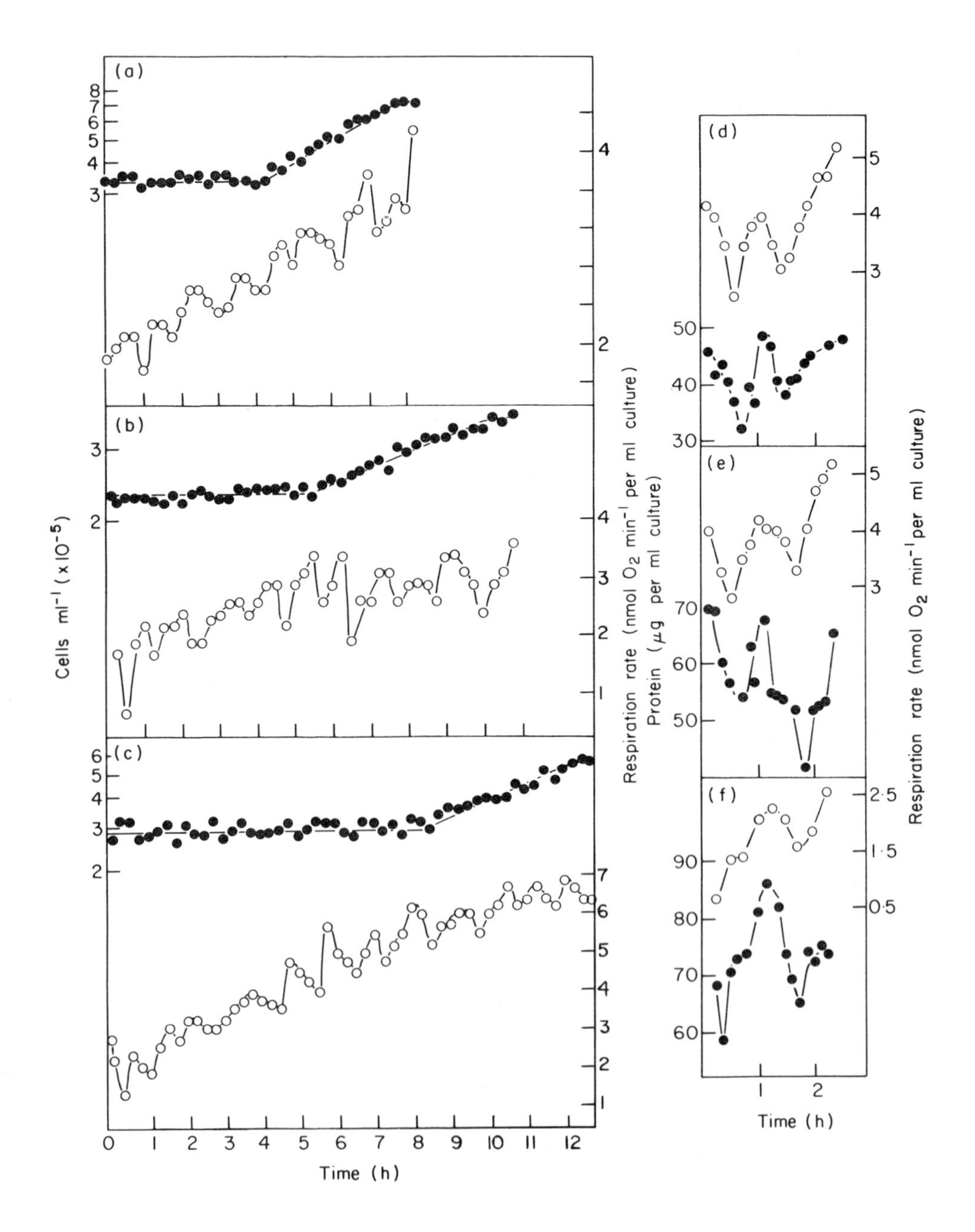
(a)
(b)
(c)
(d)
(e)
(f)
Cells ml⁻¹ (x10⁻⁵)
Respiration rate (nmol O2 min⁻¹ per ml culture)
Protein (μg per ml culture)
Respiration rate (nmol O2 min⁻¹ per ml culture)
Time (h)
Time (h)

with agents that interfere with electron transport and energy conservation, confirm that cycles of energization and de-energization of mitochondria are implicated, and that these coupled oscillators are involved in respiratory control (Edwards and Lloyd, 1978; Bashford *et al.*, 1980). However, the fundamental driving oscillator in the system is located in a transcriptional or translational feedback loop (Lloyd and Edwards, 1979, 1980a, b; Edwards and Lloyd, 1980; Lloyd and Turner, 1980), and it is the overall rate of cellular protein synthesis which determines mitochondrial energy-yielding activities. Like respiration, cellular protein content shows an hourly periodicity, and these two oscillatory components are always in phase. Moreover this period and phase relationship is identical at 25°C, 27°C and at 30°C (Fig. 1.20d–f) (S. W. Edwards and D. Lloyd, unpublished data). Oscillations in total cellular protein with periods in the range 20 min–4 h have also been demonstrated in mammalian cells in tissues ("circahoralian rhythms", Brodsky, 1975; Yarygin *et al.*, 1978, 1979; Brodsky *et al.*, 1979; Nechaeva *et al.*, 1980) and in culture (Klevecz and Ruddle, 1968; Tsilimigras and Gilbert, 1977) as well as during the early cleavages of sea urchin embryos (Mano, 1970, 1971a,b, 1975), but none of these reports cite that frequency is temperature independent *in vivo*.

Fig. 1.20 (a–c) Changes in respiration rates during synchronous growth of *A. castellanii*. Exponentially-growing cultures were centrifuged at 750 r/min for 2·4 min in the 6 × 1 l head of an MSE Mistral centrifuge. Approximately 80% of the supernatant, containing the slowest-sedimenting cells (about 10% of the original exponential culture), was decanted and grown as a synchronous culture. Samples were withdrawn at frequent intervals for measurements of respiration rates and cell counts (estimated using a Fuchs-Rosenthal haemocytometer). In (a) the temperature of incubation was 30°C, the synchrony index 0·65 and the synchronous culture contained 13% of the original exponential culture. In (b) the incubation temperature was 27°C, the synchrony index 0·63 and the culture contained 6% of the original exponentially growing culture. In (c) the incubation temperature was 25°C, the synchrony index 0·72 and the synchronous culture contained 10% of the original exponentially growing cultures. (d–e) Changes in respiration rates and protein levels in synchronously dividing cultures of *A. castellanii*. Synchronous cultures were prepared by centrifugation of exponentially growing cultures at 300 r/min for 2 min in an MSE bench centrifuge. The supernatant was carefully decanted and grown as a synchronous culture; samples were taken at frequent intervals throughout growth for estimation of cell numbers until cell division was completed. In (d) the temperature of incubation was 30°C, the synchrony index 0·7 and the generation time 8h. In (e) the temperature of incubation 27°C, synchrony index 0·72, generation time 10 h. In (f) the temperature of incubation 25°C, synchrony index 0·67, generation time 12·5 h. (○) shows respiration rate and (●) protein levels. After the completion of the cell division cycles, values of protein levels for the experiments shown in (d), (e) and (f) were 75, 130 and 132 μg/ml culture, respectively. (Results of S. W. Edwards and D. Lloyd.)

The experiments with *A. castellanii* suggest that turnover of the majority of protein species is periodic, and consists of discrete and alternating phases of synthesis and degradation (Edwards and Lloyd, 1980; Lloyd and Edwards, 1980b). This temporal compartmentation of mutually incompatible biochemical processes (occurring in identical spatial and subcellular compartments?) is timed by a temperature-compensated mechanism, may thus be a circadian subharmonic. Evidence for circadian subharmonics with periods of 3–4 cycles per day can be obtained by close examination of well-plotted circadian rhythms (Goodwin, 1963), e.g. in the case of the luminescence rhythm in *Gonyaulax polyedra* (Karakashian and Hastings, 1962). This strongly supports the suggestion that the synthesis of low frequency rhythms occurs from higher frequency components.

C. EPIGENETIC CELL CYCLE INTERACTIONS

The oscillations of energy-requiring and energy-yielding reactions in *A. castellanii* may represent interactions not only between epigenetic and circadian oscillators, but also between epigenetic events and cell division cycle timekeeping, i.e. between the growth cycle and the DNA-cell division cycle. If phase information is to be imparted to daughter cells within a population of rapidly-dividing organisms (Sweeney, 1976) then it may be necessary to make the cell division cycle time an integral number of those subcycles which are also "counted" to generate a circadian period (Fig. 1.21).

The distribution of possible generation times of mammalian cells is not continuous within the limits for each cell type but is quantized in multiples of 3–4 h (Klevecz, 1976). To satisfy a considerable list of published data it was suggested that a subcycle G_q has a traverse time equal to the period of a temperature compensated timekeeping mechanism (Fig. 1.22). Cell cycle time increases at lower temperatures, lower serum concentrations and high cell densities because the number of rounds of traverse through G_q increases. Earlier observations of a 3–4 h period of enzyme accumulation (Klevecz, 1969a; Klevecz and Kapp, 1973) total cellular protein, cellular volume (Klevecz and Ruddle, 1968) and of DNA synthesis in the S-phase (Klevecz, 1969b) were interpreted generally in terms of the expressions of activity of a cellular clock. To explain these results, it is suggested that the cell division cycle of mammalian cells is built up from multiples of a fundamental 4 h period. Differences in the length of G_1 in cells of the same culture occur as a consequence of the gated entry of cells into

S. More recently, evidence has been presented that the clock operates throughout the cell cycle, not just in G_q, but in S and G_2 as well (Klevecz *et al.*, 1978). Thus phase response to the mild perturbation effect of a transient increase in serum concentration leads to advances or delays in the time of division of Chinese hamster V79 cells; these results are in accord with the hypothesis of timekeeping by a limit

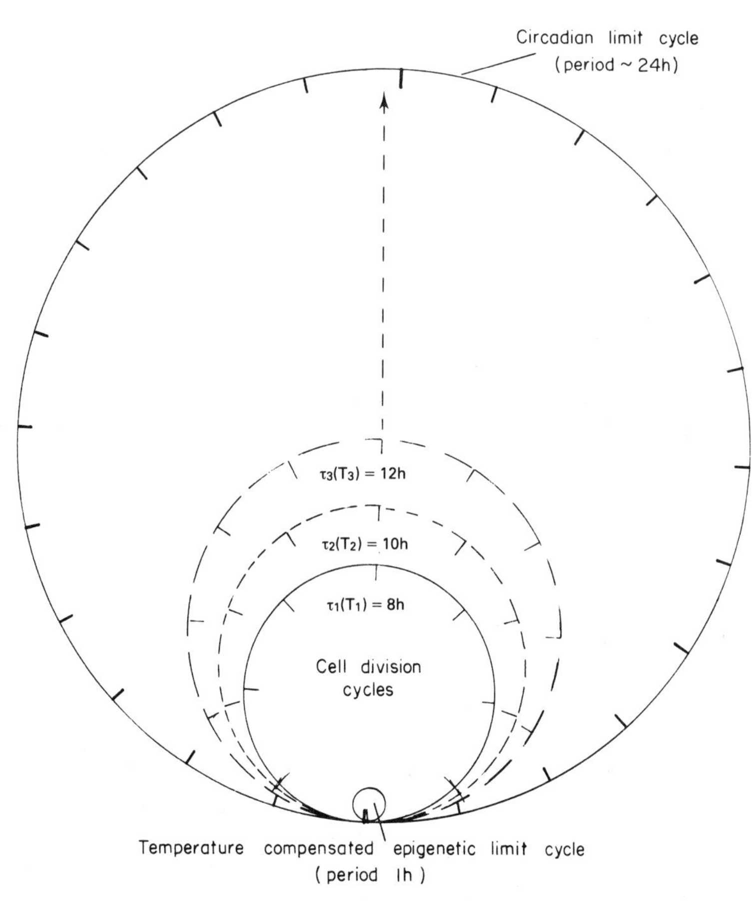

Fig. 1.21 Epigenetic-cell division cycle-circardian interaction; a hypothetical scheme based on results with *Acanthamoeba castellanii*. It is suggested that the fundamental oscillator occurs in the epigenetic time domain; this has a period of 1 h and is temperature compensated (Lloyd and Edwards, 1982). Frequency reduction by an unknown mechanism (but see Fig. 1.18 for possible alternatives) generates quantized cell division cycle times which may have values between 8 h (under optimal growth conditions) and the circadian value. The effect of temperature on cell-cycle time is also shown.

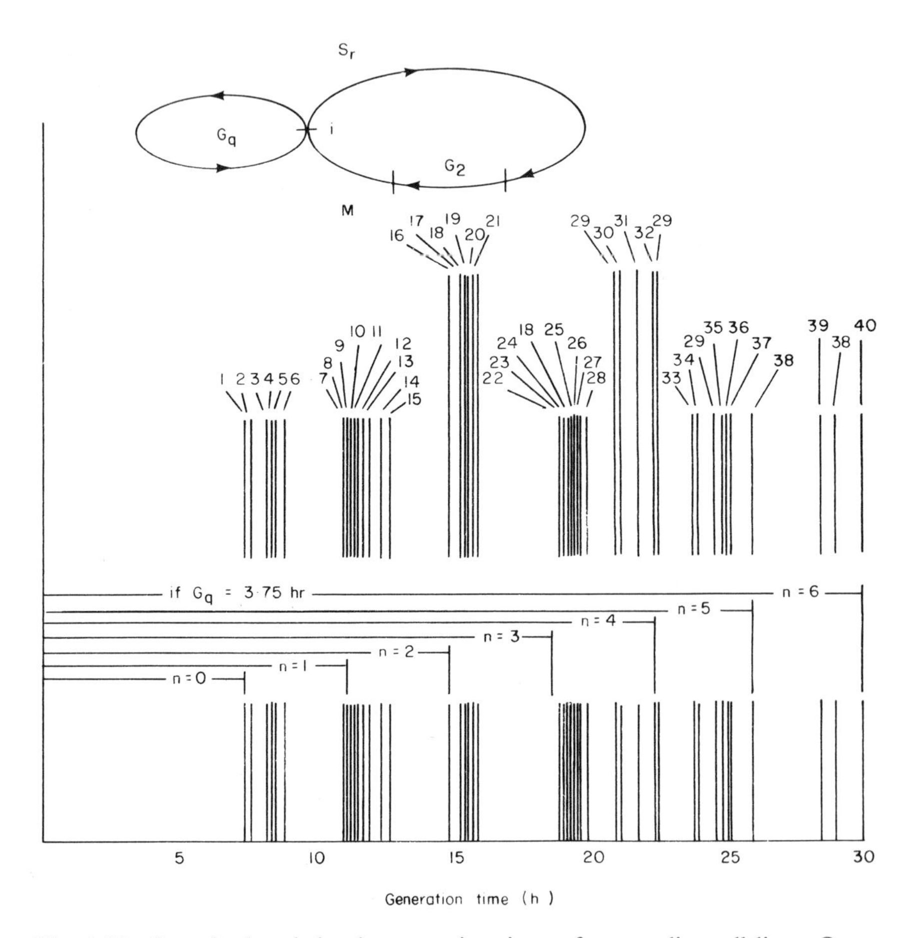

Fig. 1.22 Quantized variation in generation times of mammalian cell lines. Generation times were determined from the published data on cells synchronized by mitotic selection or from time-lapse cinematography of random cultures. The list is not exhaustive but represents a sampling of papers published between 1961 and 1976 in which the stated generation time could be directly confirmed in the data. Wherever possible modal generation times were obtained; that is, the time of maximum mitotic index rather than the centre of mass of the mitotic wave. Numbers above each line in the figure refer to the reference number in Table I of Klevecz (1976). The calculation of possible generation times uses the simple expression $T_g = nG_q + 2G_q$. T_g = generation time in hours. G_q (quantized G) = incremental increase in generation time in hours. (Reproduced with permission from Klevecz, 1976.)

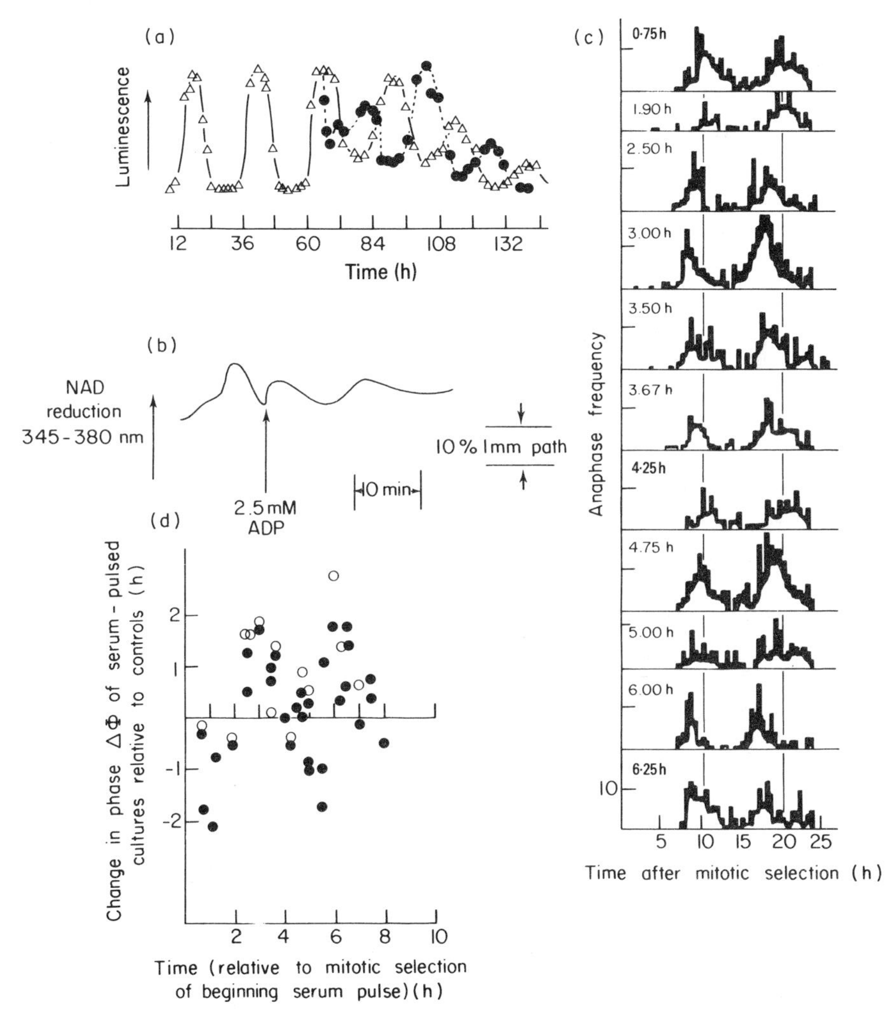

Fig. 1.23 Phase-resetting in different time domains. (a) Effect of a light pulse (dashed line) on the phase of bioluminescence in *Gonyaulax polyedra*. (Reproduced with permission from Hastings and Sweeney, 1960.) (b) Effect of ADP addition on NADH oscillations in an extract of *Saccharomyces carlsbergensis*. (Reproduced with permission from Pye, 1969.) (c) Effect of pulses of serum on anaphase frequency in V79 cells. Synchronous cultures were pulsed for 30 min with serum at the indicated times. The occurrence of cell division is plotted relative to the modal midpoint of the population of control cultures (——). For clarity not all data points are shown. (d) Phase response curve of V79 cells. Midpoints of ●, first; ○, second mitotic waves following synchronization were compared for each pair of serum-pulsed and control cultures as a function of time of beginning the serum pulse. Plus values indicate that pulsed cultures divided sooner than controls, negative values later than controls. ((c) and (d) reproduced by permission from Klevecz *et al.*, 1978.)

cycle oscillator with a 4 h period (Fig. 1.23). A biphasic response was obtained in the 8·5 h cell cycle of this cell-line. This conclusion is further confirmed by observations of phase response curves with similar periods when heat shocks or ionizing radiation were the perturbating influences (Klevecz *et al.*, 1980, 1981; King *et al.*, 1980). Pavlidis (1973) has shown that a limit-cycle oscillator will generate a phase-response curve whose repeating interval will be equal to the period of the clock.

D. ONE CELLULAR CLOCK?

Interactions between oscillations in different time domains (cell cycle—circadian, ultradian—circadian, epigenetic—cell cycle) would fit the notion that there is just one central timer, a cellular oscillator that controls the timing of epigenetic oscillations, mitosis, cell division, and circadian timing in cells. This hypothesis (Lloyd and Edwards, 1982) is an extension of the suggestion that mitotic and circadian timing share a common clock (Wille, 1979). It is suggested that the fundamental oscillator has a periodic mode characteristic for an organism or cell-line. In the ultradian mode, the period of its limit cycle may be 1 h (e.g. in *Acanthamoeba castellanii* (Edwards and Lloyd, 1978, 1980, 1981), or in mammalian cells 4 h (Klevecz, 1976; Klevecz *et al.*, 1978). For the most part, cell-cycle times will be in multiples of these fundamental frequencies. In the circadian–infradian mode, the period of the cellular oscillator becomes independent of an integral subdivision of the average generation time, and is equivalent to the circadian cell cycle period, i.e. there is a circadian oscillator (Wille, 1979). Transition from one periodic mode to another can occur in systems where periodic solutions bifurcate in at least several ways to give multiple-limit cycles (Othmer, 1975). The system can show hysteresis in flipping from one limit cycle to another depending on whether the value of the controlling parameter is increasing or decreasing (Fig. 1.24; Wille, 1979).

VIII. Concluding Remarks

It is evident that the living organism has a time-structure which is hierarchically organized and is at least as complex as its spatial structure. Present-day biochemistry to a large extent ignores this time-structure (an exception to this generalization is the field of rapid-reaction kinetics) while pursuing structural detail to submolecular

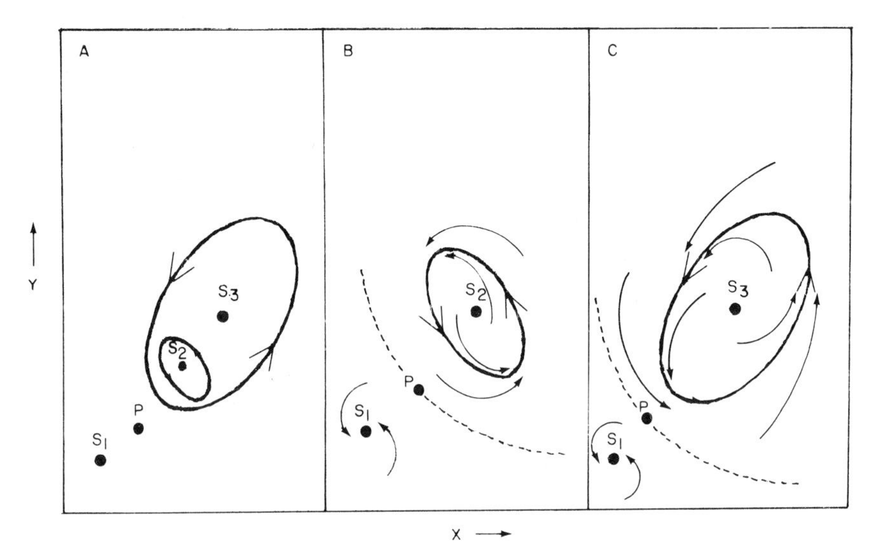

Fig. 1.24 Phase portrait of a hypothetical XY biochemical oscillator showing multiple critical points and limit cycles which result from parametric control of a hypothetical two-variable biochemical cellular oscillator. (A) The possible stable behaviours of the dynamic system as a value of the parameter, σ, are varied over a wide range. A single stationary point, S_1, exists for values of σ in the neighbourhood of σ_1. For very small changes in the values of σ near the bifurcation point, P, the system can flip from asymptotic stability at S_1 or to the low-amplitude limit cycle which orbits around the unstable focus, S_2. This possibility is shown in (B). For increasing values of the parameter σ greater than σ_2, the dynamic system has a periodic solution with a high-amplitude limit cycle oscillation around another unstable focus, S_3. The latter situation is depicted in (C). Heavy lines show the trajectories of the limit cycle path. Light arrows show the path of trajectories from the state of the system not on the limit cycle as they approach the limit cycle in time. The dotted lines extending away from P show the separatrix which divides the regions of attraction of the stable focus, S_1, from the basins of attraction of the limit cycles. (Reproduced with permission from Wille, 1979.)

dimensions. Where the dynamic aspects of protein conformation, membrane physiology, metabolic networks, and cellular growth and division *have* been examined, a rich array of temporal behaviour has been exposed, particularly the oscillatory behaviour of many of the cellular constituents, at all levels, and on all time scales (Rapp, 1979; Hirsch, 1980). In a consideration of the mechanism of evolution of circadian clocks, Hastings (1976) has suggested that:

"A byproduct of the recent ferment in theoretical population ecology is the recognition that stable steady-state (non-oscillating) operation is vanishingly rare in interactive systems of any complexity: complex feedback-regulated systems tend to instability and oscillations unless this is opposed by enormous selective pressure (May, 1973). It is not uncommon for such oscillations to exhibit much longer periods than the characteristic periods of the individual reactions involved (Stadtman, 1970). In a cellular system, cell types incapable of a true steady state operation might be the rule rather than the exception. The business of natural selection might thus be more to eliminate maladaptive oscillations, stabilize the periods (e.g. with respect to temperature) of those remaining, and couple them (e.g. by a photoreceptor) to a synchronizing environment, than to engineer a clock in cells that would otherwise not oscillate."

The hazards of overlooking cellular time-structure cannot be overstressed (Lloyd *et al.*, 1978c); it is evident that many of the dogmas currently dominating biochemistry, microbiology, and cell biology have become established from studies of exponentially growing (asynchronous) cell populations (or homogenates derived therefrom). Measurements made on heterogenous populations of this kind are time-averaged ones. Particularly suspect in this respect are claims that metabolite pools are present at "stabilized" levels, or that the accumulation of a constituent occurs "continuously". In other words observations made on biological processes always have a kinetic component, and it is imperative to avoid deriving from them "time-independent" conclusions. Weber (1976) has illustrated this point for measurements of rotational diffusion of proteins which suggest a compact shape. This conclusion is valid only for times of the order of a fluorescent lifetime; the measurements give no indication that the shape would remain the same over much longer periods.

"This situation is analogous with that involved in deciding whether a mountain is a rigid or a flexible structure. Evidently over a period of a human lifetime, of the order of tens of years, the mountain would not change appreciably its shape, but this conclusion could not be extended as valid over periods of geological time, of the order of 10^8 years. The latter time interval would correspond to 1 s on the molecular time scale if the year is 1 ns. We must therefore guard against deriving rigorous conclusions as the "molecular rigidity" from observations on the nanosecond time scale".

Similar comparisons could be made at every level within the temporal hierarchy of living systems which would illustrate the perils of disregarding the rich complexities and broad spans of its temporal organization.

Summary

Living organisms may be considered as hierarchical systems in both space and time; spatial organization has been explored more thoroughly than temporal aspects. The levels of organization within the temporal hierarchy range from ultrafast reactions on a subpicosecond time scale, through rapid reactions, metabolic, epigenetic, cell division cycle and circadian domains. A network of controls operates within and between time domains; the most studied example of molecular dynamics is a pathway of the metabolic domain. Control loops operating in the epigenetic domain are also recognized. At all levels, non-stationary state operation is likely; oscillatory behaviour has provided a mechanism for accurate timing of biological processes. No functional role for oscillations in the metabolic time domain has been definitely established, but it is suggested that some epigenetic oscillations with a temperature-compensated hourly period may represent a fundamental unit of time used both in timing of the cell division cycle and of circadian rhythms. This model can thus explain interactions between three different time domains.

2. The Cell Division Cycle: Methods of Study

"You can never step into the same river twice, for fresh water is always flowing towards you."

Heraclitus

Four types of system are available for cell division cycle studies: single cells, naturally-synchronous systems, synchronous cultures, and asynchronous exponentially-growing cultures. The general aims of such studies are to elucidate the time-course of metabolic, biosynthetic and morphological changes and the regulatory mechanisms involved during cellular growth and division. The attractions of performing experiments with an individual cell are offset by limitations of available techniques. Systems showing natural synchrony, e.g. early cleavages of sea-urchin eggs (cell division accompanied by a small mass increase), and *Physarum polycephalum* (synchronous mitoses of up to 10^8 nuclei in a common plasmodium, Guttes and Guttes, 1964) continue to provide many insights. An enormous range of procedures are now available for the preparation of synchronous cultures. We list the older well established methods for induced synchrony, and provide more detailed accounts of some newer selection methods, which are, at least for many purposes, preferable. Data from asynchronous exponentially growing cultures provide independent verification of results from synchronous cultures.

I. Some Comments on Perturbations

A major debate over the years between the induction and selection synchronists, as to what constitutes the "normal" cell cycle need not be reiterated here, but some comments on perturbations are relevant to many current studies. The deliberate use of physical or chemical agents to perturb biological systems has been, and continues to be, a major source of information on regulatory processes. To be able to make measurements on a system without disturbing it in some way is in biology, as in physics (Heisenberg, 1927), an ideal unattainable in practice. Nevertheless, progressive refinements to analytical procedures and to the methodology of selection synchrony, make studies of the minimally perturbed culture the eventual goal. What must we *not* do to organisms during our experiments? A list of instructions would read: (a) never remove them from the culture; (b) keep the culture under rigorously controlled conditions of temperature and light intensity and at constant gas tension; (c) never expose the organisms to extremes of centrifugal force, liquid shear or hydrostatic pressure; (d) carry out all procedures as rapidly as possible. Some of the recently developed methods for selection synchrony (see pp. 75–81) approach these ideal conditions more closely than has seemed hitherto possible. In those cases where organisms *must* be removed from the culture, they should be resuspended in conditioned growth medium (i.e. medium which has previously supported growth to an identical cell density), and the cell densities used should be as low as possible to minimize rate of change of composition of the medium during growth. Mitchison (1977c) has emphasized the doubts that attend the use of gradient methods for selection synchrony, and has pointed out that interpretation of data (even from improved experimental procedures) always requires the presentation of results obtained with identically-treated control (i.e. asynchronous) cultures. Again, it is almost always impossible to obtain the *ideal* control. Unequivocal evidence may require that (in addition to appropriate control experiments) events be followed through at least two successive cell cycles, and that similar results be obtained after different procedures for establishment of synchrony.

Some answers to the questions of the extent and duration of perturbation produced by altered conditions which may occur during the manipulation of cultures have been obtained for glucose-grown *Candida utilis* (Lloyd and Ball, 1979). Periods of anaerobiosis as short as 2 min produced an oscillation in respiration when the air supply was restored. Longer exposure to anoxia was followed by an overshoot in dissolved oxygen levels after switching back to a gas phase of air (Fig 2.1). Centrifugation at 2000 *g* for 2 min at 30°C gave minor disturbances over the following 30 min, although the attainment of steady state respiration was essentially monotonic. Little detectable perturbation of respiration was observed after 30 min aerobic exposure to 0°C or starvation (aeration in K-phosphate buffer at pH 6·0) for 3·5 h. These experiments suggest the limits to physiological disturbance which must not be exceeded for studies on respiration of essentially uperturbed (or minimally perturbed) cultures of the yeast. Here we observe primarily reactions of the metabolic time domain, but also see evidence (in the overshoot, Fig. 2.1) of slower adaptation (epigenetic time domain). For this system these effects can be distinguished from cell-cycle events as follows: (i) discontinuities produced by perturbation typically have periods of the order of minutes, whereas those

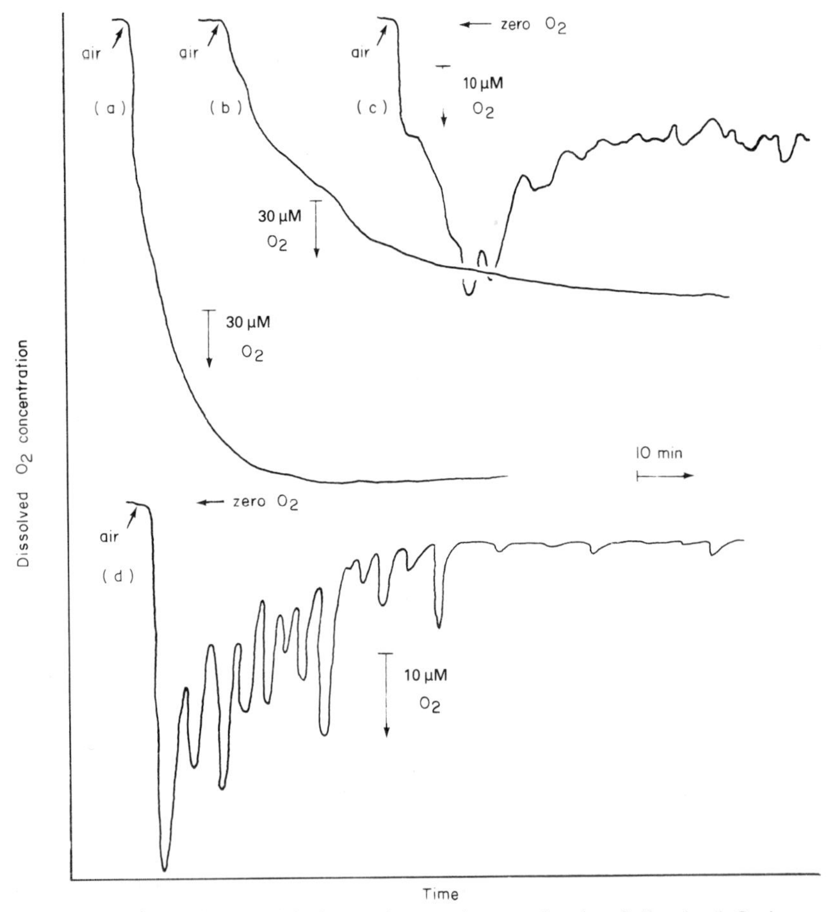

Fig. 2.1 Effects of anaerobiosis on the steady-state levels of dissolved O_2 in suspensions of *C. utilis*. Washed cell suspensions in 50 mM-potassium phosphate buffer (pH 7·2) containing 10^8 cells ml^{-1} were maintained under a gas phase of N_2 for 15 min with stirring, and then air was switched on as indicated: (*a*) no added substrate; (*b*) 50 mM-glycerol present; (*c*) 50 mM-glucose present; (*d*) Effect of anaerobiosis on the steady-state level of dissolved O_2 in a stationary-phase culture of *C. utilis* after growth with glucose. The culture contained $1{\cdot}0 \times 10^8$ cells ml^{-1}, and air was switched on after 25 min of N_2 flow. (Reproduced with permission from Lloyd and Ball, 1979.)

involving genuine cell-cycle events have longer periods (of the order of hours); (ii) perturbation often results in highly damped oscillations, whereas cell cycle-dependent oscillations are attenuated only as synchrony decays. The literature abounds with accounts of the effects of perturbating influences on cellular metabolism e.g. the effects of O_2 on glycolysis (Winfree, 1972) on cellular growth, the effects of nutritional shift-up or shift-down on bacterial growth in chemostats, (Maaløe and Kjeldgaard, 1966) and on circadian rhythms (Hastings and Schweiger, 1976). The effects of temperature shifts and nutritional deprivation on the events of the cell division cycle are well documented for several organisms (see Chapter 8), but these are usually extreme examples of environmentally-induced disturbances. Detailed information comes from the studies of phase-shifting of natural oscillators. Some of these ("soft" oscillators) are prone to disturbance by small external influences, and deviations from a limit cycle are righted only after multiple loops or oscillator periods. This is in contrast to those examples where almost immediate righting occurs ("hard oscillators"). Many experiments indicate that perturbation of oscillators leads to no permanent alterations of period or amplitude, but to a change of phase (phase resetting): this principle accounts for a diverse set of phenomena including the "set-backs" resulting from thermal shocks. The occurrence of limit-cycles at every level in the hierarchy of time domains of living systems may thus considerably reduce long-term sensitivity to external disturbances.

In the absence of general guiding principles, there is no substitute for checking individually each system of interest for its susceptibility to perturbation of the kind likely to be encountered in experimental procedures.

II. Methods for Study of Single Isolated Cells

Populations of cells sometimes display properties which are a consequence of cell–cell interactions (Gerisch *et al.*, 1975; Aldridge and Pye, 1976; Kohen and Kohen, 1977). The dispersion of cell generation-times sets the limit to the degree of synchrony attainable. Thus even in minimally-perturbed highly synchronous cultures there are at least two sources of discrepancy between the behaviour of the population and of its component individuals. Where there is a special interest in confirming that events observed in cultures are not due to cellular interactions (intercellular synchronization) or time-averaging, then experiments with single isolated cells are essential: e.g. circadian rhythms in *Gonyaulax polyedra* (Sweeney, 1960, 1963) and in *Acetabularia mediterranea* (Mergenhagen and Schweiger, 1974). Early work on the establishment of synchronous cultures by the selection of single cells, and the growth and respiration of single cells has been reviewed previously (Prescott, 1964, 1976a; Mitchison, 1971). Severe technical problems limit the range of non-invasive methods possible at present; the extent to which these methods disturb different organisms remains to be evaluated. Current developments in integrated micro-optics (Tien, 1977) may well provide a new impetus to single cell studies. Some of the methods currently available are:

(1) Time-lapse cinematography—cell growth (James *et al.*, 1975; Collyn-d'Hoaghe *et al.*, 1977);
(2) Interference microscopy—total dry mass (Mitchison, 1957);
(3) Cartesian diver technique—cell respiration or mass (Zeuthen, 1953; Chakravarty, 1976)—automated version (Oman *et al.*, 1977);
(4) Microspectrophotometry (Thorell *et al.*, 1965; Bendetti *et al.*, 1976; Bendetti and Lenci, 1977) DNA, cytochromes, chlorophyll and other chromophores; commercially-available machine from Shimadzu (MPS 50L);
(5) Microfluorimetry—redox states of nicotinamide nucleotides, (Chance *et al.*, 1973b; Kohen *et al.*, 1976) and some enzyme assays (Fig. 2.2) using fluorogenic substrates e.g. fluorescein-di-β-galactoside (Yashphe and Halvorson, 1976);
(6) Resonance Raman spectroscopy—haemoproteins (Adar, 1978)
(7) Electrophoresis of extracts of single cells—RNAs and proteins (Ko and Goldstein, 1978);
(8) Fluorescence immunological techniques—for monitoring DNA replication (Gratzner *et al.*, 1976).

III. Methods used for Monitoring Growth of Synchronous Cultures

Resistive particle counters (e.g. the Coulter Counter, Fig. 2.3) provide a rapid and accurate method for counting and sizing of cells (Harvey, 1968; Kubitschek, 1969). The large numbers counted make the data more statistically significant than cell counts as usually performed with a haemocytometer. For particles of bacterial size the elimination of background electrical interference may require screening of the instrument within a Faraday cage (Gear and Bednarek, 1972; Pickett

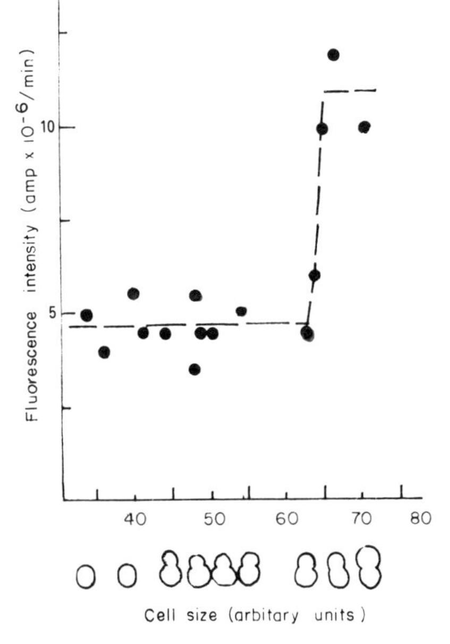

Fig. 2.2 Levels of β-galactosidase activity in single *S. lactis* cells during the cell cycle. The shape of the cells is drawn diagramatically. (Reproduced with permission from Yashphe and Halvorson, 1976.)

and Lester, 1979). Devices which automatically sample cultures for counting have been elaborated in many laboratories (e.g. Edmunds, 1965; Szyszko *et al.*, 1968; Rasmussen *et al.*, 1974) and a continuous flow counter has been constructed (Gear, 1976). Pulse height analysis the measurement of size distribution of cells requires the use of devices such as those described by Heath *et al.* (1973), Rackham (1977) or a Coulter Channelyser, and handling systems (e.g. Compucorp, Computer Design Corporation, Los Angeles, CAL). A modification of the flow system which focuses organisms into the sensing zone and reduces distortion of volume measurements is described by Kachel (1976) and its application is described by Roberts (1980). Measurement of impedance promises to give more information than resistance (Hoffman and Britt, 1979).

For budding yeasts, mild sonication facilitates separation of cells that have completed physiological division, but may otherwise remain adhered for generations (Scopes and Williamson, 1964). For some yeasts, sonication detaches even immature buds (Salmon, 1980). Alternatively visual examination in a haemocytometer requires a

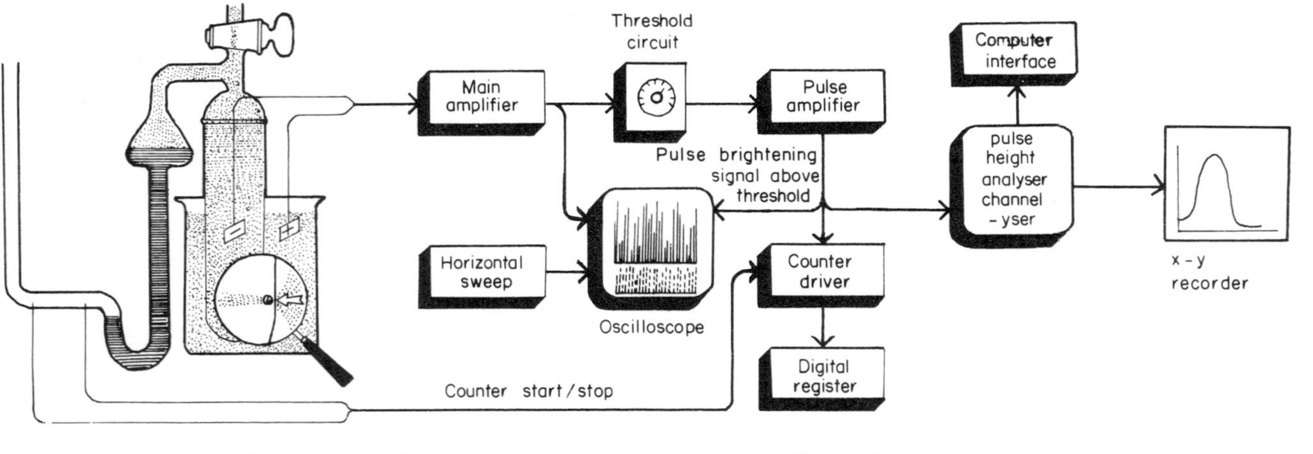

Fig. 2.3 Block diagram of the Coulter resistive particle counter, size distribution analyzer and data acquisition units.

standard convention; particles with single buds are scored as one cell, whereas those with two or three buds are scored as two cells (Williamson and Scopes, 1960). For fission-yeasts the analogous distinction between single cells and pairs of cells requires the observation of an indentation where the cell plate begins to form (Mitchison, 1970); for this organism the cell-plate index (percentage of cells engaged in the formation of transverse division plates) is a useful indicator of the degree of synchrony. Division indices (fraction of organisms which show cleavage furrows) have often been used to monitor synchrony of *Tetrahymena* (Zeuthen, 1964). Nuclear division has also been extensively used as a landmark for following synchrony; examples of methods used for yeasts, but useful also for determination of mitotic indices in other organisms, include the Giemsa stain (Williamson, 1966; Robinow and Marak, 1966) and fluorescent staining with acridine orange (Rustad, 1959) or mithramycin (Slater, 1976). Newer methods for monitoring cell cycle traverse include measurement of nuclear volumes by flow microfluorimetry (Zucker *et al.*, 1979c; Yen and Pardee, 1979). Automated recognition and counting of cells in various phases of the cell cycle is also achieved by image-processing and analysis of fields of cells observed microscopically (Brugal and Chassery, 1977; Nicolini *et al.*, 1977).

IV. Assessment of the Degree of Synchrony

Throughout this book we use the synchrony index of Blumenthal and Zahler (1962) to express the degree of synchrony of cultures prepared by selection methods. This is calculated from the equation:

$$F = (N_t/N_0 - 2^{\alpha/\beta})$$

where N_t is the number of organisms after division is completed, N_0 the number of organisms before division, α is the time taken for the organisms to divide, and β is the mean generation time (Fig. 2.4). For a culture showing instantaneous and exact doubling in cell numbers the value of F would be unity, whereas for an exponentially-growing culture its value would be zero. Values greater than 0·85 have not been encountered in practice (except where possibly an element of induction arises from the experimental procedure). Values between 0·6 and 0·85 are regarded as satisfactory. In our experience, synchronous cultures established from mid-exponential growth phase populations usually show a precise doubling (although exceptions have been

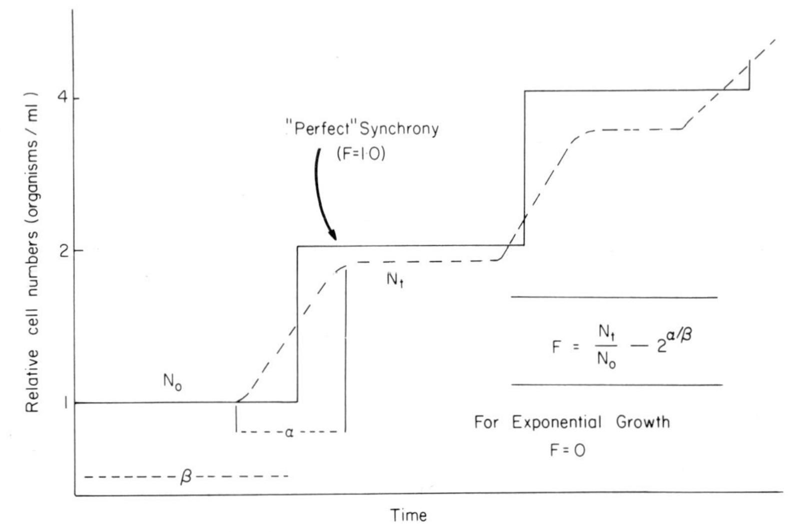

Fig. 2.4 Assessment of the degree of synchrony: the Synchrony Index of Blumenthal and Zahler (1962). For explanation see text.

noted in special cases; e.g. for *Crithidia fasciculata* which has a dyskinetoplastic subpopulation, (Edwards *et al.*, 1975) and for budding yeasts at slow growth rates (Carter and Jagadish, 1978a, see p. 369).

Other synchrony indices include those reviewed by Scherbaum (1963), Engelberg (1964) and Burnett-Hall and Waugh (1967). More recent treatments are by Anagnostopoulos (1975) and Wolosker and De'Almeida (1979). A probit transformation procedure for computer-calculated synchrony evaluation has been worked out by Hagar and Punnett (1973).

V. Induction Procedures for the Establishment of Synchronous Cultures

A. CYCLIC CHANGES OF TEMPERATURE OR PRESSURE

Temperature-induced synchronous division of *Tetrahymena pyriformis* has provided a wealth of information on the control of cell division and has been popular as a method of producing synchronous cultures of protozoa and bacteria ever since its introduction by Scherbaum and Zeuthen (1954). In the original method the temperature was shifted from 29°C to 33°C through 6 to 10 cycles with 0·5 h spent at each temperature. During this heat treatment, cell divisions are blocked, the average cell volume increases about three-fold, and the last heat shock is followed by a burst of cell division after 100 min at 24°C. Three successive division peaks then occur closer in time than in normal mitotic cycles (1·7 h as against 2·5–3 h at 24°C). Automation of this method was described by Scherbaum and Zeuthen (1955). A cold-shock method was devised for *Tetrahymena pyriformis* by Padilla and Cameron (1964) and a similar method works for *T. vorax* (Buhse and Rasmussen, 1974). Widely used for induction of synchrony in bacteria, cold shock gives distorted cell cycles (for early references see Mitchison, 1971); e.g. a two-step doubling has been observed in *Escherichia coli* K12 after a single change of temperature (Wolosker and De'Almeida, 1975, 1979).

A newer system for synchrony of *Tetrahymena* uses intervals at 34°C for periods of 30 min alternating with intervals at the optimum temperature (28°C) for periods of 160 min (Zeuthen, 1971, 1974). The time interval between shocks equals the doubling time (Fig. 2.5), and each repetitively-synchronized cell division leads to a synchronized S-period of DNA synthesis which exceeds the S-period in normal single cells by 10–20 min (accounted for by imperfect synchrony). This method can also be applied successfully to continuous-flow cultures (Zeuthen, 1974).

Cultures of *Schizosaccharomyces pombe* can be division-synchronized when treated with 5–6 heat shocks during each of which the temperature is elevated to 41°C for 30 min (Kramhøft and Zeuthen, 1971, 1975). Between heat shocks the optimal temperature for growth and division (32°C) is maintained for 110 min: an automated device is programmed to provide these conditions. A comparison of this system with synchronous cultures obtained by selection synchrony indicates normal cycles with respect to size changes, DNA synthesis and nuclear division in the heat treated cells (Kramhøft *et al.*, 1976); thus there is no indication that dissociation of the growth cycle from the DNA-division cycle occurs (as was the case for the methods originally employed for induction of synchronous cell division in *Tetrahymena*). However, the patterns of accumulation of at least two enzymes (aspartate transcarbamylase and ornithine transcarbamylase) are different after induced synchrony, showing steps rather than continuous accumulation. Mechanisms by which cyclic temperature changes induce synchronous cell division have been extensively discussed (Zeuthen, 1974; Rooney and Costello, 1977; Polanshek, 1977).

Temperature changes under constant conditions of illumination have been used to synchronize some algae, e.g. *Euglena* (Neal *et al.*,

1968; Padilla and Bragg, 1968; Terry and Edmunds, 1970), *Chlamydomonas* (Rooney *et al.*, 1971) and *Dunaliella* (Wegmann and Metzner, 1971). The colourless forms *Astasia longa* (Padilla and Cook, 1964) and *Polytomella agilis* (Cantor and Klotz, 1971) have also been synchronized by this technique as has the cyanophyte, *Anacystis nidulans* (Lorenzen and Venkataraman, 1972) and the trypanosomatid protozoon, *Crithidia oncopelti* (Newton, 1957).

Changes of hydrostatic pressure have been used to induce synchronous division in *Tetrahymena pyrifomis* (Zimmerman and Laurence, 1975). Pressure jumps to 340–680 × 10^5 N m^{-2} for 1–4 min through 5 to 7 cycles were employed with 30 min recovery periods. The division delays produced by high pressure cannot be accounted for entirely in terms of inhibited protein synthesis (Walker and Wheatley, 1979).

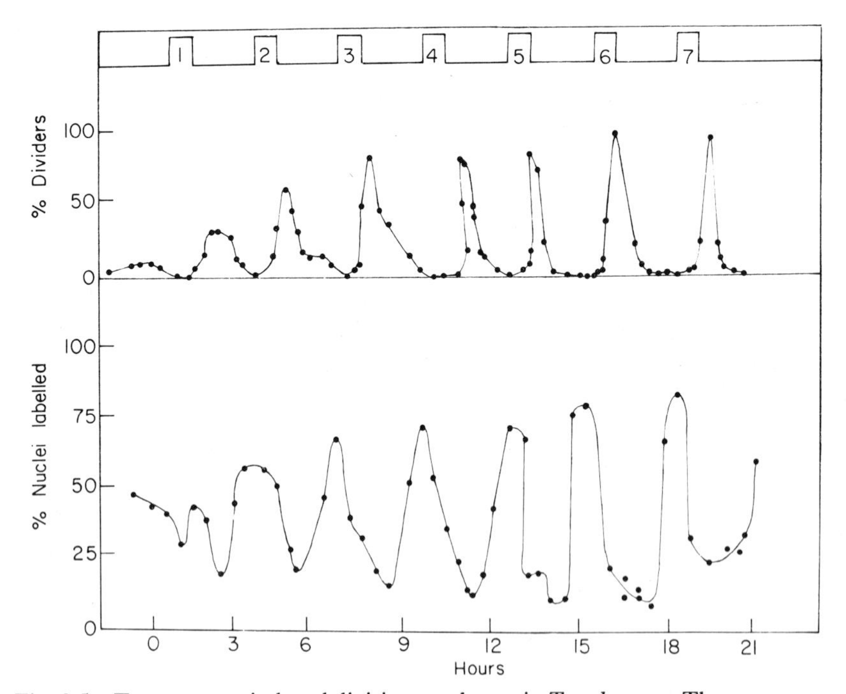

Fig. 2.5 Temperature-induced division synchrony in *Tetrahymena*. The temperature cycle is shown on top and the time is given below. The temperature was changed between 28°C (optimal, held for 160 min between shocks) and 34°C (division-blocking, held for 30 min in each of the 7 shocks shown). *Upper ordinate:* percentage of cells in division. *Lower ordinate:* percentage of cells in replication. (Indicated as % cells having a labeled macronucleus after a 15 min pulse with [^{14}C]-thymidine. Points are in the middle of the incorporation intervals). (Reproduced with permission of Zeuthen, 1974.)

B. INHIBITOR-INDUCED SYNCHRONY

Inhibitors which block cell cycle traverse at specific points may be used to induce synchronous cell division (Fig. 2.6). In the presence of the inhibitor, cells accumulate at one stage of the cell division cycle: subsequent removal (washout), or in some instances spontaneous recovery, gives synchronous release and cell-cycle progression. Table 2.1 lists some examples of the use of inhibitors for induction synchrony. These methods commonly used for mammalian cell lines, inevitably lead to perturbation of macromolecular synthesis e.g. the use of DNA synthesis inhibitors can lead to dissociation of a "growth cycle" from a "DNA cell division cycle" (Mitchison, 1971) or can obscure natural periodicities. Some methods claim to minimize metabolic interference by using short periods of blocking at suboptimal inhibitor concentrations (Thilly, 1976) but reports of obvious side-effects of cytotoxic agents e.g. of colcemid (Kato and Yosida, 1970) question whether even these procedures can give information on "natural" cell cycles. Nevertheless a great deal of information on the mechanism of cell division and its control has come from these studies.

C. HYPOXIC SHOCK AND HYPERBARIC O_2

Delayed division in *Tetrahymena* is induced by short-term exposure to anaerobiosis; Rasmussen (1963) employed two 50 min periods of anoxia separated by a 40 min growth period. This observation has been employed to devise a method to synchronize cell division by multiple anoxic shocks (Rooney and Eiler, 1967, 1969). However, more recently, it was shown for *T. pyriformis*, strain W, that a single period of anaerobiosis (4 h) is followed by synchronous cell division 105 min later ($F = 0{\cdot}81$) (Dickinson *et al.*, 1977; Gray *et al.*, 1977).

Hyperbaric O_2 leads to a delay in cell-cycle transit after DNA synthesis, but before metaphase in human diploid W1–38 cells (Balin *et al.*, 1978); this observation has not yet been exploited for the induction of synchrony.

D. NUTRIENT DEPRIVATION FOLLOWED BY REFEEDING

Synchronous cell division has been obtained in bacteria, lower eukaryotes and in higher animal and plant cell cultures by depletion and resupplementation of an essential nutrient. Bacterial examples include *Escherichia coli*, *Proteus vulgaris* and *Caulobacter crescentus* taken from

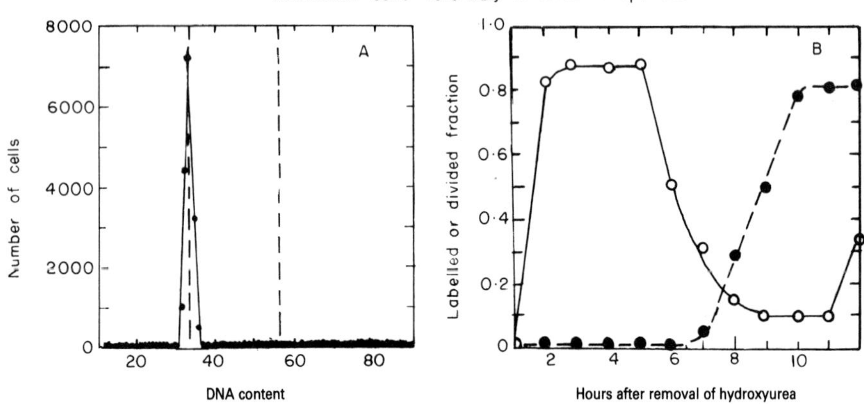
Mammalian cells reversibly arrested in G_1 phase
A
Number of cells
DNA content
B
Labelled or divided fraction
Hours after removal of hydroxyurea

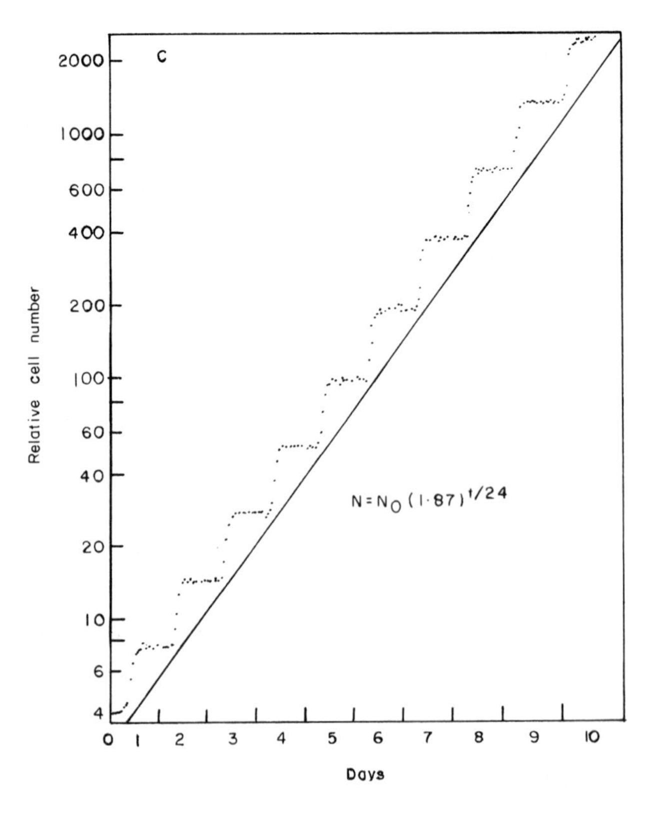
C
Relative cell number
$N = N_0 (1{\cdot}87)^{t/24}$
Days

Table 2.1 *Inhibitor-induced synchrony*

DNA synthesis inhibitors		
Thymidine (at high concentrations ~ 2 mM)	Chang appendix cells	Xeros (1962)
	S3 HeLa	Puck (1964)
	CHO cells	Petersen and Anderson (1964)
	Human kidney cell line	Bootsma *et al.* (1964)
	HeLa	Volpe and Eremenko (1973)
	Human lymphoblast	Zielke and Littlefield (1974)
	Murine mastoma	Thomas and Lingwood (1975)
	HeLa	Thilly (1976)
[^{3}H]-Thymidine (high specific radioactivity leading to selective killing)	L6OT cell line	Whitmore and Gulyas (1966)
Amethopterin (methotrexate)	HeLa	Rueckert and Mueller (1960)
	HeLa	Schindler (1960)
	HeLa	Mueller and Kajiwara (1966)
5-Fluoro-2′deoxyuridine	HeLa	Eidinoff and Rich (1959)
	HeLa	Schindler (1960)

Fig. 2.6 Evidence that cells recovering from ile$^-$-mediated G1 arrest in the presence of hydroxyurea accumulate at the G1/S boundary. (a) Cells cultivated for 30 hours in ile$^-$ medium were resuspended in fresh, complete medium containing isoleucine and hydroxyurea to 10^{-3} M for 10 h, at which time an aliquot was removed for flow microfluorometric analysis. Immediately thereafter the cells were spun down, washed, and resuspended in fresh medium without hydroxyurea. (b) At times thereafter aliquots were pulse-labelled for 15-min periods with 2 μCi/ml of [^{3}H]-thymidine, and autoradiographs were prepared for determination of the labelled fractions (open figures). Aliquots were also removed for determination of cell number with an electronic particle counter (solid figures). (Reproduced with permission from Tobey, 1973.) (c) Growth of HeLa S_3 cell culture in which synchrony was maintained by resynchronization with periodic exposure to 0·25 mM thymidine. A line describing exponential asynchronous growth in which 87% of the population is able to divide in a 24-h period has been placed on the curve to emphasize the fraction of cells dividing in each cycle. Each point is the result of duplicate counts of samples on an electronic particle counter. (Reproduced with permission from Thilly *et al.*, 1976.)

Table 2.1 *continued*

DNA synthesis inhibitors		
	L cells	Littlefield (1962)
	CHO cells	Stubblefield (1968)
	Saccharomyces cerevisiae	Esposito (1968)
2′-Deoxyadenosine	HeLa	Mueller and Kajiwara (1966)
	Schizosaccharomyces pombe	Mitchison and Creanor (1971); Sissons *et al.* (1973); Poole (1977c)
Aphidicolin	HeLa	Pedrali-Noy *et al.* (1980)
5-Aminouracil	*Vicia faba*	Smith *et al.* (1963)
	Zea mays	Clowes (1965)
	Haplopappus gracilis	Eriksson (1966)
	Various plant tissues	Novák *et al.* (1979)
2′-Deoxyguanosine	HeLa	Mueller and Kajiwara (1966)
Hydroxyurea (~ 0·5 mM)	CHO cells	Sinclair (1965, 1967)
	Higher plant (*Haplopappus*) cells	Eriksson (1966)
	Allium sativum root-meristems	Clain and Brulfert (1980)
	Mouse fibroblast (L) cells	Adams and Lindsay (1967)
	Crithidia luciliae	Steinert (1969)
	Crithidia luciliae	van Assel and Steinert (1971)
	Leishmania tarentolae	Simpson and Braly (1970)
	Tetrahymena pyriformis	Worthingon and Nachtwey (1976)
	Burkitt lymphoma cells	Miyamoto *et al.* (1976)
	Polytomella agilis	Cantor *et al.* (1978)
	Schizosaccharomyces pombe	Miyata *et al.* (1978a)

Table 2.1 *continued*

DNA synthesis inhibitors		
Hydroxyurea *(contd.)*	*Schizosaccharomyces pombe*	Miyata *et al.* (1978a, b, 1979)
Phenylethanol	*Mycobacterium phlei*	Konicek (1977)
	Rhizobium trifolii T37	Zurkowski and Lorkiewicz (1977)

Protein synthesis Inhibitor		
Chloramphenicol	*Schizosaccharomyces pombe* (glycerol grown)	Quinton and Poole (1977)

Metaphase Inhibitors		
Colchicine or colcemid	HeLa	Marcus and Robbins (1963) Kim and Stambuck (1966) Scharff and Robbins (1966) Pfeiffer and Tolmach (1967) Ohara and Terasima (1976)
	L 5178 Y cells	Doida and Okada (1967a, b)
	HeLa	Stubblefield (1968)
	malignant melanoma	Romsdahl (1968)
	Tetrahymena pyriformis	Wunderlich and Peyk (1969)
Vineblastine SO_4	HeLa	Marcus and Robbins (1963)
	HeLa	Pfeiffer and Tolmach (1967)
	Intestinal crypt (*in vivo*)	Verbin and Farber (1975)
	Tetrahymena pyriformis	Stone (1968) Sedgley and Stone (1969)

Table 2.1 *continued*

Metaplase inhibitors	
N_2O	Rao (1968)
	Cullen *et al.* (1979)
Nocodazole	Zieve *et al.* (1980)
Ansamitocin	Tanida *et al.* (1980)
G_1 Arrest	
S. cerevisiae α-factor	Hartwell (1973b)
	Chan (1977)
	Shulman (1978)

stationary-phase cultures and inoculated into fresh medium (Cutler and Evans, 1966; Nishi *et al.*, 1967; Iba *et al.*, 1975). Specific nutrient deprivation and replacement in *Lactobacillus acidophilus* R26 with thymidine (Burns, 1964), in *Escherichia coli* B/r/1 with glucose (Buckley and Anagnostopoulos, 1975), in *E. coli* with thymine (Barner and Cohen, 1956), and in *E. coli* B and *E. coli* K12 with isoleucine, leucine, threonine, methionine, histidine, proline and tyrosine (Ron *et al.*, 1975) are all conditions which yield synchrony of cell division. Amino acid auxotrophs are especially useful for this experimental approach (Matney and Suit, 1966; Stonehill and Hutchison, 1966; Chan and Cheng, 1977; Silver *et al.*, 1977b; Ron *et al.*, 1977; Allin and Guetard, 1979).

For *Saccharomyces cerevisiae*, Williamson and Scopes (1960, 1961a) developed a method which used a stationary-phase culture as the starting point. Cycles of starvation and refeeding were combined with temperature changes in this early method, but it was then found that the latter could be eliminated without loss of synchrony (Williamson, 1964a). Concentrated suspensions of *Schizosaccharomyces pombe* incubated in growth medium until nutrients are depleted and then diluted with fresh medium give synchronous cultures (Sando, 1963). This method also can be applied to *S. cerevisiae* (Tauro and Halvorson, 1966) and *Tetrahymena, Acanthamoeba* and *Dictyostelium* can also be synchronized in a similar manner (Cameron and Bols, 1975; Chagla and Griffiths, 1978; Zada-Hames and Ashworth, 1978). Algae synchronized by starvation and refeeding include *Astasia longa* starved for sulphate (Morimoto and James, 1969), *Polytomella caeca* (M. H. Cantor and D. Lloyd, unpublished) and *Polytomella agilis* (Cantor and Burton, 1975) starved for thiamine, silicon-deprived diatoms (Lewin *et al.*, 1966; Busby and Lewin, 1967; Coombs *et al.*, 1967a, b), and vitamin B_{12}-starved *Euglena gracilis* (Shehata and Kempner, 1979).

In a variety of mammalian cell lines (CHO cells, Syrian hamster and mouse L cells) isoleucine deficiency for a period of 20–24 h gives G_1 arrest (Tobey and Ley, 1970, 1971) and synchronous cell division is observed on re-addition of this amino acid. The effect appears to be specific, as depletion of other amino acids gave general inhibition of cell cycle traverse (Tobey, 1973).

E. CONTINUOUS PHASED GROWTH AND PERIODIC FEEDING

Several systems of continuous phased growth have been developed. For *Candida utilis* (Dawson, 1965, 1966, 1968; Muller and Dawson, 1968) the continous supply of medium to a chemostat is interrupted and collects during a doubling time, when its volume will equal that of the culture. It is then automatically mixed with the culture, doubling the culture volume and halving the cell density. One half of this diluted culture is removed and the cycle repeats. Claims that this system gives valid information on changes of respiration and other energy-yielding pathways during the cell division cycle (Dawson and Westlake, 1975; Dawson, 1977; Thomas and Dawson, 1977) do not take into account the disturbing effect of periodic environmental changes on the organisms. This technique has also been applied to cultures of *Azotobacter vinelandii* (Kurz *et al.*, 1975).

Division synchrony may also be maintained in a chemostat by adding the limiting nutrient periodically, the period of addition being equal to the generation time, i.e. one addition per cell-generation (Goodwin, 1969a,b,c). For *Escherichia coli* growing under phosphate-limited conditions good synchrony was obtained when pulses of phosphate (final concentration 0·18 mM) were delivered at intervals set equal to the mean generation time (calculated from the dilution rate). After several cycles of phosphate starvation the environmental periodicity induces entrainment of cell division. The oscillations of enzyme activities observed in these cultures may also result from the induction procedure. Similar methods have been used by von Meyenburg (1969) for *Saccharomyces cerevisiae*, Kjaergaard and Joergensen (1979) for *Bacillus subtilis*, and Kepes and Kepes (1980) for *E. coli*. Some conditions (e.g. periodic feeding with glucose at intervals longer than the generation time) in *E. coli* B/r/l (Anagnostopoulos, 1971; Buckley and Anagnostopoulos, 1975) and in *Schiz. pombe* (Anagnostopoulos and Salvesen, 1977) give rise to two independently dividing subpopulations.

F. CHELATING AGENTS

Both the divalent cation ionophore A23187 and the chelating agent EDTA inhibit cell division in *Schiz. pombe* strain 132 (Duffus and Paterson, 1974a,b; Ahluwalia *et al.*, 1978); when cells are released from this inhibition by inoculation into fresh medium subsequent divisions show some synchrony. Similar results have been obtained for *Kluyveromyces fragilis* (Penman and Duffus, 1975; Walker, 1978). It is suggested that Mg^{2+} plays an important role in the control of the yeast cell division cycle (Walker and Duffus, 1980) and this hypothesis is strengthened by the observation that synchrony of cell division of *Schiz. pombe* can also be induced by controlled exposure to tetrasodium pyrophosphate (40 mM, 1 h, $F_1 = 0{\cdot}68$) or citric acid (50 mM, 1 h, $F_1 = 0{\cdot}68$) (Walker and Duffus, 1979). Other chelating agents (8-hydroxyquinoline, 1,10-phenanthroline and EGTA) and Mg^{2+} sequestering antibiotics (novobiocin and lomofungin) were not effective. Chelating agents differ from DNA synthesis inhibitors as synchrony inducers, in that exposure for periods much shorter than a complete cell cycle can be employed. 1,10-Phenanthroline (but not the non-chelating 1,7 analogue) inhibits cell cycle traverse in lymphoblasts; it is suggested that Zn^{2+} is the critical divalent ion in this case (Falchuk and Krishan, 1977).

G. LIGHT–DARK SYNCHRONIZED CELL DIVISION

Light inhibits cell multiplication in many phototrophs, and synchrony can often be achieved by imposing suitable regimes of illumination interspersed with periods of darkness (Spudich and Sager, 1980). Growth conditions should be optimized to guarantee rapid cell divisions and shortest possible generation times (Lorenzen and Hesse, 1974). This approach, pioneered by Tamiya *et al.* (1953) opened a wide field of study; so rather than give individual methods, we present a short representative list of selected examples (Table 2.2) and a more detailed list for *Chlamydomonas* (Table 2.3). Some of these methods (including the original one) include a centrifugal selection step. For *Chlorella* and *Chlamydomonas*, cell multiplication can respectively involve factors as high as 20 (Lorenzen, 1964) or 16 (Lien and Knutsen, 1979) (Fig. 2.7). Continuous synchrony is obtained by continuous dilution with fresh medium (Pfau *et al.*, 1971) or by periodic dilution, once in each cycle (Lorenzen, 1964; Padilla and Cooke, 1964; Padilla and James, 1964), or to constant initial turbidity (Schmidt, 1974a,b).

Table 2.2 *Some Systems for Synchronous Cell Division Induced by Light–Dark Regimes*

Species	Reference
Anacystis nidulans	Herdman *et al.* (1970)
	Lindsey *et al.* (1971)
	Csatorday and Horvath (1977)
Chlamydomonas spp.	See Table 2.3
Chlorella ellipsoidea	Tamiya *et al.* (1953)
	Tamiya *et al.* (1961)
	Tamiya (1966)
Chlorella sorokiniana (earlier called *C. pyrenoidosa* strain 7-11-OS)	Talley *et al.* (1972)
	Schmidt (1974a,b)
	Israel *et al.* (1977)
Chlorella pyrenoidosa	Kuhl and Lorenzen (1964)
	Senger (1965)
	Howell *et al.* (1967)
	Senger and Bishop (1969)
	Lorenzen (1970)
	Wanka *et al.* (1970)
Chlorella fusca-8p (*C. pyrenoidosa* 211-8p, Cambridge)	McCullough and John, (1972)
Euglena gracilis	Padilla and Cook (1964)
	Padilla and James (1964)
	Edmunds (1964)
	Osafune *et al.* (1975a)
Navicula pelliculosa	Darley and Volcani (1971)
	Darley *et al.* (1976)
Platymonas striata Butcher	Ricketts (1977a, 1979)
Scenedesmus quadricauda	Komarek *et al.* (1968)
Scenedesmus obliquus	Senger (1970a,b)
	Myers and Graham (1975)
	Ober (1974, 1975)
Volvox carteri	Yates *et al.* (1975)

Alternating white light with yellow light through three cycles accompanied by changing the gas phase from air to air + CO_2 provides a method for synchrony of several green algae (Carroll *et al.*, 1970).

At low growth rates a number of non-photosynthetic eukaryotes show synchronous cell division with a persistent circadian periodicity after entrainment by light-dark cycles (Table 1.1) (Edmunds, 1975).

Table 2.3 *Synchronous cultures of* Chlamydomonas; *Summary of methods of synchronization, culture conditions and culturing set-ups*. (*Reproduced with permission from Lien and Knutsen, 1979*).

Reference	Species (strain)[a]	Synchronization method: LD: temp change: selection synch.	Culture conditions and culturing set-ups: Light Klx	Temp °C	% CO_2 in air[b]	Dilution: Serial[c]	Dilution: No serial[d]	Progeny average number	Culture vol (ml) LP[e]	Medium: inorg org.: N-,C-source: pH
Wetherell (1958)	*C. eugametos* Moewus	15–18: $\overline{12}$, LD	15	25	4		?	8	?[f]	inorg: NO_3^-, CO_2: 6·8
Bernstein (1960, 1964)	*C. moewusii* Gerloff	12: $\overline{12}$, LD	9	25	5		$5\times10^4\rightarrow$ $< 10^5\rightarrow$	8, 1.cycle 2.5, 2.cycle < 8	2000–4000 LP = ?	org: NH_4^+, CO_2, citrate: 6·8
Kates and Jones (1964)	*C. reinhardii* Dangeard (89.90 UTEX) *C. moewusii* (96.97 UTEX)	12: $\overline{12}$, LD	3–10	21 25	1–5		2×10^5	2–5	2500 Fernbach flask LP = ?	inorg., HSM[g]. NH_4^+, CO_2
Surzycki (1971)	*C. reinhardii* (137C)	12; $\overline{12}$, LD	6	21	5		$5\times10^4\rightarrow$ $5\times10^6\rightarrow$	4, 1.cycle 4, 2.cycle 3, 3.cycle 2, 4.cycle no division	4000 Erlenmeyer flask LP = ?	inorg.: NH_4^+, CO_2: 6·7–6·8
Mihara and Hase (1971)	*C. reinhardii* (IAM C-9-Tokyo)	12: $\overline{12}$, LD	4–6	25	2–3	2×10^6		2–4	50, 500 LP = 1·5, 2·5	inorg., 3/10 HSM[g]: NH_4^+, CO_2: 6·8
Cavalier-Smith (1974)	*C. reinhardii* (11/32d CCAP)	12: $\overline{12}$, LD	6·5	25	0 air only		3×10^3	2–5	2000 Erlenmeyer flask LP = ?	org.: NO_3^-. NH_4^+, citrate; 6·8

Schlösser (1966)	*C. reinhardii* (11-32 Gö)	12: $\overline{12}$, LD	20	34	2	$1 \cdot 56 \times 10^6$		14	300 tubes LP = 4–5	inorg.: NO_3^-, CO_2: 6·8
Senger (1975)	*C. reinhardii* (11–32 Gö)	14: $\overline{10}$, LD	15	28	3	$1 \cdot 5 \times 10^6$		10	? LP = 4–5	inorg: NO_3^-, CO_2: 6·8
Lien and Knutsen (1979)	*C. reinhardii* (11-32/90 Gö) (11/32b CCAP)	12: $\overline{4}$, LD	20	35	2	$1 \cdot 4 \times 10^6$		16	300 tubes LP = 4	inorg.: NO_3^-, CO_2: 6·8
Vaage (1973)	*C. moewussi* var. *rotunda* Tsubo (577 UTEX)	14: $\overline{10}$, LD	10	30	2	$2 \cdot 0 \times 10^6$		7–8	300 tubes LP = 4	inorg.: NO_3^-, CO_2: 6·8
Lien and Knutsen (1976)	*C. reinhardii* (CW 15—mutant)	14: $\overline{10}$, LD	12–15	30	2	$1 \cdot 0 \times 10^6$		5	300 tubes LP = 4	inorg., modified HSM[g]: NH_4^+, CO_2: 6·8

[a]CCAP = Cambridge University Culture Collection: Gö = University of Göttingen Algal Collection; UTEX = The University of Texas at Austin Culture Collection.
[b]Bubbling through culture.
[c]Serial = many cycles; each cycle starting at cells/ml.
[d]No serial = few cycles, 1 cycle starting at cells/ml.
[e]Culture = culture vessels; volume in ml: LP = light path in cm.
[f]? = no information.
[g]HSM = high salt medium.

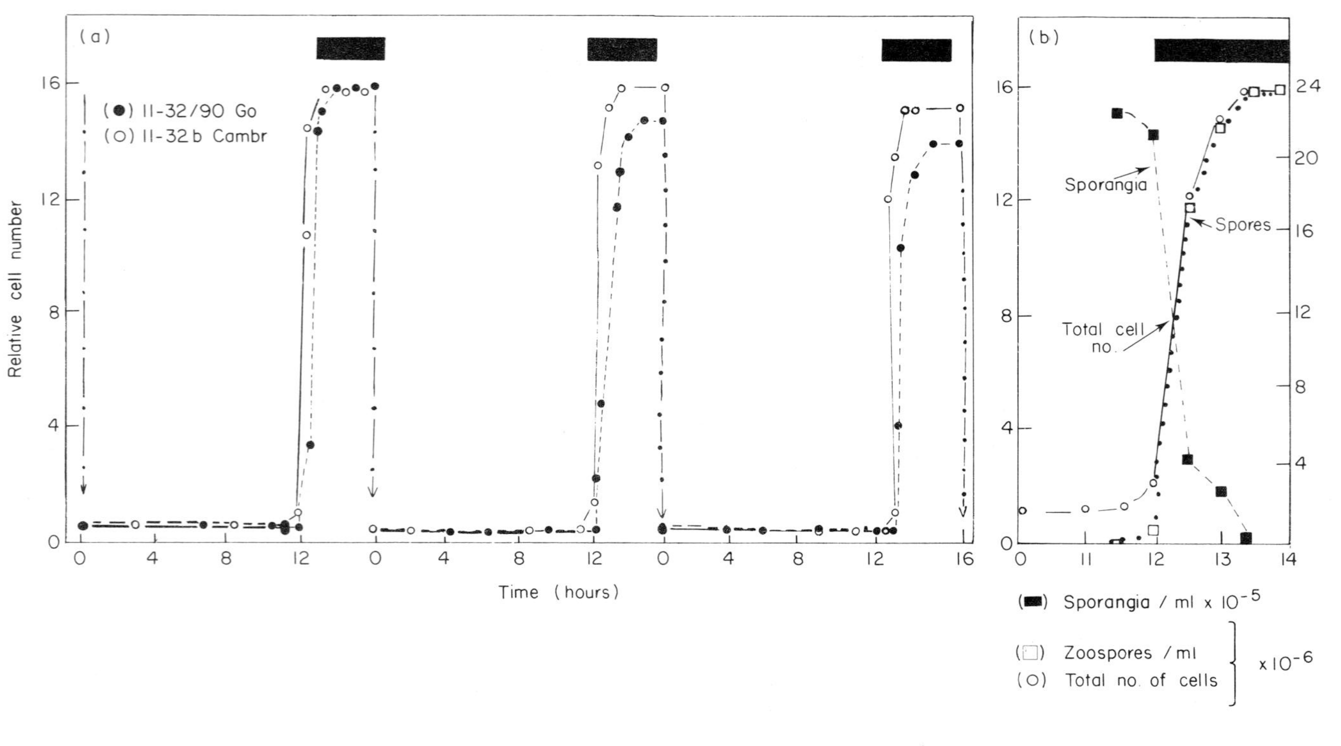

Fig. 2.7 (a) Light–dark induced synchrony of two strains of *Chlamydomonas reinhardii*. Relative cell number during three synchronous cycles measured one cycle apart: arrows indicate dilution of cultures at end of dark period to $1{\cdot}5 \times 10^6$ cells/ml (dark phases indicated by horizontal black bars). (b) Time course of sporulation with disappearance of sporangia, appearance of zoospores and increase in total cells/ml of Cambr. strain. (Reproduced with permission from Lien and Knutsen, 1979.)

VI. Selection Procedures for the Establishment of Synchronous Cultures

A. MITOTIC SELECTION

Selective detachment of mitotic cells from monolayer cultures by gentle mechanical means (Axelrod and McCulloch, 1958) provides the most satisfactory method for selection synchrony of cultured mammalian cell lines (Terasima and Tolmach, 1963a, b; Fig. 2.8). Modifications and developments of the original method of "mitotic wash-off" have provided procedures which give larger yields of synchronously dividing cell populations (using automated semi-continuous or continuous removal from rotating bottles) and improved synchrony (standardized mechanical shaking) (Lindahl and Sorenby, 1966; Petersen *et al.*, 1968; Highfield and Dewey, 1975). Some methods involve cold storage of batches of cells after wash-off (Tobey *et al.*, 1970; Yamaguchi *et al.*, 1977) but accumulation of stored cells may distort metabolic events (Ehmann and Lett, 1972). Other modifications may also be regarded as potentially perturbing, e.g. pretreatment of cultures with low calcium medium (Robbins and Marcus, 1964), use of trypsinization (Badger and Cooperband, 1976), accumulation of detached cells in metaphase by means of a colcemid block (Gaffney, 1975) or by hypertonic arrest (Wheatley, 1976). The selection process has been improved by combination of wash-off with sedimentation selection on a column of culture medium (Ooka and Daillie, 1974). An automated apparatus which selects from a random exponential culture a population of mitotic cells which are collected and dispensed into individual growth flasks, has been designed by Klevecz (1972; 1975; Fig. 2.9) and is now commercially available. This method does not rely on colcemid alignment, cooling or low-calcium pretreatment; 10^7 cells can be automatically and reproducibly prepared for every point required in the cell division cycle.

B. SIZE SELECTION BY FILTRATION

Selection synchrony of bacteria by filtration through piles of filter paper was first described for *Escherichia coli* B by Maruyama and Yanagita (1956). Using 2 layers of Whatman No. 1 on top of 18 coarser-grade papers (55 mm diameter) approximately 10^{11} cells can be selected (Maruyama, 1964). This technique was modified for use with *Alcaligenes faecalis* (Lark, 1958; Lark and Lark, 1960) and *Bacillus*

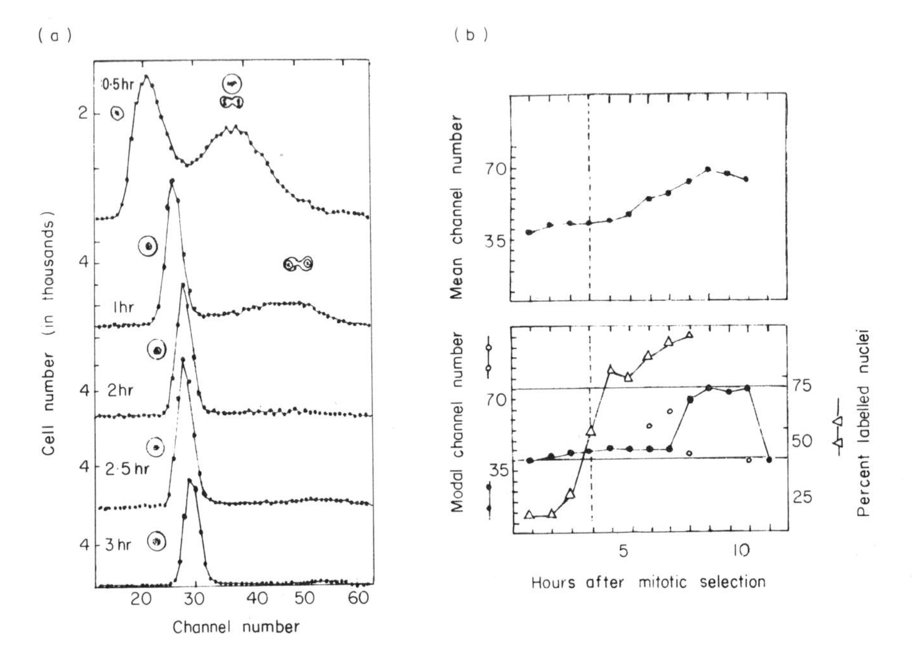

Fig. 2.8 (a) Distribution of DNA-content per cell from synchronous CHO cells. CHO cells were synchronized as described in the text and stained by the acriflavine-Feulgen method. Only the first 3 hours of the cycle are shown. Freshly detached cells are either in mitosis or are densely staining early G_1 cells [G_1(C+)]. The change in condensation states of chromatin and the division stage cells is represented in the figure. (From R. R. Klevecz and L. L. Deaven, unpublished.) (b) Mean and modal channel number (DNA content) from a flow microfluorimetric (FMF) analysis of synchronous CHO cells. Mitotic CHO cells were selectively detached and fixed for FMF analysis at hourly intervals throughout the cell cycle. Major modes are indicated by filled circles, and minor modes by open circles. The beginning of S phase was determined by autoradiographic analysis of the percent labelled nuclei (triangles) and is indicated by the vertical broken line. Mean DNA content was determined using the equation $\bar{c} = \Sigma(fc)/\Sigma f$, where $\bar{c}$ is the mean channel number or center of mass, c is any of 100 channels, and f is the number of cells in that channel. (From R. R. Klevecz and L. L. Deaven, unpublished.)

megaterium (Imanaka *et al.*, 1967). Even though modifications have been devised with a view to minimize distortion of division cycle-related events (Abbo and Pardee, 1960; Nagata, 1963; Goldberg and Chargaff, 1971), and for scale-up (Helmstetter and Uretz, 1963), the method suffers from the unavoidable problems arising from the need to concentrate cell suspensions before filtration. Direct filtration of small (100 ml) cultures of *Bacillus subtilis* though large diameter

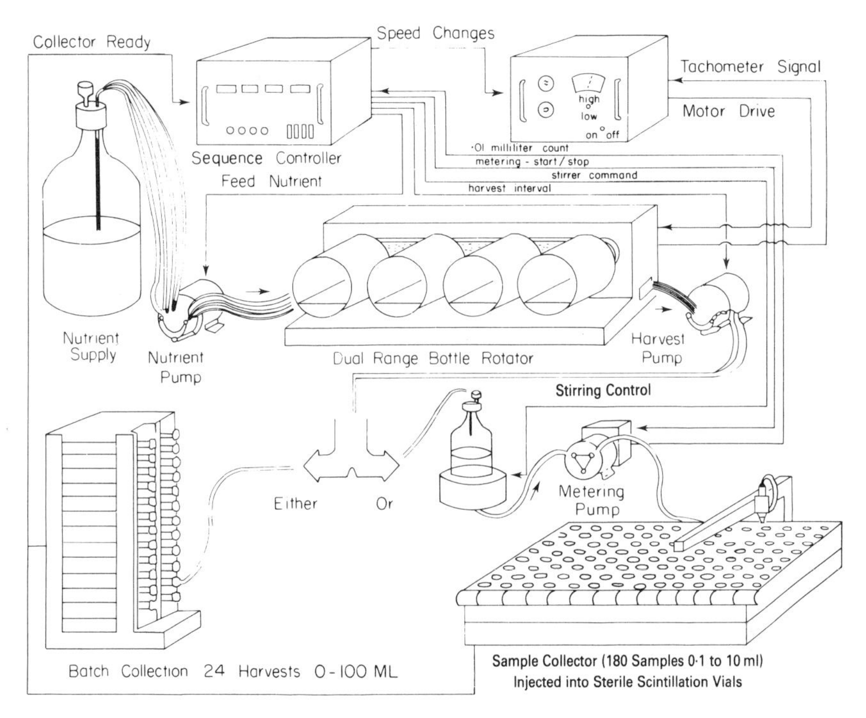

Fig. 2.9 Schematic illustration of components of Klevecz' cell-cycle analyser.

(47 mm) glass-fibre filters gives good synchrony (Sargent, 1973b, Edwards, 1980a). A similar procedure works well for the swarm-cell population of *Caulobacter* NC1B 9083 (Swoboda and Dow, 1979).

C. AGE SELECTION BY MEMBRANE ELUTION

Synchronously dividing populations of *Escherichia coli* B/r can be obtained by selecting cells of uniform age from a growing population (Helmstetter and Cummings, 1963, 1964). After binding of a culture to the surface of a membrane filter, only newly-formed daughter cells are eluted during passage of the medium back through the filter (Fig. 2.10). Collection of the eluted cells for a short time relative to the generation time gives a highly synchronous culture. A simple and inexpensive apparatus has been described by Helmstetter (1967, 1969). Samples of about 10^8 organisms can be obtained by using larger filters

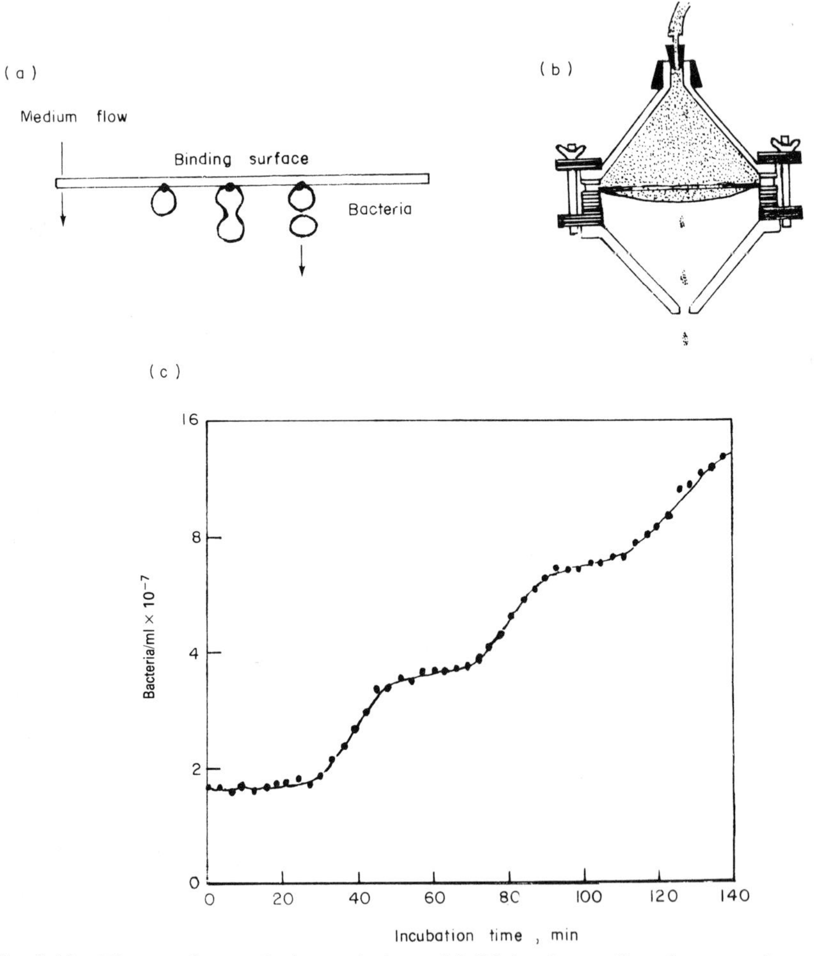

Fig. 2.10 The membrane elution technique. (a) Of the three cell cycle stages shown only the daughter cell is eluted as culture medium passes through; (b) The inverted membrane assembly; (c) Cell numbers (determined in a Coulter counter) in a sample of effluent from a membrane-bound population of *Escherichia coli* B/r growing in glucose minimal medium. The sample was collected between 38 and 40 min of elution and incubated at 37°C. (Reproduced with permission from Helmstelter, 1969.)

(Cummings, 1965; Clark and Maaløe, 1967) but collecting multiple samples from a smaller filter gives more satisfactory results (Helmstetter, 1969). A major disadvantage of this method is that it does not work well for all strains of *E. coli;* strain B/r gives the bests results of all strains tested (Helmstetter, 1969). Strains derived from B/r have been synchronized by membrane elution (Kung *et al.*, 1976; Hackenbeck and Messer, 1977a, b).

D. VELOCITY SEDIMENTATION

1. *Centrifugation through Gradients*

The rate of sedimentation of an ideal spherical particle is proportional to the square of its radius and to the density difference between the particle and the suspending fluid. Separation of exponentially-growing cell populations into different age classes on the basis of size and density differences by velocity sedimentation has become one of the most popular methods for selection synchrony. The use of density gradient methods is however a hazardous approach and one which is not to be encouraged. Any age-class may be used as an inoculum for a synchronous culture, but it is usual to use the most slowly sedimenting subpopulation, selecting about 10% of the original cells. Introduced by Mitchison and Vincent (1965) for fission and budding yeast, and for *Escherichia coli*, the method has been applied successfully to many different cell-types, even though interpretation of results has not always taken sufficient account of possible perturbative effects. The original method employed sucrose density gradients; lactose, glucose and glycerol gradients have also been used (Mitchison and Carter, 1975) as have gradients of dextran, Ficoll, Renografin and colloidal silica. The most recent modification uses Percoll for both bacteria and a yeast (Dwek *et al.*, 1980). The loading capacity of gradients is limited, and scale-up from 50 ml gradients to volumes of more than 1 litre requires the use of zonal centrifuges. Tailoring of the gradient can provide a spatial analogue of the temporal sequence of the cell cycle: thus under appropriate conditions, successive fractions of the gradient provide sub-populations at successive stages of the cell cycle (Fig. 2.11). This method of cell cycle analysis is also frequently employed (cell cycle fractionation, see also p. 89).

Examples of gradient separations of bacterial sub-populations for the preparation of synchronous cultures include those with *E. coli* (Donachie and Masters, 1966; Kung and Glaser, 1977; Mycielski *et al.*, 1977), and *Lineola longa* (Baldwin and Wegener, 1976). Similar procedures have been used for the separation of synchronously germinating spores of *Bacillus subtilis* (Siccardi *et al.*, 1975) and swarmer cells of *Rhodopseudomonas palustris* (Westmacott and Primrose, 1976). Equilibrium density separations and subsequent synchronous growth of selected fractions have been demonstrated for *E. coli* (Poole, 1977b).

Cell cycle fractionation of *E. coli* by velocity sedimentation has been described by Manor and Haselkorn (1967) who used caesium chloride gradients, and by Kubitschek (1968a,b). Optimal design for maximum resolution of the technique (Koch and Blumberg, 1976) taking into account size, shape, and density, as well as the effect of the known

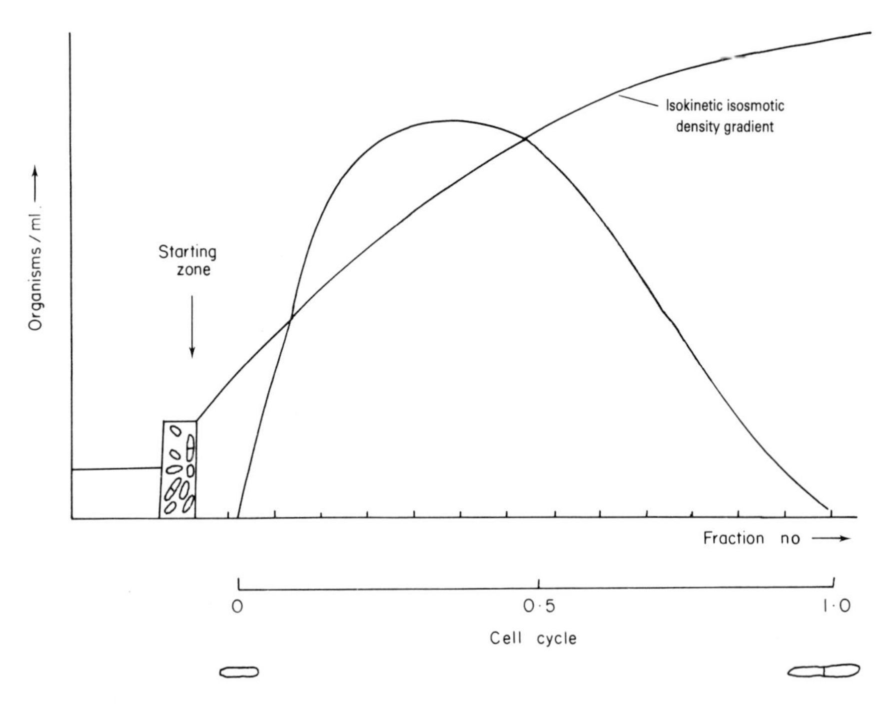

Fig. 2.11 Cell-cycle fractionation by velocity sedimentation through a density gradient: idealized conditions give a linear representation of cell-cycle traverse.

variation in cell size at division, predicts the usefulness of isokinetic gradients, in which particles sediment at constant velocity throughout the gradient (Noll, 1967). In zonal rotors equivolumetric gradients serve this function (Pollack and Price, 1971). The gradient should be iso-osmotic throughout, and the density range employed depends on the density of the cells (for bacteria $<1{\cdot}13$, Koch and Blumberg, 1976). For a recent review of centrifugal fractionation techniques including gradient media and rotors currently available see Lloyd and Poole (1979). Cell-cycle fractionations on equivolumetric gradients

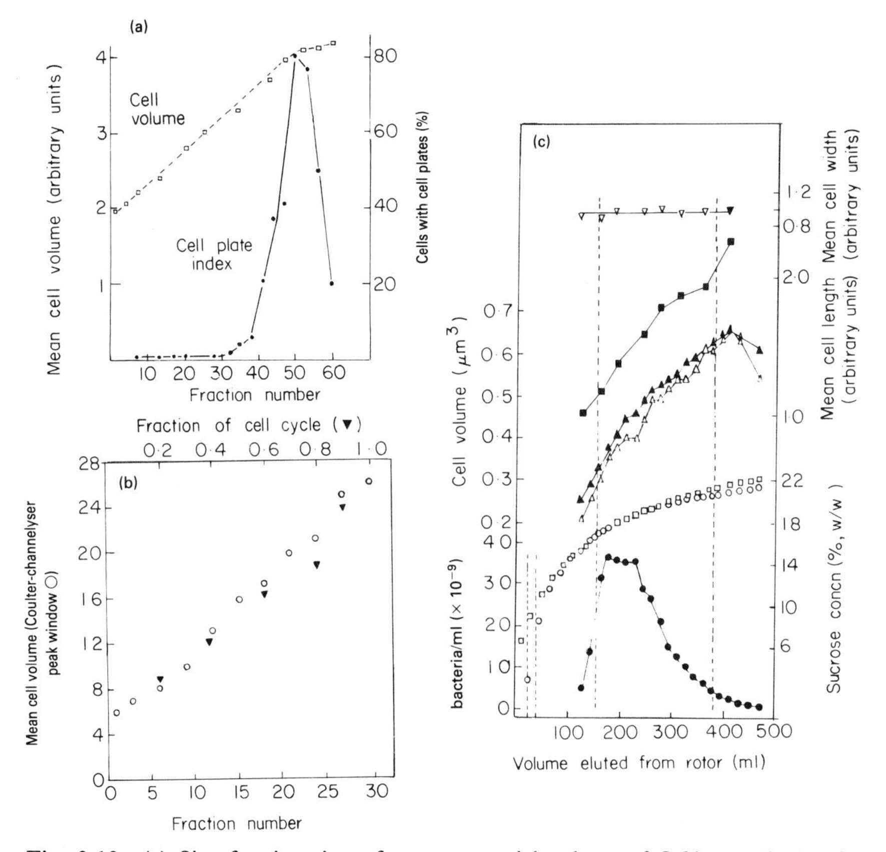

Fig. 2.12 (a) Size fractionation of an exponential culture of *Schiz. pombe* (strain $972h^-$) by zonal centrifugation. An exponential-phase culture grown in minimal medium was harvested by filtration and layered onto a chilled 15–40% sucrose gradient in an MSE Type A zonal rotor at 600 r/min. After centrifugation at 2 000 r/min until the fastest-migrating cells had moved two-thirds of the distance to the edge of the rotor, the cells were pumped out from the centre. Sixty fractions (15 ml each) were collected, and the mean cell volume of each fraction was measured with a Coulter Counter. The cell plate index (percent cells with cell plates) was measured under the microscope. (From B. L. A. Carter, unpublished data.) (b) Mean cell volumes of *S. cerevisiae* from a synchronous culture at different stages of the cell cycle compared to mean cell volumes of cells in fractions removed sequentially from a zonal rotor. Samples were taken from a synchronous culture (prepared by velocity separation on a zonal rotor) at intervals during the cell cycle (generation time of 2·5 h). Fractions from the zonal rotor were analyzed after separation of an exponential-phase culture. The mean cell volume of samples at different stages of the cycle (triangles) and of fractions from the zonal rotor (circles) was measured on a Coulter Counter. (From S. Sogin and B. L. A. Carter, unpublished data.) (c) Analysis of the cell cycle of *E. coli* by zonal fractionation into size classes of cells from an exponentially-growing culture. The integrated force-time was $3{\cdot}5 \times 10^8\ \text{rad}^2\,\text{s}^{-1}$. The measurements made in successive fractions were, cell numbers (●), sucrose concentration (○), the theoretical profile of an equivolumetric gradient (□), modal (△) and mean (▲) cell volumes, and, measured from electron micrographs, cell length (■) and diameter (▽). (Reproduced with permission from Scott *et al.*, 1980.)

have been described for *Myxobacter* (Tan *et al.*, 1974) and for *E. coli* (Scott *et al.*, 1980, Fig. 2.12).

Velocity-sedimentation selection of a slowly sedimenting sub-population of *Schiz. pombe* on linear sucrose gradients (10–60%, w/w; total vol. 540 ml) made up in defined growth medium was carried out at room temperature in an MSE HS zonal rotor (Poole *et al.*, 1973, Poole and Lloyd, 1974). Other workers have utilized the changes in density which occur in the cell cycle of yeasts to select homogeneous populations of cells after equilibrium density centrifugation in dextran (Wiemken *et al.*, 1970), Renografin (Hartwell, 1970), Urografin (Salmon, 1980), or colloidal silica (Ludox) (Shulman *et al.*, 1973; Shulman, 1978). Dextran gradients have also been used for *Candida utilis* selection (Nurse and Wiemken, 1974).

Cell cycle fractionation of *S. cerevisiae* and *S. lactis* in linear sucrose gradients (20–40%, w/w and 10 to 25%, w/w respectively, total vol. 1500 ml) was at 4°C in an A XII zonal rotor (Sebastian *et al.*, 1971; Halvorson *et al.*, 1971; Carter and Halvorson, 1973). An almost linear increase in mean cell volume with fraction number, indicated satisfactory separation primarily by size (and thus by age) in the cell cycle. Rate-sedimentation of *Schiz. pombe* in a reorienting gradient zonal rotor (Wells and James, 1972) and in an HS zonal rotor by rate- and isopycnic-sedimentation (Poole and Lloyd, 1973; Poole *et al.*, 1974; Lloyd and Edwards, 1977) gave valid analyses of the first 0·75 of the cell cycle. Cell-cycle fractionation of *Candida albicans* has also been reported by Chaffin and Sogin (1976). Synchronous cultures of *Chlorella pyrenoidosa* were established after selection by centrifuging to equilibrium in gradients of aqueous Ficoll (Hopkins *et al.*, 1970; Sitz *et al.*, 1970).

The use of sucrose gradients for selection synchrony and cell cycle fractionation of mammalian cells is widespread (Sinclair and Bishop, 1965; Morris *et al.*, 1967; Schindler and Schaer, 1973; Probst and Maisenbacher, 1973, 1975; Horáková *et al.*, 1976). Typical conditions employ sucrose gradients (25–31%, w/w), made up in growth medium and made isotonic with NaCl (7·6 to 0 g l^{-1}) in an A-type zonal rotor, and centrifugation at about 1000 r/min for about 6 min. Ficoll gradients (5–20%) have also been employed (Ayad *et al.*, 1969; Warmsley and Pasternak, 1970; Warmsley *et al.*, 1970; Lepoint, 1977; Krynicka *et al.*, 1980). Cell lines studied by these methods include Ehrlich Ascites Tumour cells, neoplastic mast cells, murine mast tumour cells and L cells. HeLa cells have been synchronized after density selection of mitotic cells on gradients of colloidal silica (Ludox) in Eagles medium with polyvinyl-pyrrolidone (Wolff and Pertoft, 1972).

2. Selection Synchrony after Density Labelling

Ciliated protozoa do not form food vacuoles during cell division; thus iron filings or tantalum particles are not taken up by dividing organisms (Hildebrandt and Duspiva, 1969). Density gradient centrifugation has been used to separate these cells from the rest of the population in order to establish synchronous cultures; this procedure has been used for *Tetrahymena pyriformis* (Wolfe, 1973) and for *Paramecium tetraurelia* (Aufderheide, 1976). A magnetic method for large scale selection of *T. pyriformis* has been devised by Dickinson *et al.* (1976).

3. Differential Sedimentation at Unit Gravity

Many investigators have used density gradients in large, stationary cylinders through which cell populations are allowed to sediment under the influence of gravity. These unit gravity devices, based on the Stayflow system of Mel (1964), are termed Stay-put chambers (Brubaker and Evans, 1969; Miller and Phillips, 1969; Ayad *et al.*, 1969). The method can be carried out at the temperature of growth or at 4°C; at the lower temperatures separation takes longer (as much as 10–15 h) due to cell shrinkage and increased viscosity of the medium. At 37°C mouse fibroblasts harvested from suspension cultures take 50 min to distribute themselves across a gradient of aqueous sucrose (2·72–10%, made iso-osmotic throughout with NaCl) (Schindler *et al.*, 1970; Shall and McClelland, 1971; Shall, 1973). Other gradient media employed include bovine serum albumin and foetal calf serum (Macdonald and Miller, 1970). Although the resolution of the method is good, and subsequent synchrony satisfactory for some purposes, some distortion of metabolic events is inevitable in these lengthy procedures. The method has also been used in combination with the selective detachment process for KB cells (Ooka, 1976). The capacity of the method is limited to about 10^9 cells; a major problem is the stability of the starting zone in such shallow gradients. Several devices have been designed to minimize disturbance while layering the sample on top of the gradient; Tulp *et al.* (1979) described the use of a flow deflector and subsequent analysis of subpopulations of murine leukemia cells. This method of selection has also been employed in conjunction with the mitotic wash-off technique (Zucker *et al.*, 1979a).

4. Differential Sedimentation in the Culture Medium

A rapid and simple method for preparing synchronous cultures of *Tetrahymena pyriformis* (Corbett, 1964) by differential centrifugation of cultures at 550 g for 6 min has been ignored while many less satisfactory procedures have been developed. This basic approach to the problem of cell selection was reintroduced for *Acanthamoeba castellanii* by Chagla and Griffiths (1978). Exponentially-growing cultures were transferred aseptically to 50 ml sterile capped centrifuge tubes and centrifuged in a swing-out rotor at 300 r/min (10 g; r_{av} 10 cm) for 2 min. The supernatant, consisting of the slowest sedimenting cells, accounted for about 10% of the original population and was carefully decanted and grown as a synchronous culture. Synchrony indices ranged from 0·59 to 0·74 with a mean of 0·67 and standard deviation of 0·04 and compare favourably with those obtained for this organism by continuous-flow centrifugation (Lloyd *et al.*, 1975). Asynchronous control cultures were obtained by remixing the tube contents after identical centrifugal treatment and growing the entire population (Edwards and Lloyd, 1978, 1980), no perturbations of respiration or macromolecular synthesis (accumulation of total protein and RNA) were detectable in these control cultures. Scale-up using 1 litre centrifuge pots (2 min at 800 r/min^{-1}, 150 g; r_{av} 21 cm) gives equally good synchrony indices (Edwards *et al.*, 1981) and assays of enzyme activities (e.g. catalase, cytochrome *c* oxidase, ATPase) F_1-ATPase inhibitor, and enzyme proteins (catalase, cytochrome *c* oxidase, cytochrome $a + a_3$, cytochrome *c*, F_1-ATPase), in asynchronous control cultures indicate absence of detectable perturbation. The small scale (50 ml) procedure has been successfully applied to *Dictyostelium discoideum* (C. Woffendin and A. J. Griffiths, unpublished data), but in this case the most slowly sedimenting subpopulation contains a proportion of non-viable cells. After discarding the supernatant, the original volume is restored with conditioned growth medium, the tube contents resuspended, and the centrifugal procedure repeated. The second supernatant gives cultures with synchrony indices in the range 0·63 to 0·78 and mean $N/No = 1{\cdot}9$. If this simple method for obtaining synchronous cultures is widely applicable, its obvious advantages may make it the ideal procedure for cultures of small volume.

5. *Continuous Flow Centrifugal Selection*

This procedure, introduced by Lloyd *et al.* (1975) for eukaryotes and by Evans (1975) for *Escherichia coli*, is widely applicable as a simple and rapid method for the preparation of large-scale synchronous cultures. Passage of an exponentially growing culture through a continuous flow centrifuge rotor under predetermined conditions of rotor speed and flow rate permits selection of the most slowly sedimenting sub-population of the culture that escapes harvesting and emerges in the rotor effluent (Fig. 2.13). In the "continuous action rotor" of MSE (Crawley, West Sussex) the maximum flow rate is 2 l min^{-1} and maximum rotor speed 18 000 r/min. A "low-speed zonal control" circuit enables accurate speed control at speeds less than 1850 r/min. Thus a wide range of conditions are available for selection of sub-populations of different organisms with very different mean sedimentation coefficients. Table 2.4 summarizes examples of applications of this method. A Sharples Laboratory Supercentrifuge (Penwalt, Camberley, Surrey) (Fig. 2.14) has been used successfully for *Alcaligenes eutrophus* (Edwards and Jones, 1977); other continuous flow centrifuges have not yet been employed for this technique, and the early work of Knutsen *et al.* (1973) who used the Rastgeldi threshold centrifuge for selection synchrony of *Chlamydomonas reinhardii* has not not been followed up.

Apart from the advantage of scale-up (we have prepared synchronous cultures of volumes up to 20 l), this method has many attractive features. The organisms are never removed from the culture medium, and at least at the highest flow rates, have a short mean residence time in the rotor (0·2 min at 2 l min^{-1} in the MSE rotor). Thus several potential sources of perturbation (anaerobiosis, nutrient deprivation or temperature changes) are minimized; the organisms do inevitably experience transient viscous shear, centrifugal forces and hydrodynamic pressure. For eukaroytes, sedimentable at rapid flow rates and low centrifugal speeds, these forces are not evidently a source of problems (some highly fragile protozoa show no loss of integrity or viability after selection), but the extreme conditions used for bacterial separations may be more suspect. As the rotors are autoclavable, aseptic procedures can easily be employed. Strict anaerobiosis can also be maintained during processing.

A disadvantage of the method is the impossibility of obtaining control asynchronous cultures which have received identical flow and centrifugal treatment; it takes some time to recover the harvested cells (as a concentrated suspension) from within the rotor. A recent claim that continuous flow centrifugation may subject organisms to metabolic shock (Walker *et al.*, 1980) is based on inconclusive evidence. The most slowly sedimenting sub-population need not be the smallest cells of the culture (Poole, 1977a,b; Poole and Pickett, 1978); this is indicated by the observation that the period between collection of the effluent and the first synchronous division is sometimes significantly shorter than the cell cycle. In these cases (e.g. for *Schiz. pombe* and *E. coli*) cell cycle dependent density fluctuations contribute to the

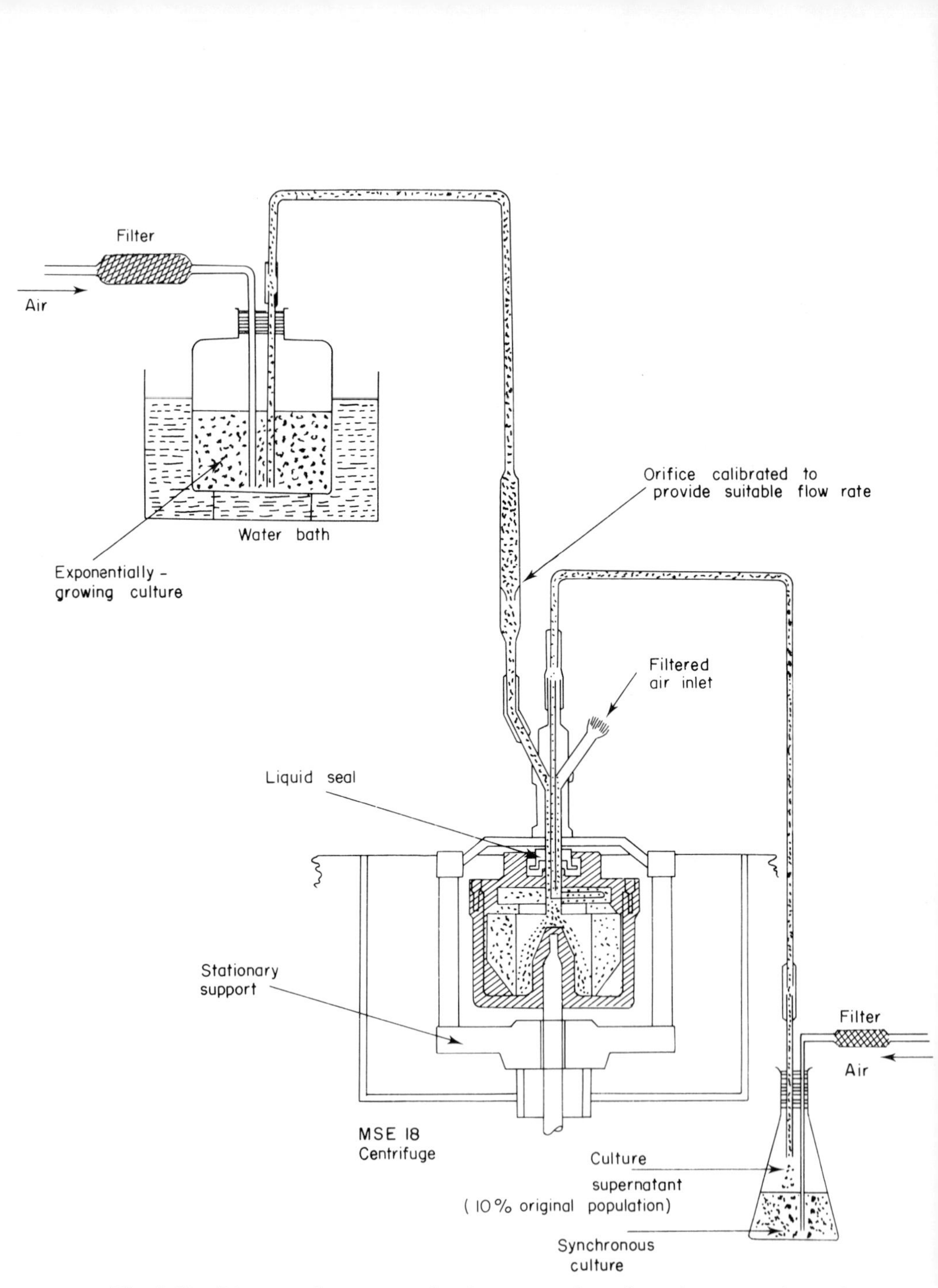

Fig. 2.13 Diagram of apparatus for the preparation of synchronous cultures by continuous-flow selection using the MSE continuous-action rotor (for details see text). (Reproduced with permission from Lloyd *et al.*, 1975.)

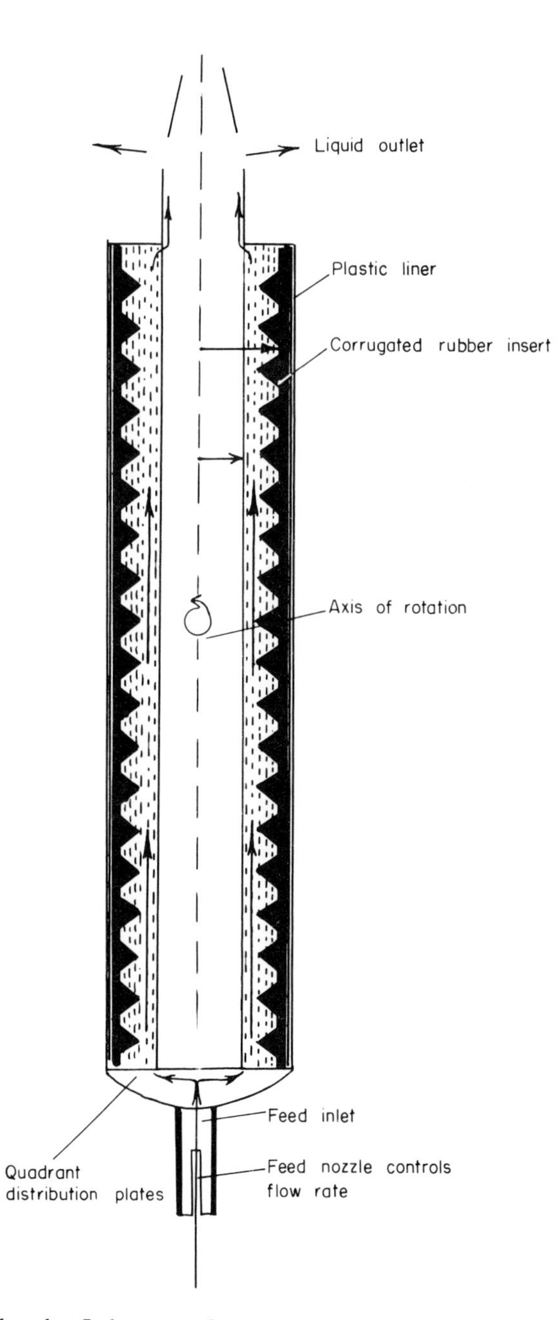

Fig. 2.14 Sharples Laboratory Supercentrifuge Rotor. This device is useful for size-selection synchrony, or fitted with a rubber insert on a plastic liner (as shown) may be used for continuous flow size selection (Lloyd *et al.*, 1977).

Table 2.4 *Selection Synchrony by Continuous Flow Centrifugation*

Organism	Flow Rate ($1\ min^{-1}$)	Rotor Speed (r/min^{-1})	S.I. (F_1, F_2)		Reference
Schizosaccharomyces pombe 972 h^- (glucose grown)	0·6	3 000	0·58	0·48	Lloyd *et al.* (1975) El'Khayat (1980) Toma (1980) Walker and Duffus (1980) Walker *et al.* (1980)
(glycerol grown)	1·0	3 000	0·82		Edwards and Lloyd (1977a)
	0·5	6 700	0·61	0·56	Poole (1977a)
(complex medium)	1·0	3 500	0·72		Bullock and Coakley (1979)
			0·74	0·63	Williams (1980)
Candida utilis NCYC 193 (glucose grown)	0·42	4 000	0·80	0·83	Lloyd *et al.* (1975)
(acetate grown)	0·42	3 500	0·72	0·72	Kader and Lloyd (1979)
(glycerol grown)	0·42	3 500	0·82	0·74	Kader and Lloyd (1979)
Tetrahymena pyriformis ST	0·42	700	0·84		Lloyd *et al.* (1975)
	1·15	2 800	0·74	0·41	Phillips and Lloyd (1978)
	1·5	1 750	0·77	0·4	Lloyd *et al.* (1978b)
	(several different conditions)		0·83		Unitt (1980)
Acanthamoeba castellanii (Neff strain)	0·38	1 000	0·78		Lloyd *et al.* (1975)
Escherichia coli W1485 (L-alanine grown)	0·2	10 500	0·69	0·69	Evans (1975)
A1002 (glucose grown)	0·16	14 500	0·64		Poole (1977b)
[a]*Alcaligenes eutrophus* H16	1·3	17 000	0·82	0·75	Edwards and Jones (1977) Edwards *et al.* (1978)

Table 2.4 *continued*

Organism	Flow Rate (1 min^{-1})	Rotor Speed (r/min^{-1})	S.I. (F_1, F_2)	Reference
Chlorella 2118p				John *et al.* (1980)
Pseudomonas aeruginosa PA01	0·22	16 000	0·75	I. R. Johnson and H. Jones (unpublished data)
[a]*Bacillus subtilis*				
(glucose grown)	0·14	16 400	0·59	Edwards and McCann, 1981 and unpublished data
(glycerol grown)	0·14	16 400	0·59	
(lactate grown)	0·14	18 750	0·67	
Clostridium pasteurianum				
ATCC 6013 (strain W5)	0·36	12 000		Clarke (1978)

[a]A Sharples Supercentrifuge was employed

age-selection process. Failure of the method in some instances e.g. for succinate-grown *E. coli* (R. I. Scott and R. K. Poole, unpublished) may be accounted for by a narrow range of sedimentation coefficients in those populations where the inverse relationships between density and size compensate for one another. This situation also prevails in some yeasts (Salmon, 1980). Selection of the smallest size-class gives poor synchronous cell division in cultures containing many small non-viable organisms; e.g. typanosome cultures often have a dyskinetoplastic subpopulation (Lloyd *et al.*, 1975).

Continuous-flow cell cycle fractionation can be achieved very rapidly without using a gradient, by means of a modified Sharples Supercentrifuge (Lloyd *et al.*, 1977). In this method, a specially-designed rubber insert retains cells harvested at successive levels within the rotor. Predetermined rotor speeds and flow rates give conditions suitable for different organisms, and size separations of *Schiz. pombe*, *S. cerevisiae*, *Candida utilis*, and *Acanthamoeba castellanii* have been obtained. Larger organisms (e.g. *Tetrahymena pyriformis*) were not amenable, as the low rotor speeds precluded stability of the column of liquid within the rotor. Nevertheless this procedure, as yet unexploited, offers a method for cycle fractionation of those fragile or highly motile organisms which cannot be separated using gradient methods. Its major advantage is that 10 litres of culture can be processed in less than 15 min.

6. *Centrifugal Elutriation (Counterflow Centrifugation)*

The principles and practical aspects of centrifugal elutriation as a method for the separation of biological particles were evolved by Lindahl (1948, 1956), but a rotor suitable for general use has only recently become available commercially (Fig 2.15). The force tending to sediment particles in a centrifugal field is counterbalanced by flowing the suspending fluid in the opposite direction (McEwen *et al.*, 1968). Fractionation of a cell population into discrete size classes is achieved by incrementally increasing the flow rate while maintaining the centrifuge rotor (4·5 ml volume) at constant speed. The progress of the process can be watched with a stroboscope. Theoretical treatment is presented by Sanderson *et al.* (1976) and Sanderson and Bird (1977). Fractionations of populations of mammalian cells and of yeasts in the JE-6 elutriator rotor (Beckman Instruments Inc., Palo Alto, California) have been reported. Mammalian cell populations fractionated by this procedure include Chinese hamster ovary cells, human lymphoma T cells, mouse L-P59 cells, and 3T3 and SV40 3T3 cells (Gerner *et al.*, 1977; Meistrich *et al.*, 1977; Mitchell and Tupper, 1977, Grdina *et al.*, 1979a, b; Piper *et al.*, 1980; Keng *et al.*, 1980; Meyn *et al.*, 1980) and cells from solid tumors (Keng and Wheeler, 1980a, b). In these studies an upper limit of about 3×10^8 cells could be fractionated with no loss of viability as assessed by the subsequent growth of fractions. Complete resolution of the population was achieved in 50 min, but if only G_1 cells were required for subsequent synchronous growth then the procedure took only 20 min. The elution medium was the growth medium, sometimes supplemented with Amphotericin B (25 μg ml^{-1}) and Gentamycin (50 μg ml^{-1}), and kept at 4°C. Rotor speeds of between 1 500 and 200 r/min (corresponding to about 600 g at the elutriating zone) and flow rates of between 9 and 40 ml min^{-1} were employed in different laboratories. The more rapid separation that are theoretically attainable at high rotor speeds and flow rates have not been reported.

For fractionation of populations of *S. cerevisiae*, Gordon and Elliott (1977) loaded the rotor with 3×10^9 organisms at 9 ml min^{-1} and at 3 000 r/min and used increments in pumping speed of 1 ml min^{-1} up to 27 ml min^{-1}. The elution was with water at 4°C. Excellent separation was obtained of small unbudded cells, budded cells in which nuclei had not migrated to the necks of buds, cells with nuclei in the process

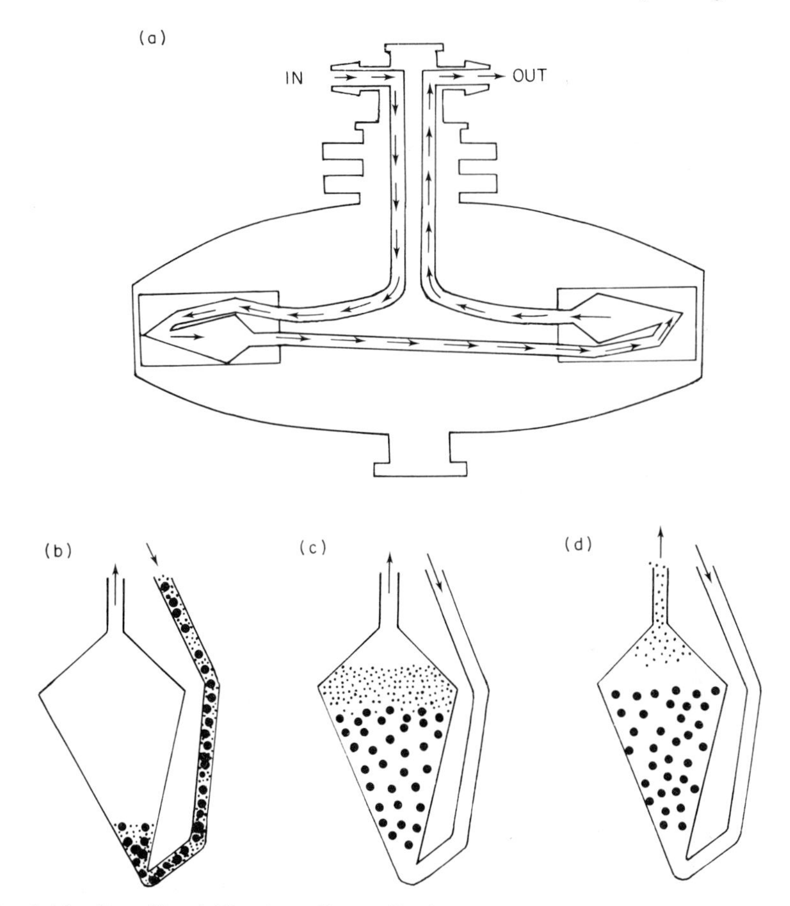

Fig. 2.15 Centrifugal Elutriator Rotor (Beckman. J.-21B). (a) The rotor in cross-section showing pathway for fluid flow through the 4·5 ml vol. chambers. (b)–(d) Sequence of operation: (b) Sample suspended in medium is pumped into chamber; (c) Sedimentation tendency of particles balanced by counter flow; (d) Flow increased; slowly-sedimenting particles elutriated out of chamber.

of migration, and cells with nuclei partitioned between mother cells and buds. Despite the fact that a low temperature was used to minimize cell growth during the procedure (which took 2–3 h), and that the cells spent this time in the absence of nutrients, a reasonable degree of synchrony was observed when subpopulations were reincubated in growth medium. It was pointed out that if subsequent establishment of synchronous cultures (rather than cell-cycle fractionation) was the primary aim, selection of the smallest cells could be achieved more rapidly, and that use of warm growth medium would give less metabolic disturbance. These precautions have been observed by Creanor and Mitchison (1979) who have obtained good synchrony of *Schiz. pombe* ($F_1 = 0{\cdot}65$). With this organism maximum rotor loading occurred at 7×10^9 cells and took between 10 and 40 min. The initial population for synchronous growth was a total of 2×10^8 cells. Control asynchronous cultures using cells exposed to centrifugal elutriation showed less perturbation (as measured by the rate of incorporation of [^{3}H]-leucine) than those for velocity-sedimentation separations on sucrose gradients in tubes.

It is evident that centrifugal elutriation is becoming a popular method for cell cycle fractionation, although this extended procedure is of doubtful value for most purposes. Some advantages over density-gradient based methods are: (i) a long effective path-length; and (ii) its use of any medium of uniform density (and hence osmolarity). It employs low centrifugal fields, but suspended particles are subjected to some viscous shear forces which are potentially damaging. Coriolis or streaming effects tend to make removal of small particles less efficient than large particles. As a technique for separation of a sub-population prior to establishment of synchronous cultures, centrifugal elutriation suffers by comparison with the continuous-flow centrifugation or simple differential centrifugation methods. The elutriation method takes longer, and must therefore be regarded as potentially more perturbing. During the procedure, organisms are accumulated in the rotor at such high cell densities that they must become nutrient depleted and anaerobic even though perfused by fresh growth medium. Further work is necessary for an evaluation of the extent of these problems. Only small cultures are obtainable, the device is costly, has to be sterilized chemically, and the present limitation on rotor speed excludes application to prokaryotes.

VII. Use of Exponentially-Growing Cultures for Cell Cycle Analyses

A. CLASSIFICATION OF ISOTOPICALLY-LABELLED INDIVIDUAL CELLS

An alternative approach to analysis of cell division cycle-related sequences is to classify subpopulations within an exponentially-growing culture. There are two ways in which this may be achieved: (1) classification into age classes, and (2) continuous observation of a single known age-class within the growing culture. Historically, both types of experimental approaches have been used to investigate the period of DNA synthesis during the cell division cycle. In the first type, an asynchronous exponentially dividing population is pulse labelled with [^{3}H]-thymidine and examined by autoradiography to determine isotopic incorporation into DNA as a function of cell size (Woodward *et al.*, 1961). Alternatively the number of labelled metaphase figures observed as a function of time subsequent to pulse labelling gives a method of measurement of the relative durations of G_1, S and G_2 (Painter and Drew, 1959).

All ages of cells are represented in an asynchronous exponentially-growing culture, but small newly-divided cells are twice as frequent as large cells about to divide. The distribution of ages in such a population is ideally shown in Fig 2.16 (Cook and James, 1964). Age-size relationships for *Escherichia coli*, *Sterigmatomyces halophilus* and *Schizosaccharomyces pombe* have been investigated by Bugeja *et al.* (1980).

The proportion of cells in a population that are observed in mitosis (the mitotic index *M*) is related to the proportion of the cell cycle time

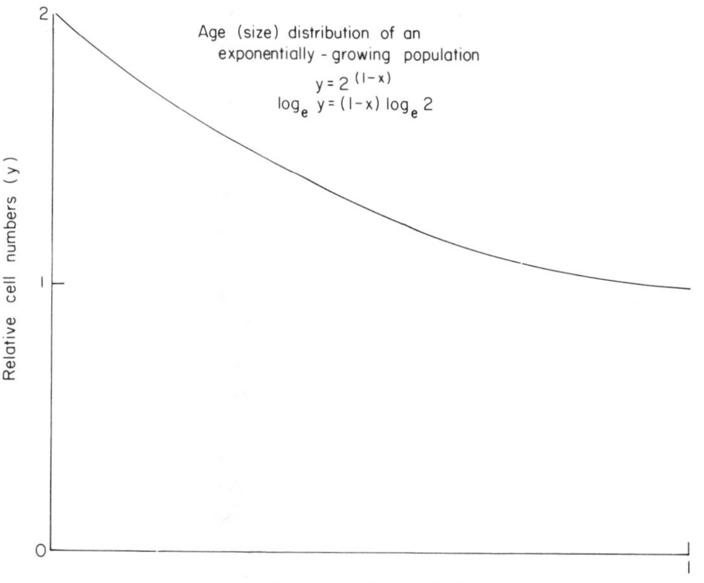

Fig. 2.16 Age (size) distribution of an exponentially-growing culture (ideal case). (Reproduced with permission from Mitchison, 1971.)

spent in mitosis t_m by the equation:

$$\frac{t_m}{T} = \frac{\log_e(M+1)}{\log_e 2}$$

where T is the cell cycle time. Cleaver (1967) and Mitchison (1971) have reviewed the use of isotopes in determining the phases of the DNA cycle; an example is the determination of the S-phase for yeast (Williamson, 1965). Recent accounts of these methods include those by Gerecke *et al.* (1976), Barranco *et al.* (1977a, b), Gray *et al.* (1977) and Gould (1979).

B. DETERMINATION OF TRANSITION POINTS

A transition point (defined as that stage of the cell division cycle beyond which an inhibitor does not affect division in the ongoing cycle but will affect division in the subsequent cycle, Hamburger and Zeuthen, 1957) may be determined using asynchronous cultures (Howell *et al.*, 1975). In this case measurement of residual cell division (N/No, where No is the population when the inhibitor is added, and N is the final population attained) for an organism which gives two daughter cells on division, adequately locates the transition point (X_{TP}) which can be calculated from the equation:

$$X_{TP} = 1 - \frac{\log_e(N/No)}{\log_e 2}$$

C. FLOW MICROCYTOFLUORIMETRY AND CELL SORTING

High speed flow methods of cytometry and cytofluorimetry (Dittrich and Gohde, 1969; Van Dilla *et al.*, 1969), are now widely used to determine frequency distribution curves for cellular components within randomly-dividing cellular populations of mammalian cells. Thus fluoresence measurements of DNA content of individuals can be made at rates of $>10^3$ cells s^{-1} (Kamentsky *et al.*, 1965) and this method is rapidly replacing autoradiography for the determination of the proportion of a population in S-phase. A number of different dyes (e.g. acriflavine, ethidium bromide, propidium iodide, mithramycin) have been used (Crissman *et al.*, 1975). The rapidity and simplicity of this technique using mithramycin staining of mammalian cells fixed in aqueous ethanol, permits the monitoring of cycle kinetics during an experiment (Fig. 2.18) (Crissman and Tobey, 1974). The propidium iodide method gives results in good agreement with those obtained from autoradiography (Fried *et al.*, 1976) and has been successfully used for *Saccharomyces cerevisiae*, although the DNA content is two-hundredfold lower than that of a typical mammalian cell (Johnston *et al.*, 1980). Recently the method has also been applied to bacteria (Paau *et al.*, 1977; Bailey *et al.*, 1977; Hutter and Eipel, 1978b). Introduction of this technique into microbiology has been slow considering its obvious potential (Hutter and Eipel, 1979); other examples of its application to work on lower eukaryotes are those with *Euglena* (Falchuk *et al.*, 1975), *S. cerevisiae* (Slater *et al.*, 1977) and *Tetrahymena* (Phillips and Lloyd, 1978).

Fluorescent staining methods for polysaccharide (periodic acid, Schiff reaction), and for protein (fluorescein isothiocyanate) have been used (Crissman *et al.*, 1975; Freeman and Crissman, 1975) and fluorescein diacetate hydrolysis gives the distribution of carboxylesterase within a population of tumour cells (Watson *et al.*, 1977).

In the flow cytofluorimeter, cells in liquid suspension are passed in single file through a sensing zone, thereby generating signals which are detected and analysed. The flow system uses the laminar sheath-flow technique, and excitation of fluoresence is by selection of appropriate lines of an argon–ion or helium-neon laser (Fig. 2.17a). Photomultipliers and/or solid-state photosensors, measure intensity of emitted fluoresence, axial light loss, and wide or narrow forward-angle scatter of individual cells. Each detector has three modes of measurement: pulse height analysis, pulse area analysis and pulse width analysis, and "windows" enable selection of subpopulations for differential counting and subpopulation frequency distribution. Data presentation is by a frequency-distribution histogram or by two or three dimensional "cytograms"; and mathematical treatments of data have been reported (Fried *et al.*, 1976; Gray, 1976; Beck, 1978; Kim *et al.*, 1978; Kosugi *et al.*, 1978; Fried *et al.*, 1980). Physical sorting of cell populations analysed in this way can be achieved as cells exit from the flow chamber in a liquid jet by ultrasonic vibration to 30 kHz (Fig. 2.17b, Horan and Wheeless, 1977; Schaap *et al.*, 1979). Cells are isolated in tiny droplets and those of interest are electrostatically charged and deflected. Quantitative multiparameter flow analysis and sorting of cells provides the ability to obtain measurements on single cells at high rates and to sort cells for further analysis; evidently the emergence of this new technology (equipment is available from Ortho instruments, Westwood, MASS., or Coulter Electronics, Hialeah, FLA) will revolutionize cell cycle research from the fundamental to the diagnostic

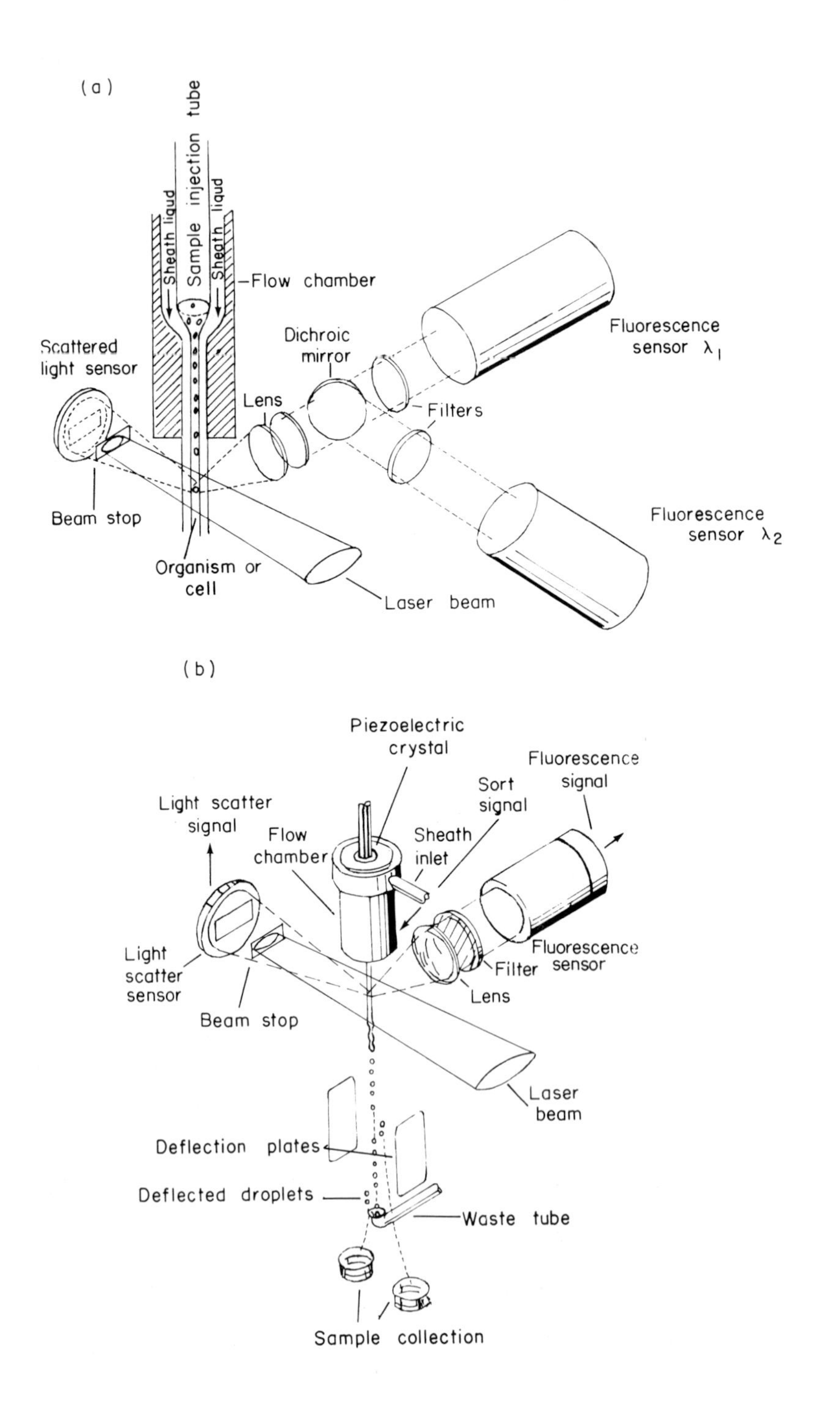
(a)
Sample injection tube
Sheath liqud
Sheath liqud
Flow chamber
Scattered light sensor
Dichroic mirror
Fluorescence sensor λ1
Lens
Filters
Beam stop
Fluorescence sensor λ2
Organism or cell
Laser beam
(b)
Piezoelectric crystal
Fluorescence signal
Sort signal
Light scatter signal
Flow chamber
Sheath inlet
Light scatter sensor
Fluorescence sensor
Filter
Lens
Beam stop
Laser beam
Deflection plates
Deflected droplets
Waste tube
Sample collection

levels (see Journal of Histochemistry and Cytochemistry, volume **24**, No. 1, 1976 and volume **27**, No. 1, 1979).

D. LASER FLYING SPOT TWO DIMENSIONAL FLUORIMETRY

Another technique which measures the intensity of fluorescence from individual organisms and sorts these into channels is the laser flying spot 2-D microfluorometer of Chance *et al.*, 1978 (Fig. 2.19). Primarily used to analyse two dimensional spatial heterogeneity in tissues, one report (Bashford *et al.*, 1980) demonstrates its potential usefulness for the determination of time-dependent changes in monolayers of cells. In this study oscillations in redox states of flavoproteins and nicotinamide nucleotides in growing cells of *Acanthamoeba castellanii* and *Schizosaccharomyces pombe* were followed over periods of up to 2 h using appropriate lines from an argon ion laser for excitation.

E. CELL CYCLE FRACTIONATION

Cell cycle fractionation procedures permit physical separation into sub-populations of large number of organisms from exponentially-growing cultures; these methods are based on size or density changes during the cell cycle and density gradients and centrifugation conditions employed have been briefly referred to in the section relating to methods for selection synchrony. Although the techniques are basically similar, some further comments are pertinent. The basic philosophy underlying cell cycle fractionation is to subject the exponentially-growing culture to some treatment so that all the cells receive an identical exposure before instantaneous arrest (by chemical fixation or rapid cooling), and then to separate into fractions of increasing size (or density) by centrifugation. In this kind of experiment the physiological state of the cells at time of treatment can be precisely and

Fig. 2.17 (a) Generalized flow analyser showing sample and sheath fluid paths, cell stream, laser illumination and light scatter, and fluorescence sensors. Sheath fluid flows coaxially around the sample injection tube and serves to centre cells for measurement. Low-angle light scatter and fluorescence at two wavelengths are measured as cells traverse the laser excitation beam; (b) Generalized flow sorter. Ultrasonic energy is used to break the cell stream into uniform droplets. Cells flowing through the instrument are isolated in these tiny droplets. Based on measurements recorded on cells in flow, droplets containing cells of interest are charged. As the droplet stream passes through an electrostatic field, the charged droplets are deflected right or left, carrying the sorted cells. (Reproduced with permission from Horan and Wheeless, 1977.)

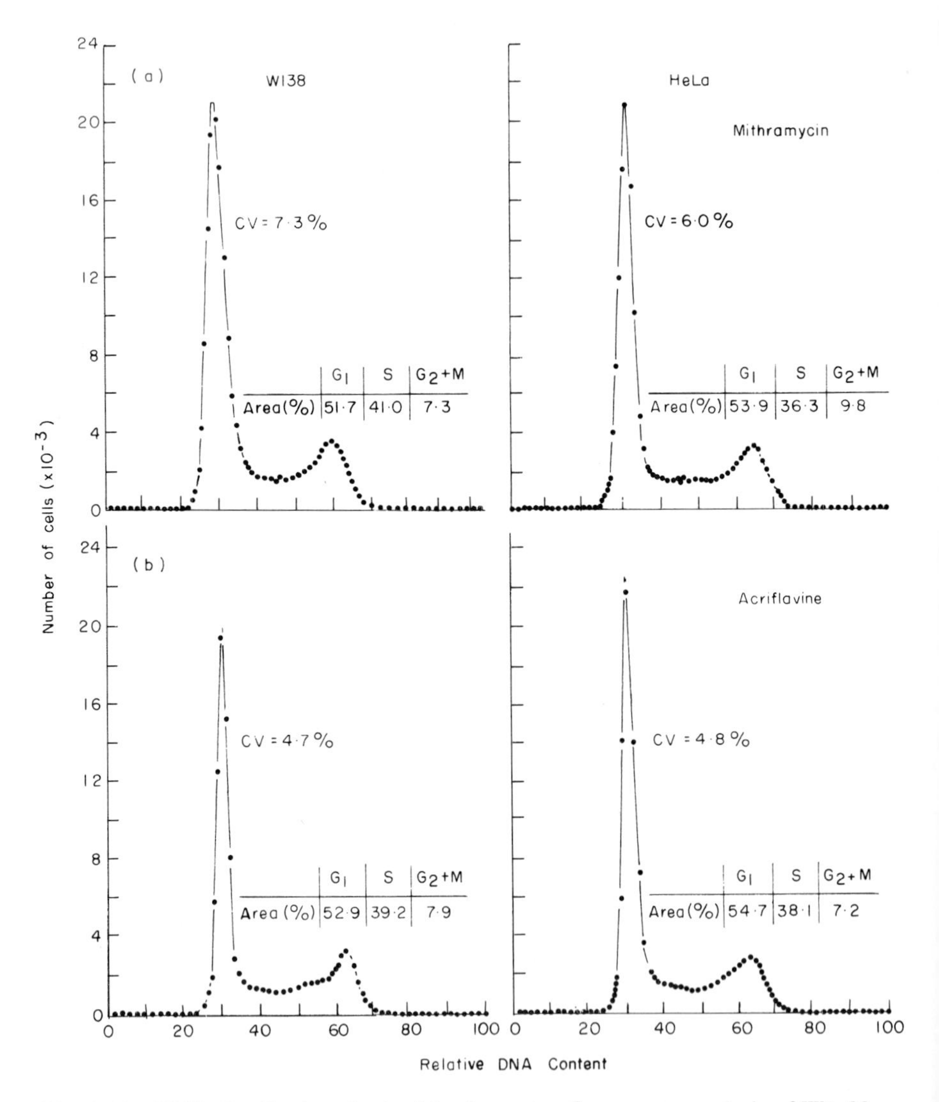

Fig. 2.18 DNA distribution obtained by flow microfluorometer analysis of W1–38 and HeLa cells stained with either (a) mithramycin, or by the (b) acriflavin-Feulgen procedure. Coefficients of variation (CV) for the G_1 distributions and the percentage of cells in G_1, S, or $G_2 + M$ were obtained by computer-fit analyses of the DNA distribution curves. (Reproduced with permission from Crissman and Tobey, 1974.)

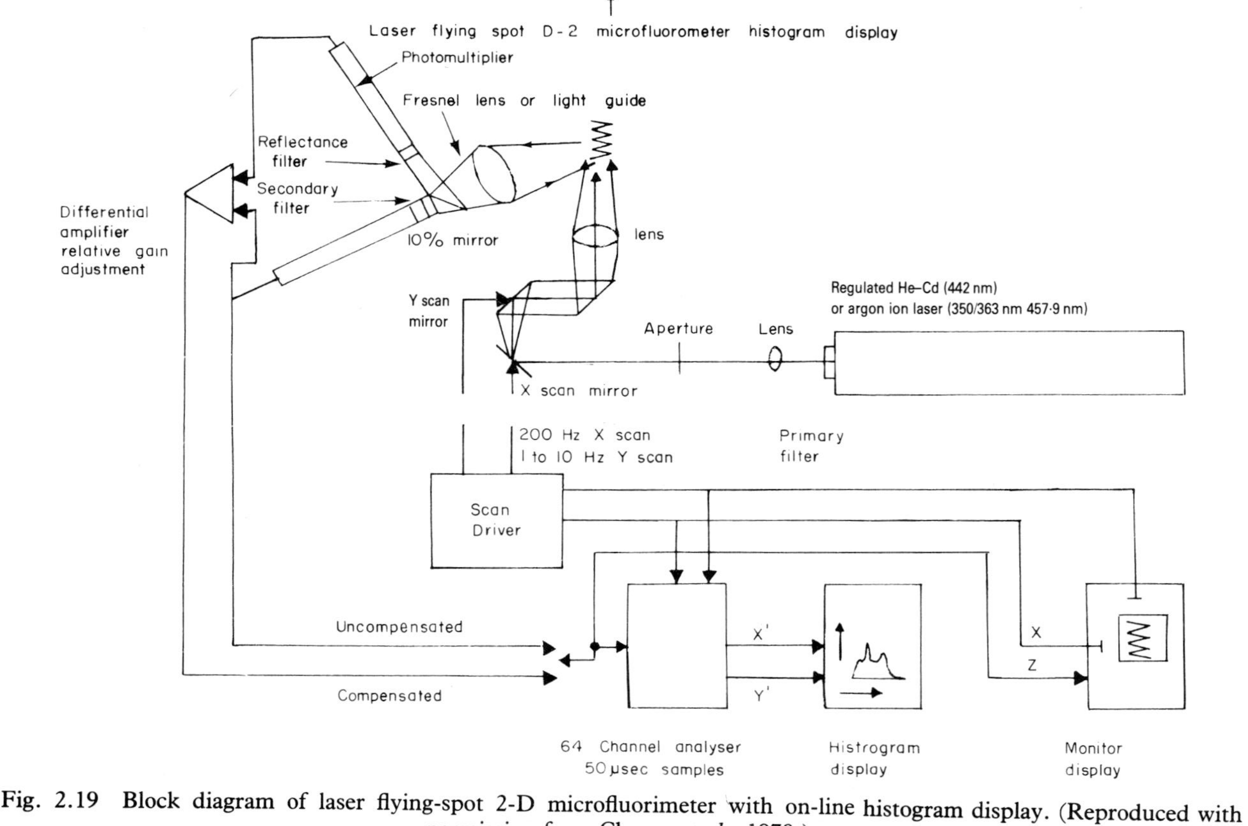

Fig. 2.19 Block diagram of laser flying-spot 2-D microfluorimeter with on-line histogram display. (Reproduced with permission from Chance *et al.*, 1978.)

reproducibly defined. A wide range of applications has been reported: e.g. susceptibility to mutagenesis through the cell division cycle (Dawes and Carter, 1974), chemical analysis of different size classes (Poole *et al.*, 1974), incorporation of a pulse of a labelled precursor for macromolecular synthesis (Kubitschek *et al.*, 1967), measurement of the rate of enzyme induction at successive stages of the cell cycle after a pulse of inducer (Sebastian *et al.*, 1973), and levels of enzyme activities in culture fractions (Poole and Lloyd, 1973). Although it is often maintained that data obtained from cycle fractionations are less likely to be distorted as a consequence of physiological or biochemical disturbance than those obtained from synchronous cultures, this is not unquestionably so. The critical factor is the rate of arrest. This should be achieved as rapidly as possible in the bulk culture (e.g. by circulation of cooling water at 0°C through an immersed cooling coil). Subsequent rigorous control of temperature during harvesting, loading, centrifugation and recovery of fractions is of crucial importance. All these procedures should be carried out as rapidly as possible and (where appropriate) extracts prepared immediately. Although these precautions will not prevent disturbances to reactions with time constants in the metabolic time domain, epigenetic reactions will (at least for the most part) be effectively frozen. Even so the investigator should be aware of the potential hazards of proteolytic degradation—even at low temperatures (Pringle, 1975) and the use of a mixture of protease inhibitors is recommended for some control experiments (Edwards *et al.*, 1981). It is also worth remembering that protein synthesis can proceed slowly in intact cells even at 0°C, and in some investigations protein synthesis inhibitors (e.g. cycloheximide and chloramphenicol, Cartledge and Lloyd, 1972) are necessary to give complete arrest. Thus while cell cycle fractionation provides an invaluable confirmatory adjunct to selection synchrony, both techniques should be used in parallel.

Cell cycle fractionation is not a suitable technique for cell separation prior to measurement of pool sizes of intermediary metabolites or respiration rates. A slight inconvenience of the technique is that it does not separate cells strictly on the basis of age in the division cycle. Thus it is necessary to size the separated sub-populations (usually by measurements of mean cell volumes using a resistive particle counter fitted with multichannel particle size analyser). Even on iso-osmotic gradients (especially after lengthy zonal centrifuge-runs during which cells are subjected to high hydrostatic as well as centrifugal forces) at least some organisms (even some of those with, and particularly those without, cell walls) will show volume changes. Even the sudden cooling (or chemical fixation) during the metabolic arrest step may alter cell volume (MacKnight and Leaf, 1977). Despite all these cautionary remarks, cell cycle fractionation is a useful high-resolution method which provides large quantities of material for biochemical analyses.

F. CLASSIFICATION BY AGE

The cell division cycle of *Escherichia coli* B/r in asynchronous cultures has also been studied by a modification of the membrane elution technique of Helmstetter and Cummings (1963, 1964). The rate of synthesis of a macromolecule is determined by pulse-labelling the cells of a culture with a labelled precursor, binding them to a membrane filter, and then measuring the radioactivity in newborn cells which are eluted continuously from the filter. This technique, described in detail by Helmstetter (1969) can be used to study the effects of any environmental change (e.g. presentation of radioactive precursors, inducers, radiation, chemical mutagens, antibiotics and nutritional shifts) and has been invaluable in determining the relationship between chromosome replication and the division cycle (Helmstetter and Cooper, 1968; Cooper and Helmstetter, 1968).

Summary

The cell division cycle can be studied in single cells, naturally-synchronous systems, synchronous cultures and in asynchronous exponentially-growing cultures (cell cycle fractionation and classification by age). For many purposes, environmental perturbations must as far as possible be minimized in order to distinguish between cell-cycle dependent phenomena and direct consequences of experimental procedures. Parallel experiments with identically-treated asynchronous cultures are mandatory. Perturbation does, on the other hand, afford well-established approaches to the preparation of synchronous cultures; induced synchrony provides information on control mechanisms involved in the cell division process. Recently introduced methods for selection synchrony offer considerable advantages over density-gradient based procedures. Major developments in automated analytical cytology (especially flow microcytofluorimetry) facilitate cell cycle analyses.

3. Biosynthesis of Macromolecular Cell Components: DNA

"We lack not rhymes and reasons
As in the whirligig of time
We circle with the seasons."

Alfred Lord Tennyson
Will Waterproof's Lyrical Monologue, 1842.

PART 1
SYNTHESIS OF DNA DURING THE EUKARYOTIC CELL CYCLE

I. Introduction

Of all the studies of the cell cycle, no single area has received as much attention as that attempting to understand the structure of DNA and the mechanisms by which this macromolecule is replicated. The literature contains a vast number of reports on these topics and is regularly and extensively reviewed. In view of the great importance of DNA and its key role in the regulation of cell division, this effort is more than adequately justified. Despite many recent advances, particularly in our understanding of the structure of DNA, identification of the components of the multi-enzyme complex required for replication and the mechanisms by which the molecule is duplicated, there are still some fundamental questions unanswered. For many years now it has been established that DNA synthesis occurs during a restricted portion of the cell cycle, the S-phase, which is preceded by a gap (G_1) and followed by another gap (G_2) where no synthesis occurs (Howard and Pelc 1953); the G_2 phase is then followed by mitosis (M) (Fig. 3.1a). In general, these subdivisions of the cell cycle are still applicable but as we shall see later, there are some instances where these delineations do not fully describe the observed patterns of synthesis. In eukaryotes, the bulk of the genetic material is located

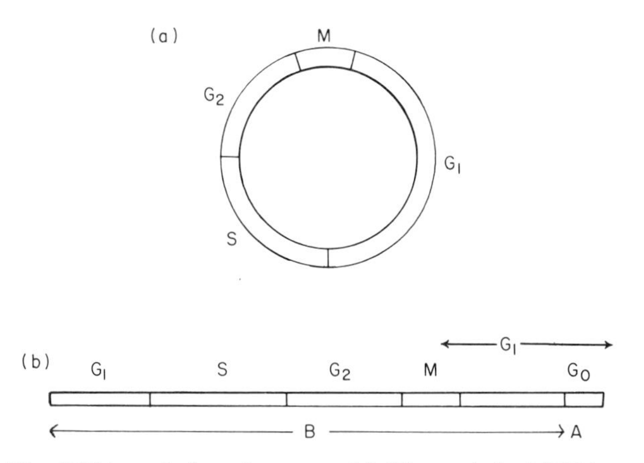

Fig. 3.1 The DNA cycle in eukaryotes. (a) The period of DNA synthesis (S) is preceded by a gap, G_1 and followed by another gap, G_2. Mitosis (M) completes the cycle; (b) The cell cycle may consist of two portions A and B, as shown. State A represents G_0, the variable portion of G_1. For details see text.

within the nucleus but there may also be extranuclear DNA e.g. mitochondrial DNA (mt DNA), or chloroplast DNA (cp DNA), which may or may not be replicated at the same time as that of the nucleus. Replication of nuclear DNA, which is semi-conservative (Taylor *et al.*, 1957), must be precise and must occur before cell division; in asexual reproduction (which we will be concerned with here) each daughter cell must contain a copy of the genetic material (except where mutations may have arisen) identical to that of the mother cell. While the molecular processes involved in the DNA synthesis are becoming elucidated, little is known about the events which occur in G_2; even less is known about G_1 where initiation of DNA synthesis is triggered. The factors which initiate replication, especially when differentiated cells undergo proliferation as in tumour development, are of prime importance and largely unidentified. Strict control must also be exercised so that the genome is only replicated once per cycle. In many areas of research, valuable information on aspects of eukaryotic processes has resulted from studying less structurally complex prokaryotic systems. This approach is not readily applicable to the synthesis of DNA, however, which appears to be fundamentally different in the two cell types. Alternative, simpler systems in which the synthesis of eukaryotic DNA has been studied are organellar DNA and DNA of mammalian viruses. Of particular importance has been the use of papovaviruses as model systems (see Fareed and Davoli, 1977).

A comprehensive survey of the literature and an extensive and detailed account of the development of our current understanding of all aspects of DNA replication would probably encompass an entire volume and will not be attempted here. Instead these sections will attempt to summarize the major aspects of this topic, including genome structure and proposed mechanisms of replication, and will include only illustrative references. For a more complete account the reader is referred to some of the recent reviews by Edenberg and Huberman (1975), Baserga and Nicolini (1976), Callan (1976), Horgen and Silver (1978), Pardee *et al.* (1978), Sheinin *et al.* (1978b), Wintersberger (1978), De Pamphilis and Wassarman (1980) and Petes (1980). Also certain of the Cold Spring Harbor Symposia on Quantitative Biology are dedicated entirely to this topic (e.g. vol. **43**, part 2 (1978)).

II. G_1 and G_0 Periods

The G_1 period, which precedes DNA synthesis, is the phase when cells are "triggered" into the phase of DNA replication. Events in G_1 may be sequential and arrest of one event may delay subsequent processes (Naha *et al.*, 1975). In mammalian cells the S, G_2 and M phases usually last about 6–8 h, 2–6 h and 1 h, respectively, but considerable variations are found in the duration of G_1 (Pardee *et al.*, 1978). The length of G_1 phase may also vary between individuals of a population of cells of the same type (Wintersberger, 1978). Also, when cells cease proliferation, either as a result of inhibition of growth or upon differentiation, they are usually arrested in G_1 or have an extended G_1 period (Dell' Orco *et al.*, 1975). In view of this, it has been postulated (Smith and Martin, 1973) that the cycle may be divided into two fractions, A and B, where B is fixed and includes S, G_2, M and part of G_1 (Fig. 3.1b). State A represents the variable portion of G_1 and may be termed G_0; the variations in duration of G_1 may therefore rest in the variability of state A. There is evidence to suggest that upon the restoration of optimal growth conditions, cells escape from their arrested state at the same point in G_1 (Pardee, 1974a). It is proposed that cells may enter G_0 after they pass a "restriction point" in G_1 as they differentiate or as a response to sub-optimal growth conditions, and thus retain their viability. Alternatively, cells may pass this "restriction point" and undergo replication (Pardee, 1974a).

The initiation of DNA synthesis occurs in G_1 and this then must also be the portion of the cycle when the components necessary for replication must either be synthesized or activated. It is interesting to note therefore, that in some cell types, particularly eukaryotic micro-organisms, the G_1 phase is extremely short or absent i.e. a new round of DNA synthesis begins immediately after mitosis or cell division. Some reports of cases where G_1 has been shown to be absent include certain mammalian cell lines (Robbins and Scharff, 1967), in the micronucleus of *Tetrahymena* (Gorovsky *et al.*, 1977) or *Euplotes* (Prescott, 1966) in *Physarum polycephalum* (Nygaard *et al.*, 1960; Kessler, 1967), *Amoeba proteus* (Ord, 1968; Prescott and Goldstein, 1967) and the fission yeast *Schizosaccharomyces pombe* (Bostock *et al.*, 1966; Bostock, 1970). Presumably in these instances, the necessary preparations required for the processes of DNA synthesis occur in the G_2 phase and it is not a prerequisite that these preparations follow mitosis (see Chapter 8).

III. Chromatin Structure and Requirements for Chromatin Replication

It is unrealistic to consider replication of DNA alone during the cell cycle of eukaryotes since this DNA is complexed with histones and non-histone proteins. Replication of the entire complex requires a multi-enzyme system which must also be synthesized and activated. Understanding of the detailed structure of chromatin and of the individual components that comprise the complex is essential if we are to elucidate the mechanisms by which the genome is replicated. The structure of chromatin undergoes marked changes during interphase (Nicolini *et al.*, 1975; Hildebrand and Tobey, 1975) notably during mitosis and chromosome condensation.

A. CHROMATIN STRUCTURE

There are a number of recent reviews on the detailed structure of chromatin (e.g. Baserga and Nicolini, 1976; Horgen and Silver, 1978; Sheinin *et al.*, 1978b; Lilley and Pardon, 1979); the basic structure of this complex is shown in Fig. 3.2. The duplex of DNA is coiled in a precise manner and associated with histones to form a repeating pattern of nucleoprotein particles called nucleosomes (McGhee and Felsenfeld, 1980). Each nucleosome is composed of histones and DNA in a ratio of approximately 1:1 (Wintersberger, 1978). There are five major histones present in virtually all cells and tissues which are termed H1, H2A, H2B, H3 and H4 (Horgen and Silver, 1978; Isenberg, 1979). There are two molecules each of H2A, H2B, H3 and H4 which form an octameric complex (Thomas and Kornberg, 1975) that represent a core particle around which about 140 base pairs are wrapped (Jorcano and Ruiz-Carrillo, 1979; Isenberg, 1979). The

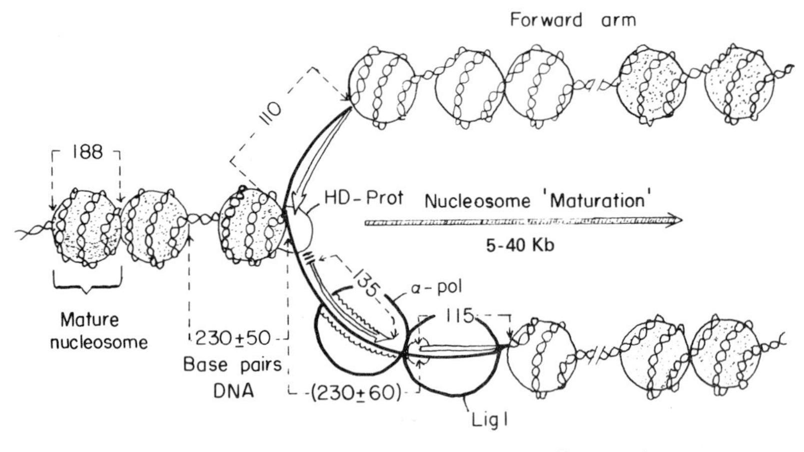

Fig. 3.2 Replication fork of a eukaryotic chromosome. Only those components for which there is substantial experimental evidence are represented: mature (heavy shading) and immature (light shading) nucleosomes, an RNA-primed Okazaki fragment |||⇒) on the retrograde arm, helix destabilizing (HD) protein, DNA polymerase (α-pol) and DNA ligase (Lig-1). The sizes of components are drawn to reflect their molecular weights. Numbers of base pairs shown are for SV40 chromosomes since most of the available data comes from viral genomes. (Reproduced with permission from DePamphilis and Wassarman, 1980.)

histone octamer has a two-fold axis of symmetry and the overall shape of a left-handed helical spool on which are wound two turns of a flat super-helix of DNA (Klug *et al.*, 1980). Nucleosomes are joined by a spacer or linker region of 40–60 base pairs, presumably coiled and associated with one molecule of H1; this linker region is more sensitive to nucleolytic attack and may not normally be in an extended formation as in the "beads on a string" visualization (Woodcock, 1973) but may be an integral part of the nucleosome (Horgen and Silver, 1978). Specific phosphorylation reactions occur on histone H1 which may lead to condensation of chromatin fibres into chromosomal structures at metaphase (Lake, 1973; Bradbury *et al.*, 1973, 1974b). There may be differences in amino acid sequences in histones of the same types in different species.

Also associated with chromatin are non-histone proteins, which are mostly acidic, difficult to isolate and heterogenous (Elgin and Weintraub, 1975). Up to 450 non-histone proteins have been detected on sodium dodecylsulphate polyacrylamide gels (Wintersberger, 1978) and these proteins presumably perform structural, enzymatic, and regulatory roles within the chromatin. They may act as positive regulators in gene expression, in histone and chromatin modifications, in processing of rRNA, or be involved in chromosomal protein metabolism.

The 2 μm plasmids of *Saccharomyces cerevisiae* are organized into structures very similar to the chromatin of yeast chromosomes (Nelson and Fangman, 1979). This yeast shows some variations in this basic chromatin structure. While yeast chromatin may be partially digested to give nucleosomes of 160 base pairs (11S) which are similar to those of higher eukaryotes (Nelson *et al.*, 1977), there are differences in the composition of histones present. While histones H2A, H2B, and H4 are detected (Franco *et al.*, 1974), H1 is absent and a histone analogous to H3 is present which has a quite different electrophoretic mobility (Brandt and Von Holt, 1976). Structural changes in chromatin normally observed during the cycle (e.g. Collins *et al.*, 1977) are not found in this organism notably the absence of condensation of chromatin into chromatids during mitosis (Gordon, 1977).

B. SYNTHESIS OF HISTONES DURING THE CELL CYCLE

There is general agreement that histones are synthesized during the S-phase and that inhibition of DNA synthesis also results in inhibition of histone synthesis (e.g. Moll and Wintersberger, 1976). More recently it has been suggested that histones may also be synthesized in G_1 (Groppi and Coffino, 1980). However, there is a possibility that at certain stages of the cycle of some cells, pools of free histones are present, since inhibition of protein synthesis in the middle of the S-phase does not inhibit DNA-chain elongation (see later). Histone synthesis *in vitro* is unaffected by inhibitors of RNA synthesis (Gallwitz and Mueller, 1969). Transcription of Group A histone mRNA (7–9S RNA) is restricted to the S-phase (Borun *et al.*, 1967) and non-histone chromosomal proteins are thought to be responsible for regulating the transcription of the regions of genome containing the genes for histone synthesis (Stein *et al.*, 1975). In the ciliate *Tetrahymena pyriformis* which contains an amitotic macronucleus and a mitotic micronucleus (Gorovsky *et al.*, 1977), histone H1 is present in the macronucleus but absent in the micronucleus (Gorovsky and Keevert, 1975).

After synthesis, histones are subject to various post-synthetic modifications such as acetylation, methylation, phosphorylation etc. which may alter the charge of the molecules; these changes occur at specific stages of the cycle and may involve modifications at specific sites on the histone molecules (Elgin and Weintraub, 1975) and are implicated in the control of cell division (see Chapter 8).

C. COMPONENTS OF THE MULTIENZYME COMPLEX NECESSARY FOR CHROMATIN REPLICATION AND REPAIR

A number of enzymes and proteins are required for the normal repair mechanisms and replication of chromatin. Probably the most extensively studied enzymes of this complex are the DNA polymerases (see Chambon, 1975; Loeb, 1974; Holmes and Johnston, 1975; Weissbach, 1977 and Wintersberger, 1978 for reviews).

Table 3.1 summarizes the properties of the DNA polymerases commonly found in eukaryotic cells. Some important features of these polymerases are:

(i) their location within different subcellular compartments (mitochondria contain a specific mitochondrial DNA polymerase);
(ii) their ability to polymerize DNA only in the $5' \rightarrow 3'$ direction (see later);
(iii) all accept "activated" DNA (i.e. single-stranded DNA) as a template. DNA polymerase α is the major polymerase present in eukaryotic cells and uses short sequences of RNA (Dressler, 1975) as a primer (see later).

A DNA polymerase with $3' \rightarrow 5'$ exonuclease activity (similar to an enzyme found in prokaryotes) has been isolated in bone marrow cells (Byrnes *et al.*, 1976). This enzyme catalyses the removal of $3'$ terminal nucleotides from DNA and probably acts as a "proof-reading" mechanism removing mis-matched nucleotides. Different DNA polymerases show distinctive cell cycle-dependent changes in activity (Spadari and Weissbach, 1974).

Other enzymes that may be required for synthesis and repair include:

(i) Endonucleases that introduce single-strand nicks into double-stranded DNA (Mechali and De Recondo, 1975; Otto and

Table 3.1. *Nomenclature for eukaryotic DNA polymerases (from Weissbach, 1977).*

DNA polymerase	Molecular weight	Inhibition by *N*-ethylmaleimide	Salt effect
α	120 to 300 k	+	Inhibited by NaCl concentrations >25 mM
β	30 to 50 k	−	Stimulated by 100–200 mM NaCl but inhibited by 50 mM–PO_4^{3-}
γ	150 to 300 k	+	Stimulated by 100–250 mM KCl and 50 mM–PO_4^{3-}
Mitochondrial	150 k	+	Stimulated by 100–200 mM KCl

Knippers, 1976; Lavin *et al.*, 1976; Wang and Furth, 1977). An enzyme introducing a single-stranded DNA into the molecule may allow the strand to rotate relative to the helix axis, thus allowing swivelling of coiled sections (Champoux and Dulbecco, 1972);

(ii) Unwinding enzymes (Keller, 1975; Poccia *et al.*, 1978);
(iii) DNA ligases (Söderhäll and Lindahl, 1976);
(iv) DNA-dependent ATPases, which require single-stranded DNA (Hachman and Lezius, 1976; Otto, 1977);
(v) DNA binding proteins (Hoch and McVey, 1977) which bind single-stranded DNA and maintain this extended configuration (Otto *et al.*, 1977);
(vi) Ribonuclease H, which hydrolyses RNA:DNA hybrids (Tashiro *et al.*, 1976) or other ribonucleases (Banks, 1974).

IV. Initiation of DNA Synthesis

There is much evidence in the literature to suggest that the initiation of DNA synthesis is under positive control. A cell must make available all the necessary precursors that are required for the assembly of new chromatin e.g. deoxyribonucleotides for DNA, histone precursors etc. There must be activation or synthesis of those components present in the multi-enzyme complex constituting the replicative and repair apparatus which is essential for the assembly of new chromatin.

However, it is obvious that there must also be a tighter control mechanism operating, that is responsible for the initiation of replication, to ensure that, for example, replication occurs only once per cycle and that cells do not continue proliferation once they have fully differentiated. This inducer has been the focus of much research, but it has so far remained elusive. Several approaches have been adopted in this area.

One approach is to study the factors which may stimulate or accompany the transition when resting cells (in G_0) enter the S-phase. Growth factors present in serum are capable of stimulating cells to proliferate (Pardee *et al.*, 1978) although the addition of Ca^{2+} to resting cultures stimulates DNA replication even in serum-free cultures (Dulbecco and Elkington, 1975). The rate of transport of low molecular weight compounds has also been suggested as the primary growth regulatory mechanism in mammalian cells (Holley, 1972). Changes in rates of DNA synthesis closely coincide with changing rates of exogenous thymidine-incorporation at certain concentrations (Miller *et al.*, 1979). Changes in the surface membrane fluidity have also been implicated in control of growth transformed cells (Pardee, 1975). A more recent proposal for this positive growth regulator is diadenosine $5'$–$5'''$–P^1, P^4-tetraphosphate (Ap_4A), which is present in rapidly growing mammalian cells at a level of 20×10^{-4} of that of ATP, but $0{\cdot}1 \times 10^{-4}$ of the ATP level of slowly growing cells (Rapaport and Zamecnik, 1976). Levels of Ap_4A decrease when inhibitors of protein synthesis or DNA synthesis are added (Rapaport and Zamecnik, 1976) and the addition of this nucleotide to resting mammalian cells results in initiation of DNA synthesis (Grummt, 1978) *in vitro*.

Protein synthesis appears to be essential for the initiation of DNA synthesis since addition of protein synthesis inhibitors to cells in G_1 prevents replication (Williamson, 1973). Furthermore, when quiescent cells are stimulated, protein synthesis increases (Hassell and Engelhardt, 1976). However, this protein synthesis may be required for the formation of an uncharacterized initiation factor, for the synthesis of the enzymes responsible for DNA replication, or for the production and assembly of precursors of chromatin. Although the activities of DNA polymerases, DNA ligases (Wintersberger, 1978), and thymidine kinases (Gröbner and Sachsenmaier, 1976; Gröbner, 1979; Wright and Tollon, 1979) increase during the period of DNA synthesis, it seems unlikely that any of these are rate-limiting in G_1-phase. Regeneration of rat liver is accompanied by modifications (e.g. phosphorylation, acetylation, ADP-ribosylation) of nuclear proteins (Caplan *et al.*, 1978) and the phosphorylation of chromatin proteins has also been implicated in the initiation of DNA synthesis (Bradbury *et al.*, 1974a; Defer *et al.*, 1979).

Deoxyribonucleoside triphosphates in a variety of cell types show cyclic variations during growth and division (see Chapter 8), and thymidine nucleotides in cells not engaged in DNA synthesis are degraded (Schaer and Maurer, 1977). In *Physarum polycephalum*, these nucleotides increase before and after the initiation of DNA synthesis but decrease during the S-phase (Fink, 1975), while in HeLa cells, pyrimidine deoxyribonucleotide biosynthesis is thought to be triggered prior to, or at the same time as the initiation of DNA replication (Bray and Brent, 1972). In Chinese hamster cells the dCTP pool is closely correlated with the rate of DNA synthesis perhaps suggesting a regulatory function (Skoog *et al.*, 1973). However, if Chinese hamster cells are made permeable to exogenous deoxyribonucleotides by the addition of lysolecithin, the supply of these precursors alone does not limit the entry of G_1 cells into S-phase (Miller *et al.*, 1978).

There is strong evidence from cell fusion experiments to suggest that the signal for initiation is cytoplasmic in origin (see Rao and Sunkara, 1978 for a review). Differentiated cells are reversibly supressed in their capacity for DNA or RNA synthesis. If these cells are fused with cells normally synthesizing DNA, then in the heterokaryon, DNA and RNA synthesis occurs in *both* types of nuclei (Harris *et al.*, 1960); the inactive cell does not supress synthesis in the active cell. In another series of experiments, nuclei from differentiated cells (liver, brain and blood cells) synthesized DNA when injected into unfertilized eggs of the same species (*Xenopus laevis*), again implicating a cytoplasmic factor present in cells synthesizing DNA, which then stimulates resting nuclei as quickly as 90 min after injection (Graham *et al.*, 1966). This cytoplasmic factor is not species-specific or even class-specific since injection of mouse liver nuclei into *Xenopus* eggs produce the same effect (Graham *et al.*, 1966). Fusion of HeLa cells at different stages of the cell cycle has shown that regulation of DNA synthesis is under positive control (Rao and Johnson, 1970). Fusion of G_2 cells with either G_1 or S-phase cells does not prevent the initiation or continuation of synthesis (i.e. there is no inhibitor present in G_2 cytoplasm); the G_2 nuclei do not enter S-phase suggesting that G_2 nuclei cannot be stimulated into S-phase until mitosis has occurred (Rao and Johnson, 1970). If cells from two species with different G_1 durations are fused, then the heterokaryons initiate DNA synthesis after a time corresponding to that of the shorter G_1, but the duration of the S-phase is unchanged (Graves, 1972). These studies suggest that initiation is effected by a non-species-specific cytoplasmic factor, which does not affect the duration of the S-phase; once initiated, replication can continue in the absence of this factor. Results from experiments of cell fusion in early, mid and late G_1 phases suggest that the levels of this inducer(s) gradually accumulate during G_1, are maximum in S-phase and decline in G_2 (Rao and Sunkara, 1978); DNA synthesis can only occur when the inducer is present above a critical level, coinciding with the beginning of S-phase (Fig. 3.3).

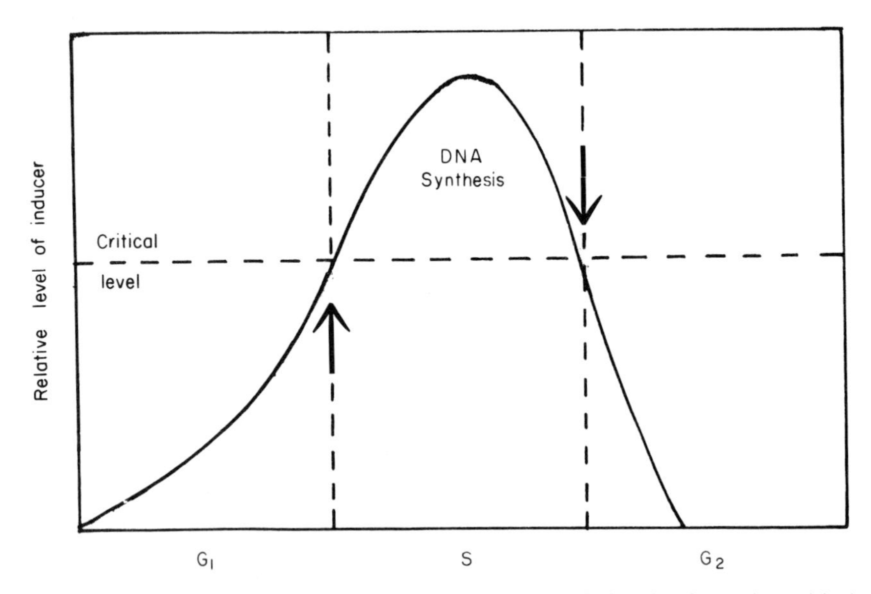

Fig. 3.3 Levels of inducer during the HeLa cell cycle. When levels reach a critical value (↑), DNA synthesis is initiated. When levels fall below this critical level (↓), DNA synthesis stops. (Reproduced with permission from Rao and Sunkara, 1978.)

V. Mechanisms of Chromosomal Replication

Much of our understanding of the replication of chromosomes has come from the technique of DNA fibre autoradiography, first developed by Cairns (1963, 1966). Using this technique, whereby newly-synthesized DNA may be pulse-labelled with high or low doses of tritiated thymidine, Huberman and Riggs (1968) proposed a model for replication of DNA fibres of Chinese hamster or HeLa cells, which is probably valid for most (if not all) eukaryotic chromosomes (Fig. 3.4). In this model, each chromosome may be envisaged as being divided into a number of replicating units, or replicons of varying size

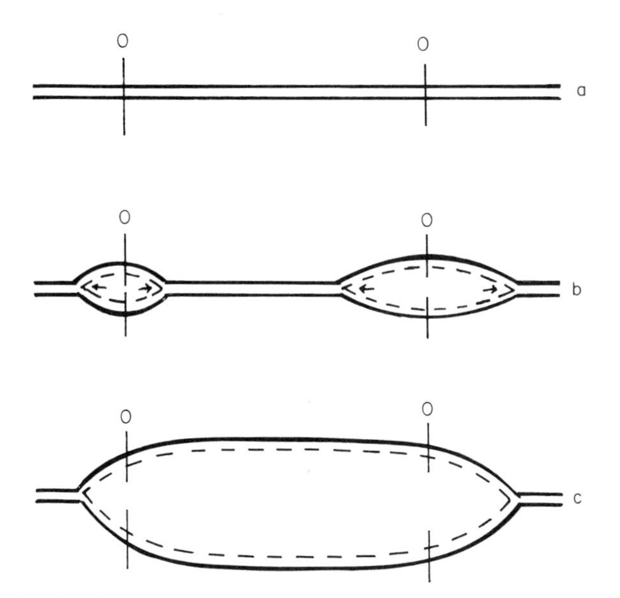

Fig. 3.4 Bidirectional replication of eukaryotic DNA. Solid horizontal lines in (a) represent the DNA duplex and O represents an "origin" of replication in two adjoining replication units; (b) DNA synthesis is initiated at the two origins and proceeds bidirectionally with newly-synthesized DNA represented as dashed lines. Horizontal arrows indicate the direction of fork movement. In (c) replication is completed in both units (modified from Huberman and Riggs, 1968).

in which DNA replication is initiated at an "origin". Replication then proceeds bidirectionally at the two forks, giving rise to a series of "bubbles" which can be visualized using fibre autoradiography. In their original model, Huberman and Riggs (1968) tentatively suggested that replication proceeded until a "terminus" was reached in the DNA molecule, whereby replication ceased. There is no evidence to support this and indeed there is some evidence to the contrary (see later); replication probably continues until adjacent replicons converge.

The size of replicons varies within different cells and even within individual chromosomes. In HeLa and Chinese hamster cells, replicons are about 30 μm long (Huberman and Riggs, 1968), in yeast from 3–80 μm, although mostly between 15–20 μm and 30–35 μm (Petes *et al.*, 1974) and in most cell types probably range from 15–120 μm, averaging about 50 μm (Edenberg and Huberman, 1975; Callan, 1976). The rate of fork movement in yeast is about 0·7 μm/min/replication fork (Petes and Williamson, 1975b), in mammalian cells about 2·5 μm/min or less (Huberman and Riggs, 1968), and generally within the range 0·1–2·0 μm/min, i.e. 2–40 $\times$ 10^5 daltons or 1–15 $\times$ 10^3 nucleotides/min (Sheinen *et al.*, 1978b).

Eukaryotic chromosomes therefore contain a large number of replicons and it has been found that active replicons are often distributed non-randomly on chromosomes or in clusters. While the distances between adjacent replicons may vary in different species (Hand and Tamm, 1974) or are similar in the same species (Hand, 1975a), clusters of replicons appear to initiate replication synchronously i.e. replication may occur in bursts over topographically contiguous regions. This phenomenon of synchrony of initiation is important when considering the sequence of events occuring within the S-phase (see later). Much effort has been involved in the identification of these origins since they are replicated at the same time in the S-phase in different cycles, suggesting that they are unique, specific sequences of nucleotides. Experiments with viruses have shown that these origins in fact contain inverted repetitive sequences of bases or palindromes (Fig. 3.5a), which may then form 1 or 2 hairpin loops per strand due to intrachain pairing (Fig. 3.5b). In SV40, this palindrome sequence at or near the origin is 28 nucleotides long, with 2 complementary sequences of 8 nucleotides located 2 nucleotides away from an axis of symmetry (Jay *et al.*, 1976). Similar palindromes have been identified at the initiation sites of adenovirus 2 (Padmanabhan *et al.*, 1976) and AAV (Straus *et al.*, 1976). These hairpin loops represent discontinuities in the double helical structure which may then be recognized by proteins. The single-stranded sequence at the loop may also serve as a template for RNA or DNA primers, a site for nuclease cleavage or a site for unwinding protein (Pardee *et al.*, 1978). While no evidence has been presented in favour of specific termination points for replication (Hand, 1975b; Blumenthal *et al.*, 1973), there is in fact no necessity for such a mechanism to operate; termination of replication may occur when two adjacent replicons converge.

Evidence against the existence of termination points, however, comes from work with tumour viruses (see Fareed and Davoli, 1977 for review). These viruses contain a circular, double-stranded DNA molecule (3–3·6 $\times$ 10^6 daltons) and replication is initiated at a specific

(a)

—A-A-T-T-C-A-G-G-A-G-C-T-C-T-G-A-A-T-T—

—T-T-A-A-G-T-C-C-T-C-G-A-G-A-C-T-T-A-A—

(b)

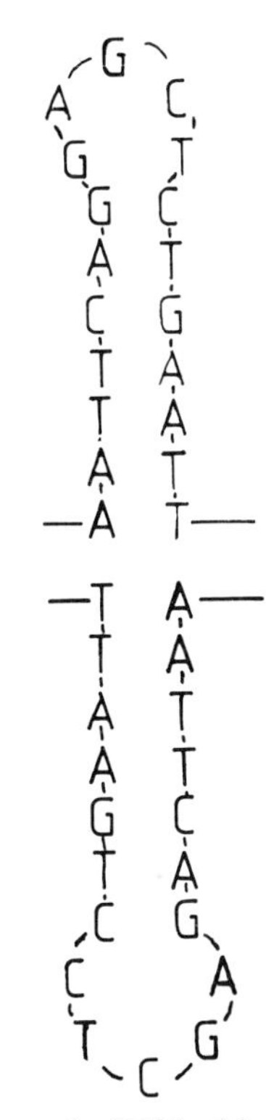

Fig. 3.5 Palindrome sequence of bases in DNA. (a) Shows sequences of bases in a portion of the two strands of the DNA duplex. This palindrome sequence can potentially form a hairpin-loop configuration as indicated in (b).

origin (Fig. 3.6a). Replication proceeds bidirectionally at two replication forks which meet 180° opposite the origin. However, in a mutant containing a deletion i.e. a smaller circular DNA molecule (Fig. 3.6b), replication again terminates 180° from the origin, at a nucleotide sequence quite different from that in the wild type. This observation strongly argues against the existence of fixed termination points of specific nucleotide sequences.

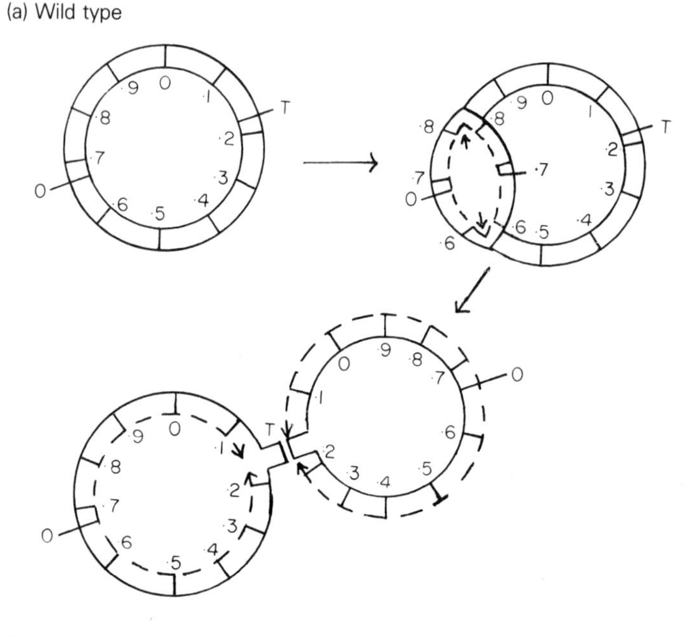

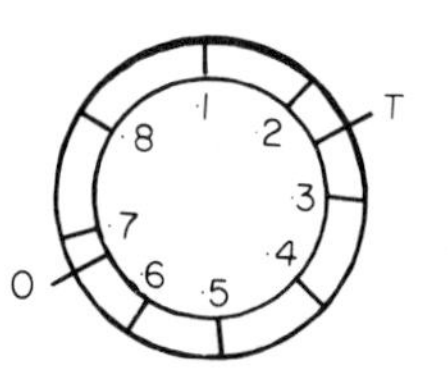

Fig. 3.6 DNA replication in tumour viruses. In (a) a circular, double-stranded viral DNA molecule is divided into ten segments. DNA replication is initiated at the origin 0, and proceeds bidirectionally, as indicated by the arrows. Replication is terminated at T which is 180° opposite the origin. Dashed lines represent newly-synthesized strands. In a mutant containing a deletion (b), termination of replication also occurs 180° opposite the origin (modified from Wintersberger, 1978).

VI. Molecular Events at the Replication Fork

For replication to occur at the growing fork as indicated, one chain must be synthesized in the $5' \rightarrow 3'$ direction and the other in the $3' \rightarrow 5'$ direction. However, no known DNA polymerase can synthesize DNA in the $3' \rightarrow 5'$ direction. It is envisaged, therefore, that one chain is synthesized in the $5' \rightarrow 3'$ direction, perhaps continuously, while the other chain is synthesized discontinuously in the same direction resulting in the formation of short pieces of DNA, analogous to "Okazaki fragments" in bacteria although somewhat smaller (Okazaki *et al.*, 1968; Ogawa and Okazaki, 1980), which are later joined together by DNA ligases. Also, all DNA polymerases require RNA primers (Dressler, 1975) (i.e. short pieces of RNA) which are 9 ± 1 nucleotides in length (De Pamphilis and Wassarman, 1980).

Metabolism of "Okazaki fragments" in eukaryotic chromosomes is comprehensively described in the review by De Pamphilis and Wassarman (1980). In Fig. 3.7, initiation of synthesis of "Okazaki fragments" occurs randomly at sites within a stretch of single-stranded DNA in an initiation zone. Synthesis begins after RNA primers, utilizing either deoxyribo- or ribonucleotides, are extended by DNA polymerase α, extending the primer in the $5' \rightarrow 3'$ direction. Excision of these RNA primers then may occur in a two step mechanism, irrespective of whether the "Okazaki fragment" is extending or is mature (approx. 135 nucleotides). After excision of primers the gap is then filled by an α-polymerase (plus other proteins) and eventually DNA ligases form a phosphodiester link when the final nucleotide is inserted, thus completing the chain.

Accompanying DNA replication, there must also be specific processes that unwind nucleosomal DNA, and then reassemble mature nucleosomes behind the replication fork. These processes are represented in Fig. 3.2 (De Pamphilis and Wassarman, 1980). A specific unwinding protein (helix-destabilizing protein) destabilizes the double helix and maintains single-stranded DNA in an extended conformation. After semi-conservative replication of DNA, the new duplexes must be associated with histone octamers, and this may occur in one of several ways (Fig. 3.8). There are conflicting reports in the literature concerning the association of newly-synthesized and old histones in the two chains of DNA after replication, although individual octamers contain either all new or all old histones. Newly-synthesized histones may be randomly dispersed over the chromosome (Jackson *et al.*, 1975) and pre-existing histones must migrate to become associated with post-replicational DNA (Jackson *et al.*, 1976). However, other patterns of distribution of newly-synthesized histones have been reported (Tsanev and Russev, 1974; Seale, 1975). Newly-replicated DNA in *Physarum polycephalum* is more easily degraded than "old" DNA, but becomes nuclease-resistant in less than 10 min after synthesis (Jalouzot *et al.*, 1980).

Although discontinuous DNA replication must occur in one strand of DNA, such discontinuity of synthesis is not necessary in the other strand. There is some argument as to whether replication is in fact

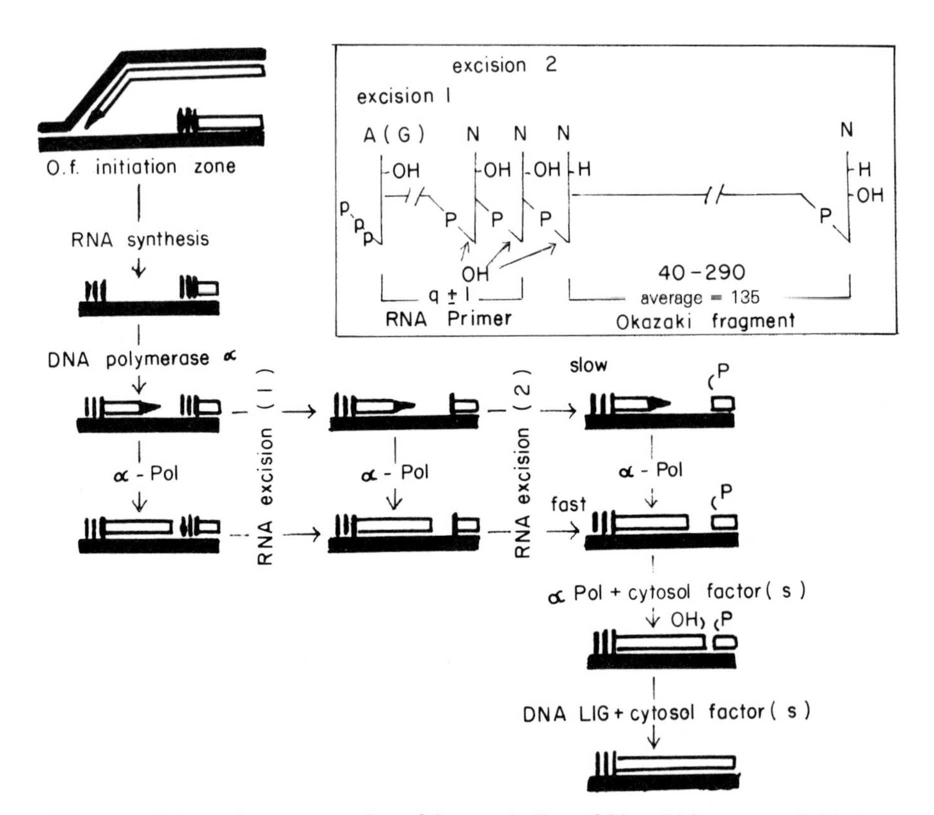

Fig. 3.7 Schematic representation of the metabolism of Okazaki fragments. Initiation of Okazaki fragments (O.f.) occurs stochastically within an "initiation zone" via *de novo* synthesis of RNA primers (|||) complementary to the retrograde template (—). Within this zone, initiation events are promoted at preferred DNA sequences. The maturation of Okazaki fragments (▭▶) is represented on the vertical coordinate in four distinct steps: synthesis of RNA primers, elongation of DNA by DNA polymerase α, incorporation of the final deoxyribonucleotides (gap-filling), and joining of Okazaki fragments to growing DNA chains (ligation). Excision of RNA primers is represented on the horizontal coordinate in two distinct steps: excision of the bulk of the RNA which is independent of Okazaki fragment synthesis, and removal of the p-rN-p-dN-$(pdN)_n$ junction which is facilitated by concomitant DNA synthesis. Excision of all RNA primers is presumed to take place concurrently. A single example is illustrated for simplicity. Inset shows the structure of RNA-primed Okazaki fragments with sizes in nucleotides. N represents A, G, C, or T. Cleavage sites during alkaline hydrolysis are indicated by arrows. Excision 1 and 2 refer to regions excised (solid bars) during replication; possible extensions of these regions are indicated by broken bars. (Reproduced with permission from DePamphilis and Wassarman, 1980.)

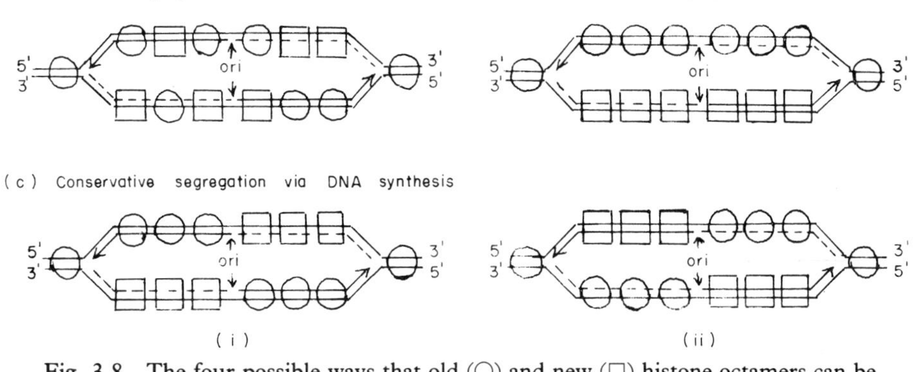

Fig. 3.8 The four possible ways that old (○) and new (□) histone octamers can be distributed between the two arms of a replication fork. DNA synthesis occurs in the $5' \rightarrow 3'$ direction (arrowheads) continuously (solid line) on the forward side and discontinuously (broken line) on the retrograde side, beginning at the origin (ori) of DNA replication. (Reproduced with permission from DePamphilis and Wassarman, 1980.)

discontinuous in *both* strands and evidence in favour of this proposal comes from a variety of sources (Sussenbach *et al.*, 1975; Blumenthal and Clark, 1977; Krokan *et al.*, 1977). These discontinuities in DNA replication may be divided into three phases: (i) rapid joining of short (4S) segments to give an intermediate of 25S in 2–3 min, (ii) continuous growth of 25S segments into 50–70S fragments (which may represent completed replicons) in 15 min and (iii) discontinuous joining of completed replicons to form larger strands after about 2 h (Friedman *et al.*, 1974). There is also evidence to suggest that replication may be unidirectional in some instances (Sussenbach *et al.*, 1975; Robberson *et al.*, 1975; Blumenthal and Clark, 1977; Fareed and Davoli, 1977).

VII. The S-phase

A. GENERAL PATTERNS OF DNA SYNTHESIS

Table 3.2 lists some representative examples of patterns of DNA synthesis in a variety of different cell types. In general, higher eukaryotes (mammalian cells) contain more DNA per nucleus than lower eukaryotes (e.g. yeast) and therefore on this basis alone may be expected to have longer S-phase duration. This, however, is an over-

Table 3.2. *Some examples of timings of DNA synthesis in eukaryotic cells*

Cell type	Method of Analysis (S = synchronous culture)	Comments	Reference
(a) Mammalian Cells			
L-929	Microspectrophotometry	$G_1 = 8$ h, $S = 6$ h, $G_2 = 5$ h	Zetterberg and Killander (1965a)
	Mitotic detachment (S)	$G_1 = 6$ h, $S = 9$ h, $G_2 = 5$ h	Gaffney and Nardone (1968)
Mouse ear skin	*In vivo*	$S = 30$ h	Blenkinsopp (1968)
W138	Mitotic detachment (S)	3 peaks of DNA synthesis in S-phase	Klevecz and Kapp (1973)
(b) Yeasts			
Schizosaccharomyces pombe	Sucrose gradient centrifugation (S)	$S = 10$ min, no G_1	Bostock (1970)
Saccharomyces cerevisiae	Centrifugation	Initiation of DNA synthesis at bud emergence	Matur and Berry (1978) Sena *et al*. (1975)
	Centrifugal elutriation	Peak of DNA synthesis between 0.1 and 0.4 of cycle	Elliott and McLaughlin (1978)
Candida utilis	Sucrose gradient centrifugation (S)	Mid-point of doubling of DNA at 0.94 of cycle	Wain *et al*. (1976)
(c) Protozoa			
Tetrahymena pyriformis	Function of cell size (cytophotometry)	DNA replicated from 55–135 min after cell division (26–65% of the cycle)	Cleffman *et al*. (1979b)
(amicronucleate strain)	Continuous flow size selection (S)	S phase occupied 40–50% of the cycle, G_1 40% and G_2 10–20%	Phillips and Lloyd (1978)

	Micro-manipulation of dividing cells (S)	DNA synthesis in first half of interphase	Prescott (1960)
Paramecium caudatum	Autoradiography	Macronuclear DNA synthesis high in second part of cycle	Rao and Prescott (1967)
Paramecium aurelia	Autoradiography	[^{3}H]-thymidine incorporation into macronuclear DNA throughout second half of interphase	Kimball and Perdue (1962)
Physarum polycephalum	Natural synchrony	DNA synthesis stops within 5 min of the end of anaphase—No G_1	Kessler (1967)
Acanthamoeba castellanii	Low speed centrifugation in growth medium (S)	DNA synthesized between 0.37 and 0.6 of the cycle	Edwards and Lloyd (1978, 1980)
(d) Algae			
Chlorella	Light: Dark (S)	Large increase in incorporation of precursor into DNA in late light phase	Takabaya *et al*. (1976)
Chlamydomonas reinhardii	Light: Dark (S)	DNA synthesis increases at onset of dark phase	Schor *et al*. (1970)
Platymonas striata	Light: Dark (S)	DNA synthesis occurred in 6 h preceding cell division (in light)	Ricketts (1977b)
Navicula pelliculosa	Light: Dark (S)	DNA synthesis from 3–6 h in light	Darley *et al*. (1976)
Cylindrotheca fusiformis	Light: Dark, but grown subsequently in continuous light (S)	DNA synthesis from 3–9 h	Paul and Volcani (1976)

simplification as we shall see later. In general, DNA synthesis occurs during a restricted portion of the cycle (i.e. the S-phase) with no synthesis of nuclear DNA outside this period. Outside the S-phase there may, however, be cytoplasmic DNA synthesis (see later) or uptake and incorporation of precursors into nuclear DNA (repair processes).

There are some exceptions to these observations. In *Amoeba proteus*, some DNA may be replicated in interphase and not all nuclear DNA of the mother is transferred to the progeny during mitosis (Makhlin *et al.*, 1979). This unusual observation requires further investigation. The dinoflagellates are unusual organisms and have some bacteria-like properties, including, in *Prorocentrum micans*, continuous nuclear DNA synthesis during most of the cycle (Filfilan and Sigee, 1977). In the photosynthetic dinoflagellate, *Amphidinium carterae* (synchronized by a LD regime), DNA synthesis occurs during an S-phase lasting 6–9 h, but also at times outside this period (Galleron and Durrand, 1979). In photosynthetic organisms (e.g. *Anacystis nidulans*), DNA synthesis is dependent on illumination conditions and only occurs in normal or reduced light (Ssymank *et al.*, 1977) while in others (e.g. *Chlamydomonas reinhardii*), synthesis of DNA increases sharply at the onset of the dark phase (Schor *et al.*, 1970).

DNA synthesis is often measured by incorporation of tritiated thymidine, a suitable DNA precursor which labels DNA in those cells containing a thymidine kinase. In some cells e.g. V79 fibroblasts, perturbations of growth may occur at certain concentrations of this labelled precursor (Ehmann *et al.*, 1975). Other procedures which affect the normal sequences of DNA replication and division include heat shocks in *Tetrahymena pyriformis* (Zeuthen, 1978), where extending the final heat shock to 5 h results in 2 phases of DNA synthesis followed by 2 cell divisions. These experiments may provide clues to the mechanisms controlling DNA synthesis, cell growth and division.

B. TEMPORAL ORGANIZATION OF THE S-PHASE

There is much evidence to suggest that the S-phase is a highly ordered but discontinuous series of events. Wide variations in the length of the S-phase occur; in *Schizosaccharomyces pombe* the S-phase lasts 10 min (Bostock, 1970) while in mouse ear skin the S-phase lasts about 30 h (Blenkinsopp, 1968). Differences in the amount of DNA to be replicated do not alone account for these variations. Also, there are vast differences in the duration of the S-phases of embryos and adult cells of the same species; DNA replication in embryos sometimes occurs in a much shorter time—up to 100 times faster than in the adult (see Callan, 1973). In different mammalian cell lines with considerable variations in DNA-replication times (S-phases ranging from 6–25 h), there is little variation in the rate of chain growth from one cell to another (Painter and Schaefer, 1969). These observations are explained by the fact that the overall rate of replication during S-phase is not governed by the rate of chain elongation but by the degree of synchrony of initiation of replicons, i.e. the number of replicons initiated at any one time. Thus, in early embryos there is a high degree of synchronization of all replicons compared to adult cells, resulting in a shorter S-phase. The structure of heterochromatin has been implicated in this mechanism whereby structural changes in this complex may result in changes in the distributions of compacted and non-compacted regions of the chromosome, which then may or may not be amenable to activation (Blumenthal *et al.*, 1973).

There are also variations in the rate of DNA synthesis at different times during the S-phase. In *S. cerevisiae* (Williamson, 1974) and *Physarum polycephalum* (Hall and Turnock, 1976) 80–90% of the total DNA is replicated within the initial 10–25% of S-phase, while in mammalian cells the rate of fork movement increases threefold from early S- to late S-phase; highest rates of replication occur in mid S-phase (Housman and Huberman, 1975). In Chinese hamster cells, rates of fork displacement are constant through most of the S-phase, in HeLa cells the highest rates are in mid and late S-phase, while in human diploid fibroblasts the lowest rates are in mid S-phase (Kapp and Painter, 1979). Also, in the three mammalian cell lines, thymidine uptake is periodic during the S-phase and may show 2, 1 and 3 peaks, respectively, in uptake, possibly dependent on the duration of the S-phase (Kapp and Painter, 1977). W138 cells also show three peaks of thymidine uptake (Kapp and Klevecz, 1976), which are paralleled by three peaks of DNA synthesis during an S-phase of 12 h (Klevecz and Kapp, 1973). Damage to early replicating DNA markedly supresses the entire S-phase, whereas damage to mid- or late-replicating DNA does not affect early replication in the following cycle (Hamlin, 1978). This evidence suggests that there is a sequential triggering of replicons which provides an ordered progression through S. While replication sites in eukaryotic chromosomes are not associated with the nuclear membrane as in prokaryotic chromosomal replication (Fakan *et al.*, 1972; Wise and Prescott, 1973), the initiation of DNA replication occurs at sites within the nucleus, or at sites close to the membrane (O'Brien *et al.*, 1973). Euchromatin is replicated in early S-phase at sites distributed through the nucleus, whereas heterochromatin is replicated later in S-phase at or near, the nuclear membrane (Williams and Ockey, 1970; Erlandson and De Harven, 1971; Huberman *et al.*, 1973).

The requirement for protein synthesis in the initiation of replication has previously been mentioned. In mammalian cells, addition of inhibitors of protein synthesis at the start of a pulse label in DNA fibre autoradiograph experiments, decreases the frequency and synchrony of initiation of replication and reduces the incidence of bidirectional fork movement (Hand, 1975b). However, in *S. cerevisiae*, protein synthesis is required for the initiation, but not the continuation, of DNA synthesis (Hereford and Hartwell, 1973). Inhibition of protein synthesis in mid-S has no effect on DNA replication in this organism (Williamson, 1973). This may be due in part to the initiation of most replicons in early S-phase (Moll and Wintersberger, 1976), but also suggests that there may be pools of histones already available for assembly into chromatin.

VIII. Nucleolar DNA Synthesis

Nucleolar DNA codes for ribosomal RNA, and this (rDNA) is replicated independently of nuclear DNA in rat liver (Wintzerith *et al.*, 1975). In rat kangaroo cells, rDNA is duplicated in late S (Giacomoni and Finkel, 1972). In *Tetrahymena pyriformis*, nucleolar DNA replication occurs during a fraction of the macronuclear S-phase (Charret, 1969) probably at the onset of replication (Anderson and Engberg, 1975), whereas in *Amoeba proteus*, nucleolar DNA is late replicating and occurs in G_2 (Minassian and Bell, 1976).

In *Physarum polycephalum*, about 10% of the total amount of radioactively-labelled DNA precursor incorporated is found in the cytoplasm, and some of this incorporation occurs outside the S-phase (Holt and Gurney, 1969). This incorporation during G_2 was found to occur within the nucleolus (Guttes and Guttes, 1969; Guttes and Telatnyk, 1971; Ryser *et al.*, 1973).

During oogenesis in *Xenopus laevis* there is a massive increase in the genes coding for rRNA resulting in gene amplification. This gene amplification arises from replication of rDNA by a rolling circle mechanism (Fig. 3.9) which results in semi-conservative replication (Hourcade *et al.*, 1973). Replication is first initiated in one strand of the circular duplex of rDNA and later occurs on the single-stranded molecule that is formed as the circle unfolds.

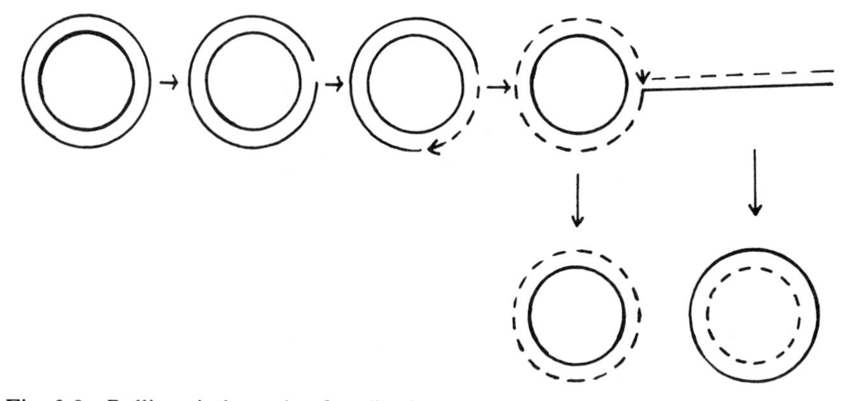

Fig. 3.9 Rolling circle mode of replication of rDNA occurring during gene amplification in amphibian oogenesis (modified from Wintersberger, 1978).

IX. Mitochondrial DNA Replication

A. GENERAL PATTERNS OF SYNTHESIS DURING THE CELL CYCLE

Mitochondrial DNA (mt DNA) is present in many copies per cell although the amount present varies with growth conditions and ploidy (Williamson, 1970), and the species (Lloyd and Turner, 1980). Mitochondria also contain their own specific DNA polymerases (Weissbach, 1977) and hence may be expected to replicate their DNA to some extent independently of nuclear DNA synthesis. Some studies suggest that in budding yeast, mt DNA is replicated periodically in a stepwise fashion and at a different time in the cell cycle from nuclear DNA (Smith *et al.*, 1968; Cottrell and Avers, 1970; Dawes and Carter, 1974). Any proposals for correlating these patterns with mitochondrial growth and division must take into account the presence of a mitochondrial reticulum in yeast (see Chapter 5). In *Euglena gracilis* however, mt DNA replicates simultaneously with nuclear DNA replication (Calvayrac *et al.*, 1972) and in kinetoplastidae the timings of syntheses of nuclear and kinetoplast DNA are closely correlated (reviewed by Simpson, 1972).

There are many reports, however, which suggest that mt DNA does not replicate synchronously with nuclear DNA (e.g. Leff and Lam, 1976) and in fact is replicated continuously. In *S. cerevisiae*, continuous replication represent replication of individual duplexes over the entire *et al.*, 1975; Cottrell, 1979); although it has been suggested that the discrepancies between these and earlier reports may result from the fact that incorporation of labelled precursor into mt DNA does not parallel replication (Cottrell, 1977). Continuous synthesis of mt DNA for most or all of the cell cycle has also been observed in *P. polycephalum* (Guttes *et al.*, 1967; Braun and Evans, 1969) and in *T. pyriformis* (Parsons, 1965; Charret and André, 1968; Parsons and Rustad, 1968), although in the latter organism an increased rate of replication occurs when the macronuclear DNA is also replicated (Cameron, 1966). However, whether all these observations of continuous mt DNA replication represent replication of individual duplexes over the entire cycle, or over a restricted part of the cycle, but initiated asynchronously, cannot be distinguished.

B. MECHANISMS OF MT DNA REPLICATION

There is general agreement that mt DNA replicates semi-conservatively (e.g. Banks, 1973; Cummings, 1977; Mattick and Hall, 1977). In rat tissues, it is suggested that mt DNA can replicate in one of two modes: (i) sections of the molecule are completely replicated sequentially by synthesis first along one section of one strand followed by synthesis along the equivalent portion of the complementary strand or (ii) synthesis on one strand must be at least 80% complete before initiation occurs on the other strand (Wolstenholme *et al.*, 1973). In addition, there may be pauses in replication; at least 44% of the contour of mt DNA molecule bears discrete sites at which synthesis may be arrested (Koike and Wolstenholme, 1974).

Perhaps the best understood mechanism of mt DNA replication is that of mouse L cells (Robberson *et al.*, 1972; Kasamatsu and Vinograd, 1973; Berk and Clayton, 1974; Bogenhagen *et al.*, 1978). The mt DNA duplex comprises a heavy strand (H-strand) and a light strand (L-strand) and initiation of synthesis of a new H-strand occurs in one direction only (Fig. 3.10). Elongation of this new DNA strand proceeds until about 450 ± 80 nucleotide bases of the H-strand are displaced and a loop is visible in electronmicrographs (the D-loop). Synthesis of the L-strand is not initiated until about 60% of the H-strand has replicated and eventually two new duplexes, termed α and β molecules are produced (Fig. 3.10). It is thought that replication may proceed continuously.

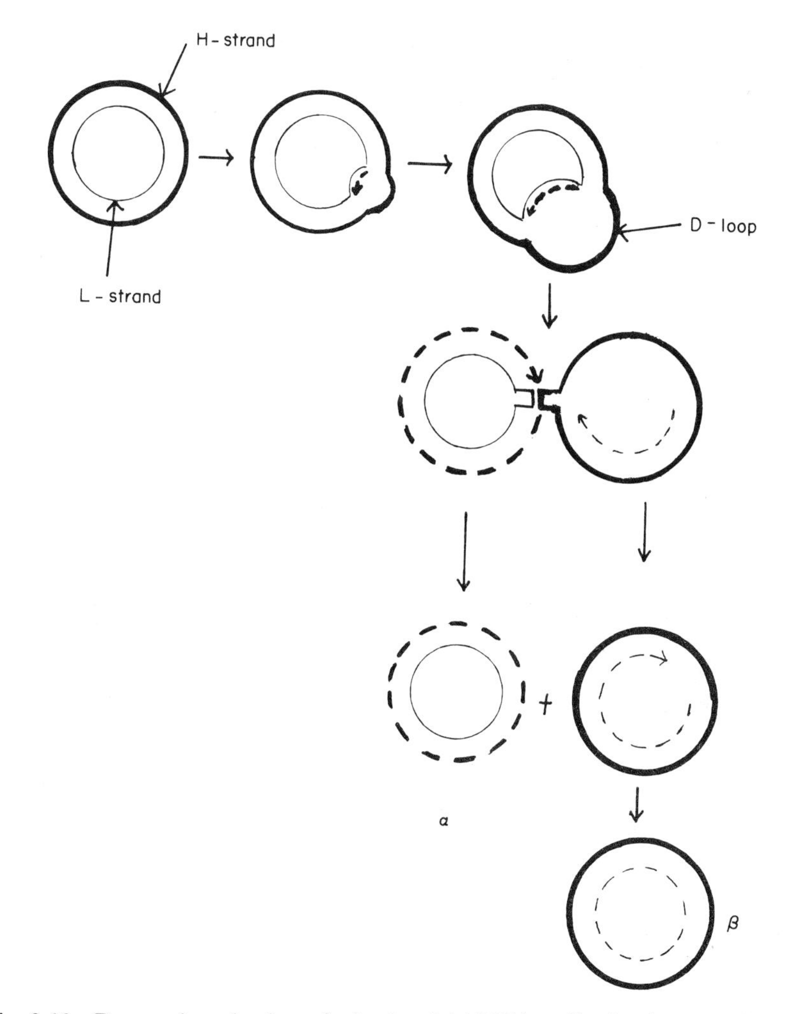

Fig. 3.10 Proposed mechanism of mitochondrial DNA replication in mouse L cells. Heavy (H) strand, light (L) strand and displacement loop (D-loop) as indicated. For details see text (modified from Wintersberger, 1978).

X. Chloroplast DNA Synthesis

Chloroplast DNA (cp DNA) in photosynthetic eukaryotes presents a further complication to studies of nuclear and extra-nuclear DNA. In *Chlamydomonas reinhardii*, the chloroplast genome contains sufficient DNA for 200–300 gene products (Howell and Walker, 1977);

replication is semi-conservative (Chiang and Sueoka, 1967) and does not appear to occur in synchrony with nuclear DNA (Grant *et al.*, 1978). Incorporation of labelled precursors into cp DNA occurs at all stages of the cycle of *Euglena gracilis* but cytoplasmic (mitochondrial or chloroplast) shows two peaks of synthesis, the second occurring at or about the time of nuclear division (Cook, 1966; Brandt, 1975).

PART 2
SYNTHESIS OF DNA DURING THE PROKARYOTIC CELL CYCLE

I. Introduction

At the beginning of the 1970s, it was apparent that there were major differences in the molecular mechanisms whereby eukaryotic and prokaryotic DNAs were replicated. It was established that most eukaryotic DNA was located within the nucleus; this DNA was complexed with histones and organized linearly into duplexes comprising many chromosomes. Synthesis was generally restricted to a specialized portion of the cycle, the S-phase, and replication of chromosomes occurred bi-directionally at multiple replication sites.

Prokaryotic DNA, however, was known to be attached to the cell membrane, not associated with histones and present as a single, circular chromosome within the cytoplasm. Replication, which was semi-conservative, was not limited to a particular portion of the cycle and was continuous in fast-growing bacteria. It was believed that replication proceeded unidirectionally with a speed of 20–30 $\mu m \, min^{-1}$ at a single replication fork (Cairns, 1963), ending at a terminus which was situated on the chromosome close to the origin.

The last ten years have seen major developments in this area, in particular the identification of the components of the replication machinery by biochemical and genetic analyses. The latter approach has involved the isolation of a series of temperature-sensitive mutants, which are defective in DNA replication or repair. The use of bacteriophage or other viral DNA molecules as model systems has also proved invaluable. Whereas the *E. coli* chromosome has about 5×10^6 base pairs (Wickner, 1978a), *E. coli* phages e.g. ϕX174 and fd, have single-stranded DNA containing about 5000–6000 bases (Kornberg, 1977; Tomizawa and Selzer, 1979). Since the synthesis of DNA in these viruses is dependent on the enzymes of the host cell, studies on the mechanisms of replication of these DNAs must give information relevant to the understanding of events occurring during replication of the bacterial chromosome.

This section will briefly attempt to summarize our current understanding of the processes involved in prokaryotic DNA replication and again will include only illustrative references. For more detailed accounts, the reader is referred to the more recent reviews on this topic e.g. Gefter (1975), Jovin (1976), Cozzarelli (1977), Kornberg (1977, 1978), Alberts and Sternglanz (1977), Champoux (1978), Wickner (1978a,b), Kolter and Helinski (1979), Tomizawa and Selzer (1979), Ogawa and Okazaki (1980) and in particular, the Cold Spring Harbor Symposia on Quantitative Biology vol. **43**, part 1. The following is divided into three major sections to describe; (1) some general features of prokaryotic DNA replication, including possible controlling factors; (2) studies with bacteriophage and viral DNA replication; and (3) how these latter studies have helped to elucidate some of the mechanisms which occur at the chromosomal replication fork.

II. DNA Replication in Prokaryotes

A. GENERAL FEATURES

Replication of DNA during rapid growth of bacteria is continuous for most or all of the cell cycle (Table 3.3). Under certain conditions (i.e. bacteria growing at the same temperature, but with different generation times), replication of DNA may be initiated before the previous round of synthesis has been completed i.e. the chromosome contains a number of replication forks, initiated at the origin at different times. These observations are accommodated in a model proposed by Cooper and Helmstetter (1968) to explain the relationship between DNA replication and the cell division cycle in *E. coli* B/r. In this model, if the generation time is 60 min or less, there are two events of fixed duration: C, which equals 40 min and is the time taken for one round of DNA replication to be completed after initiation at the origin, and D, which equals 20 min and is the time interval between completion of a round of replication and cell division. Thus, if the generation time is 60 min, DNA replication begins at time zero, is completed after 40 min and there is then a gap of 20 min before cell division. If the generation time is 50 min, this gap lasts 10 min, while if the generation time is 40 min, continuous DNA synthesis results. This model explains the fact that rapidly-growing cells may contain multiple replication forks, since with generation times of less than 40 min, new rounds of replication are initiated before existing rounds are completed. Many experimental data fit this model, even in cells growing with generation times of up to 90 min (Buckley and Anagnostopoulos,

1976). The C period has been shown to decrease continuously with the growth rate but approaches a constant value of 37 min at high growth rates (Churchward and Bremer, 1977). A constant C period, at different growth rates, has also been reported in *E. coli* K12 (Chandler *et al.*, 1975).

B. REGULATION OF REPLICATION

As with eukaryotic DNA synthesis, little is known about the factors that control the initiation of replication in prokaryotes. The involvement of protein and RNA synthesis in the initiation process has been suggested e.g. in *E. coli*, the addition of chloramphenicol inhibits initiation, but two separate processes are implicated which are inhibited by high and low concentrations of this compound (Ward and Glaser, 1969). Once initiated, however, further replication may occur in the absence of further protein synthesis (Rosenberg, B. H. *et al.*, 1969). These observations have been interpreted as suggesting that initiation of replication is under a negative control whereby a replication repressor is synthesized periodically, while an antirepressor protein is synthesized continuously; derepression of initiation occurs when the antirepressor protein accumulates above a certain level (Rosenberg, B. H. *et al.*, 1969). Transcription or synthesis of an RNA species is also required for initiation (Messer, 1972) and it may be that transcription is required at a time when protein synthesis is no longer necessary (Lark, 1972). Presumably transcription is required for the synthesis of an RNA primer (see later), but other unknown factors and controls are also required for initiation (Lark and Renger, 1969; Tippe-Schindler *et al.*, 1979); starved *E. coli* cells can be stimulated into DNA synthesis by the addition of spermidine (Geiger and Morris, 1980). In a temperature-sensitive mutant of *Bacillus subtilis*, initiation of new rounds of replication can occur in the absence of protein synthesis, but initiation is inhibited by rifampin, suggesting that transcription is required (Laurent, 1973).

A positive control of initiation has been proposed in the "Replicon Model" of DNA synthesis (Jacob *et al.*, 1963), whereby replication is correlated with the synthesis of a specific activator. Alternatively, a negative control mechanism has been postulated in the "Autorepressor" model (Sompayrac and Maaløe, 1973) or in the postulated existence of an "anti-initiator" protein which controls the frequency of chromosome initiation (Blau and Mordoh, 1972).

Table 3.3. *DNA replication during the cell cycles of some prokaryotes*

Organism	Method of Analysis (S = synchronous culture)	Comments	Reference
Escherichia coli	Filtration (S)	Rate of DNA synthesis is continuous through the cycle in rapidly growing organisms.	Abbo and Pardee (1960)
		Continuous DNA synthesis.	Rudner *et al*. (1965)
	Inoculation of stationary phase cells into fresh medium (S)	Total DNA synthesized continuously; rate of synthesis doubled in mid-cycle.	Cutler and Evans (1967a)
4 strains of *E. coli*	Sucrose gradient centrifugation (S)	DNA synthesis (C-period) occupied 55–65% of the cycle during slow growth.	Gudas and Pardee (1974)
Bacillus subtilis	Filtration (S)	Chromosome initiation at beginning of the cycle.	Sargeant (1975a)
Proteus vulgaris	Inoculation of stationary phase cultures into fresh medium (S)	DNA per cell increases continuously.	Cutler and Evans (1966)

Myxococcus xanthus	Autoradiography	Chromosome replication initiated at 0·12 of cycle and is continuous for 80% of cycle.	Zusman and Rosenberg (1971)
	Function of cell length	DNA synthesis for 85% of the cycle, independent of generation time (from 5–11 h).	Zusman *et al.* (1978)
Rhodopseudomonas sphaeroides	Inoculation of stationary phase cultures into fresh medium (S)	Total DNA increases continuously but rate of incorporation of precursors is discontinuous.	Ferretti and Gray (1968)
Anacystis nidulans	Light: Dark (S)	DNA synthesis just before cell division.	Asato (1979)
		DNA synthesis occurs from mid cycle and is completed at cell division.	Bagi *et al.* (1979)
Caulobacter crescentus	Plate selection method (S)	DNA synthesis in both swarmer and stalked cells takes 50 min.	Iba *et al.* (1977a)
		DNA synthesis from 50–150 min (g = 200 min).	Evinger and Agabian (1979)

The association of the chromosome with the cell membrane may be involved in the control of timing of the initiation of DNA synthesis (Marvin, 1968; Helmstetter, 1974a,b).

Experimental evidence in favour of this suggestion has led to the proposal of a model (Fig. 3.11; Helmstetter, 1974b) in which the

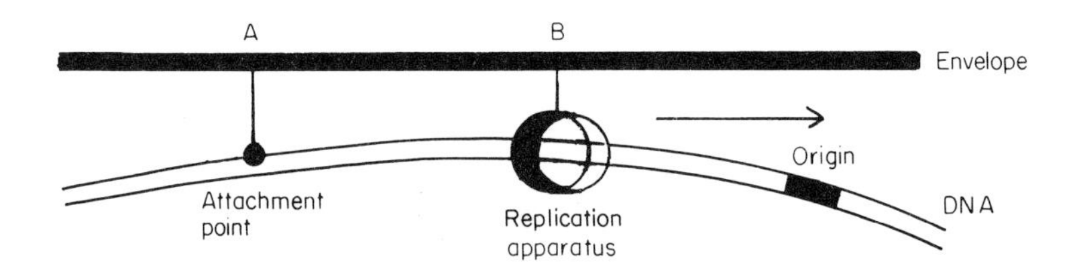

Fig. 3.11 Proposed model for chromosomal attachment and DNA synthesis. The thick horizontal line represents the cell envelope which is attached to one strand of the bacterial chromosome by the attachment point at A. The replication apparatus is attached to the cell envelope at B, but is not attached to the chromosome. Thus elongation of the envelope between A and B causes the replication apparatus to move along the chromosome and initiate replication when it reaches the origin. (Reproduced with permission from Helmstetter, 1974b.)

chromosome is attached to the cell envelope at a fixed point (A) and the replication apparatus is fixed at another point (B). Elongation of the envelope between A and B causes the replication apparatus to move along the chromosome, initiating replication when it reaches the origin. The timing of initiation is thus controlled by the rate of envelope growth. Multiple replication forks which may occur at different growth rates can be envisaged in Fig. 3.12 (Helmstetter, 1974b): in Class 0 cells, the replication apparatus (RA) reaches the origin before septum cross wall formation, in Class 1, the RA reaches the origin after septum-crosswall formation has begun and in Class 2 the RA arrives after division is completed. The evidence for localized membrane growth required by this model is, however, not unequivocal and is discussed in more detail in Chapter 5. Recently, Craine and Rupert (1979) have also argued against the necessity of a membrane/DNA interaction for initiation or continuation of DNA synthesis.

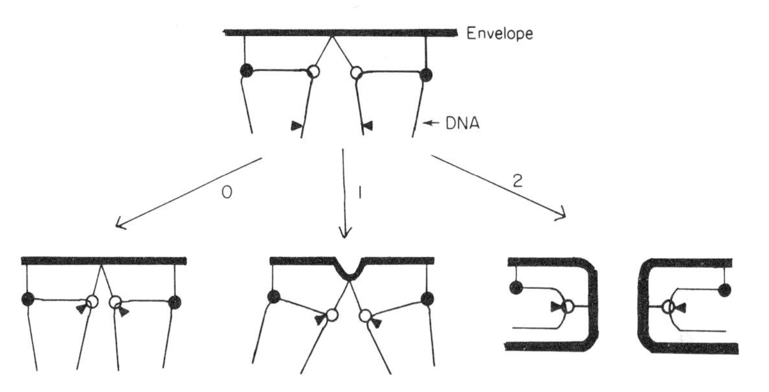

Fig. 3.12 Chromosome attachment and cell division in bacteria. The bacterial DNA is attached to the envelope at the attachment point (solid circle). As the envelope elongates, the replication apparatus (open circle) approaches the origin (▶), and initiation of replication may occur before or after septum formation, as shown. For details see text. (Reproduced with permission from Helmstetter, 1974b.)

C. REQUIREMENTS FOR REPLICATION

Biochemical and genetic studies have revealed in *E. coli* a variety of proteins that are required for both initiation and elongation of DNA fragments into new molecules (Table 3.4). There is confusion in the nomenclature of some of these replication proteins; more work is necessary to further identify and characterize some of these, especially the elongation factors X, Y, Z (Wickner and Hurwitz, 1974) and the i, n and n′ proteins (Kornberg, 1978). For phage or virus replication, other gene products, coded for by viral DNA, may also be required and will briefly be mentioned later. In addition to these proteins, ribonucleotide triphosphates (rNTP) are required for primer formation and deoxyribonucleotide triphosphates (dNTP), Mg^{2+} and ATP are required for polymerization. Some of the functions and properties of the major proteins are described:

(i) DNA polymerase I. Three different enzymes all with the same catalytic properties and capable of polymerizing dNTP in the $5' \rightarrow 3'$ direction, but with some different functions, are found in *E. coli*. DNA polymerase I has $5' \rightarrow 3'$ exonuclease activity which may excise ribonucleotides (Westergaard *et al.*, 1973) or thymine dimers (Grossman *et al.*, 1975) and also has $3' \rightarrow 5'$ exonuclease activity which may function as a "proof-reading" mechanism excising mis-matched nucleotides (Kornberg,

Table 3.4. E. coli *replication proteins*

Protein	Approx mol. wt ($\times 10^{-3}$)	Function	Reference
DNA binding protein	75–80	Binds to single-stranded DNA.	Sigal *et al.* (1972)
Replication factors X, Y, Z Proteins i, n, n′	X ≈ 45 Y ≈ 55 i ≈ 80 n ≈ 30 n′ ≈ 75	Required for prepriming reactions. Y and n′ are single-stranded DNA-dependent ATPases.	Wickner and Hurwitz (1974, 1975b) Kornberg (1978)
dnaC	25–29	Interacts with *dnaB* protein in a prepriming mechanism.	Wickner and Hurwitz (1975a)
dnaB	250	"Mobile replication promoter" required in prepriming and priming reactions.	McMaken *et al.* (1977) Meyer *et al.* (1978)
dnaG (primase)	60–65	Priming protein, synthesizing ribo-, deoxy- and mixed oligonucleotides.	Wickner *et al.* (1973) Rowen and Kornberg (1978a,b)

DNA polymerase III holoenzyme *polC* (α) (polymerase)	180	DNA synthesis by elongation of primed chain.	Nusslein *et al.* (1971) Gefter *et al.* (1971)
DNA elongation factor I (β)	40		Livingston and Richardson (1975)
DNA elongation factor III (γ or δ)	30–60 (?)		Livingston *et al.* (1975)
dnaZ	125		Wickner, S. (1976)
ε	27		Wickner and Hurwitz (1976)
τ	83		McHenry and Kornberg (1977)
θ	9		Meyer *et al.* (1978)
DNA polymerase I	109	May add nucleotides to chain after primer excision.	Gefter (1975)
DNA ligase	74	Seals nascent DNA chains.	Lehman (1974)
DNA gyrase	400	Supertwisting.	Gellert *et al.* (1976, 1977)
NalA	210		Sugino *et al.* (1977)
Cou	190		
rep	65–67	Protein required for helicase action.	Scott *et al.* (1977) Eisenberg *et al.* (1977)
dUTPase	64	Prevents uracil incorporation into DNA.	Shlømai and Kornberg (1978)

1969). This enzyme is thought to be present at levels of about 400 molecules per cell and is probably involved in the maturation of duplicated DNA into high molecular weight structures (Gefter, 1975).

(ii) DNA polymerase II—about 10 molecules per cell (Gefter, 1975) is specifically stimulated by the *E. coli* DNA binding protein (Sigal *et al.*, 1972; Molineux and Gefter, 1975b), elongates primed DNA (Sherman and Gefter, 1976) and is probably involved in polymerase activity of chains less than 50 nucleotides in length (Gefter, 1975).

(iii) DNA polymerase III also has both $3' \rightarrow 5'$ and $5' \rightarrow 3'$ exonuclease activity (Livingston and Richardson, 1975) and elongates primed DNA in the form of a holoenzyme complex (Wickner and Kornberg, 1974b) containing *dnaZ* protein and elongation factors I and III (Wickner and Hurwitz, 1976).

(iv) The *E. coli* DNA binding protein. This has also been termed the unwinding protein, binds to single-stranded DNA, is essential for initiation and elongation and stimulates the activity of DNA polymerase II (Sigal *et al.*, 1972; Weiner *et al.*, 1975). Binding of this protein to DNA prevents nuclease activity associated with the polymerases (Molineux and Gefter, 1975a).

Other proteins which are important for replication of prokaryotic DNA may have similar functions to those found in the eukaryotic DNA replication complex. Those proteins unique to prokaryotic DNA replication will be described later.

D. CHROMOSOME ELONGATION

There is general agreement that the bacterial chromosome contains a single origin of initiation of replication. In *E. coli*, the origin has been mapped at about 70 min on the genetic map (Bird *et al.*, 1972) and a 38 K base region is thought to contain the origin, probably near the 1·3K base *Hind*III fragment (Marsh and Worcel, 1977). A 422 base pair region has recently been mapped and is believed to contain the origin of replication (Sugimoto *et al.*, 1979; Meijer *et al.*, 1979); the nucleotide sequence in this region contains a high degree of repetitiveness due to inverted and direct repeats. Single origins have been located in *Salmonella typhimurium*, close to the *ilvA* gene (Nishioka and Eisenstark, 1970; Fujisawa and Eisenstark, 1973), and in *Bacillus subtilis* close to the *purA16* gene (Harford, 1975). In these latter two bacteria, replication was shown to be bidirectional (see

Gyurasits and Wake (1973) for *B. subtilis*) as demonstrated in *E. coli* (Masters and Broda, 1971; McKenna and Masters, 1972; Prescott and Kuempel, 1972) which has a terminus located near the *rac* locus (Kuempel and Duerr, 1978). However, in a thymine-requiring strain of *E. coli*, it has been shown that replication is bidirectional at high thymine concentrations but unidirectional at low thymine concentrations (Edlund *et al.*, 1976).

Initiation of replication, as in eukaryotic DNA, requires an RNA primer (Wells *et al.*, 1972; Sugino *et al.*, 1972) and replication on both strands must occur in both the $5' \rightarrow 3'$ and $3' \rightarrow 5'$ directions. No known DNA polymerase can polymerize DNA in the $3' \rightarrow 5'$ direction; it has been shown that synthesis in both chains occurs in the $5' \rightarrow 3'$ direction, (but discontinuously in one chain) by the formation of nascent fragments of DNA or "Okazaki fragments" (Okazaki *et al.*, 1968; Ogawa and Okazaki, 1980) which are then excised of primer and joined by a ligase (see Part 1 for details). Nascent DNA fragments (Richardson, 1969; Goulian, 1971) are found in a variety of cell types (see Ogawa and Okazaki, 1980) and accumulate when DNA polymerase (Kuempel and Veomett, 1970; Okazaki *et al.*, 1971) or DNA ligase is inhibited (Hosoda and Matthews, 1968; Newman and Hanawalt, 1968). While DNA replication must be discontinuous on one strand, it is uncertain whether replication in the other strand is also discontinuous.

III. Replication of Phage DNA

A. GENERAL FEATURES

Many of the commonly-used bacterial phage and viral DNAs are single-stranded DNA molecules, which must be primed before any synthesis can occur. This priming may involve the synthesis of a short RNA sequence or by "nicking" a strand to expose a 3′–hydroxyl group to serve as a primer. As we shall see, priming occurs at a specific origin of replication and elongation of this primer may occur in a variety of ways depending on the type of DNA that is to be replicated. Invariably, the replication proteins of the host cell are required for phage DNA replication and again depending on the nature of the DNA, various combinations of these proteins, perhaps together with some phage DNA gene products are necessary.

Before any duplication can occur, circular single-stranded phage DNA (SS) must be converted into a duplex replicative form (RF).

The viral or phage strand is designated the (+) strand and the complementary strand which is synthesized, the (−) strand. Important in this stage is the binding of the (+) strand with the *E. coli* DNA binding protein to all parts of the molecule except the primer attachment site (Geider and Kornberg, 1974). The binding protein is not itself recognized by RNA polymerases (Molineux *et al.*, 1974). The primer which is formed is generally RNA (but there may be exceptions) which is linked to the DNA via a 3′:5′–phosphodiester bond and later specifically removed by an excision mechanism possibly involving RNase H (Keller, 1972). In some cases specific pre-priming reactions must occur which involve protein–protein and protein–DNA interactions. The sequences of events leading to the production of an RF structure do not generally require phage DNA gene products.

The RF molecule that is produced may then be duplicated to form more RF structures (Tomizawa and Selzer, 1979), in a mechanism requiring a phage gene product which functions in nicking the RF molecule at a specific point on the (+) strand. A complementary (−) strand may be synthesized from this new (+) strand (as above) or phage gene products may block (−) strand synthesis by encapsulating the newly synthesized (+) strand. Replication of this (+) strand may occur by a rolling circle mechanism (Gilbert and Dressler, 1968; Godson, 1977).

In the following sections, descriptions of replication of DNA of some of the best studied phages will be given, but for more complete accounts of various phages, plasmids etc. the reader is referred to the Cold Spring Harbor Symposia on Quantitative Biology vol. **43**, part 1 (1978).

B. MECHANISMS LEADING TO PRODUCTION OF REPLICATIVE FORMS

1. *Phage G4*

Replication of Phage G4 DNA is initiated by a mechanism that also operates in St-1, α3 and ϕK replication (Wickner, 1978b), and may proceed in a number of stages (Fig. 3.13; Kornberg, 1978; Wickner, 1978a,b). The *E. coli* DNA binding protein initially attaches to the single-stranded viral DNA. This binding may be important for maintaining DNA in the correct conformation for primer synthesis, melting small duplex DNA regions which may otherwise block elongation, affecting the properties of the replication proteins by protein–protein interactions and blocking transcription by RNA polymerase (Wickner, 1978a). The *dnaG* protein or primase (Rowen

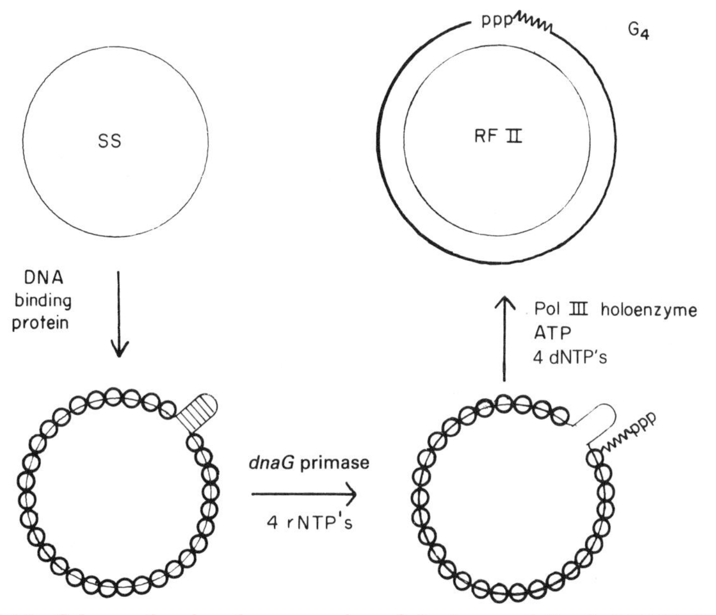

Fig. 3.13 Scheme showing the conversion of single strand G4 viral DNA (SS) into RFII DNA. (Reproduced with permission from Kornberg, 1978.)

and Kornberg, 1978a) then, in the presence of ADP (or ATP) and dNTPs or rNTPs (Wickner, 1977) initiates synthesis of a primer complementary to the viral DNA (Martin and Godson, 1977; Fiddes *et al.*, 1978) at a unique sequence of DNA (Sims and Dressler, 1978). This contains potential hairpin loop structures (Bouché *et al.*, 1978; Godson, 1978; Sims and Dressler, 1978; Sims *et al.*, 1978) and a primer which may be a 29 nucleotide RNA molecule (Rowen and Kornberg, 1978b) or less than 6 nucleotides containing substituted deoxyribonucleotides (Kornberg *et al.*, 1978). This primer is formed 5 nucleotides from the base of a hairpin loop at an adenine residue (Godson, 1978).

This primer is then extended by DNA synthesis effected by the DNA polymerase III holoenzyme which contains the *polC* gene product, the *dnaZ* protein and two proteins termed elongation factors (EF) I and III (Wickner and Hurwitz, 1976; Wickner, 1977). This process may be divided into a number of stages (Wickner, 1978b):

(i) The *dnaZ* protein and DNA EFIII form a protein complex independent of nucleotides, DNA and the other proteins required for elongation;

(ii) This complex catalyzes the transfer of DNA EFI to primed SS DNA in an ATP- or dATP-dependent reaction;
(iii) DNA polymerase III binds to the primed DNA/EFI complex (not to primed DNA alone);
(iv) this complex of DNA polymerase III, EFI and primed DNA catalyzes the polymerization of dNTPs.

DNA polymerase III and the *dnaZ* protein are also required for *E. coli* chromosome replication.

2. *Phage φX174*

Formation of a complementary strand of φX174 DNA requires a complex prepriming reaction, but since *dnaB* and *dnaC* proteins participate in this mechanism, the system may provide a more useful analogy for the events occurring in *E. coli* chromosome replication (Kornberg, 1977). Before priming by *dnaG* protein or primase can occur, five proteins must interact to form a complex. The three other proteins that are required are termed replication factors X, Y and Z (Wicker and Hurwitz, 1974) or factors i, n and n′ (Meyer *et al.*, 1978). Initially, replication factors Y and Z bind to viral DNA previously coated with *E. coli* DNA binding protein (Fig. 3.14; Wickner, 1978b). The *dnaB* protein, which is a ribonucleoside triphosphatase (Wickner *et al.*, 1974) forms a complex with the *dnaC* protein in an ATP-dependent reaction, in the absence of DNA and other proteins. The *dnaB* protein is then transferred from this complex to the DNA/binding protein/replication factors Y and Z complex in a reaction which requires ATP and replication factor X. A new complex is then formed which contains DNA binding protein, *dnaB* protein and factors Y and Z but *not* factor X or *dnaC* protein. The *dnaG* then initiates primer synthesis (Wickner *et al.*, 1972b) presumably at a site exposed by, or associated with, this complex. Elongation of this primer may then occur by the DNA polymerase III holoenzyme complex to form an RFII complex which then requires DNA polymerase I and ligase to complete synthesis of the complementary (−) strand (Kornberg, 1978).

3. *f1, fd and M13 phage*

The method of formation of a complementary (−) strand is similar in these three phages but quite different to those previously described, in that initiation is inhibited by rifampicin, indicating that an RNA polymerase of the host cell is required for primer formation (Brutlag

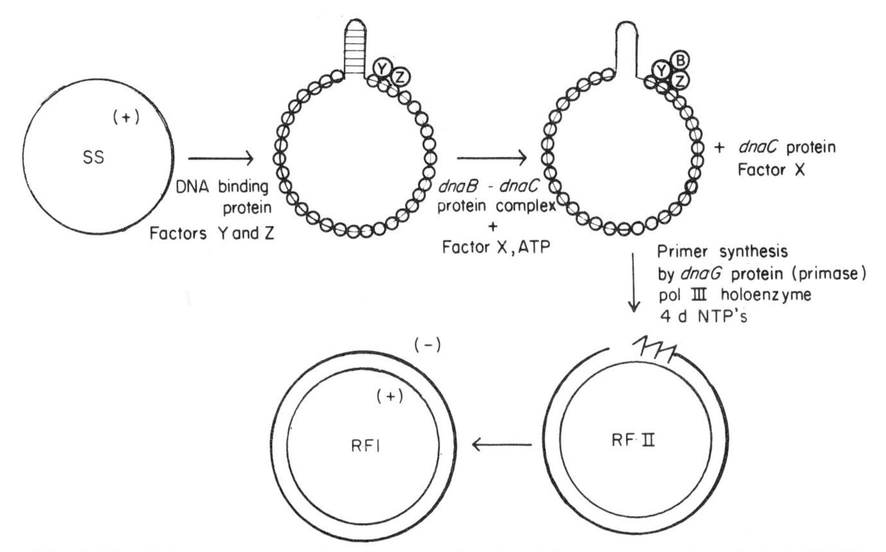

Fig. 3.14 Scheme representing the stages involved in the conversion of viral ϕX174 DNA (SS) into RFI. Replication factors X, Y and Z (analogous to proteins i, n and n′, see text) interact with binding protein (○) and *dnaB* protein (B) as shown. (Redrawn from Wickner, 1978b and Kornberg, 1978.)

et al., 1971; Wickner *et al.*, 1972a; Wickner and Kornberg, 1974a). *E. coli* DNA binding-protein binds the single-stranded DNA (Schaller *et al.*, 1976) but RNase H and two other proteins, termed discriminatory factors α and β are also required before primer synthesis can begin (Vicuna *et al.*, 1977a,b). In the presence of RNA polymerase and 4 rNTPs, primer synthesis begins at an origin which is a region of DNA possibly containing hairpin-loop structures. In fd, the primer site is a region of about 30 nucleotides and the RNA polymerase associates with this hairpin structure (Geider *et al.*, 1978) starting primer synthesis with pppApGpGpG and terminating where the polymerase meets binding protein, some 30 nucleotides later (Schaller, 1978). In M13, the origin contains two potential hairpin structures of about 30 nucleotides and primer synthesis begins at the base of one hairpin (Suggs and Ray, 1978). The polymerase proceeds to transcribe the DNA until it encounters the binding protein where primer synthesis terminates; DNA polymerase III holoenzyme then extends this primer (Suggs and Ray, 1978). Completion of this newly-synthesized complementary strand requires another DNA polymerase and a ligase (Tabak *et al.*, 1974). Hairpin regions are also found at the origin in f1 (Horiuchi *et al.*, 1978).

4. *Other Phages and Plasmids*

There are many other phages which are used as models for bacterial DNA replication but will not be described in detail here. These phages and plasmids also require a number of combinations of host proteins. Replication of phage λ may proceed bidirectionally from a single origin (Schnös and Inman, 1970) and may contain single-stranded connections at branch points of replication (Inman and Schnös, 1971). There are some differences in the requirements for different proteins for DNA replication than those given previously (Wickner, 1978b). In T7, DNA synthesis involves the direct elongation of 1 daughter strand at a growing point (Wolfson and Dressler, 1972) and the RNA primer (Ogawa and Okazaki, 1979) may be a tetraribonucleotide (pppApCp$^{C}_{A}$pN-) or a pentaribonucleotide (pppApCp$^{C}_{A}$pNpN″) where N is mainly A and C (Seki and Okazaki, 1979). The primer in T4 is 6–8 nucleotides long (Liu *et al.*, 1978). In *E. coli* replication is unidirectional with a terminus close to the origin (Sakakibara and Tomizawa, 1974) and requires *de novo* RNA synthesis but not protein synthesis (Tomizawa *et al.*, 1974). In the R-plasmids there may be a single origin (Silver *et al.*, 1977a,b) or multiple origins (Perlman and Rownd, 1976), while in P2 there is thought to be a single origin, with unidirectional fork movement (Schnös and Inman, 1971). In *B. subtilis* phage SPO1, there are two origins, one of which may replicate bidirectionally, while the other replication fork proceeds unidirectionally (Glassberg *et al.*, 1977). In the phages ϕ80, λ, 434 and ϕ21 the origins may form hairpin structures or "clover leaf" structures (Grosschedl and Hobom, 1979), which may assist recognition by proteins at initiation.

C. REPLICATION OF DOUBLE-STRANDED DNA

The replication of DNA from RF molecules of ϕX174 will be given as an example to illustrate the relevent events. For replication of phage DNA from this RF molecule at least two additional proteins are required. One of these is a phage gene product, the *cisA* protein which specifically nicks the (+) strand by breaking a phosphodiester bond (Francke and Ray, 1971; Henry and Knippers, 1974). The other protein that is required is the *E. coli rep* gene product (Scott and Kornberg, 1978; Kornberg *et al.*, 1978). The product of this *rep* gene separates the duplex in advance of replication and utilizes 2 ATPs per base-pair melted. This *cisA* gene product also functions in replication (Fig. 3.15) by supporting fork movement by strand separation and,

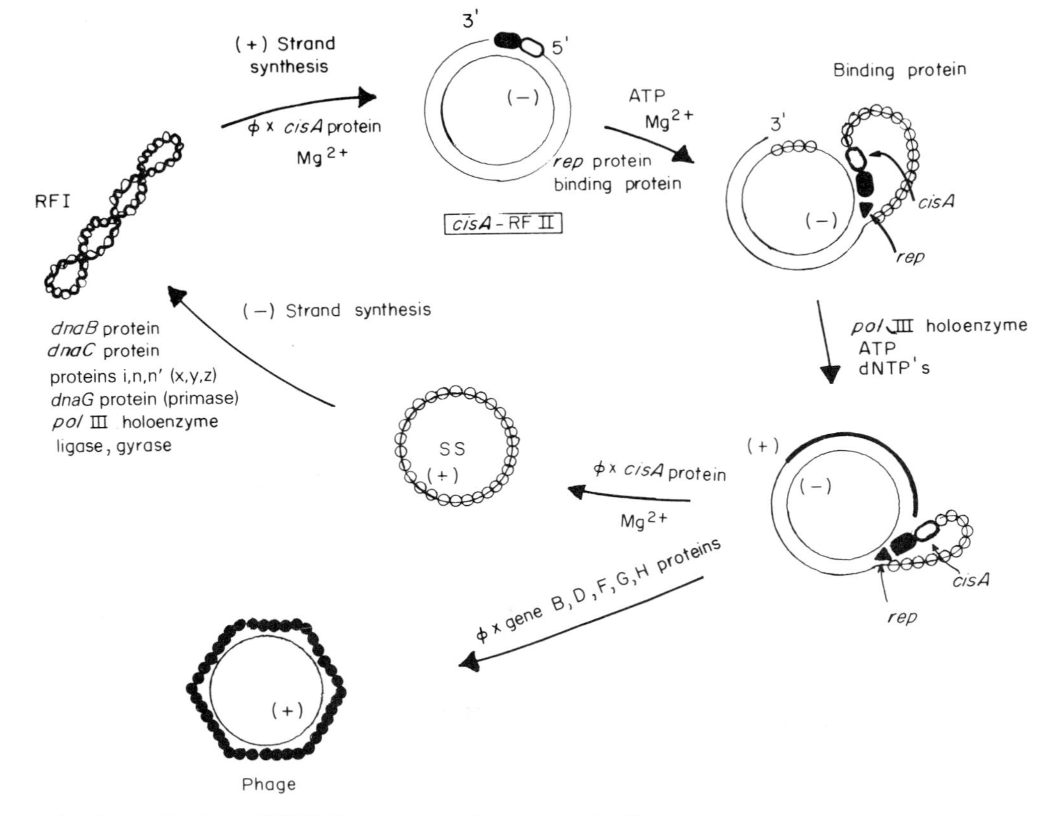

Fig. 3.15 Scheme for the replication of RF DNA and related events in the life cycle of phage ϕX. For details see text. (Reproduced with permission from Kornberg, 1978.)

if DNA binding protein is present, a new (+) strand forms by a rolling circle mechanism (Eisenberg *et al.*, 1977; Sims *et al.*, 1978). Generation of new (+) strand in this way may be continuous, whereas complementary strand synthesis is discontinuous i.e. requiring fresh initiation on each circle (Kornberg, 1978). The (+) strand produced may be used as a template for new (−) strand synthesis as before, or may combine with viral gene products to form new, intact phage which may then be liberated.

IV. Events at *E. coli* Replication Forks

As we have seen in the same cell there are a number of different pathways, whereby DNA may be replicated requiring different combinations of proteins. What then are the mechanisms of DNA synthesis at the bacterial chromosomal replication fork?

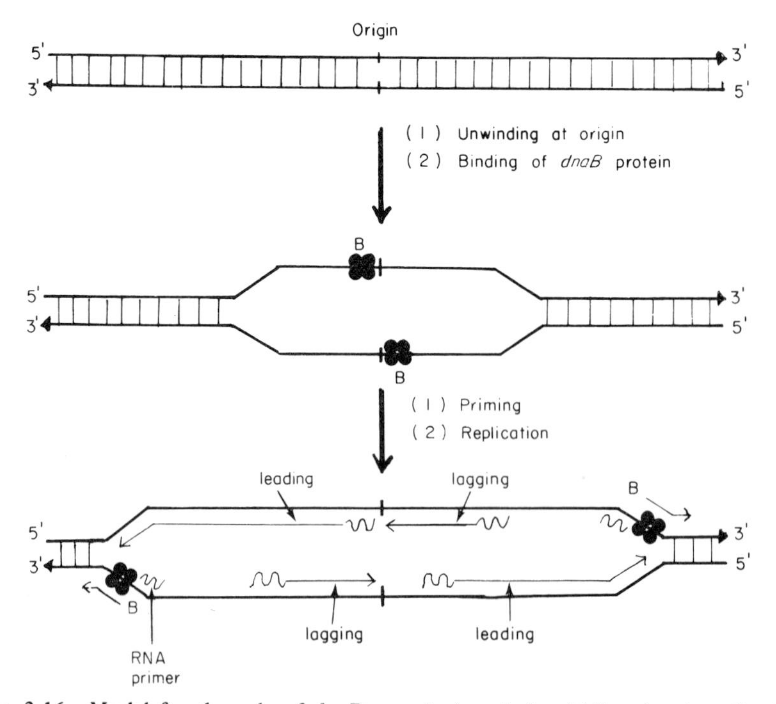

Fig. 3.16 Model for the role of *dnaB* protein in priming bidirectional replication at the chromosomal origin in *E. coli*. (Reproduced with permission from Kornberg, 1978.)

The *dnaB* protein is envisaged as being a "mobile replication promotor" (McMacken *et al.*, 1977) i.e. it may give rise to many primers on a single length of preprimed template. It is proposed that this mechanism may result in initiation of replication of *E. coli* DNA and also in the formation of "Okazaki fragments" (Fig. 3.16; Kornberg, 1977). The following proposals and model in Fig. 3.17 are taken from Gefter (1975), Kornberg (1978) and Ogawa and Okazaki (1980).

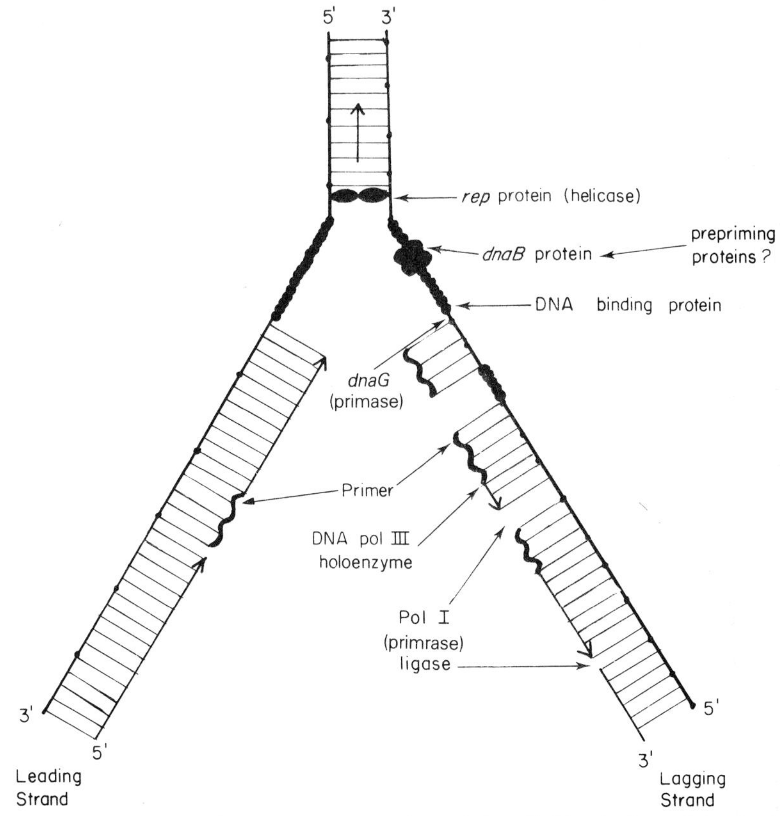

Fig. 3.17 Proposed events at the replication fork of *E. coli*. Unwinding of the duplex is facilitated by the *rep* protein (helicase) and the extended conformation may be maintained by the *E. coli* DNA-binding protein. Primer synthesis by the *dnaG* protein, in conjunction with the *dnaB* protein (plus other pre-priming reactions) may occur at intervals (indicated by small circles on both the leading and lagging strands) on the chromosome. Primer extension by the DNA polymerase III holoenzyme complex results in the formation of Okazaki fragments. After primer excision the fragments are joined by DNA polymerase I (pol I) and ligase. Replication on the leading strand may also be discontinuous, but primers initiated less frequently. (Redrawn and modified from Gefter, 1975; Kornberg, 1978; and Ogawa and Okazaki, 1980.)

DNA unwinding occurs at the origin and the *dnaB* protein (together with prepriming proteins) binds after the attachment of the *E. coli* DNA binding protein. Sites for initiation of RNA primers by *dnaG* protein (primase) are present on both strands approximately every 130 nucleotides. Once initiated the leading strand may be replicated continuously, or discontinuously with longer nascent fragments than those produced on the lagging strand. The helicase action of the *rep* protein progressively opens the duplex and produces single-stranded DNA which may be stabilized and protected from nucleolytic attack by the attachment of the binding protein. The "mobile replication promoter" action of the *dnaB* protein may allow initiation of primer synthesis by *dnaG* protein (primase) which is then elongated by the DNA polymerase III holoenzyme. Primers may then be excised by RNase H (or other mechanisms) and DNA polymerase I (or II) may fill these gaps, which are finally joined by ligase.

V. Incorporation of Thymidine into DNA and Proposed Sites of Action of Inhibitors of Macromolecular Synthesis

The incorporation of exogenous thymidine into (particularly eukaryotic) DNA, in those organisms possessing a thymidine kinase, provides a mechanism whereby rates of DNA synthesis may be followed. The proposed pathway for the incorporation of thymidine is shown in Fig. 3.18. Also shown are the proposed sites of action of some of the commonly-used inhibitors of DNA synthesis, and also inhibitors of RNA and protein synthesis.

Summary

Eukaryotic nuclear DNA replication occurs only during a restricted portion of the cell cycle, the S-phase. Nuclear DNA is complexed with proteins (histones) to form chromatin which is a duplex organized linearly into chromosomes, contained within a membrane-bounded nucleus.

Much work over the past decade has resulted in a detailed understanding of the structure of chromatin and the identification of the components of the multi-enzyme complex that is necessary for repair and replication. The major advances in this area have resulted from

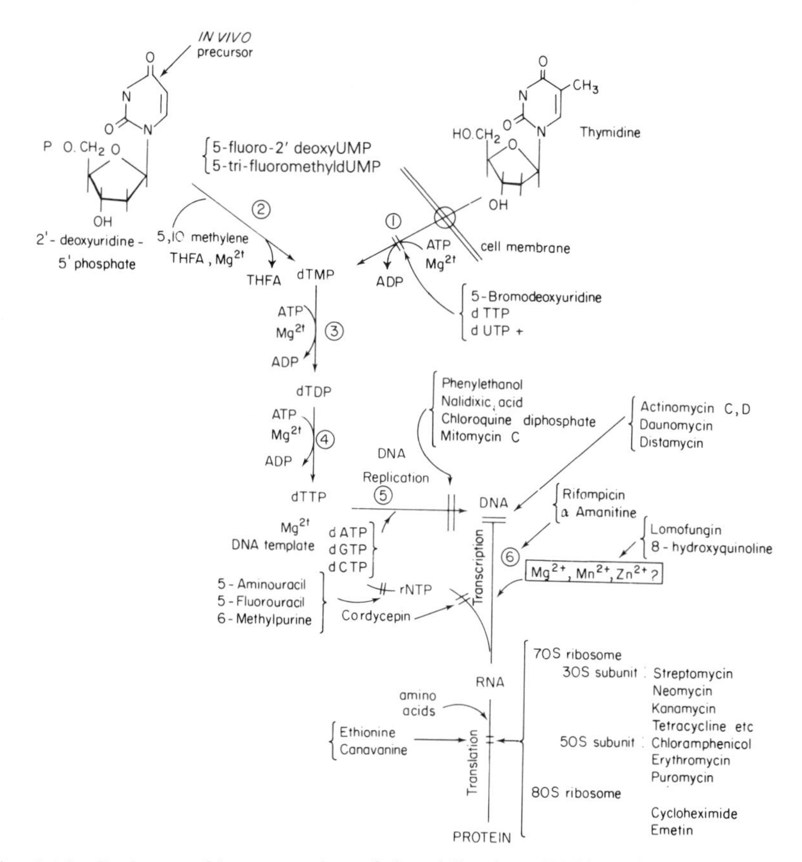

Fig. 3.18 Pathway of incorporation of thymidine into DNA and proposed sites of action of inhibitors of macromolecular synthesis. Abbreviations: THFA—tetrahydrafolic acid; enzymes—(1) Thymidine kinase; (2) Thymidylate synthetase; (3) TMP kinase; (4) Nucleoside diphosphate kinase; (5) DNA polymerase; (6) RNA polymerase. Proposed sites of inhibition are represented by →‖ and stimulation is represented as (+).

experiments using the structurally simple mammalian viruses, especially papavoviruses, as model systems.

There are, however, still some fundamental questions which remain unanswered especially concerning the initiation of DNA synthesis. The identification of the initiation factor(s) has remained elusive, but it is thought to be of cytoplasmic origin.

Organellar DNA appears to be replicated by mechanisms which are quite different from nuclear DNA, although there may be instances when these processes are temporally coordinated.

As with eukaryotes, much of the recent progress in the understanding of the mechanisms of prokaryotic DNA replication has resulted from experiments with structurally simpler viral or bacteriophage DNA. The isolation of a series of temperature-sensitive mutants, defective in DNA replication or repair has helped to identify the gene products necessary for these processes. Again, the signal or initiation factor for replication is unknown.

The molecular events which occur at the bacterial replication fork have largely been elucidated but work with viral and bacteriophage DNAs have shown that a number of different pathways, each requiring different combinations of proteins, exist within a single organism.

4. Biosynthesis of Macromolecular Components: RNA, Proteins, and other Cellular Constituents

"Dost thou love life, then do not squander time,
for that is the stuff life is made of."

Benjamin Franklin
"The Way to Wealth" July 5th 1757

The cell growth and division cycle involves accumulation of the necessary cellular constituents in a time- and spatially-ordered sequence. Measurement of total cell volume may be considered to be a complete estimate of cellular growth since all the constituents of the cell are accounted for (i.e. protein, nucleic acids, carbohydrates, lipids, low molecular weight pools etc.) including water, the largest single component. Changes in volume may be direct indications of dry mass changes, or may simply reflect variations in the water content, which may be considerable in highly osmotically-active cells (e.g. protozoa or mammalian cells). For this reason the assessment of cell growth should not rest with volume measurements alone but should ideally include estimations of macromolecules. It is only by investigating many cellular processes that we can hope to achieve an understanding of the complex network of temporally and spatially sequenced events which constitute the cell cycle. This requires both bulk measurements and studies of individual processes and components.

I. Cellular Protein

A. TURNOVER OF PROTEINS

Protein constitutes the largest fraction of cellular macromolecules and consequently many measurements of total cell protein during the cell cycle have been reported and interpreted as representing the pattern on cellular growth. Bulk measurements of total protein, like that of any cellular component however, only represent the accumulated or quasi steady-state value of the component at the time of sampling. Since Schoenheimer (1942) coined the phrase "the dynamic state of body constituents" it has become appreciated that the components of cells are constantly changing and that turnover of cellular components is an important life process. Accumulation of a protein, like that of any cellular constituent, depends on both its rate of synthesis and rate of breakdown. Synthesis of a functional protein may depend on a number of steps including transcription, polypeptide chain initiation and elongation and assembly into intact, functional components. Once assembled, these proteins are then subject to the normal processes of breakdown by mechanisms that are rather poorly understood. A vast literature covers protein turnover and proteolysis (e.g. Pine, 1972; Goldberg and Dice, 1974; Schimke and Katunuma, 1975; Goldberg and St John, 1976; Ribbons and Brew, 1976) and a detailed account of the mechanisms involved and methods of study will not be given here. Instead, we will focus on aspects of protein turnover which are pertinent to, and often overlooked in, studies of the cell cycle.

As well as degradation of proteins to individual amino acids, related phenomena include loss of protein by secretion, protein transport from one cellular compartment to another, degradation of extracellular proteins and inactivation of enzymes by covalent modifications (Goldberg and Dice, 1974). The mechanisms responsible for protein turnover in mammalian cells have analogies in bacterial systems (Goldberg and St John, 1976) although a major difference is the absence of a lysosomal system in prokaryotes. Overall rates of protein degradation appeared to be low in rapidly-growing bacterial cells. Mammalian cells are subject to higher rates of turnover—about 40% of rat liver protein is turned over every day (Dean, 1978) and all proteins so far investigated are degraded. The breakdown of enzymes follows first order kinetics i.e. newly synthesized proteins are as likely to be degraded as older proteins and the degradation rates of individual enzymes vary over a wide range (Table 4.1) from 10 min for ornithine decarboxylase (Russell and Snyder, 1969) to 19 days for isoenzyme 5 of lactate dehydrogenase (Fritz *et al.*, 1973). Different components within the same organelle can have different half-lives, which suggests complex modes of degradation and its control (Goldberg and Dice, 1974).

Whilst protein turnover can be detected in growing cells, certain conditions can accelerate this process. For example, in *E. coli*, breakdown of proteins during exponential growth occurs at a relatively low

Table 4.1. *Half-lives of some rat liver enzymes (from Dean, 1978)*

Enzyme	Half-life
Ornithine decarboxylase (soluble)	10 min
δ-Aminolevulinate synthetase (mitochondria)	60 min
Tyrosin aminotransferase (soluble)	1·5 h
Tryptophan oxygenase (soluble)	2 h
Hydroxymethylglutaryl CoA reductase (endoplasmic reticulum)	2–3 h
Alanine-aminotransferase (soluble)	0·7–1·0 days
Glucokinase (soluble)	1·25 days
Catalase (peroxisomal)	1·4 days
Acetyl CoA carboxylase (soluble)	2 days
Glutamate-alanine aminotransferase	2–3 days
Cytochrome *c* reductase (endoplasmic reticulum)	3–4 days
Arginase (soluble)	4–5 days

level but increases about fourfold upon nutrient deprivation (Pine, 1972) or as cells attain stationary phase (Goldberg *et al.*, 1975). The possible functions of protein turnover during growth may be summarized as follows (Goldberg and Dice, 1974):

(1) Removal of abnormal proteins arising from mutations, errors in gene expression, denaturation or chemical modifications. This may be more important in slow growing cells where such proteins may accumulate rather than become diluted out during rapid growth (Schmidt, 1975).

(2) Continuous turnover of proteins must significantly increase the ability of the organism to adapt to environmental changes. The more rapidly an enzyme is degraded, the faster its intracellular concentration can change in response to external stimuli. In fact, it is found that the key or rate-limiting enzymes of metabolic pathways are turned over very rapidly (Goldberg and Dice, 1974). Also important in this respect is the requirement for recycling of cellular material e.g. *E. coli* starved of nitrogen can still synthesize β-galactosidase in the presence of the inducer but cannot do so when protein degradation is prevented (Goldberg, 1971; Prouty and Goldberg, 1972).

(3) Protein constitutes an energy store which may be utilized during conditions of energy deprivation.

There appears to be a relationship between protein structure and degradation rate, and in this respect protein conformation, molecular weight and charge are important factors (Goldberg and St. John, 1976). This degradation requires energy and, in eukaryotes, can occur within the lysosomal system. However, the lysosomal system cannot account for all the breakdown e.g. in mammalian cells about 70% of intracellular protein degradation is non-lysosomal (B. Poole, personal communication). Especially difficult to understand are the variations in half-lives of different enzymes within the same organelle. Also, prokaryotes are lacking such compartmentalized locations of hydrolytic enzymes and some peptidases have pH optima in a range far outside that found within the lysosome. An interesting observation has resulted from work with pepstatin (Dean, 1975) an inhibitor of Cathepsin D. The inhibitor does not normally enter cells but can be introduced to perfused liver via liposomes. Once inside the cell, this compound accumulates within lysosomes and inhibits protein degradation. In perfused liver, levels of protein turnover are greater than the basal level *in vivo*; pepstatin reduces protein turnover to this basal level; only the accelerated turnover is inhibited. This leads to the attractive suggestion that lysosomal activity is largely responsible for accelerated turnover, which may arise from nutrient or energy deprivation, or upon differentiation and that some other mechanism may be involved in normal turnover in actively growing cells.

Although the general opinion is that high rates of biosynthesis and low rates of degradation characterize rapidly growing cells (Goldberg and St. John, 1976), there are a number of cases where this is not so. In rat fibroblasts, 20% of newly synthesized proteins are degraded within the first 2 h of synthesis, whereas many proteins have a half-life of 1–2 days (Poole and Wibo, 1973). There is also a more rapid turnover of body protein in young humans than in non-growing or aged adults (Young *et al.*, 1975) and proteolysis is faster in muscle of young growing rats than in adults (Millward *et al.*, 1975). More recent estimations of turnover rates of proteins in *S. cerevisiae*, which consider post-incorporation and recycling of the label, show that the half-life of products of mitochondrial protein synthesis is 20 min in exponentially-growing cells (Bakalkin *et al.*, 1978) and 60 min in early stationary phase cultures (Kalnov *et al.*, 1979) (Fig. 4.1). Furthermore, protein degradation can be inhibited by certain proteinase inhibitors. It is suggested that some mitochondrial proteins are produced in excess of the requirement for full catalytic activity and are not integrated into the functional membrane, but proteolysed (Galkin *et al.*, 1979a,b). Turnover rates of total cell protein, measured isotopically in growing

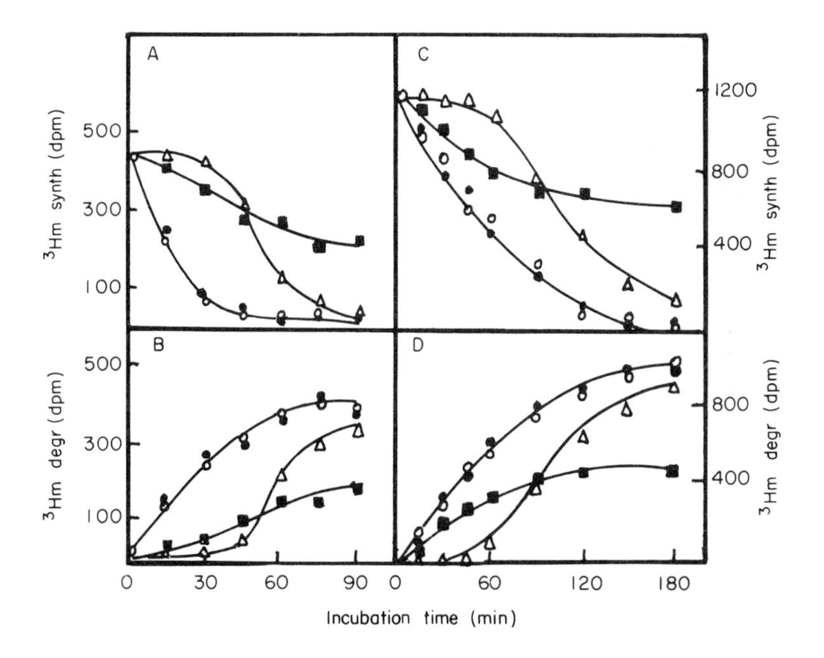

Fig. 4.1. Breakdown of mitochondrial translational products in yeast. A and B in exponentially growing cells, C and D in stationary phase cells. A and C show the decrease in the content of pulse-labelled mitochondrial proteins. B and D show the accumulation of the acid-soluble products of proteolysis. Additions: none (o); phenylmethyl sulphonyl fluoride (△); leupeptin (□); pepstatin (●). ^{3}Hm synth. represents the amount of [^{3}H]-leucine incorporated into mitochondrial translation products and retained in mitochondrial protein. ^{3}Hm degr. is the amount of [^{3}H]-leucine accumulated in the acid-soluble fraction of mitochondria owing to the breakdown of the pulse-labelled mitochondrial translation products. (Reproduced with permission from Kalnov *et al.*, 1979.)

cultures of *Acanthamoeba castellanii* are between 6 and 14% breakdown per hour (D. Lloyd, S. L. Kalnov, A. S. Zubatov, L. Novikova, and V. N. Luzikov, unpublished results) and from 1·2–2·13% per hour in growing cultures of *Chlorella fusca* var. *vacuolata* (Richards and Thurston, 1980). In the latter organism an unidentified fraction (8% of the total protein) is degraded at a rate of 20% per hour and since the maximum specific growth rate is 0·087 per hour, replacement of degraded protein must account for one tenth of the total protein synthesized (Thurston and Richards, 1980).

In monkey kidney epithelial cells, it has been suggested that there are two main mitochondrial populations with apparent half-lives of 1·6 h and 6 days (Knecht *et al.*, 1980). Any cell cycle dependence of such phenomena has not yet been demonstrated but offers an attractive system for study. The implications of these findings are not yet widely appreciated by workers in this field; perhaps patterns of accumulation or synthesis of macromolecules that have been reported are somewhat oversimplified.

1. *Total Protein Levels during the Cell Cycle*

Three approaches have been used: (a) measurement of total (bulk) protein, usually by chemical methods; (b) rate of incorporation of radioactive amino acids into polypeptides; or (c) radioautographic analysis of pulse labelled cells. More recent studies involve the examination of individual proteins on two-dimensional polyacrylamide gels.

For the majority of cells, protein appears to increase continuously during the cell cycle. The increase may be exponential, linear or linear with a rate doubling at a specific point in the cycle. In practice, it is often very difficult to distinguish between these three patterns of increase if total quantity is estimated (Fig. 4.2a). Instead the *rate* of synthesis (Fig. 4.2b) shows the distinction between the different patterns of increase much more clearly. The most commonly used

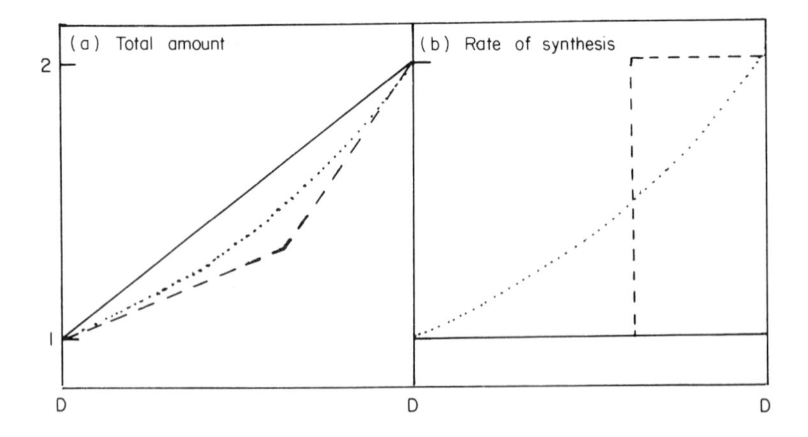

Fig. 4.2 Predicted patterns of synthesis of macromolecules during the cell cycle. (a) measurement of total amount; (b) measurement of the rate of synthesis. (—) linear synthesis, (– –) linear synthesis with a rate change, ... exponential increase. D = cell division.

technique for measuring the rate of synthesis is pulse labelling with a radioactively-labelled precursor but it must be ascertained whether the precursor taken up by the cell has been incorporated into the macromolecule or remains in a pool (Mitchison, 1971).

(a) *Eukaryotes*

(i) *Mammalian cells.* In mouse L cells, the total macromolecular dry mass, measured by interferometry, increases continuously (Killander and Zetterberg, 1965) and pulse labelling with radioactive amino acids suggests a continuous, increasing rate of incorporation (Zetterberg and Killander, 1965b). In HeLa cells (Scharff and Robbins, 1965) and Chinese hamster lung cells (Robbins and Scharff, 1967), pulse labelling shows a steadily increasing rate, but in P815Y the rate of synthesis increases in G_1 and S, but decreases in G_2 (Warmsley and Pasternak, 1970). Constant rates of increase followed by a sharper increase in the S-phase have been reported after colcemid treatment of rat hepatoma cells (Martin *et al.*, 1969) and in Chinese hamster Don C cells (Stubblefield *et al.*, 1967). The rate of histidine incorporation during mitosis in Chinese hamster cells was about a quarter of the interphase value (Prescott and Bender, 1962), whereas uridine incorporation into RNA was reduced to about one fiftieth. However, other reports show a different pattern. While there was about a 50% reduction of phenylalanine incorporation into human amnion cells (Konrad, 1963), no reduction was found during mitosis of Chinese hamster cells (Konrad, 1963; Taylor, 1960) or sea urchin eggs. (Gross and Fry, 1966). Whilst the rate of RNA synthesis is usually diminished during mitosis in these cells (see later), protein synthesis at this time may be possible because of long lived mRNA species present in the cytoplasm. In V79-8 cell line mutants which may or may not contain a G_1 phase, the presence of G_1 is associated with decreased rates of total protein synthesis (Liskay *et al.*, 1980).

Incorporation of radioactively-labelled amino acids into cytoplasmic proteins or nuclear proteins is activated at the end of the G_1 phase in Ehrlich tumour cells (Kazymin and Sherban, 1980). Analysis of about 700 polypeptides by two-dimensional gel electrophoresis showed that no polypeptide was synthesized only at a particular stage of the HeLa cell cycle (Bravo and Celis, 1980); about 100 of the most abundant polypeptides were synthesized constantly through the cycle, but some showed cell cycle-specific increases. In these cells six proteins varied in their rates of labelling by more than fourfold while others showed small but significant changes (Milcarek and Zahn, 1978).

A quite different pattern of accumulation of total protein was observed after colcemid treatment of Chinese hamster cells (Klevecz and Ruddle, 1968). Total cell protein (as well as mean cell volume and activities of lactate dehydrogenase and glucose-6-phosphate dehydrogenase) increased periodically with three maxima per cell cycle. These authors suggested that periodic enzyme synthesis at discrete stages of the cell cycle was part of a more generalized phenomenon affecting the entire population of proteins, probably initiated by increased RNA synthesis. Simultaneous measurements of the rates of protein synthesis and degradation during the cell cycle of NHIK 3025 cells, synchronized by mitotic selection, showed that protein degradation is extensive under these conditions; rates of protein degradation were approximately 1·5% per hour while rates of protein accumulation were from 3·5–4·5% per hour (Rønning *et al.*, 1979).

(ii) *Yeasts*. Continuous protein synthesis has been reported in synchronously dividing cultures of *S. cerevisiae* (Williamson and Scopes, 1961b) and *S. cerevisiae* × *S. dobzhanskii* (Gorman *et al.*, 1964), after re-feeding starved, stationary phase cultures (Williamson and Scopes, 1961a). In the fission yeast *Schiz. pombe*, the amount of radioactively-labelled protein precursor incorporated as a function of cell size suggests an exponential rate of increase through the cycle (Mitchison and Wilbur, 1962). Using a dual-label technique of incubating *S. cerevisiae* for 3 h with [^{3}H]-yeast protein hydrolysate followed by 10 min with [^{35}S]-methionine, the rate of synthesis of individual proteins can be estimated (Elliott *et al.*, 1979). After cell cycle fractionation of labelled cells by centrifugal elutriation and two dimensional gel electrophoresis, all but two out of 110 proteins showed constant ratios of counts though the cycle, suggesting that processes of degradation and modification of proteins were not of general occurrence in this organism (Elliott and McLaughlin, 1978, 1979). In synchronously-dividing cultures of *Candida utilis* prepared by continuous-flow size-selection, total protein (Fig. 4.3) does not accumulate smoothly, but instead levels oscillate (three maxima per cycle), doubling overall during one cycle (Lloyd *et al.*, 1981); control cultures indicate that these discontinuities in protein accumulation do not result from pertubations (Fig. 4.4).

(iii) *Protozoa*. Early work suggested that the increase in total protein was linear in *Tetrahymena* (Prescott, 1960). In induction-synchronized cultures of *T. pyriformis* the rate of protein synthesis decreased slightly at cell division (Kuzmich and Zimmerman, 1972; Zimmerman and Laurence, 1975), while in synchronously-dividing cultures prepared

Table 4.2. *Enzyme activities during the cell cycles of eukaryotes*

Cell type	Method of Analysis (S = synchronous culture)	Comments	Reference
MAMMALIAN CELLS			
HeLa	Double thymidine block (S)	α-2-fucosyltransferase (a gene product associated with blood group H) activity maximal in mid S and early G_2 intermediate in G_2 and mitoshs, and low in G_1.	Kuhns (1978)
		Activity of protein kinase associated with microtubules maximal at mitosis while that in cytosol is constant in S, G_2 and decreases in mitosis.	Piras and Piras (1975)
	Methotrexate block (S)	Poly(ADP-ribose) polymerase activity in cytoplasm parallels rate of DNA synthesis	Roberts *et al.* (1975)
		Nuclear protein kinase activity doubled in S and G_2 and decreased at mitosis.	Phillips *et al.* (1979)
	Thymidine followed by N_2O block (S)	dUTPase activity low in G_1, rises in S and G_2 and is maximal just before cell divison.	Mahagaokar *et al.* (1980)

Table 4.2. *continued*

Cell type	Method of Analysis (S = synchronous culture)	Comments	Reference
	Mitotic detachment (S)	Alkaline phosphatase most active in G_1 and mitosis, 5′-nucleosidase activity highest in mid G_1 and mitosis, Ca^{2+}-ATPase activity only detected in S phase. No detectable changes in Mg^{2+}-ATPase activity.	Vorbrodt and Borun (1979)
		Alkaline phosphatase activity (but not protein) highest in mitosis.	Lucid and Griffin (1979)
Chinese hamster cells	Mitotic detachment (S)	L-ornithine decarboxylase activity peaks in G_1, decreases in S and exhibits a second maximum just before mitosis.	Heby *et al.* (1976)
		Increase in ornithine decarboxylase activity in mid-S phase.	Russell and Stambrook (1975)
		Nuclear protein kinase activity/cell increases more than twofold from G_1 to S, concomitant with 70% increase in nuclear non-histone protein. Cytoplasmic cAMP-dependent protein kinase increases twofold through G_1.	Costa *et al.* (1977)
		Plasminogin activator (a neutral protease) secreted only at G_2/M.	Aggeler *et al.* (1978)

	Colcemid (S)	Activities of lactate dehydrogenase and glucose-6-phosphate dehydrogenase increase as three peaks/cycle.	Klevecz and Ruddle (1968)
	Colcemid plus mitotic detachment (S)	High turnover rate of cAMP-dependent protein kinases $t_{\frac{1}{2}}$ 30 min.	Haddox *et al.* (1980)
	Serum depletion and refeeding (S)	Induction of ornithine decarboxylase highest in late G_1/early S.	Meloni *et al.* (1980)
	Temperature shift of temperature-sensitive mutant (S)	dUTPase and uracil-DNA glycosylase activity minimum in G_0 and maximum in S.	Duker and Grant (1980)
Leydig I-10	Double Thymidine block (S)	Variations in activities of cAMP dependent and independent protein kinase activities of different subcellular fractions.	Christensen *et al.* (1979)
Erythroleukemic mouse spleen cells	Isoleucine starvation/feeding (S)	Poly (A) polymerase activity low in G_1 doubles in early S-phase.	Adolf and Swetly (1978)
KB	Double thymidine block (S)	Fumarase and lactate dehydrogenase synthesized continuously.	Bello (1969)
Mouse L	Double thymidine block (S)	Nuclear endoribonuclease increases in activity 4-fold, exoribonuclease 30-fold and 5′-nuceotidase 16-fold from G_1–G_2, while poly (A) polymerase activity is continuous.	Muller *et al.* (1977)
Burkitts lymphoma		Two peaks of lactate dehydrogenase activity.	Trenfield and Masters (1980)

Table 4.2. *continued*

Cell type	Method of Analysis (S = synchronous culture)	Comments	Reference
Neoplastic Mast	Cell cycle fractionation	NADPH—cytochrome *c* reductase, uridine diphosphatase, succinate—cytochrome *c* reductase, cytochrome *c* oxidase increase continuously.	Warmsley *et al.* (1970)
P815Y or NIL-2HSV		5′-nucleotidase and K^+/Na^+ ATPase activities are constant but adenyl kinase activity increases periodically.	Graham *et al.* (1973)
Mouse fibroblasts	Stimulation of resting cells	Thymidylate synthetase activity stimulated as cells enter S-phase.	Navalgund *et al.* (1980)
		Dihydrofolate reductase increases at beginning of S-phase.	Hendrick *et al.* (1980)
YEASTS			
Saccharomyces cerevisae	Selection of cells from sucrose gradient (S)	Sucrase and acid phosphatase increase as steps at 0·4 and 0·6 respectively of the cycle. Alkaline phosphatase increases as two steps/cycle, at 0·12 and 0·45.	Matur and Berry (1978)
		Exo-and endo-1-3,β-glucanase activities increase as steps, at or after bud emergence and from mid to late S-phase, respectively.	Delrey *et al.* (1979)

	Re-inoculation of stationary phase cultures (S)	Cytochromes *c*, *b* and $a + a_3$ increase continuously but cytochrome *c* oxidase activity increases as 1 step/cycle.	Cottrell *et al.* (1975)
	Cyclic heat treatment (S)	Ribonucleotide reductase peaks at onset of S-phase coinciding with bud emergence.	Lowdon and Vitols (1973)
	G_1 arrest	Ornithine decarboxylase activity (usually) low in G_1.	Kay *et al.* (1980)
		Increase in levels of total RNAase in G_1.	McFarlane (1980)
Saccharomyces mutant 1710	Differential centrifugation (S)	Small (carbohydrate-free) invertase increases as 1 step/cycle while large isomer (glycoprotein bound) increases continuously.	Gallili and Lampen (1977)
Schizosaccharomyces pombe	Selection from sucrose gradient (S)	Ornithine transcarbamylase, aspartate transcarbamylase and alcohol dehydrogenase increase as "steps".	Sissons *et al.* (1973)
		Acid phosphatase activity per ml increases as 1 step/cycle but specific activity shows 1 peak/cycle.	Miyata and Miyata (1978)
		glucose grown: All enzymes assayed increased periodically as two peaks per cycle, except cytochrome *c* oxidase which showed a single peak. Confirmed by cell cycle fractionation.	Poole and Lloyd (1973)

Table 4.2. *continued*

Cell type	Method of Analysis (S = synchronous culture)	Comments	Reference
		glycerol grown: Cytochrome *c* oxidase activity and levels of cytochromes increase as steps, but other respiratory enzymes increases as peaks.	Poole and Lloyd (1974)
		18 enzymes show continuous increases while TMP kinase is periodic.	Mitchison (1977c)
		Sucrase activity/ml shows one step/cycle, but specific activity/peak/cycle.	Miyata *et al.* (1980)
	Continuous flow size-selection (S)	Oligomycin-sensitive ATPase activity increases periodically as peaks in glucose-containing medium but steps in glycerol-containing medium. Confirmed by cell cycle fractionation.	Edwards and Lloyd (1977a)
	Cell cycle fractionation	Periodic accumulation of cytochromes c_{548}, b_{554}, b_{560}, b_{563} and $a + a_3$. Cytochrome P-420 decreased while cytochrome P-450 increased during the first three quarters of the cycle.	Poole *et al.* (1974)
		Sensitivity of ATPase to eleven different inhibitors shows cell cycle-dependent variations.	Lloyd and Edwards (1977)

	Hydroxyurea (S)	Specific activity of acid phosphatase shows 1 peak/cycle but activity/ml culture increases continuously.	Miyata and Miyata (1978)
	Deoxyadenosine (S)	Potential of sucrase and maltase increases in steps in early S-phase.	Sissons *et al.* (1973)
ALGAE			
Chlorella sorokiniana (=*C. pyrenoidosa* 7-11-05)	Light/Dark (S)	Aspartate transcarbamylase and dihydroorotase increase as single steps/cycle, out of phase, in NO_3^{2-} containing medium, or in phase in NH_4^+-containing medium.	Dunn *et al.* (1977)
		Linear accumulation of glutamate dehydrogenase (in presence of inducer) with rate change early in S-phase.	Israel *et al.* (1977)
C. pyrenoidosa. 7-11-05	Light/Dark followed by equilibrium centrifugation. Grown in continuous illumination (S)	Ribulose bisphosphate carboxylase, dihydroorotase, dTMP kinase, dCMP deaminase and aspartate transcarbamylase increase as steps. Phosphoribosylglycinamide synthetase increases linearly with a rate increase.	Schmidt (1975)
		Isocitrate lyase synthesized only in dark. NH_4^+-induced synthesis of NADPH—specific glutamate dehydrogenase only in light.	Schmidt (1974b)
211-8b		Nitrite reductase inducible throughout cycle. Potential increases in S.	Schmidt (1974b)

Table 4.2. *continued*

Cell type	Method of Analysis (S = synchronous culture)	Comments	Reference
C. vulgaris and *C. pyrenoidosa*	Light/Dark (S)	Increase in activities of nitrite and nitrate reductase in 1st hour of light.	Tischner (1976)
C. sorokiniana		NADP-glutamate dehydrogenase inducible throughout cycle and potential doubles in early S-phase.	Turner *et al.* (1978)
Chlamydomonas reinhardii		Ferredoxin-NADP reductase, ribulose bisphosphate carboxylase and phosphoribulokinase increase in light period.	Armstrong *et al.* (1971)
		Ribulose bisphosphate carboxylase activity increases linearly in light and remains constant in dark.	Iwanij *et al.* (1975)
	or selection of zoosproes (S)	Aryl sulphatase derepressible at all stages of the cycle but peaks at 8 and 16 h.	Schreiner *et al.* (1975)
Ankistrodesmus braunii	Light/Dark (S)	Activity of NADP-glyceraldehyde-3-phosphate dehydrogenase greater in first half of light phase.	Theiss-Seuberling (1975)
Porphyridium purpureum		Activities of ADPG-transferase and phosphorylase increase linearly in light. Amylase activity increases at end of light.	Sheath *et al.* (1977)

Euglena gracilis		Succinate dehydrogenase and fumarase activities constant in light, double in early dark and then remain constant. Increase affected by light/dark transition.	Davis and Merrett (1974)
		Ornithine decarboxylase activity peaks in G_1, after S-phase and in G_2. In vitamin B_{12}-deficient growth only G_1 peak retained.	Lafarge-Frayssinet *et al*. (1978)
PROTOZOA			
Physarum polycephalum	Natural mitotic cycle	Ornithine decarboxylase activity increased only in S-phase.	Sedory and Mitchell (1977)
		Ca^{2+}-ATPase activity maximal at mitosis.	Petzelt *et al*. (1980)
		cAMP phosphodiesterase activity decreases at mitosis, protein kinase maximal at S and early G_2.	Trakht *et al*. (1980)
		RNA polymerase B activity highest in S-phase.	Pierron and Sauer (1980)
Tetrahymena pyriformis	Heat shock (S)	Succinate dehydrogenase, cytochrome *c* oxidase and malate dehydrogenase activities/ml increase exponentially, but rotenone-insensitive NADH-cytochrome *c* reductase activity increases periodically.	Cowan and Young (1978)

Table 4.2. *continued*

Cell type	Method of Analysis (S = synchronous culture)	Comments	Reference
		Thymidine kinase activity maximal in S-phase	Fink (1980)
		Acid RNAse activity increases sharply at the end of heat shock but no increase observed in free running synchrony.	Tarnowka and Yuyama (1978)
		Adenylate cyclase activity/cell highest at division, but affected by temperature.	Kassis and Zeuthen (1979)
	Heat shock (S) or tantalum loading (S)	Nucleoside phosphotransferase activity increases continuously.	Bols and Zimmerman (1977)
Acanthamoeba castellanii	Low-speed centrifugation in growth medium (S)	Catalase activity and immunologically-determined catalase protein increase periodically (seven maxima/cycle)	Edwards *et al.* (1981)
		ATPase activity (and the naturally occuring ATPase inhibitor) show seven maxima per cycle and sensitivities to 4 different inhibitors show cyclic variations.	S. W. Edwards and D. Lloyd (unpublished observations)
		Total amounts of cytochrome *c* and $a + a_3$, as well as the activity of cytochrome *c* oxidase accumulate discontinously as seven maxima per cell cycle.	S. W. Edwards and D. Lloyd (unpublished observations)

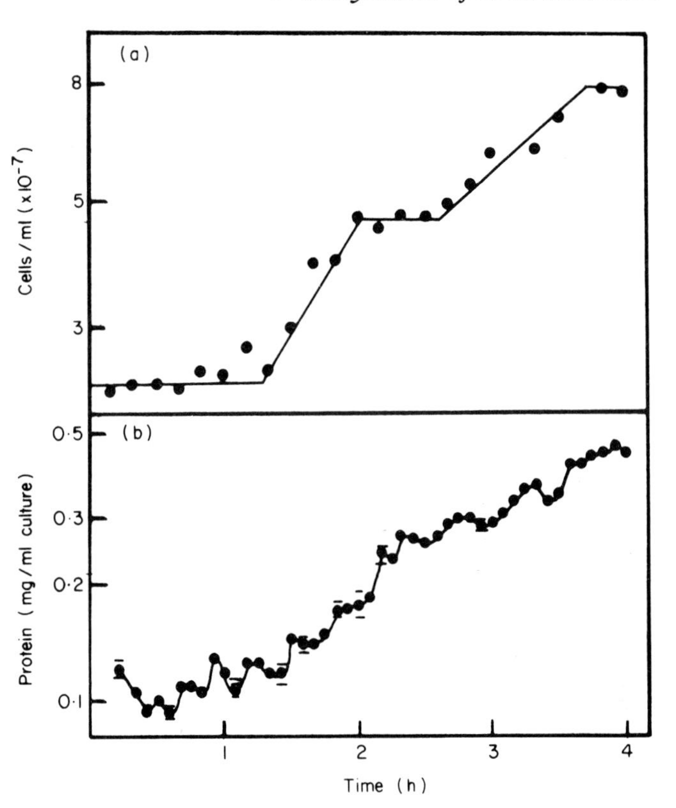

Fig. 4.3 Changes in total protein levels during synchronous growth of *C. utilis* in a glycerol-containing medium. An exponentially-growing culture was passed through a continuous-flow rotor at 420 ml/min, while the rotor speed was 2 500 r/min (Lloyd *et al.*, 1975). The effluent (containing 10% of the original population) was grown as a synchronous culture. (a) Cell numbers (b) total protein, where each point is an average of three protein measurements with ranges of values obtained (where these are greater than the diameter of the symbol). (Reproduced from Lloyd *et al.*, 1981.)

by continuous-flow size-selection a continuous increase in total protein was observed (Phillips and Lloyd, 1978). However, when measured by cytophotometry and expressed as a function of cell size rates of protein synthesis were highest in G_1 and G_2 but decreased in early S (Cleffmann *et al.*, 1979a,b). Synthesis of RNA was approximately the inverse of this and thus the RNA : protein ratio decreased to a minimal value in mid S, irrespective of whether the timing of S was altered by heat shock. These authors suggested that DNA replication may be

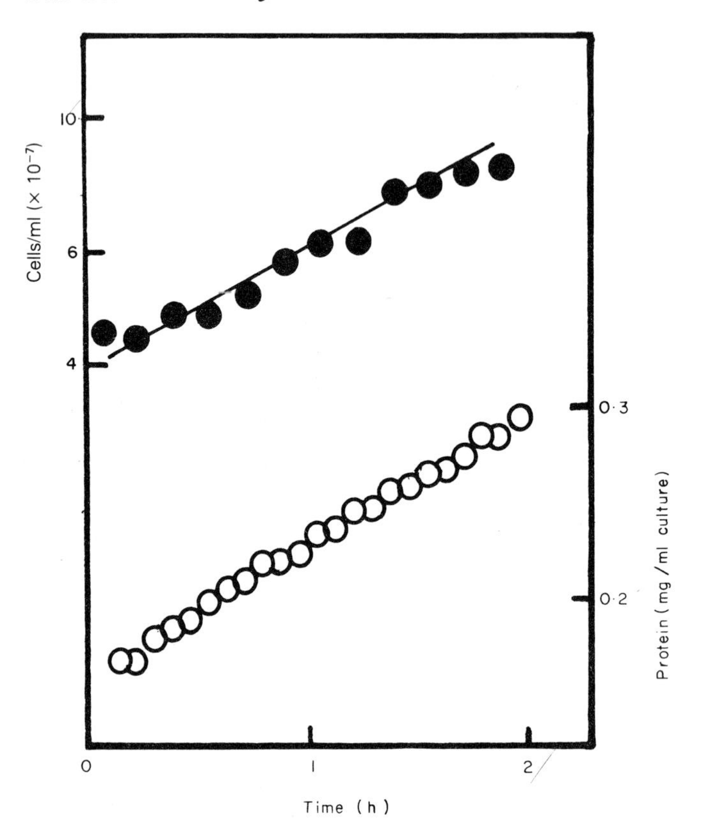

Fig. 4.4 Changes in protein levels and cell numbers in a control asynchronous culture of *C. utilis*. An exponentially-growing culture was centrifuged at 5 000 *g* for 1 min. After this, the pellet and supernatant were remixed and returned to the growth vessel. Cell numbers and total protein were measured as in Fig. 4.3. (Reproduced from Lloyd *et al.* 1981.)

under the negative control of the relative RNA content. In *Paramecium aurelia* there is an increasing rate of protein accumulation which is highest in S (Woodard *et al.*, 1961). In synchronously-dividing cultures of *Crithidia fasciculata*, after sedimentation-velocity size-selection (Edwards *et al.*, 1975), total protein increases continuously, doubling during a cell cycle.

Total protein in the naturally-synchronous mitotic cycle of the acellular slime mould, *Physarum polycephalum* shows little increase just prior to and just after mitosis (Mittermayer *et al.*, 1966). The rate of incorporation of ^{3}H-lysine shows two peaks (Fig. 4.5) varying

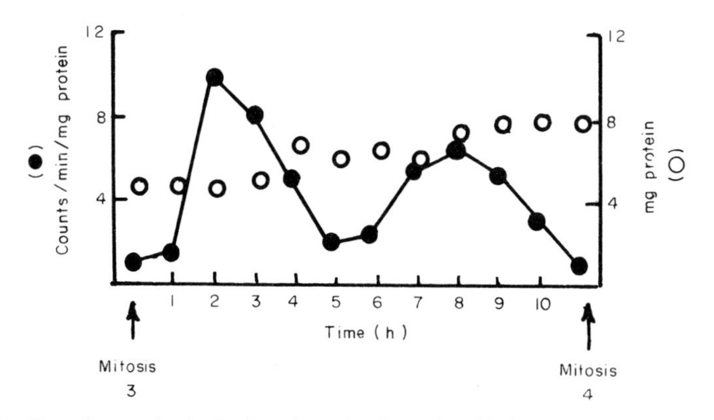

Fig. 4.5 Protein synthesis during the mitotic cycle of *Physarum*. Open circles show total protein measured by the Biuret method, closed circles show the incorporation of 10 min pulses of [^{3}H]-lysine. (Reproduced with permission from Mittermayer *et al.*, 1966.)

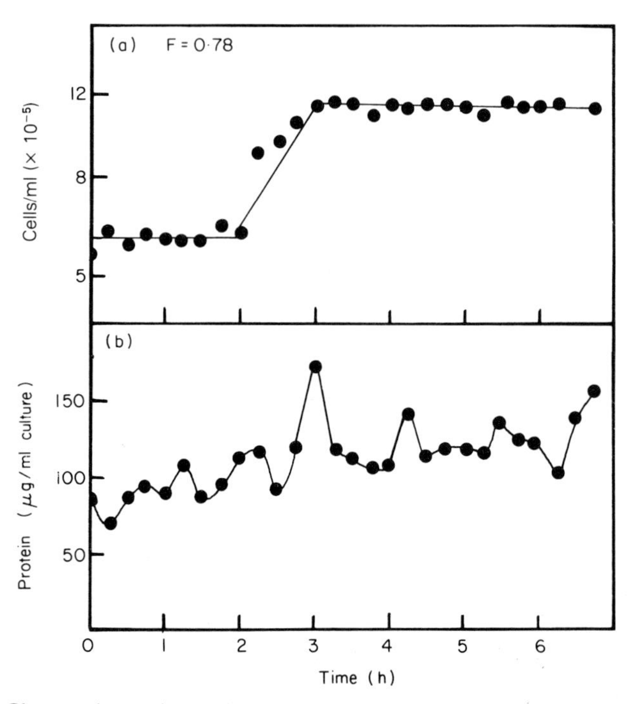

Fig. 4.6 Changes in total protein content in synchronously-dividing cultures of *D. discoideum*. The synchronous culture contained 13% of the original exponential culture. (a) Cell numbers and (b) total protein estimated by the Lowry method. Each point is the mean of two estimations. (Unpublished results of C. Woffendin and A. J. Griffiths.)

maximally about tenfold during one cycle. A continuous increase in protein synthesis was reported in *P. polycephalum*, with an increased rate in G_2 phase (Birch and Turnock, 1977). In the cellular slime mould *Dictyostelium discoideum*, total protein accumulates discontinuously (C. Woffendin and A. J. Griffiths, unpublished data), but doubles overall during one generation time (Fig. 4.6). In control cultures subject to the same experimental conditions smooth, continuous increases in cell numbers and total protein were observed. Evidently macromolecular synthesis is under a complex control mechanism in this organism.

Total protein accumulation in *Amoeba proteus* followed a similar pattern to "reduced weight" (Prescott, 1955), increasing for most of the cycle but with no net increase just before cell division. In *Acanthamoeba castellanii*, total protein accumulated discontinuously rising to seven maximal values in a generation time of 8 h (Edwards and Lloyd, 1980). Levels doubled overall during one cell cycle and the

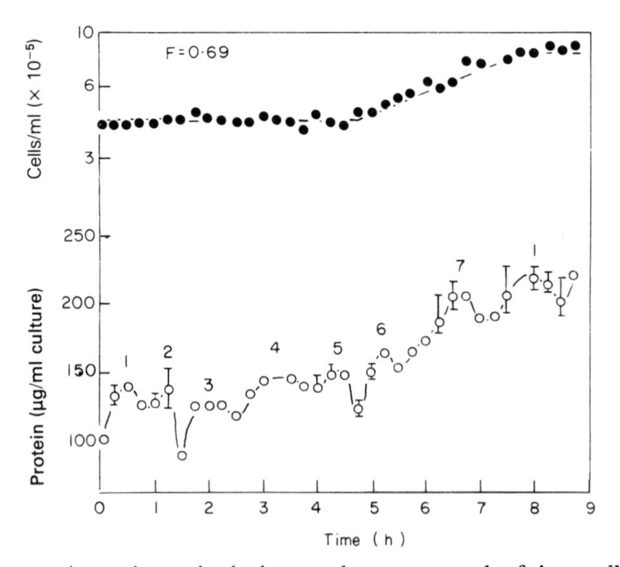

Fig. 4.7 Changes in total protein during synchronous growth of *A. castellanii*. The synchronous culture contained 6% of the exponential culture and shows (●) cell numbers with synchrony index F and (○) changes in total protein levels as μg/ml culture. Cell suspensions were washed twice with 10 vol. 50 mM $MgCl_2$ (pH 7·4) before estimation of total protein. Each point is the mean value of 3 separate protein estimations on each sample and the range of values obtained are indicated where these are greater than the diameter of the symbol. (Reproduced with permission from Edwards and Lloyd, 1980.)

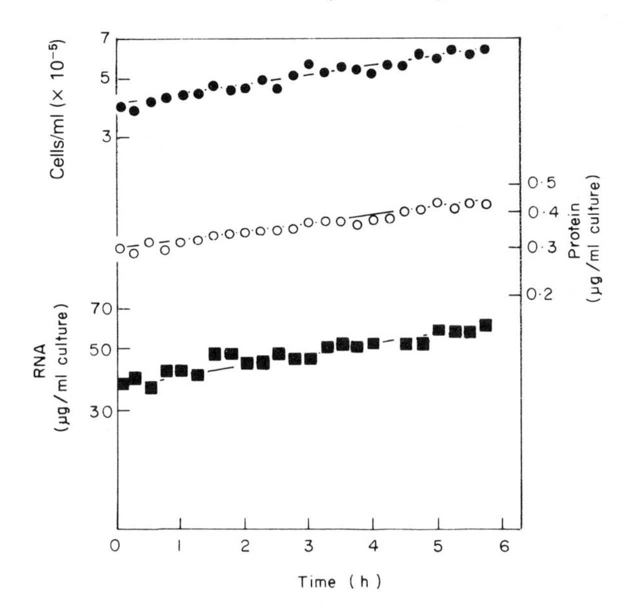

Fig. 4.8 Changes in cell numbers, RNA and protein levels after centrifugation and resuspension of exponentially-growing cultures of *A. castellanii*. An exponential culture was transferred to 4 sterile 50 ml stoppered tubes and centrifuged at 300 r min^{-1} for 2 min in the 4 × 50 ml head of an MSE bench centrifuge. After centrifugation the pellet and supernatant were resuspended and returned to the growth vessel. At 15 min intervals, samples were removed for cell counts (●) and measurements of RNA (■) and protein (○). (Reproduced with permission from Edwards and Lloyd, 1980.)

maximum amplitude (peak-trough, % minimal value) was 54% (Fig. 4.7). Oscillatory accumulation of protein was not observed in control, asynchronous cultures subject to identical procedures (Fig. 4.8), indicating that periodic synthesis and degradation is not a consequence of perturbation. The phase relationship (Figs 4.9 and 4.10) of timings of maxima of respiration rate, protein, ADP (all in phase), and RNA and ATP/ADP ratios (in phase with each other but out of phase with protein) suggest that the cell cycle of *A. castellanii* is temporally ordered into subcycles of biosynthesis and degration and that turnover of protein and RNA is extensive even in conditions of rapid growth. The phase relationship of protein and ATP/ADP ratios (high protein when ATP/ADP levels are low and vice versa) indicate that the changing biosynthetic requirements of the cell result in the ATP/ADP changes, rather than the latter controlling biosynthesis. Further work is necessary to determine whether this phenomenon is of widespread

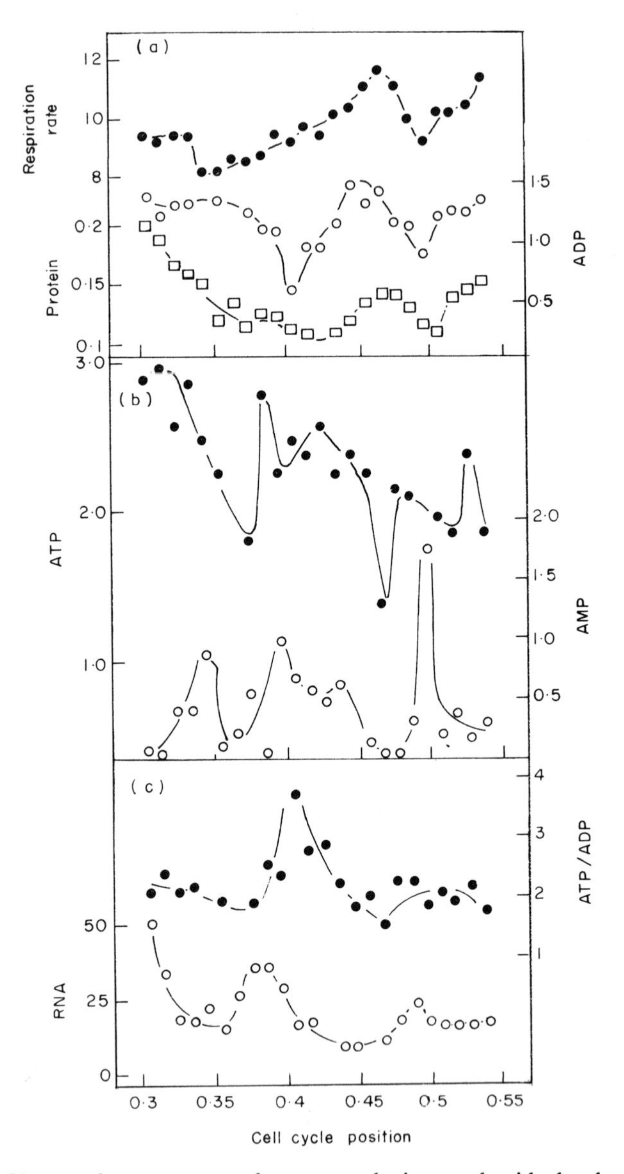

Fig. 4.9 Changes in oxygen uptake rates, adenine nucleotide levels, RNA and protein levels during the cell cycle of *A. castellanii*. A synchronous culture containing 13% of the exponential culture was prepared and growth followed for 9 h. After 2·5 h growth, samples were removed at intervals for 2 h for measurements of oxygen-uptake rates, adenine nucleotides, RNA and protein levels. (a) Shows (●) respiration rates (nmol O_2/min/ml culture), (○) ADP concentration (nmol/ml culture) and (□) total protein (mg/ml culture). (b) shows (●) ATP levels and (○) AMP levels, both as nmol/ml culture. (c) shows (●) ATP/ADP ratio and (○) RNA levels, (μg/ml culture). Data are presented in relation to the proportion of the cell cycle traversed at the time of sampling, assuming a linear representation of the cell cycle from 0–1·0. 0·1 of the cell cycle time represents 0·825 h. (Reproduced with permission from Edwards and Lloyd, 1980.)

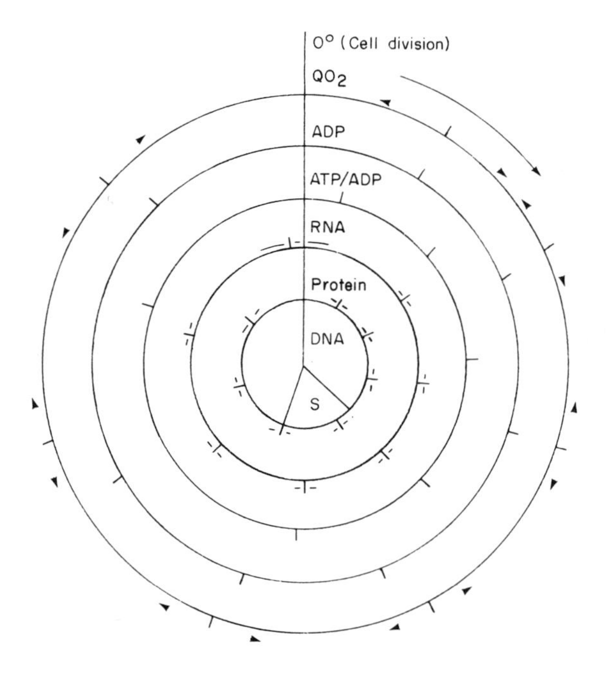

Fig. 4.10 Timings of cell cycle events in *A. castellanii*. The cell cycle is represented as a circle from 0° to 360°. 0° represents time zero in a synchronous culture and 360° the time when all the cells have completed division. Maxima of cell cycle events are thus shown as phase angles with respect to their timings in synchronous cultures. Maxima of Q_{O_2} values, total protein and RNA are shown as means, plus range of values obtained in 3 experiments. (Reproduced with permission from Edwards and Lloyd, 1980.)

occurrence, although similar observations have been made on a related (*D. discoideum*) and an unrelated (*C. utilis*) organism (see earlier).

(iv) *Algae*. In cultures of algae synchronized by alternating cycles of light and dark, the light phase is generally the period of cell growth while cell division accompanies or follows the transition to darkness. While cell division is well synchronized in these cultures, results may equally well be interpreted as either being cell cycle-dependent or illumination cycle-dependent. Consequently suitable control experiments must be carried out to distinguish between these possibilities.

In *Chlamydomonas reinhardii*, protein synthesis increases linearly in the light phase but remains constant in the dark phase (Iwanij *et al.*, 1975). The conclusion that the rate of protein synthesis in this organism is not constant comes from experiments where cells were pulse labelled with [$^{35}SO_4^{-2}$] or [^{3}H]-arginine (Howell *et al.*, 1977). Of 100 different polypeptides examined, 20 varied in their rate of labelling during the cycle by 3–12-fold. The overall variations in rates of polypeptide chain elongation were twofold, polypeptide chain initiation 25-fold, and overall rates of protein synthesis >25-fold (Baumgartel and Howell, 1977). However these authors point out that the rate of synthesis of proteins is greatly affected by the illumination conditions.

In light–dark synchronized cultures of *Chlorella*, protein increases continuously in the light but is constant in the dark, paralleling changes in cell volume (Takabaya *et al.*, 1976). During synchronous growth of *Chlorella pyrenoidosa* in continuous light after selection of cells by equilibrium sedimentation, total protein increases exponentially but remains constant just after DNA replication (Schmidt, 1975). Again, it is apparent that in this organism, rates of protein synthesis are very dependent on the intensity of illumination (Talley *et al.*, 1972). In *Cylindrotheca fusiformis* two-dimensional gel electrophoresis revealed that 208 out of 600 polypeptides detected showed changes in the rates of labelling during the cell cycle in light/dark synchronized cultures (Okita and Volani, 1980) but rates of labelling were affected by the illumination conditions.

(b) *Prokaryotes*. Total protein levels during synchronous growth of *Escherichia coli* obtained after filtration have been shown to increase exponentially (Abbo and Pardee, 1960). This result contrasts with that of Kubitschek (1968a) who found a constant rate of incorporation of radioactive precursors into protein, thus indicating a linear increase during the cycle. Nishi and Kogoma (1965) found that in this organism total protein increased through most of the cycle but was constant at the end of the cycle, even though incorporation of labelled leucine was continued. There was also no reduction in the capacity for enzyme induction at this time. It was proposed that protein synthesis was continuous throughout but that turnover increased at the end of the cycle. Continuous accumulation of total protein was reported during the cell cycle of *E. coli* both in synchronously-dividing cultures prepared after centrifugation though a sucrose gradient (Scott *et al.*, 1980) and on cell cycle fractionation (Scott *et al.*, 1980; Scott *et al.*, 1981) although the pattern of increase was unclear. However, Lutkenhaus *et al.* (1979) using two-dimensional gel electrophoresis at different times in the cycle found that no differential rates of synthesis of individual proteins (out of 750 polypeptides examined) could be shown.

Synchronous cultures of *Rhodopseudomonas spheroides* prepared by re-inoculation of stationary phase cultures into fresh growth medium showed an exponential increase in total protein (Ferretti and Gray, 1968) while an indirect protein estimation in a synchronous culture of *Lactobacillus acidophilus* (prepared by thymidine starvation and refeeding) suggested a linear increase (Burns, 1961). An exponential increase was observed in *Salmonella typhimurium* (Ecker and Kokaisl, 1969), the rate of protein synthesis being related to cell size.

B. ENZYMES

In constrast to the commonly-observed continuous accumulation of total protein, individual proteins, notably enzymes, show more complex patterns of accumulation through the cell cycle. The literature contains extensive reviews on patterns of enzyme synthesis during the cell cycles of a wide variety of cell types and proposals of possible control mechanisms responsible for the molecular basis of gene expression. Before such models can be viewed critically it must be established whether events observed during synchronous growth are genuinely associated with the cell cycle, or are artefacts resulting from metabolic perturbations caused by the method employed for synchrony. For this reason, many of the patterns of enzyme synthesis found in published lists (e.g. Halvorson *et al.*, 1971b; Mitchison, 1971) require scrutinization and re-examination as methods to prepare synchronous cultures have improved (see Chapter 2). Unfortunately, much of the earlier work, especially with mammalian cells, has been performed on synchronously-dividing cultures established by induction techniques and for reasons given earlier it is very doubtful if these cultures give useful information concerning the temporal sequence of events associated with the undisturbed cell cycle. Even with some of the selection methods currently employed, the possibility that perturbations have not been introduced cannot be overlooked or taken for granted. Ideally cells must not be removed from ideal growth conditions since transient periods of anaerobiosis (less than five minutes) are sufficient to induce high frequency metabolic respiratory oscillations (Pye and Chance, 1966), in yeast (Lloyd and Ball, 1979) or bacteria (Harrison, 1970). In these examples, detection of this phenomenon was only possible because a continuous readout of oxygen levels was obtained and where possible such techniques should be employed in cell cycle studies in order to detect the fine time-dependent structure.

Final acceptance of many of the cited examples of periodic enzyme synthesis requires further consideration as follows:

Table 4.3. *Enzyme activities during the cell cycles of prokaryotes*

Organism	Method of Analysis (S = synchronous culture)	Comments	Reference
Escherichia coli	Temperature induction (S)	Cytochrome *b* increased as 1 step/cycle.	Ohki (1972)
	Cell cycle fractionation	Specific activities of carboxypeptidase I, β-galactosidase and amidase constant but that of carboxypeptidase II rose at division.	Beck and Park (1976, 1977)
	Isoleucine starvation/feeding (S)	D-alanine carboxypeptidase activity maximum before cell division. Transpeptidase activity was maximal immediately after cell division.	Mirelman *et al.* (1978)
	Membrane elution	Cell wall-bound murein hydrolases oscillated whilst transglycosylase increased exponentially.	Hakenbeck and Messer (1977b)
	Filtration (S)	β-galactosidase induced continuously.	Abbo and Pardee (1960)
	Cell cycle fractionation or separation of cells on gradient (S)	ATPase activity increased discontinuously with two peaks/cycle at 0·37 and 0·8	Scott *et al.* (1980)
	Cell cycle fractionation	Cytochrome *o* accumulated continuously.	Scott *et al.* (1981)

		All cytochromes resolved in α-region by fourth order finite difference analysis accumulate continuously in succinate-grown cells.	Poole *et al.* (1980b) Scott and Poole (1982).
Rhodopseudomonas spheroides	Dilution of stationary phase cultures (S)	Succinate dehydrogenase and NADH-dehydrogenase activity increased as steps but cytochromes *b* and *c* synthesized continuously.	Wraight *et al.* (1978)
		Enzymes involved in synthesis of tetrapyrolle of bacteriochlorophyll increased discontinuously.	Ferretti and Gray (1968)
Bacillus subtilis	Filtration (S)	Succinate dehydrogenase activity increased as a step, while that of NADH dehydrogenase and malate dehydrogenase increases continuously.	Sargent (1973a)
Alcaligenes eutrophus	Continuous flow size-selection (S)	ATPase activity increases as 2 peaks/cycle with maxima at 0·4 and 0·9 and minima at 0·1 and 0·6.	Edwards *et al.* (1978)
Azotobacter vinelandii	Continuous phased cultures (S)	Nitrogenase activity rose gradually during most of cycle.	Kurz *et al.* (1975)
Myxobacter AL-I	Culture fractionation	Succinate dehydrogenase activity increases as a step.	Hartmann *et al.* (1977)
Nocardia restricta	Inoculation of stationary phase cultures into fresh medium (S) or sucrose gradient centrifugation	Adenylate cyclase activity high during DNA replication; cAMP-phosphodiesterase activity increases after DNA replication.	Lefebvre *et al.* (1980)

(a) The event or periodic marker should be absent in control cultures subject to experimental procedures identical to those used to produce the synchronous culture;
(b) The event should be repeated in subsequent cycles;
(c) Independent methods of analysis (e.g. synchronous cultures and cell cycle fractionation) should give essentially the same result.

The following section is not intended to provide an extensive and comprehensive account of enzyme synthesis during the cycle as reviews on this subject can be found elsewhere e.g. (Halvorson *et al.*, 1971b; Schmidt, 1974b, 1975; Halvorson, 1977; Klevecz, 1977; Mitchison, 1977a). Instead, some examples of different patterns of enzyme accumulation during the cell cycles of a wide range of cell types will be presented, as well as a brief survey of current models for regulation of enzyme activity. It is apparent that no single model fits all the existing data.

1. *Patterns of Enzyme Synthesis*

Synthesis of an enzyme during the cell cycle can follow one of a number of patterns (shown in Fig. 4.11). Gene transcription may or may not be continuous and similarly enzyme synthesis may be periodic or continuous.

Continuous enzyme synthesis may occur linearly, with a rate increase at a particular point in the cycle or exponentially, while periodic enzyme synthesis and/or degradation may result in "steps" (for stable enzymes) or "peaks" for unstable enzymes. The decline phase in the latter case is assumed to result from enzyme inactivation or breakdown. In practice it is difficult to distinguish between linear and exponential increases since there is maximally a difference of only 3% between an exponential curve that doubles in rate over a cycle time and the straight line of best fit (Mitchison, 1971). In the literature, examples of all these patterns of increase can be found in a variety of cell types and the published lists are updated by the examples shown in Tables 4.2 and 4.3. For periodic enzyme synthesis, there may be multiple "steps" or "peaks": increases (or decreases) in enzyme activity have often been assumed to represent increases (or decreases) in enzyme protein. Only in a few cases have attempts been made to correlate enzyme synthesis with enzyme protein. This is important because enzyme activity may be dependent on the presence or absence of specific activators or inhibitors, but there is evidence that in some cases increasing enzyme activity requires *de novo* synthesis since addition of protein synthesis

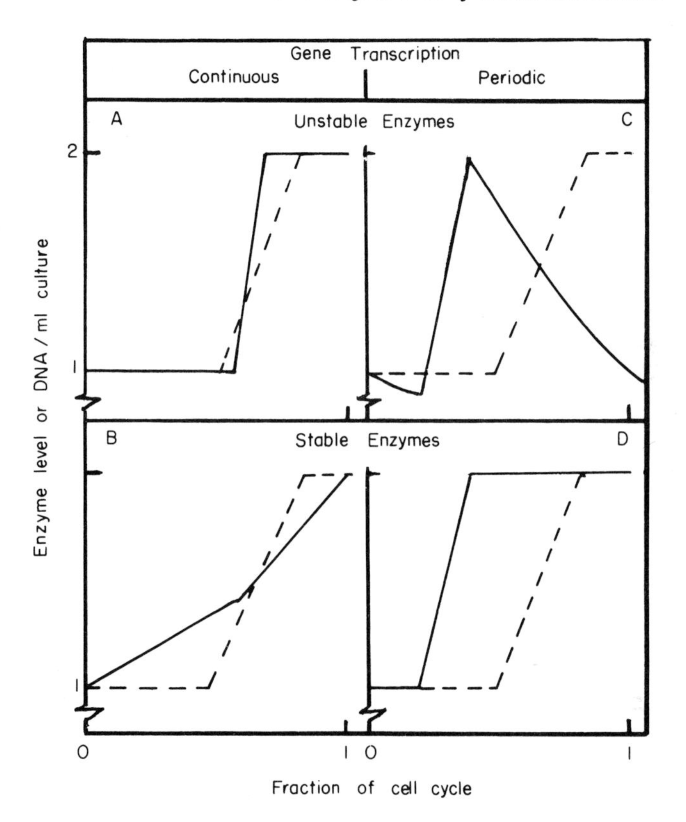

Fig. 4.11 Predicted patterns for enzyme synthesis *in vivo*. Gene transcription may be continuous or periodic and enzymes stable or unstable. — enzyme levels and - - - DNA levels, respectively. mRNAs are assumed to be non-labile and post-transcriptional factors non-limiting throughout the cell cycle. (Reproduced with permission from Schmidt, 1974b.)

inhibitors prevents an increase in activity (e.g. Lowdon and Vitols, 1973; Davis and Merrett, 1974). In synchronously-dividing cultures of *A. castellanii*, the accumulation of catalase was discontinuous so that during a generation time of 8 h catalase activity oscillated to give seven maxima per cell cycle (Edwards *et al.*, 1981). These changes in enzyme activity were paralleled by changes in immunologically-determined catalase protein (Fig. 4.12) and were absent in control cultures subjected to the same experimental conditions (Fig. 4.13). In contrast, oscillating ATPase activity in *E. coli* (Scott *et al.*, 1980) is accompanied

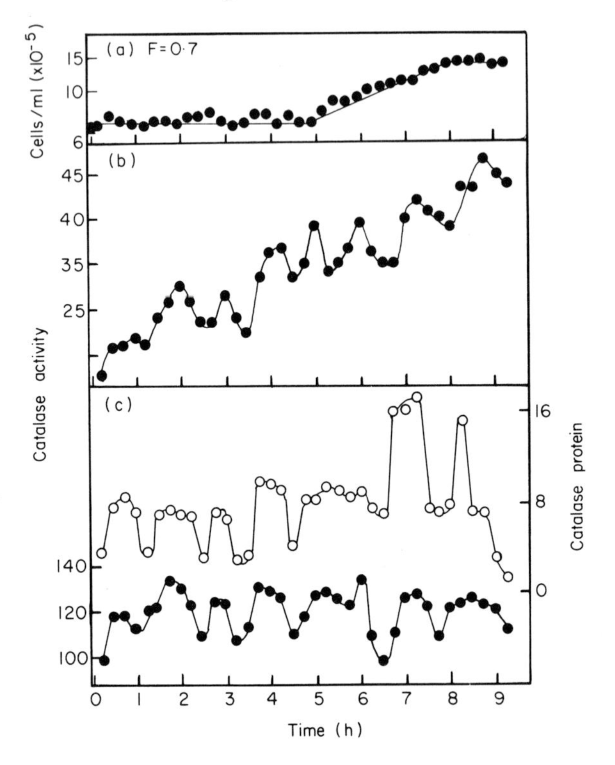

Fig. 4.12 Changes in catalase activity and catalase protein during synchronous growth of *A. castellanii*. An exponentially-growing culture ($4{\cdot}5 \times 10^6$ cells/ml) was centrifuged at 800 r min^{-1} for 2 min in the 6 × 1 l rotor of an MSE Mistral centrifuge. The supernatant contained 7% of the original culture. (a) Shows cell numbers, (b) shows catalase activity (mU/ml culture, ●) and (c) shows catalase activity (mU/mg protein, ●) and immunologically determined catalase protein (m*U*/mg protein, ○). (Reproduced with permission from Edwards *et al.*, 1981.)

by a continuous increase in immunologically-determined F_1 protein (see Chapter 6).

It is important to be able to describe the pattern of enzyme *synthesis* precisely, so that the validity of the various models proposed for the temporal control of gene expression can be evaluated. In most cases however, enzyme *accumulation* is measured and it is often assumed that this represents enzyme synthesis in the absence of turnover. As discussed earlier, this may be an over-simplification, as turnover (at least in some organisms) may be extensive. For interpretation of patterns of enzyme accumulation it is also important to ascertain the S-phase of the synchronous cultures so as to correlate increased enzyme

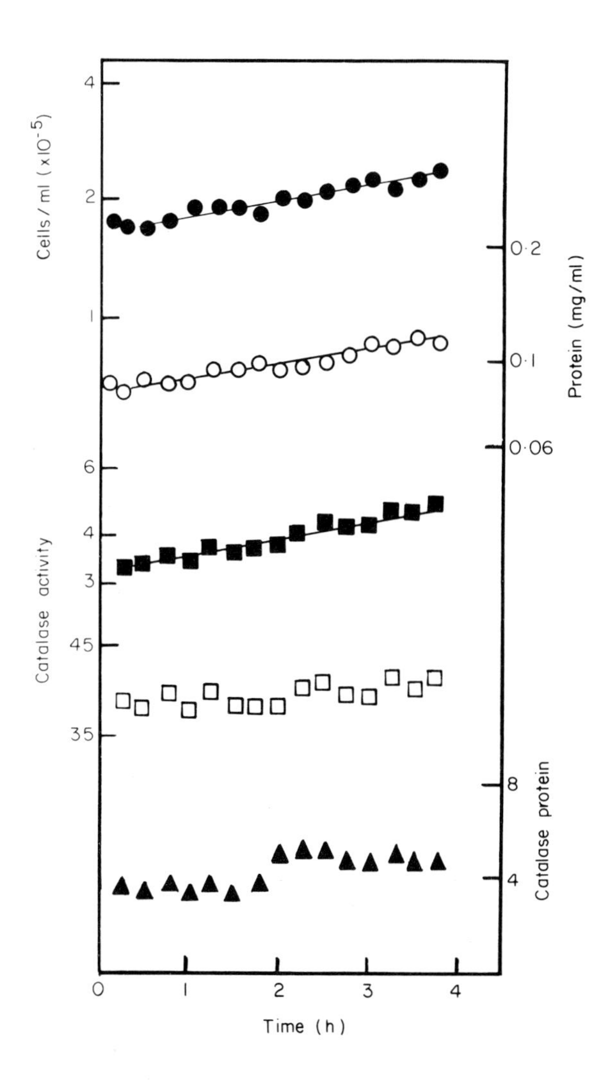

Fig. 4.13 Changes in cell numbers, protein levels, catalase activity and catalase protein after centrifugation of exponentially-growing cultures of *A. castellanii* (a synchronous control, c.f. Fig. 4.12). An exponential culture was centrifuged at 800 r min^{-1} for 2 min in the 6 × 1 l head of an MSE Mistral Centrifuge. After this the pellet and supernatant were re-mixed and returned to the growth vessel. At intervals, samples were removed for cell counts (●), protein (○), catalase activity expressed as m*U*/ml culture (■) and m*U*/mg protein (□) and immunologically-determined catalase protein, in *U*/ml culture (▲). (Reproduced with permission from Edwards *et al.*, 1981.)

amount with gene dosage. There is general agreement that in eukaryotes enzymes involved in DNA synthesis are synthesized periodically just prior to the onset of the S-phase, but other periodic enzymes may or may not be associated with specific stages of the cycle.

Also to be considered is whether enzyme activity as measured, is fully induced or repressed, or in a state somewhere between the two extremes (autoregulated). The differences in levels of some enzymes between the two extreme states is very dramatic, especially in prokaryotes where changes in enzyme levels from basal to fully induced can be as much as one thousand fold. This difference is likely to be an order of magnitude lower in eukaryotes (see Halvorson *et al.*, 1971b). These differences in prokaryotes and eukaryotes may be of importance in the control mechanisms postulated below.

Two approaches to the study of enzyme synthesis during the cell cycle can be adopted. First, cells can be grown synchronously in the absence of inducer (or presence of repressor) and the capacity for enzyme induction at different times in the cycle measured (enzyme potential). Secondly enzyme synthesis may be measured during synchronous growth in the presence of the inducer.

2. *Enzyme Potential*

Certain enzymes will be present at very low levels in cells cultured in the absence of an inducer (or presence of repressor). The capacity for induction of an enzyme can be measured by removing cells at different times during synchronous growth (in the absence of inducer) and incubating them under conditions of full induction or derepression. The potential for that enzyme is the maximum rate of enzyme accumulation under these conditions (Kuempel *et al.*, 1965; Schmidt, 1974). Usually enzyme potential is not limited to a particular cell cycle stage i.e. there is unrestricted potential (Halvorson *et al.*, 1971b). In prokaryotes usually remains constant for a portion of the cycle and then doubles sharply at a particular cell cycle stage (see Mitchison, 1971). There is some evidence in prokaryotes that the time of the doubling in enzyme potential is also that time when the structural gene for the enzyme is duplicated e.g. sucrase potential in *Bacillus subtilis* doubles at the same time as the sucrase transforming ability of the DNA (Masters and Pardee, 1965). There are also examples of unrestricted potential, doubling over a short period in eukaryotes, notably in *Chlorella* where the doubling occurs in the S-phase (Knutsen, 1965; Schmidt, 1974; Turner *et al.*, 1978). It is believed that in this case there may be a delay between gene replication and transcription. Before induced enzyme accumulation can be said to reflect availability of a structural gene for transcription, inducibility must be shown to be dependent upon both *de novo* RNA and protein synthesis (Schmidt, 1974). In *Schiz. pombe*, after induction synchronization, the stepwise increases in enzyme potential are in early S-phase (Sissons *et al.*, 1973); whilst in cultures synchronized by a selection method, the steps occur sometime after S phase (Mitchison and Creanor, 1969).

3. *Oscillatory Repression*

The model of oscillatory repression has been developed by a number of workers to account for periodic enzyme accumulation (Masters and Pardee, 1965; Goodwin, 1966, 1969c; Masters and Donachie, 1966; Donachie and Masters, 1969). Prerequisites for the application of this model are that there is end-product repression, that the system is partially derepressed (or induced) and that genes are available for transcription at all times. This hypothesis states that under favourable conditions, stable oscillations will be produced in a system where enzyme product represses synthesis of that enzyme by negative feedback. Thus, when a critical concentration of product is reached, enzyme synthesis is repressed; when the concentration of the product decreases during growth and falls below a critical concentration, enzyme synthesis again will be triggered. Oscillatory induction may also be envisaged as producing oscillations in enzyme levels by a similar mechanism. It has been proposed (Goodwin, 1966c) that a cell cycle dependent event couples these oscillations to the cycle so that events are repeated at the same time in successive cycles. If the enzyme is completely repressed (basal levels) or fully derepressed (or induced) continuous synthesis will be observed. Most data on prokaryotic enzyme synthesis fit this model; e.g. in *B. subtilis* sucrase activity is periodic when partially induced and basal levels show a continuous increase (Masters and Donachie, 1966). In eukaryotes there is little available evidence in support of this theory. Ribulose-1,5-diphosphate carboxylase accumulation in *Chlorella pyrenoidosa* is, however, continuous when cells are cultured under conditions of high light intensity, but shows a step pattern at lower illumination (Malloy and Schmidt, 1970). This finding may be interpreted as resulting from oscillatory repression by a photosynthetic product acting as a corepressor (Schmidt, 1974a, b). There are however, many more examples from studies with eukaryotes where the evidence does not fit the model (Halvorson, 1977).

The theory has recently been re-evaluated by Tyson (1979) who considers that the period of the oscillation in a negative feedback loop is much longer than the half-life of the most stable element in the loop. Thus a severe constraint is placed on the theory that stepwise increases in enzyme activity are caused by periodic repression of enzyme synthesis.

4. *Linear Reading*

This mechanism (also termed sequential transcription) has been proposed by Halvorson and coworkers based on experiments primarily with *Saccharomyces* spp. (see Halvorson *et al.*, 1971b; Halvorson, 1977). In this model, a gene is considered available for transcription only during a specific period of the cell cycle and genes are transcribed in the same order as their linear sequence on the chromosome. The model assumes that mRNAs are unstable and that there is a constant interval between the transcription of a gene and its translation and conversion to active enzyme (Halvorson *et al.*, 1971b). The model makes a number of predictions, some of which have been realized.

(a) The amount, but not the time, of enzyme expression is under environmental control. Basal and induced levels of α-glucosidase in *S. cerevisiae* vary by about fifty-fold but the increase in enzyme activity during the cycle occurs at the same cycle stage under both conditions (Halvorson *et al.*, 1966).

(b) For non-allelic genes there is only one period of enzyme synthesis per structural gene per cell cycle. Multiple steps may therefore result from multiplicity of non-allelic genes for the same enzymes. For example in a *S. dobzhanskii* × *S. fragilis* hybrid there are two antigenically distinct β-glucosidases and activity increases as two steps per cell cycle. This may be interpreted as representing two non-allelic genes transcribed at different points in the cycle (Gorman *et al.*, 1964; Halvorson *et al.*, 1966). Continuous enzyme synthesis may in fact be multiple steps (see later).

(c) Periodic enzyme synthesis should be observed at all times during the cell cycle.

(d) For genes on the same chromosome, the time interval between the expression of the gene in the cycle should be related to their linkage distance. Some evidence for this comes from Cox and Gilbert (1970). In *S. cerevisiae*, the distance on the chromosome between the genes for two enzymes are related to the time interval between the two steps in enzyme activity during the cycle.

The model has recently been modified (Halvorson, 1977) so that

(i) ordered transcription can be overridden by inducer and (ii) not *all* proteins in *S. cerevisiae* are subject to control by this mechanism. However, many data do not fit this model, especially those instances of continuous enzyme activity and unrestricted potential. Schmidt (1974b) has proposed that in this model the term periodic transcription should be replaced by "periodic or sequential processing of precursor mRNAs (i.e. different hn RNAs) or periodic transport of specific mRNAs from the nucleus during the cell cycle". It may be that sequential transcription acts only over restricted regions of the genome.

5. *Critical Point Control*

This control of enzyme synthesis was proposed by Mitchison and Creanor (1969) and suggests that the genome produced during the S-phase is unavailable for transcription until a critical point is reached (some time after DNA replication). In *Schiz. pombe*, the time of doubling in activities of sucrase, alkaline phosphatase, acid phosphatase and sucrase potential is in G_2 suggesting that there is a delay between "chemical replication" and "functional replication". This observation has been restricted to only a few examples.

6. *Post-transcriptional Controls*

Whereas in prokaryotes, enzyme synthesis is regulated mainly at the transcriptional level, in eukaryotes a number of mechanisms may be involved in the post-transcriptional control of enzyme synthesis. These include:

(a) Processing of precursor mRNA (e.g. hn RNA) into RNA;
(b) Transport of mRNA with poly A segments from nucleus into cytoplasm;
(c) mRNA degradation in nucleus and cytoplasm;
(d) Binding of regulator proteins to mRNA;
(e) Other forms of masking of mRNA;
(f) Attachment of mRNA to ribosomes requiring initiator proteins;
(g) Dependency of fate of polypeptides on class of ribosomes on which they are synthesized;
(h) Differential rates of translation of different polypeptides.

7. *Post-translation Controls*

These include mechanisms responsible for "*enzyme instability*":

(a) Loss of stabilizing ligands;
(b) Loss of enzyme protein by proteolytic action;
(c) Loss by release into the medium;
(d) Loss of activity resulting from change in ionic environment etc.;
(e) Change in turnover rate by change in activity of proteolytic enzymes;
(f) Abortive folding;
(g) Storage in vesicles e.g. lysosomal enzymes

or *enzyme activation*:

(a) Conformational change;
(b) Covalent modification e.g. phosphorylation, hydroxylation, glycosylation, cleavage;
(c) Combination of oligomeric subunits;
(d) Assembly into a membrane;
(e) Integration into multienzyme complexes.

Considerations which must also be made in the interpretation of patterns of enzyme accumulation include:

(a) Are the *in vivo* conditions accurately represented by the standard conditions used in the *in vitro* enzyme assay?
(b) Do pool sizes of substrates, products and effectors vary during the cycle?
(c) Are the enzymes involved in transport mechanisms synthesized periodically?

The mechanisms that are responsible for the temporal control of gene expression during the cell cycle are far from completely resolved. In prokaryotes, where enzyme synthesis is regulated primarily at the transcriptional level and responses to changing environmental conditions must be rapid, it may be that oscillatory repression is the important mechanism. In eukaryotes more complex mechanisms may be involved when e.g. subcellular compartmentation is considered. Also, it may be that some lower unicellular eukaryotes, which also need to respond rapidly to environmental changes, may have different mechanisms to higher eukaryotic cells which, in tissues, for example, are exposed to a more constant, homeostatically-controlled environment.

II. RNA

There are a number of RNA species found within a cell and in all cases ribosomal RNA (rRNA) constitutes the largest single species, in some cases accounting for up to 85% of the total RNA. This type of RNA is relatively stable and found associated with ribosomes. Transfer RNA (tRNA) is also found outside the nucleus while messenger RNA (mRNA) which is relatively unstable, is synthesized inside the nucleus and migrates to the cytoplasm. Messenger RNA may also contain long sequences of polyadenylic acid [poly (A)–mRNA]. A species of RNA found exclusively within the nucleus is heterogenous nuclear RNA (hnRNA). The role of this component is not entirely certain, but it is not related to tRNA or rRNA and contains pre-mRNA transcripts and other processing products, or RNAs which may function differently to mRNA (see review by Perry, 1976).

Total cellular RNA may be measured by chemical methods which give an estimate of *all* RNA species present at a given time. This, however, only gives a measure of accumulated RNA and does not account for variations in rates of synthesis and degradation that may occur. To measure the rate of synthesis of RNA it is usual to follow the rate of labelling of RNA species after exposure to radioactively-labelled precursors. Usually labelled uridine is used for this purpose, although adenine can be employed if measurements of DNA are made simultaneously. Again with such labelling procedures, rates of uptake of precursors must be interpreted with caution as variations in pool sizes may invalidate the assumption that uptake is directly proportional to synthesis. Ideally rates in incorporation of labelled precursors into the macromolecule under investigation should be followed to eliminate this possibility. Also to be considered in these experiments is the rate of entry of precursor molecules into the cell and there is some evidence that in rat hepatoma cells permeation of uridine and choline is the rate-limiting step in incorporation into macromolecules (Plagemann and Roth, 1969).

In pulse labelling experiments the length of the pulse determines which RNA species is labelled e.g. short pulses are incorporated into rapidly labelled RNA (e.g. mRNA, hnRNA) while longer pulses are incorporated into the more stable RNA species (e.g. rRNA). Problems of turnover are also very important and the half-lives of individual RNA species vary greatly. In yeast, where transcription of rRNA accounts for 60% of the total, 30% of all RNA synthesized is unstable (Shulman *et al.*, 1977). In mammalian cells, however, the unstable fraction (mostly hnRNA) may account for 75% of the total RNA (Soeiro *et al.*, 1968). The half-lives of mRNAs are also short and vary between individual messengers, cell types and cultural conditions. For example half-lives for mRNAs may be 1 h in mammalian cells (Borun *et al.*, 1967) or as short as 2 min in *B. subtilis* where 54% of the total label incorporated into RNA is found in the unstable fraction (Zingales and Colli, 1977). In *S. cerevisiae*, the half-life of poly(A)-RNA is 21 min (Venkov *et al.*, 1977) while that of invertase mRNA is 30–35 min when derepressed, but 45–50 min when fully repressed (Elorza *et al.*, 1977). Also, in *Neurospora crassa* as the growth rate increases, so the number of ribosomes per genome increases (Alberghina *et al.*, 1975). In yeast, however, the fraction of ribosomes functioning in protein synthesis falls from 84% in fast growing cells to 50% in slow growing cells (Waldron *et al.*, 1977).

After transcription, all RNA species undergo processing (see Perry (1976) for review), which may include:

(a) Nucleolytic reactions e.g. cleavage, trimming of large precursors;
(b) Terminal additions of nucleotides e.g. 3′-polyadenylation of mRNA, 5′-capping of mRNA;
(c) Nucleoside modification e.g. ribosome methylation.

In eukaryotes, a further consideration is that organelles such as mitochondria and chloroplasts contain their own RNA species (rRNA and possible mRNA) and this small but significant quantity must be accounted for in total cellular measurements. Over recent years, more elaborate techniques have been employed to attempt to study the patterns of individual RNA components during the cell cycle.

1. *Patterns of RNA Synthesis (accumulation) during the Cell Cycle*

Details of reports of the synthesis of both total RNA and individual RNA species during the cell cycles of a wide variety of cell types are given in Tables 4.4–4.7. Again it must be pointed out that much work has employed the use of induction-synchronized cultures, particularly treatment with DNA synthesis inhibitors, and the validity of these experiments as representing the undisturbed cell cycle must be questioned. Even so, it is apparent that no single mode of synthesis is found universally in all cell types.

(a) *Mammalian cells*. The general pattern that emerges from Table 4.4 is that RNA synthesis is continuous through the entire cell cycle except during mitosis, when synthesis stops (e.g. Feinendegen *et al.*, 1960; Prescott and Bender, 1962; Scharff and Robbins, 1965). More detailed studies have shown that RNA synthesis proceeds up to the end of prophase, ceases in metaphase (e.g. Feinendegen and Bond, 1963;

Table 4.4. *Changes in RNA during the cells cycles of mammalian cells in culture*

Cell type	Method of Analysis (S = synchronous culture)	Comments	Reference
HeLa	Autoradiography	RNA synthesis stops during mitosis.	Feindengen *et al.* (1960)
		$[^3H]$-Uridine incorporation stops completely during mitosis, $[^3H]$-histidine incorporation supressed.	Prescott and Bender (1962)
Chinese hamster cells		Most of initially-labelled RNA is bound to chromosomes. RNA synthesis until late prophase; label gradually becomes associated with cytoplasmic RNA. RNA synthesis resumes at telophase.	Feindengen and Bond (1963)
	Mitotic detachment (S)	Rate of RNA synthesis increases three-fold during interphase but no synthesis in mitosis.	Terasima and Tolmach (1963a)
		RNA synthesis lowered at mitosis.	Scharff and Robbins (1965)
		Incorporation of $[^3H]$-uridine maximal at 18 and 24 h and decreases in mitosis.	Kim and Perez (1965)
		Rate of synthesis increases slightly in G_1 and doubles in first half of S-phase. No pool fluctuations.	Pfeiffer and Tolmach (1968)

Table 4.4. *continued*

Cell type	Method of Analysis (S = synchronous culture)	Comments	Reference
		Synthesis of whole cell, nuclear pellet, nuclear supernatant, messenger, transfer and cytoplasmic rRNA occurs throughout interphase, doubling in rate in early S-phase.	Pfeiffer (1968) Hodge *et al*. (1969)
		Increased mitochondrial RNA synthesis in S and G_2.	England *et al*. (1974)
	Hydroxyurea (S)	Protein-bound poly (ADP ribose) polymerase maximum at mitosis and in mitotic chromosomes.	Tanuma *et al*. (1978)
	Double thymidine block (S)	Polysomal 7-9S RNA (mRNA for group A histone polypeptides) associated with polysomes in S-phase.	Borun *et al*. (1967)
		hnRNA synthesized in G_1, G_2 and S.	Pagoulatos and Darnell (1970)
		Polysomal units, monomers and polyribosomes increase in G_1 and decrease in S. Polysome formation predominant in G_1 and M.	Eremenko and Volpe (1975)

(Also Chinese hamster cells)	Selective detachment after thymidine block (S)	Nuclear RNA synthesis ceases in mitosis but ribosomal precursors (4S and 32S RNA) persist. Resumption of protein synthesis after mitosis is not dependent on RNA synthesis.	Fan and Penman (1971)
		Multiple RNA polymerases persist through mitosis and are involved in initiation of RNA synthesis in early telophase.	Simmons *et al.* (1973)
	Metaphase arrested	Little nuclear RNA synthesis in metaphase, but mitochondrial RNA synthesis occurs in mitosis.	Fan and Penman (1970b)
		Synthesis of 5S and 4S RNA in metaphase reduced to 74 and 33%, respectively of interphase rate.	Zylber and Penman (1971)
HeLa S_3	Double thymidine block (S)	Methylation of 40S subunit of cytoplasmic ribosome occurs mostly in late G_1, but methylation of 60S ribosome is mostly in early S-phase.	Chang *et al.* (1978)
Chinese hamster	Autoradiography	Chromosomal and nucleolar RNA synthesis low in G_1, increases in S and remains constant in G_2. Synthesis drops at end of prophase and ceases in metaphase.	Crippa (1966)
		Rate of RNA synthesis decreases in prophase.	Konrad (1963)
	Mitotic detachment (S)	No RNA synthesis at mitosis.	King and Barnhisel (1967)

Table 4.4. *continued*

Cell type	Method of Analysis (S = synchronous culture)	Comments	Reference
		Incorporation into 18S RNA and total RNA increases continuously in interphase.	Enger and Toby (1969)
		Rate of progression through cycle may be correlated with number of ribosomes.	Darzynkiewicz *et al.* (1979)
	Mitotic detachment plus colcemid (S)	In rich medium, RNA synthesis shows gene dosage effect but synthesis is continuous in minimal medium or rich medium containing high uridine concentrations.	Stambrook and Sisken (1972)
	Colcemid (S)	Incorporation into most rapidly sedimenting RNA greater in S and G_2 than in G_1 may show gene dosage effect but also fluctuations which may result from periodic synthesis of a particular class of RNA.	Klevecz and Stubblefield (1967)
	Detachment after colcemid treatment (S)	Polysomes absent in metaphase but reassembled after mitosis. A threefold increase in rate of protein synthesis after mitosis occurs independently of *de novo* RNA synthesis.	Steward *et al.* (1968)

L-cells	Mitotic detachment (S)	Rate of RNA synthesis low in G_1 and increases in S.	Fujiwara (1967)
	Autoradiography	Cytoplasmic RNA is 70–85% of total and accumulated throughout interphase.	Zetterberg (1966)
	Double thymidine block (S)	Poly (A) polymerase activity constant in G_1 and G_2 but degradative activity greater than synthesizing activity in S.	Muller *et al*. (1977)
Mouse fibroblasts	Function of relative cell age	Rate of RNA synthesis 2–2·5× higher at end of interphase. Increase in rate at the same time as doubling of DNA.	Zetterberg and Killander (1965a,b)
Erythroleukaemic mouse cells	G_1 arrested	Activity of poly (ADP ribose) polymerase increases 3–4-fold in G_1.	Rastl and Swetly (1978)
	Thymidine ± hydroxyurea (S)	Globin mRNA synthesis first detected in G_1.	Gambari *et al*. (1978)
L-929	Mitotic detachment (S)	RNA synthesis continuous through interphase. Peaks of nucleolar and non-nucleolar RNA synthesis in late S and G_2.	Gaffney and Nardone (1968)
Mouse lymphoma	Thymidine and colcemid (S)	No nuclear RNA synthesis at mitosis.	Doida and Okada (1967b)
Ehrlich ascites tumour cells	Autoradiography	Uptake of [^{14}C]-uridine continuous except in late prophase and telophase.	Baserga (1962)

Table 4.4. *continued*

Cell type	Method of Analysis (S = synchronous culture)	Comments	Reference
Baby hamster kidney	Aminopterin block and thymidine release (S)	Low molecular weight RNA synthesized at or just after DNA synthesis (3 h after release).	Clason and Burdon (1969)
	Double thymidine block (S)	DNA-like RNA is constant fraction of RNA synthesized at all times. RNA synthesis continuous through interphase.	Bello (1968)
		Some of the same DNA-like RNA species synthesized at 3 and 12 h of cycle.	Bello (1969)
P815Y	Cell cycle fractionation	Continuous synthesis of RNA, but rate may drop in G_2.	Warmsley and Pasternak (1970)

Crippa, 1966; Kessler, 1967), and then resumes again in telophase (Feinendegen and Bond, 1963). There may also be a reduction of synthesis in metaphase e.g. synthesis of 5S and 4S RNA are reduced to 74 and 33% respectively, of the interphase rates (Zylber and Penman, 1971). However, synthesis of mitochondrial RNA does not decline during mitosis (Fan and Penman, 1970b), and mRNAs (Steward *et al.*, 1968; Hodge *et al.*, 1969) and multiple RNA polymerases, which are involved in RNA synthesis in early telophase (Simmons *et al.*, 1973), persist during mitosis.

There is some debate as to whether the rate of RNA synthesis, which usually doubles during interphase, is dependent on a gene dosage effect. Some evidence points to a continuous increase in the rate of synthesis (e.g. Scharff and Robbins, 1965; Enger and Toby, 1969) while other observations suggest a gene dosage effect (e.g. Zetterberg and Killander, 1965b; Crippa, 1966). Stambrook and Sisken (1972) found that if Chinese hamster cells were grown in a rich medium, the rate of RNA synthesis doubled in S, whereas in a minimal medium or a rich medium containing high concentrations of uridine, a continuous increase in rate of synthesis was observed. This was thought to be brought about by incorporation of precursor into a variable pool, an observation which must be considered in all uptake measurements.

(b) *Eukaryotic micro-organisms*. Tables 4.5 and 4.6 list reports of RNA measurements during the cell cycles of eukaryotic micro-organisms. Again it is apparent that no single control mechanism fits all the data from different cell types. A pattern of RNA synthesis similar to that found in mammalian cells is observed in the naturally-mitotic cycle of *Physarum polycephalum* i.e. synthesis decreases or stops in mitosis (Kessler, 1967). Synthesis of tRNA and rRNA also stops at this time (Fink and Turnock, 1977), but continuous RNA synthesis is found during most of interphase (Braun *et al.*, 1966; Mittermayer *et al.*, 1964). However, there is also a decrease in synthesis in mid-interphase (Fig. 4.14; Mittermayer *et al.*, 1964), which corresponds to that time in the cycle when the rate of protein synthesis decreases (Mittermayer *et al.*, 1966). RNA synthesis then resumes at the end of anaphase (Kessler, 1967).

Continuous synthesis of RNA throughout the entire cycle has been reported for fission yeast (e.g. Mitchison *et al.*, 1969) or budding yeast (e.g. Sogin *et al.*, 1974). However, in *Schiz. pombe*, the rates of synthesis of rRNA and poly (A)–mRNA double just after DNA synthesis, which may suggest a gene dosage effect (Wain and Staatz,

Table 4.5. *Changes in RNA during the cell cycles of protozoa*

Organism	Method of analysis (S = synchronous culture)	Comments	Reference
Physarum polycephalum	Naturally-synchronous mitotic cycle (S)	[^{3}H]-uridine (5 min pulses) not incorporated in nuclear RNA during metaphase or anaphase, but low labelling evident if pulses are of 10 min duration. RNA synthesis resumes within 5 min of end of anaphase.	Kessler (1967)
		Steady decline in the number of polysomes relative to number of ribosomes after mitosis.	Mittermayer *et al.* (1966)
		Rate of rRNA synthesis low immediately after nuclear division and increases 5–6-fold during interphase.	Hall and Turnock (1976)
		tRNA and rRNA synthesized continuously except in mitosis.	Fink and Turnock (1977)
Tetrahymena pyriformis	Micromanipulation of dividing amicronucleate cells (S)	[^{14}C]-adenine incorporation into RNA lower in first half of interphase.	Prescott (1960)
	Heat shock (S)	Low rRNA in early part of cycle.	Hermolin and Zimmerman (1976)

		Synthesis of messenger-like RNA containing poly (A) is required for division after heat shock.	Yuyama (1975)
	Heat shock or starvation/ feeding (S)	Replication of genes for rRNA and 5S RNA follow pattern for bulk macromolecular genes.	Tonnesen and Andersen (1977)
		Genes coding for rRNA (rDNA) replicated in early S-phase.	Andersen and Engberg (1975)
		Increase in rate of rRNA synthesis may show gene dosage effect.	Keiding and Andersen (1978)
		Rate of rRNA synthesis dependent on population density. At high cell density, cells secrete a factor repressing synthesis.	Andersen and Nielsen (1979)
	Continuous flow size-selection (S)	Continuous increase of total RNA.	Phillips and Lloyd (1978)
	Cytophotometry	RNA per cell increase in S and G_2 greater than in G_1. RNA: protein decreases from division mid S, increases late S and then remains constant.	Cleffmann *et al.* (1979 a,b)
Euplotes	Autoradiography	No RNA synthesis at DNA synthesis.	Prescott and Kimball (1961)
Euplotes eurystomus	Micromanipulation of dividing cells (S)	Rate of RNA synthesis of micronucleus increases 2–4-fold during G_1, but declines early S-phase. Large amount of stable, non-migrating RNA in macronucleus.	Evenson and Prescott (1970)

Table 4.5. *continued*

Organism	Method of analysis (S = synchronous culture)	Comments	Reference
Paramecium aurelia	Isolation of single cells to initiate synchronous culture (S)	Rate of synthesis of cytoplasmic RNA gradually increases and is maximal just before cell division.	Woodard *et al.* (1961)
	Autoradiography	Precursors initially incorporated into nuclear RNA, then into cytoplasmic RNA. Rate of incorporation per unit volume constant in interphase.	Kimball and Perdue (1962)
Paramecium caudatum	Autoradiography	Constant incorporation of RNA precursors into macronucleus throughout cycle. Incorporation into micronucleus restricted to S phase.	Rao and Prescott (1967)
Crithidia fasciculata	Sedimentation-velocity size-selection (S)	Total RNA increases continuously.	Edwards *et al.* (1975)
Acanthamoeba castellanii	Low speed centrifugation in growth medium (S)	Total RNA accumulates discontinuously.	Edwards and Lloyd (1980)
Dictyostelium discoideum		Total RNA accumulates discontinuously.	C. Woffendin and A. J. Griffiths unpublished results

Table 4.6. *Changes in RNA during the cell cycles of yeasts and algae*

Organism	Method of analysis (S = synchronous culture)	Comments	Reference
Saccharomyces cerevisiae	Cell cycle fractionation	Rate of rRNA synthesis increases exponentially and total RNA accumulated continuously.	Sogin *et al.* (1974)
	Ludox centrifugation	Continuous synthesis of ribosomal proteins.	Shulman *et al.* (1973)
	Cell cycle fractionation	Specific activities of nuclear RNA polymerase I constant through cycle while that of nuclear RNA polymerase II increases at beginning of G_2.	Sebastian *et al.* (1974)
	Zonal centrifugation	Rate of synthesis of poly (A)-containing RNA is continuous.	Hynes and Phillips (1976)
	Velocity-sedimentation	RNA polymerase I (possibly rRNA polymerase) constant. RNA polymerase II (possibly mRNA polymerase) increases for initial 1/3 of cycle then decreases.	Carter and Dawes (1975)
		Rate of rRNA synthesis constant.	Fraser and Carter (1976)
	Centrifugal elutriation	Amount of RNA increases exponentially.	Elliott and McLaughlin (1978)

Table 4.6. *continued*

Cell type	Method of Analysis (S = synchronous culture)	Comments	Reference
Schizosaccharomyces pombe	Function of cell length	Continuous increase in RNA.	Mitchison and Walker (1959)
		Rate of incorporation of precursor continuous but may be constant at end of cycle.	Mitchison and Lark (1962)
	Selection from sucrose gradient (S)	rRNA and rapidly-labelled nuclear RNA synthesis increases exponentially.	Mitchison *et al.* (1969)
		Rates of synthesis of rRNA and poly (A)-mRNA double just after DNA replication. Doubling in rate of synthesis of total RNA independent of [uridine] in range 0·03–820 μM.	Wain and Staatz (1973)
		poly $(A)^+$-mRNA increases as 1 step per cycle in each of three mutants of different cell size. Cell size-related control of the rate of synthesis maintains the mean value at a constant proportion of cell mass.	Fraser and Moreno (1976) Fraser and Nurse (1978) Barnes *et al.* (1979) Fraser and Nurse (1979)

Chlorella pyrenoidosa	Equilibrium sedimentation after entrainment by light/dark shifts but grown in continuous light (S)	Total RNA increases exponentially but remains constant just after DNA replication.	Schmidt (1975)
	Light : Dark (S)	Continuous increase of incorporation of label into RNA paralleling increase in cell volume.	Takabaya *et al*. (1976)
C. ellipsoidea		Rate of synthesis of chloroplast RNA is higher than that of cytoplasmic RNA in early stages of the cycle.	Hirai *et al*. (1979)
Chlamydomonas reinhardii		20–30% of chloroplast ribosomes are bound to thylakoid membrane in light.	Chua *et al*. (1976)
Platymonas striata		RNA synthesis only in light.	Ricketts (1977a,b)
Navicula pelliculosa		RNA only increases in light.	Darley *et al*. (1976)

1973). In *Chlorella*, RNA synthesis increases exponentially, but remains constant just after DNA replication (Schmidt, 1975). In *Paramecium caudatum*, there is continuous incorporation of RNA precursor into the macronucleus at all times in the cycle, but incorporation into the micronucleus is restricted to the S-phase (Rao and Prescott, 1967).

A more complex pattern of accumulation of total RNA was observed during the cell cycle of *Acanthamoeba castellanii* (Edwards and Lloyd, 1980). In a generation time of 8 h, RNA levels fluctuated (with a maximum peak-to-trough amplitude of 90%) to give seven maxima per cycle (Fig. 4.15). These oscillations were not apparent in asynchronous control cultures (Fig. 4.8) and provide further evidence that the cell cycle of this organism is temporally-ordered into subcycles of biosynthesis and degradation (see p. 164). Similar changes in total RNA are observed in the cell cycle of *Dictyostelium discoideum* (C. Woffendin and A. J. Griffiths, unpublished results) where levels vary with a maximal amplitude of 80% (Fig. 4.16).

(c) *Prokaryotes.* Measurements of RNA during the cell cycles of prokaryotes are listed in Table 4.7. Since these organisms are structurally simpler than eukaryotes and sometimes exhibit continuous DNA synthesis, one might expect that RNA synthesis during their cell cycles would be an attractive system for study. Despite this, there are surprisingly few reports to date. The general picture is that RNA synthesis is continuous through the cycle.

III. Membranes and Cell Walls

Cell membrane and wall growth must accompany overall cell growth throughout the cycle. However, at cell division a greater and more abrupt increase in membrane area must occur.

The mode of development of surface membranes and cell walls is described in detail in Chapter 5. Reports of the accumulation of some enzymes associated with these structures are presented in Tables 4.2 and 4.3 whilst Tables 4.8 and 4.9 summarize results of studies of the syntheses of lipids in both eukaryotes and prokaryotes. In the case of surface structures, turnover may be a particularly important contribution to the measured extent of accumulation. For example, in *Amoeba proteus*, the turnover rate of membranes is as great as 0·2% per min (Wolpert and O'Neill, 1962).

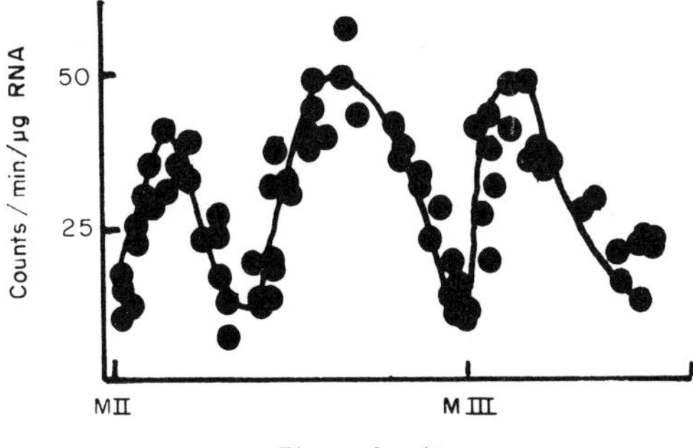

Fig. 4.14 Rate of RNA synthesis in *Physarum*. Incorporation of 10 min pulses of [^{3}H]-uridine into RNA at different times during the mitotic cycle MII and MIII are 2nd and 3rd mitoses, separated by 8–10 h. (Reproduced with permission from Mittermayer *et al.*, 1964.)

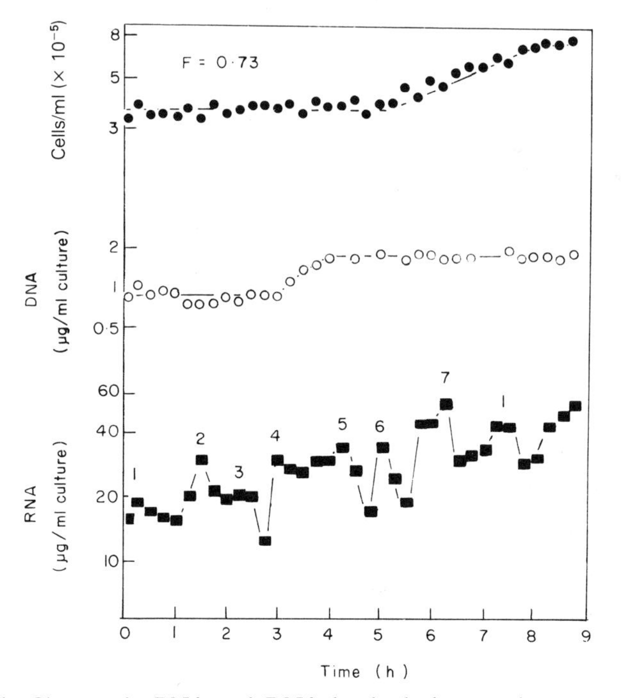

Fig. 4.15 Changes in RNA and DNA levels during synchronous growth of *A. castellanii*. The synchronous culture contained 12% of the exponential culture. (●) Cell numbers with synchrony index F, (○) DNA and (■) RNA. After the final hot trichloracetic acid extraction the supernatant was assayed for RNA, while the insoluble fraction was assayed for DNA. Positions of seven maxima indicated (1–7) (Reproduced with permission from Edwards and Lloyd, 1980.)

Table 4.7. *Changes in RNA during the cell cycles of prokaryotes*

Organism	Method of analysis (S = synchronous culture)	Comments	Reference
Escherichia coli	Cell cycle fractionation	Rate of synthesis of stable RNA varies linearly with size of cell.	Manor and Haselkorn (1967)
	Filtration (S)	Exponential increase in total RNA.	Abbo and Pardee (1960)
		Two "bursts" of RNA synthesis (probably rRNA) when 40 and 80% of DNA replicated.	Rudner *et al.* (1965)
		Periodic RNA synthesis.	Rudner *et al.* (1964)
	Stationary phase cultures re-inoculated (S)	Entire genome continuously transcribed but different sections show characteristic changes.	Cutler and Evans (1967b)
Salmonella typhimurium	Function of cell size	Delay in RNA synthesis after cell division. Linear increase in rate of synthesis.	Ecker and Kokaisl (1969)

Myxococcus xanthus	Autoradiography	Stable and unstable RNA synthesized throughout the cycle. Rate of synthesis of stable RNA increases in two steps (0·15 and 0·75 of the cycle) while that of unstable RNA increases just before cell divisison; $t_{\frac{1}{2}}$ = 4 min (gen. time = 390 min).	Zusman and Rosenberg (1971)
Caulobacter crescentus	Plate selection method (S)	$t_{\frac{1}{2}}$ mRNA is 0·4–5·8 min.	Iba *et al.* (1978)
Anacystis nidulans	Light/Dark (S)	Rate of incorporation of label into RNA is maximal in mid-cycle. Total RNA increases as 1 step/cycle.	Asato (1979)
Rhodopseudomonas spheroides	Stationary phase cells re-inoculated (S)	Rate of incorporation of precursors into RNA is discontinuous.	Ferretti and Gray (1968).

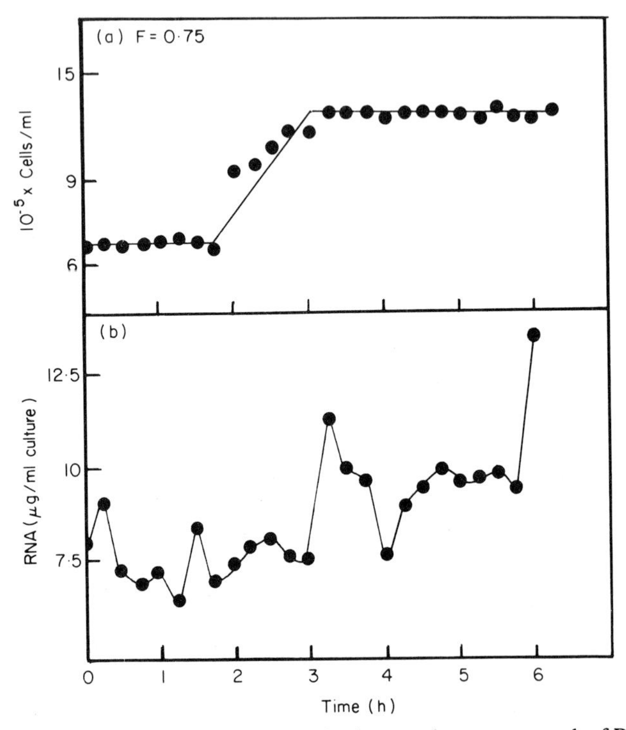

Fig. 4.16 Changes in total RNA content during synchronous growth of *D. discoideum*. (a) Cell numbers; (b) RNA by the orcinol method. (Unpublished results of C. Woffendin and A. J. Griffiths.)

Summary

There is an increasing body of evidence which suggests that turnover of cellular constituents is an important feature of growth under optimal conditions. This is particularly true for mammalian cells or eukaryotic micro-organisms with fairly long generation times. Protein degradation can be detected in cells which have high rates of protein synthesis and may be a large proportion of the total protein accumulated. In some cells e.g. *Acanthamoeba castellanii* these processes of synthesis and degradation are temporally sequenced into periods where protein-synthesizing reactions predominate followed by periods of the cycle where protein-degrading reactions predominate. This temporal switching may be necessary to isolate these opposing processes in *time* rather than in space i.e. in separate subcellular compartments. These

Table 4.8. *Accumulation of other cellular components during the cell cycles of eukaryotes*

Organism or cell type	Method of analysis (S = synchronous culture)	Comments	References
Lymphocytes	Cell cycle fractionation	2–3× increase in glucocorticoid receptor sites per cell in S and post-S cells.	Crabtree *et al.* (1980)
HeLa	Double thymidine block (S)	Degree of tubulin phosphorylation 2–3× higher in S and mitosis.	Piras and Piras (1975)
		Polyamine content affected by synchronization method.	Goyns (1980)
P815Y	Cell cycle fractionation	Continuous phospholipid synthesis, but rate of synthesis drops in G_2.	Warmsley and Pastnernak (1970)
		Carbohydrate: protein high in mid G_1 and G_2, low in S.	Graham *et al.* (1973)
3T3 Mouse fibroblasts	Serum stimulation of arrested cells	^{31}[Pi]incorporation into phosphatidylethanolamine and phosphatidylcholine maximal in G_1 but incorporation into phosphatidylinositol increases steadily up to S-phase.	Dubois and Rampini (1978)
Neoplastic Mast cells	Cell cycle fractionation	Rate of incorporation of labelled choline and inositol into nuclear, mitochondrial or microsomal phospholipids follows that of total phospholipids i.e. continuous through most of intermitotic period.	Warmsley *et al.* (1970)

Table 4.8. *continued*

Organism or cell type	Method of analysis (S = synchronous culture)	Comments	References
C1300 Mouse neuroblastoma cells	Mitotic detachment (S)	Diffusion coefficients for membrane lipids increases in G_1.	De Laat *et al.* (1980)
Chlorella ellipsoidea	Light/Dark (S)	Starch accumulated in light phase, but levels decreased in dark.	Takeda and Hirokawa (1978)
Chlamydomonas reinhardii		Decline in synthesis of 35 000–45 000 mol.wt. proteins in late light and early dark phases. Little or no tubulin produced in late dark phase—mid-light phase. Tubulin synthesis accounts for 5–10% of total protein synthesized during cell division.	Weeks and Collis (1979); Weeks *et al.* (1977)
		Chlorophyll content increases in mid-light phase and remains constant in dark. Chloroplast thylakoid membrane undergoes a series of alterations i.e. insertion of chlorophyll and cytochromes at different stages of the cycle.	Schor *et al.* (1970)
		Chloroplast components (chlorophyll, cyt_{553}, cyt_{563} ferredoxin, ferredoxin–NADP	Armstrong *et al.* (1971)

reductase) all increase after 4 h growth in light; no increase in dark.	
Incorporation of newly-synthesized polypeptides into unstacked thylakoid membranes is continuous but highest in G_1.	Bourguignon and Palade (1976)
Chlorophyll increases as a step from 4–8 h in light.	Iwanij *et al.* (1975)
Chlorophyll synthesis is maximal from 6–12 h in light. Variations in the rate of labelling of four major polypeptides of chloroplast membrane in the light.	Beck and Levine (1974)
All major carotenoids increased through most of light phase. Lutein and violaxanthin increases precede that of β-carotene. Drop in total carotenoids at 9 h.	Francis *et al.* (1975)
Phosphatidylglycerol synthesized between 3–4 h light, sulpholipid labelled between 7–9 h light, galactolipid shows maximal labelling at beginning of light and 7 h. Chlorophyll is synthesized maximally after 7 h light.	Beck and Levine (1977)

Table 4.8. *continued*

Organism	Method of analysis (S = synchronous culture)	Comments	Reference
		Major cytoplasmically-synthesized polypeptides of thylakoid membrane were synthesized almost exclusively during latter half of light phase.	Michaels *et al.* (1980)
Scenedesmus obliquus		Starch content low at 8 h, but 4× greater at 16 h.	Senger and Bishop (1979)
		Chlorophyll synthesis is from 2–8 h.	Myers and Graham (1975)
Candida utilis	Phased cultures (S)	Some fluctuations in levels of triacylglycerols, free steroids, phospholipids and esterified sterols, but phospholipid:sterol molar ratio varies from 8·5 (at 0·5 h) to 5·7 (at 2·5 h).	Hossack *et al.* (1979)
Platymonas striata	Light/Dark (S)	Synthesis of carbohydrates, chlorophyll *a* and *b*, carotenoids and phospholipids only in the light; maximum rates of synthesis at end of light.	Ricketts (1977b)
Navicula pelliculosa		Carbohydrates, lipids, total carotenoids, chlorophyll *a* and *c* all increase in light. In the dark, carbohydrates decrease, lipids	Darley *et al.* (1976)

		remain constant, chlorophyll *a* increases at reduced rate, chlorophyll *c* remains constant then increases.	
Euglena gracilis	Re-inoculation of stationary phase cultures (S)	Major polyamines maximal at end of mitosis.	Villanueva *et al.* (1980)
Porphyridium purpureum		Cellular starch levels increase in light and decline to original levels during the dark when cells divide.	Sheath *et al.* (1977)
Tetrahymena pyriformis	1 heat shock/cycle (S)	Mitochondrial and microsomal lipids labelled maximally just before cytokinesis, but lipids of cilia and ciliary supernatant labelled maximally during division. Protein incorporation into membranes preceeds lipid incorporation, i.e. selective membrane extension during the cycle.	Baugh and Thompson (1975)
Saccharomyces cerevisiae	Not specified	Total phospholipid and phosphatidylethanolamine increase as steps while phosphatidylinositol and phosphatidylserine increase as peaks at bud formation. May be periodic synthesis of phospholipids followed by continuous accretion into membranes.	Cottrell *et al.* (1978)

Table 4.9. *Accumulation of other cell components during the cell cycles of prokaryotes*

Organism	Method of analysis (S = synchronous culture)	Comments	Reference
A. LIPIDS			
Escherichia coli	Temperature-induced synchrony (S)	Phospholipids increases continuously but with stepwise loss of radioactivity from phosphatidylglycerol due to turnover.	Ohki (1972)
	Methionine starvation and refeeding (S)	Incorporation of [2-^{3}H]- or [1-^{14}C] glycerol into lipid/mg dry weight increases at cell division.	Daniels (1969)
	Cell cycle fractionation on sucrose gradients	Rate of incorporation into lipids is highest in large cells, lowest in small cells.	Daniels (1969)
	Membrane elution (S)	Stepwise changes in the rate of synthesis of phosphatidylethanolamine and total phospholipid.	Pierucci (1979) (see also Fig. 5.20)
		Exponential increase in rate of phospholipid synthesis.	Churchward and Holland (1976a)
		Rates of synthesis of phospholipids oscillate.	Hakenbeck and Messer (1974, 1977a)
	Dilution of stationary phase cultures (S)	Rate of synthesis of phospholipids is constant.	Bauza *et al.* (1976)
	Selection of cells from sucrose gradient (S)	Incorporation of free lipoprotein into outer membrane is maximal just before cell division.	James and Gudas (1976)

Rhodopseudomonas spheroides	Reinoculation of stationary phase cells (S)	Continuous incorporation of polypeptides into cytoplasmic membrane but discontinuous phospholipid incorporation giving rise to variations in cytoplasmic membrane density.	Lueking *et al.* (1978) Fraley *et al.* (1978) (see also Chapter 5)
Bacillus megaterium	Threonine starvation and refeeding (S)	Rate of incorporation of [2-^{3}H]glycerol into lipids/mg dry weight increases at cell division.	Daniels (1969)
Bacillus licheniformis	(S)	Cyclic variations in rate of [^{32}P] incorporation.	Lubochinsky and Burger (1969)
Anacystis nidulans	Light/Dark (S)	Rate of incorporation of ^{32}P into phospholipid is maximal at cell division and mid cycle.	Asato (1979)
Caulobacter crescentus	Differential centrifugation (S)	Rate of synthesis of phospholipids declines during the cycle.	Galdiero (1973b)
		Patterns of membrane phospholipid syntheses vary in different cell types.	Mansour *et al.* (1980)
B. OTHERS			
Bacillus subtilis	Filtration (S)	Phospholipid synthesis is continuous.	Sargent (1973a)
Caulobacter crescentus	Plate selection method (S)	Rates of synthesis of total protein and half of "major proteins" increase in G_1 and S and are constant in G_2. Rates in swarmer and stalked cells different.	Iba *et al.* (1978)
E. coli	Cell cycle fractionation	Continuous increase in four e.p.r. detectable signals.	Poole *et al.* (1981) (See also Chapter 5)

rapidly-degraded proteins have not been identified but may include abnormal proteins (i.e. products resulting from errors in translation or transcription) or key rate-limiting enzymes of metabolic pathways where regulation of activity and hence pathway flux may be controlled.

Some of the problems of protein turnover also (potentially) apply to RNA turnover, although the vast majority of cellular RNA is ribosomal, a somewhat more stable species than the short lived mRNAs or Hn RNA. However, if protein turnover is extensive, then one may expect a rapid and extensive turnover of at least some species of RNA.

The patterns of enzyme synthesis through the cell cycles of a variety of cell types that have been reported include periodic and continuous accumulation. Only a few reports have attempted to correlate enzyme activity and enzyme protein. Enzyme turnover is also important here since what is usually assayed is enzyme accumulation, which is a balance between enzyme synthesis and enzyme degradation. The possibility that periodic enzyme synthesis may reflect perturbations may be a consequence of the procedures employed for establishment of synchronous growth cannot be neglected, even in some cases where selection synchrony has been used.

5. Development of Sub-cellular Structures

"Remorselessly bound to the forward direction of time".

J. Bronowski
The Ascent of Man

I. Introduction

The most obvious manifestations of morphogenesis and organelle development that occur during the cell cycle are the growth in cell volume and surface area in interphase. At the subcellular and sub-organellar levels, however, equally dramatic and important changes occur, only some of which, notably those of the nucleus, have been studied at the molecular level. This chapter describes the development of boundary cell layers and organelles during the division cycles of both eukaryotic and prokaryotic cells.

The structural complexity of both lower and higher eukaryotic cells makes them a rich field of study. Cycles of organelle growth and division, tightly coordinated with the growth and division of the cell, have been described in detail for many organelles, for example nuclei, the oral apparatus of *Tetrahymena*, and mitochondria.

The degree to which cycles of organelle growth and division are controlled with respect to population size and intracellular distribution is unknown. In an interesting exercise to study the possibility that there is no cellular control, Shonkwiler (1977) has shown that the likelihood of a cell, normally having 4–8 mitochondria, arising devoid of mitochondria either by "extinction" or by maldistribution of organelles, is vanishingly small. Using these values for mitochondrial numbers per cell, chance alone appears to be adequate in explaining the inclusion of mitochondria in both daughter cells at division. However, as we shall see (p. 218), certain cell types have fewer mitochondria than this. Under these circumstances, the probability will increase that a cell may arise devoid of mitochondria. A reassessment of the extent of cellular control over organelle duplication appears timely.

Although it is likely that such developmental cycles will be firmly established for most, if not all, organelles, there exist instances where there is little evidence for a duplication cycle and where the possibility of *de novo* synthesis in each cell cycle must be considered. Thus, in *Chlorella*, division stages of the single microbody were never observed by Atkinson *et al.* (1974) and it was considered likely by these authors that both microbodies and pyrenoids arose *de novo* in daughter cells. There is, however, some evidence for division of microbodies in *Candida utilis*; high voltage electron microscopy of thick sections has revealed small, well-contrasted dense bodies in a "dumbbell" configuration (M. T. Davison, personal communication).

For many organelles, there appears to be only scattered information in the literature on their development in the cell cycle, a situation that has changed little since Mitchison's (1971) review of the field, and such cases will not be described here. The development of the endoplasmic reticulum (Moor, 1967), Golgi apparatus and dictyosomes (e.g. Nilshamma and Walles, 1974; Ueda and Noguchi, 1976; Menge and Kiermayer, 1977; Okamura *et al.*, 1977; Noguchi, 1978; Quintart *et al.*, 1980), flagella and basal bodies (e.g. Cavalier-Smith, 1974; Oakley and Bisalputra, 1977; Matthys-rochon, 1979), microplasmadesmata (Giddings and Staehelin, 1978) and intracellular symbiotic algae (Weis, 1977) have all been studied but integrated descriptions of development at the biochemical, morphological and physiological levels are not available.

Virtually nothing is known of the interactions between organelles during cell growth and division. Although examples of physical contact between organelles may be cited, for example between chloroplast and mitochondrion (Osafune *et al.*, 1972a,b), between nucleus and vacuole (Severs *et al.*, 1976), and between individual mitochondria in algae (Calvayrac *et al.*, 1972), their significance is unknown. A powerful approach to this problem might be the analytical, rather than preparative (Lloyd and Poole, 1979), fractionation of homogenates prepared at successive stages of the cell cycle, thus providing information on the physical characteristics of organelles as well as material for biochemical analyses. With few exceptions (e.g. Quintart *et al.*, 1979) this approach has not been exploited.

The interest in bacterial membranes and organelles lies perhaps not so much in their (apparent) structural simplicity as in their functional complexity. The bacterial cytoplasmic membrane is involved in respiration and ATP synthesis, transport of low molecular weight nutrients into the cell and of high molecular weight wall precursors out of the cell, as well as in biosynthesis and the segregation of DNA at division. The understanding of how the components responsible for this functional diversity are synthesized, either simultaneously or sequentially, and assembled to yield the fully-functional membrane remains a great challenge.

II. Development of Eukaryotic Structures

A. NUCLEI

1. *Changes in Nuclear Volume*

In a great many cell types (but not all; see Ross and Mel, 1972), the growth in mass and volume in the nucleus differs from growth of the whole cell in showing most of its increase late in the cycle, especially just before division (Mitchison, 1971; Zucker *et al.*, 1979a). Since about 80% of nuclear dry weight is protein (Altman and Katz, 1976), growth in nuclear volume presumably reflects the net accumulation of nuclear proteins. Michaels *et al.* (1977) have used Coulter analysis of isolated nuclei to demonstrate a dramatic enlargement of nuclei in late S and G_2. This expansion was accompanied by a significant decline in the number of binding sites for concanavalin A per nucleus, suggesting a masking or destruction of sites. However, in a recent study of a human cell line, Steen and Lindmo (1978) sized nuclei and found growth to be restricted to S. Nuclear volume measurements may be used to monitor cycle position and traverse of synchronized erythroleukaemia cells (Zucker *et al.*, 1979b).

2. *The Nuclear Envelope*

The well-known association of DNA with the nuclear envelope in animal cells was originally attributed to the membrane being the site of initiation of DNA synthesis (Comings and Kafekuda, 1968). Careful re-examination of this question (for references, see Prescott, 1976a), however, has led to the conclusion that this association has no special significance for the control, initiation or execution of DNA replication.

During mitosis, the nuclear envelope of many eukaryotic cells fragments, sometimes only partially (Gull and Trinci, 1974), releasing a large amount of nuclear material into the cytoplasm. In many cases, there is evidence that the dispersed proteins return to the reforming nucleus in late telophase and early interphase (e.g. Prescott and Goldstein, 1968). It now seems that the protein components of the envelope may be synthesized throughout the cycle and the envelope reassembled from preformed components after mitosis, rather than being synthesized *de novo* (Sieber-Blum and Burger, 1977). Some polypeptide components of the nuclear matrix (a network extending from the nuclear periphery to the interior; Aaronson and Blobel, 1975) survive nuclear dispersal and subsequent reformation during mitosis (Hodge *et al.*, 1977).

There has been some interest in the behaviour of the nuclear pore complexes that are to be found in the nuclear envelope (Wunderlich *et al.*, 1976). The number of pores has been shown to rise in G_1 and/or in S phase (Scott *et al.*, 1971; Maul *et al.*, 1972; de la Torre *et al.*, 1979), a fact that has been taken to suggest that they are involved in DNA replication. In *S. cerevisiae*, the density of nuclear pores, expressed with respect to membrane area and nuclear volume, have been described (Jordan *et al.*, 1976, 1977; Severs, 1977; Fig. 5.1), but the significance of the observed fluctuations is not understood. That portion of the nuclear surface that lies adjacent to the large vacuole remains devoid of nuclear pores (Fig. 5.2, Severs *et al.*, 1976). In contrast to the constant pore diameter reported by Severs and Jordan (1975), Willison and Johnston (1978) have reported abnormally wide pores in G_1-arrested cells. Nuclear pore frequency also increases during interphase in *Physarum* (Pendland and Aldrich, 1978).

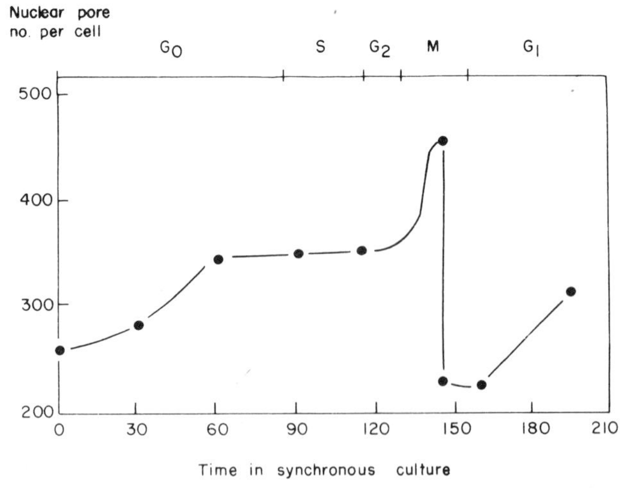

Fig. 5.1 Number of nuclear pores per nucleus during the first cycle of synchronous growth in *Saccharomyces cerevisiae*. The positions of the cell-cycle phases were established from measurements of budding and cell division. (Reproduced with permission from Jordan *et al.*, 1976.)

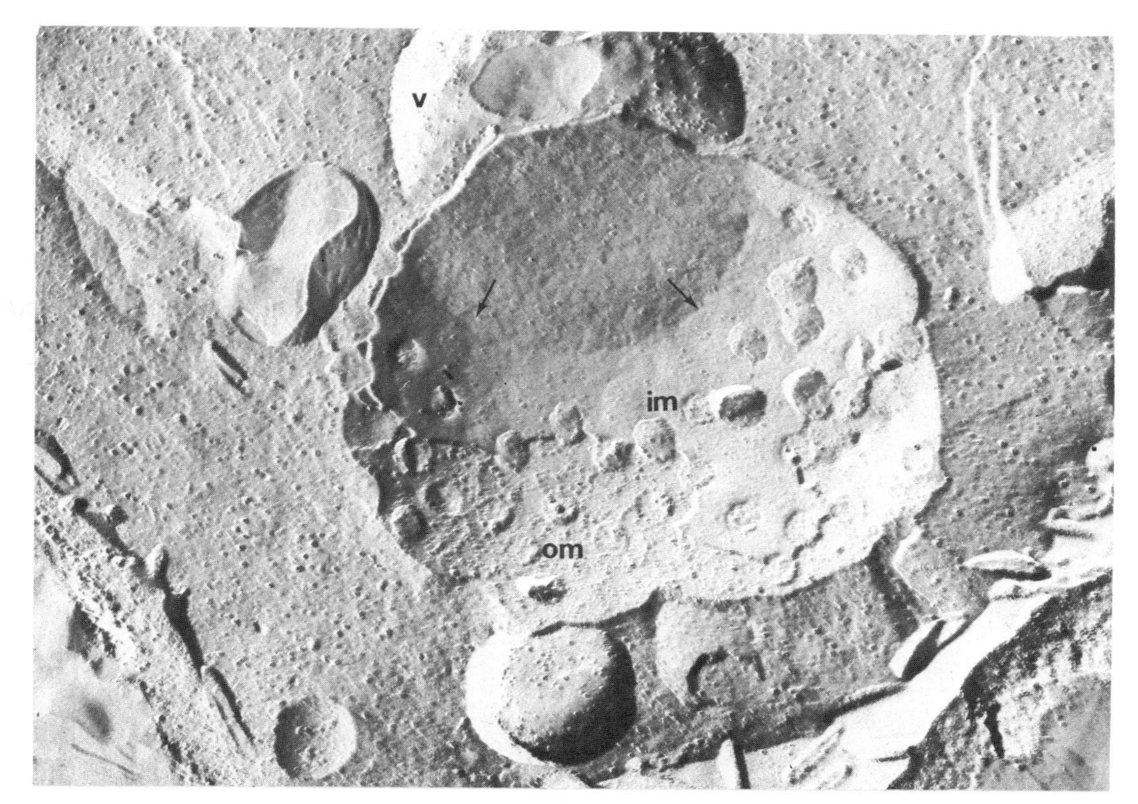

Fig. 5.2 Nuclear pores in a "G0" cell of *Saccharomyces cerevisiae*. The nucleus has been fractured concavely to reveal the A-face of the outer membrane (om) and B-face of the inner membrane (im). A small vacuole (V) extends below the upper portion of the nuclear membrane causing it to accommodate to the vacuole shape (see arrows). Nuclear pores are absent in this region of nucleus–vacuole interaction, though common elsewhere (×41 000). (Reproduced with permission from Severs *et al.*, 1976.)

Embedded in the nuclear envelope of yeast is the spindle pole body, a structure that has attracted particular attention because of its possible interplay with cell bud emergence and cell cycle control (Chapters 7, 8). Figure 5.3 and Byers and Goetsch (1975a,b) show the behaviour of spindle pole bodies in the budding cycle of *S. cerevisiae*. Unbudded

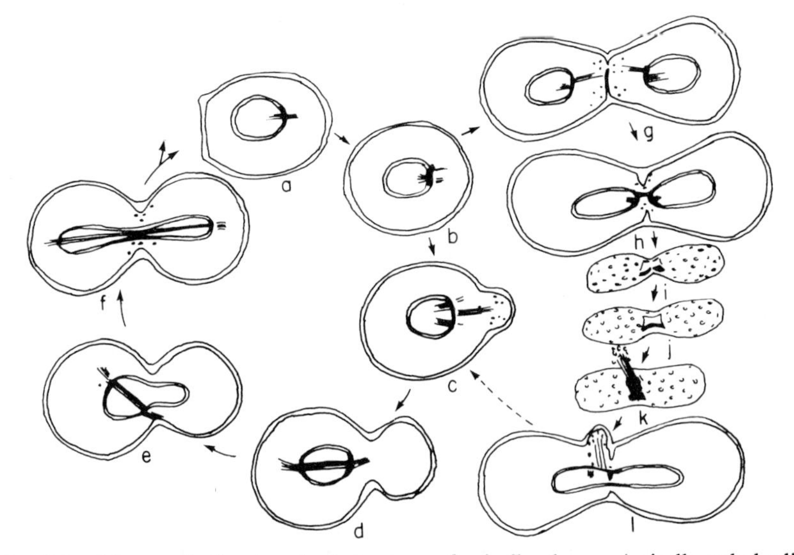

Fig. 5.3 Diagram indicating the behaviour of spindle plaques (spindle pole bodies) and satellites (both stippled), half-bridges (bold lines), and microtubules (straight lines) during the budding cycle, and the conjugation process of *Saccharomyces cerevisiae* in ideal cross-sectional views (except perspective views of nuclear fusion during conjugation in i–k). (Reproduced with permission from Byers and Goetsch, 1975).

cells possess a single body with microtubules of the intranuclear fibre abutting its intranuclear surface. Spindle pole body duplication is coincident with S phase and results in a double structure, which persists during the start of budding, eventually separating into two distinct bodies to form the poles of a complete intranuclear fibre. After migration of the nucleus to the neck, fibre elongation occurs, forcing the poles of the nucleus into mother and bud before cytokinesis. The behaviour of the spindle pole bodies in the *cdc* (cell division cycle) mutants and the clues that these studies have given to cell cycle controls are discussed in Chapter 7.

3. *Nucleoli*

Nucleoli are dense-staining, Feulgen-negative bodies that constitute one of the most conspicuous features of interphase nuclei. Nucleoli increase in size during the cell cycle (González and Nardone, 1968), with the most prominent growth occurring in the granular portion, during the first half of interphase (Sacristán-Gárate *et al.*, 1974a). In diploid human fibroblast cells, the number of nucleoli falls from 3·62 per nucleus at the beginning of the interphase to 2·51 in G_2, but the mean "projection area" increases, due to fission (Schnedl and Schnedl, 1972), whilst in *Physarum polycephalum* nucleus and nucleolar areas increase in parallel (Matsumoto and Funakoshi, 1978).

Concurrent with the shutdown of RNA synthesis and the decrease in the rate of protein synthesis at mitosis, the nucleolus begins to break down (Erlandson and De Harven, 1971). Nucleolar material has often been described to spread out as a film on the surface of mitotic chromosomes, perhaps accounting for the RNA that is associated with chromosomes at division (Prescott, 1970). The nucleolus is reconstituted in late telophase, probably associated with the return of ribosomal precursor RNA to the postmitotic nucleus (Prescott, 1976a). In *Euglena*, the nucleolus (or endosome; for references, see Moyne *et al.*, 1975) does not disappear during mitosis.

4. *Chromatin and its Organization*

DNA and its associated proteins are highly organized in nu bodies or nucleosomes and, at higher levels, in nucleofilaments and fibrils (Back, 1976) that probably have functional significance in development (for references, see Setterfield *et al.*, 1978). Chromatin conformation and the levels of phosphorylation of nuclear protein change through the mammalian cell cycle (Pederson, 1972; Nicolini *et al.*, 1975; Hildebrand and Tobey, 1975; Baserga and Nicolini, 1976; Chapter 8). This is reflected in differences in the Feulgen staining and fluorescent characteristics of interphase nuclei (Sawicki *et al.*, 1974; Moser *et al.*, 1975). In an attempt to quantify such changes in intact cells, Kendall *et al.* (1977) have measured the DNA content of Feulgen-stained nuclei and chromosomes by scanning densitometry. In addition to confirming earlier observations (Nicolini, 1975) on chromatin changes in the cell cycle, geometric parameters (Fig. 5.4) showed modulations of chromatin morphology that were not necessarily correlated with DNA content. For instance, chromatin area changed between late G_1 and early S even though DNA content was practically constant.

Expression of a temperature-sensitive mutation in mutants tsA 159

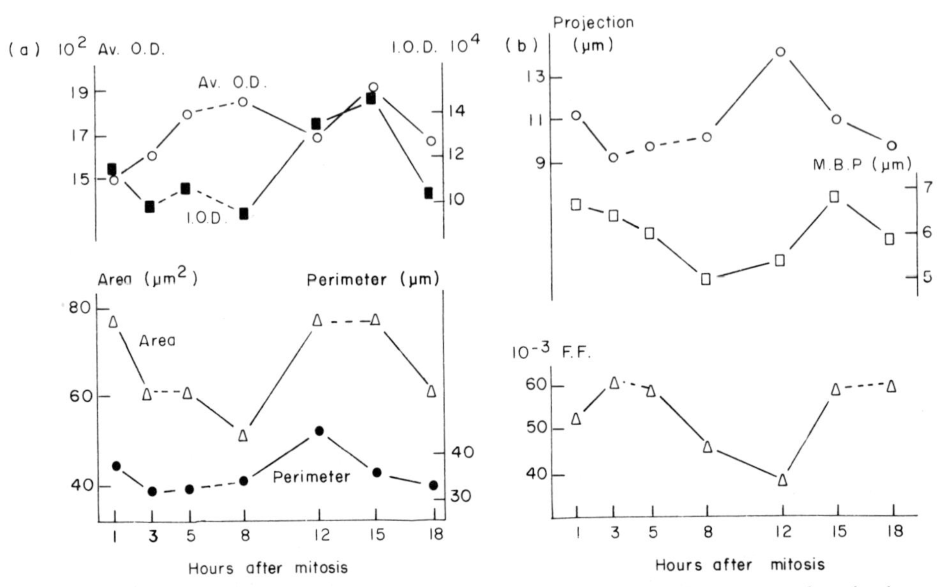

Fig. 5.4 Nuclear morphometry during the cell cycle. HeLa cells were synchronized by mitotic detachment and Feulgen-stained. Geometric and densitometric parameters of nuclei during the cell cycle were obtained using the automated image analyser Quantimet 720-D. (a) Mean integrated optical density (I.O.D.), area, perimeter, and average optical density (Av.O.D.) (obtained by dividing the I.O.D. by the area) for each nuclear image of cells at various time intervals after mitosis. The dashed lines mean the adjacent values were not significant at the 2·5% level, while the solid lines connecting two mean values indicate that the differences between such means were significant at the 5% level. (b) Mean values of projection, form factor (F.F. a measure of shape modulation), and Mean Bounded Path (M.B.P.) of cells at various intervals after detachment. Data at the higher thresholds included area, projection, and perimeter, but only the base value data at threshold of 0·04 O.D. are reported. As noted in A, dashed lines mean that adjacent values were *not* significant in two-tailed Student's *t*-tests at the 2·5% level, while solid lines connecting two means, indicate a significant difference. (Reproduced with permission from Kendall *et al.*, 1977.)

and ts C1 of mouse L cells (see Chapter 7) leads, not only to defects in DNA synthesis and cell development, but also substantial and reversible changes in the gross organization of interphase chromatin (Setterfield *et al.*, 1978). These results further support the idea that condensed chromatin normally undergoes an ordered cycle of transient, localized disaggregation and reaggregation associated with replication (Rao and Johnson, 1974). The fluctuations in spin lattice relaxation times of intracellular water that have been observed by n.m.r. spectroscopy during the HeLa cell cycle may be due in part to conformational changes in chromatin (Beall *et al.*, 1976). No such changes were observed in HTC cells (Müller *et al.*, 1980).

In *Euglena*, the nucleus reveals distinctive changes in chromatin condensation (Bertaux *et al.*, 1976; Gillott and Triemer, 1978), which seem to be correlated with the modification of nucleic acids by methylation (see Magnaval *et al.*, 1979). In *S. cerevisiae*, chromatin is dispersed during interphase and nuclear division and does not condense into discrete chromatids (Peterson and Ris, 1976; Gordon, 1977). These reports are in direct conflict with the earlier claim by Wintersberger *et al.* (1975) that condensed chromosomes are seen during a substantial part of the yeast cycle.

5. *Mitosis*

Mitosis may be regarded as the distributive phase of the cycle (Mazia, 1977), in which the sister chromatids are segregated between the daughter cells. In higher eukaryote cells, two types are found: (a) that of animals where centrioles are associated with spindle organization, and (b) the acentric mitosis of higher plants, where there are no obvious pole-associated structures. In the lower eukaryotes, there is an enormous diversity of mechanisms, all of which are outside the scope of this book. The reader is referred to the recent reviews by Heath (1980), Dodge and Vickerman (1980) and Mazia (1977).

B. MITOCHONDRIA

The partial autonomy that mitochondria exhibit with respect to their genetic constitution and biosynthetic capacities makes a study of their biogenesis and population dynamics during the cell cycle of special interest. In particular, we may ask how mitochondrial growth and division is coupled to that of the nucleus and other organelles, and whether the individual mitochondria in a single cell (in those cells that contain more than one; see below) exhibit synchrony in their biogenesis. Lloyd (1974) has described the early work on the growth and division of mitochondria in micro-organisms.

Interpretation of some reports describing the appearance of partitioned (Tandler and Hoppel, 1972) and "dividing" dumbell shaped mitochondria in a variety of cell types (e.g. Bahr and Zeitler, 1962; Hawley and Wagner, 1967; Osumi and Sando, 1969) is frustrated by the relatively recent realization that, in many cells, but not all (Heywood, 1977), one or a few highly-branched and ramifying mitochondria are present (for references, see Lloyd and Turner, 1980).

Clearly, randomly-selected thin sections of cells may give misleading information on organelle size and numbers. More recent work has exploited the techniques of serial sectioning and high voltage electron microscopy (HVEM) to provide information on mitochondrial morphology, and changes thereof, associated with the cell cycle. An alternative means of study, that of directly counting and sizing isolated organelles, has been advocated repeatedly for rat liver mitochondria (Gebicki and Hunter, 1964; Glas and Bahr, 1966; Gear and Bednarek, 1972; Schmidt *et al.*, 1977) and more recently for mitochondria isolated from *Tetrahymena pyriformis* (R. K. Poole, unpublished results). As described in Section II.A1, this approach has yielded useful information on nuclei during the cell cycle.

1. *Yeasts*

Cottrell and Avers (1971) examined thin sections of *S. cerevisiae* removed at intervals from synchronous cultures prepared essentially by the method of Williamson and Scopes (1962), and found that "all cells contained elongate and bizarrely shaped mitochondria". There was an approximately linear rate of increase in the number of mitochondrial profiles (from 5 to 10) per cell section during the interval corresponding to one cell cycle. The increases in mitochondrial DNA and several respiratory enzymes (Cottrell and Avers, 1970, 1971) and cardiolipin (Greksák *et al.*, 1977) were stepwise, however. On the basis of these results, Cottrell and Avers concluded that a mechanism other than growth and division was responsible for increase in the mitochondrial population. In contrast, Osumi *et al.* (1968) and Osumi and Sando (1969) claimed to find an abrupt increase (at about 0·5 of the cycle) in the number of mitochondria per cell in synchronized cultures of *Schiz. pombe*. The lack of statistical support for these measurements and the finding by Davison and Garland (1977) that *Schiz. pombe*, by 0·6 of a cell cycle, contains an extensive and largely continuous mitochondrial reticulum suggest that these results should be re-evaluated. In this organism, mitochondria appear to be juxtaposed with nuclei at the time of nuclear division (McCully and Robinow, 1971; Johnson *et al.*, 1973).

The most detailed study of mitochondrial form and volume in budding yeast is that of Stevens (1977). Computer-aided, three-dimensional reconstructions were used to show that rapidly-growing cells contain only a few mitochondria, usually less than ten. One mitochondrion of each cell is highly branched, much larger than the others and represents more than half the volume of the total cellular "chondriome" in rapidly growing cells. Out of 15 budding cells examined, this giant mitochondrion extended into the bud in 5 of them. A similar observation was made by Davison and Garland (1977). In 8 others, there was clearly no connection between the organelles in the mother and bud and, in 1 cell, a much smaller mitochondrion extended into the bud. In two cells in which the nucleus was elongated into the bud, the mitochondrion was not (Fig. 5.5). Variation in mitochondrial numbers and forms was not specifically related to any particular stage in the cell cycle; small numbers of highly-reticulated forms were found during all stages of bud formation.

2. *Protozoa*

A detailed study of mitochondrial proliferation in synchronized *Tetrahymena pyriformis* was made by Kolb-Bachofen and Vogell (1975). Only equatorial sections through cells were considered; morphometric analysis revealed that a synchronous mitochondrial division occurred during late S phase, even though the ratio of total mitochondrial area to cytoplasmic area remained constant throughout the cycle. An organelle cycle with three morphologically-characterized phases was proposed (Fig. 5.6). The results are similar to those predicted by Lloyd *et al.* (1971); this model proposed that mitochondria divide when they reach a critical size and that mitochondrial and nuclear divisions are co-ordinated to maintain the mitochondrial population within appropriate limits.

In *Paramecium*, the problem of interpreting electron micrographs of thin sections was circumvented by direct observation of mitochondria in living cells slightly compressed under a coverslip, or by measurement of organelle length after release of the mitochondria from compressed cells (Perasso and Beisson, 1978). Mean mitochondrial lengths doubled during the first quarter of the cell cycle and decreased progressively until the next cell division. Mitochondrial diameter and ultrastructure appeared to be invariant through the cycle.

In certain trypanosomatids, a single branched mitochondrion has been demonstrated (Paulin, 1975) by building three-dimensional models from micrographs taken with the HVEM and from stereopair analysis. Occasional bifurcations in the tubular extensions of the organelle were interpreted as possibly being early stages in the replication of the organelle. In other trypanosomatids, kinetoplast replication precedes karyokinesis in the cell cycle (Cosgrove, 1971).

A relatively simple cycle of synchronous mitochondrial growth and division has been observed in the plasmodia of *Physarum polycephalum*

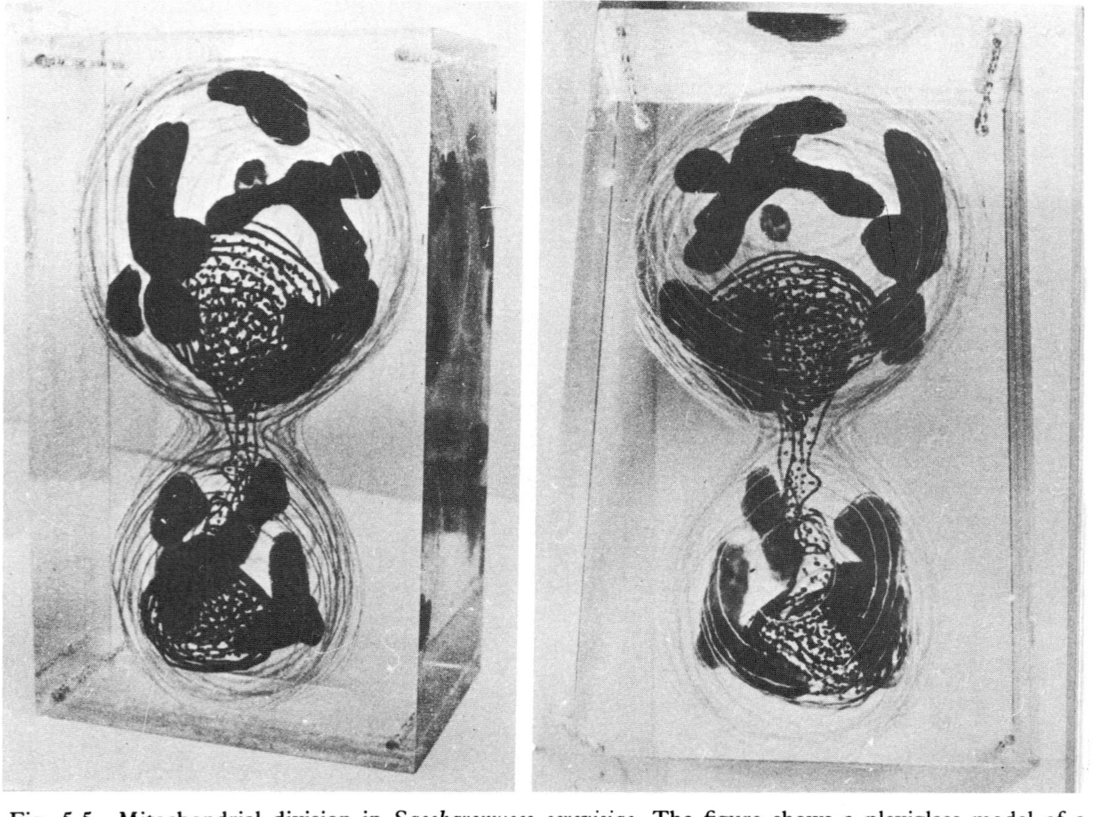

Fig. 5.5 Mitochondrial division in *Saccharomyces cerevisiae*. The figure shows a plexiglass model of a budding cell (strain R1) viewed from 2 sides. Mitochondria in black, nucleus in dotted black. (Reproduced with permission from Stevens, 1977.)

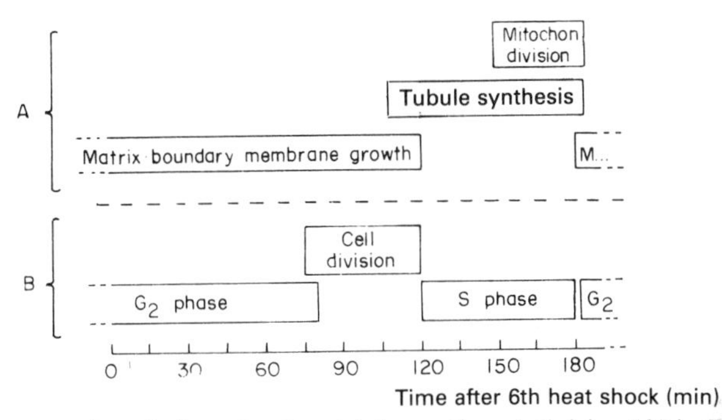

Fig. 5.6 A proposed cycle for mitochondrial growth and division (A) in *Tetrahymena pyriformis*, based on morphometric data, and its relation to the phases of the cell cycle (B). (Reproduced with permission from Kolb-Bachofen and Vogell, 1975.)

by Kuroiwa *et al.* (1978). The mitochondrial nucleoid divides just before division of the mitochondrion (Kuroiwa *et al.*, 1977).

3. *Algae*

Temporary formation of giant mitochondria occurs during interphase in *Euglena gracilis* under photo-autotrophic (Osafune, 1973; Osafune *et al.*, 1975a) and heterotrophic growth conditions (Calvayrac *et al.*, 1972, 1974; Ledoigt and Calvayrac, 1979). The transient, giant mitochondria divide by formation of a membraneous partition, separating the matrix into two or more chambers (Osafune *et al.*, 1975b,c). Formation of the giant mitochondria is accompanied by a striking decrease in respiration rate, which is restored when the giant mitochondria divide into smaller forms (Osafune *et al.*, 1975a). Recent three-dimensional reconstructions of mitochondria (Pellegrini and Pellegrini, 1976), however, reveal the presence of a single mitochondrial reticulum throughout the cycle (Pellegrini, 1980). Fig. 5.7 shows the relationship of mitochondrial development to that of other organelles in lactate-grown cells.

A remarkably similar cycle of mitochondrial morphology has been observed during the cell cycle of *Chlamydomonas reinhardii*. Examination of random thin sections (Osafune *et al.*, 1972a,b) and of serial sections (Osafune *et al.*, 1975d) has shown that giant mitochondria are formed from the fusion of smaller ones, accompanied by a decrease in respiration rate. Before formation of the giant forms, the mitochondria temporarily show close association with the chloroplast (Osafune *et al.*, 1972a,b). During cytokinesis and prior to the equal partitioning of mitochondrial volume between daughter cells, the total

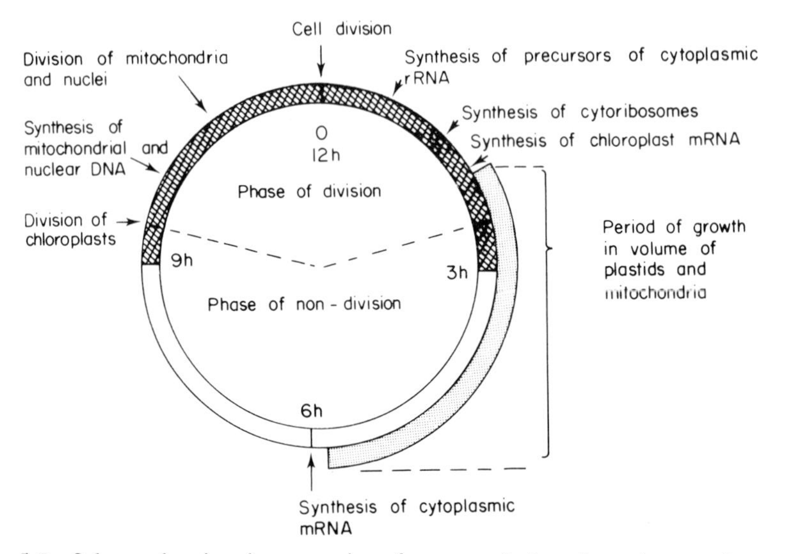

Fig. 5.7 Scheme showing the succession of events and of syntheses in a synchronous culture of *Euglena gracilis* growing in a medium containing lactate and in the presence of light. (Reproduced with permission from Ledoigt and Calvayrac, 1979.)

mitochondrial volume of the incipient cell remains nearly constant, whilst the number of mitochondria increases significantly (Osafune *et al.*, 1976). In the closely-related alga *Polytoma papillatum*, a single mitochondrion is formed by fusion of the numerous mitochondria present in cells following division (Gaffal and Kreutzer, 1977).

In contrast, the unicellular green alga *Chlorella fusca* var. *vacuolata* possesses a single mitochondrion throughout its cell cycle. The organelle forms an extensive reticulum, which is partitioned among the daughter cells by the outward progress of septa in the plane of the phycoplast microtubules (Atkinson *et al.*, 1974). The ratio of inner to outer mitochondrial membrane area remains constant throughout the cell cycle, although mitochondrial volume oscillates (Forde *et al.*, 1976). In contrast to the synthesis of succinate dehydrogenase and cytochrome *c* oxidase, which are synthesized only in the last third of the cycle (Forde and John, 1973, 1974), growth of the inner membrane occupies an extended period of the cycle. It was concluded (Forde *et al.*, 1976) that mitochondrial growth involves the intercalation of periodically synthesized respiratory enzymes into membranes made earlier in the cycle, with consequent fivefold changes in the occupancy of active enzyme molecules in the membrane.

4. *Higher Animal and Plant Cells*

Several measurements on cells of higher animals and plants have added support to the early light microscopic observations (e.g. Frederic, 1958) that the number of mitochondria decreases during mitosis (Schnedl, 1974; Sacristán-Gárate *et al.*, 1974b; Dewey and Fuhr, 1976). In contrast, Ross and Mel (1972) showed that, in Chinese hamster cells, the mean total mitochondrial volume and number of mitochondria per cell both showed a significant decrease in G_1 and G_2 as compared with S. These results and those of Posakony *et al.* (1977) with HeLa cells do not support a simple model in which a cycle of mitochondrial growth and division is synchronized with the cell cycle (Okamura *et al.*, 1977).

A study of mitochondrial membranes during the cell cycle of P815Y mast cells revealed that the accumulation of lipid and protein was continuous through the cell cycle (Warmsley *et al.*, 1970). The specific activity of mitochondrial structures involved in energy coupling, assayed with a fluorescent probe, remained constant during the cell cycle, as did the specific activity of cytochrome *c* oxidase and succinate-cytochrome *c* reductase. Evidence that HeLa cell mitochondria grow by the random insertion of the components into pre-existing membrane, and then divide, has been presented by Storrie and Attardi (1973).

5. *Mitochondrial Segregation*

If we accept that, in the normal division cycle of diverse cell types, mitochondrial formation occurs through a "duplication cycle" of growth and division, at least one other important question remains to be answered. This relates to the mode of segregation of daughter organelles between the two daughter cells formed at division. The problem of segregation of cellular structures has been most closely studied with respect to cell walls of both eukaryotes and prokaryotes where covalent and other labels have yielded conflicting data on the sites of deposition of newly-synthesized wall components. Apart from studies on the morphology of organelles at division, examples of which are given elsewhere in this chapter, little attention has been paid to the segregation of biochemically functional units at division, although it is tacitly assumed that this is equal with respect to the progeny.

Galdiero (1973a) studied the distribution of membranes between mother and bud cells of *S. cerevisiae* after pulse labelling an exponentially-growing culture. Synchronous cultures prepared from such a culture by size selection were sampled at intervals and the cells vigorously shaken in a Waring blender to separate buds from parent cells. Membranes isolated from either the purified buds or parent cells exhibited a specific radioactivity (of both [^{3}H]-leucine and [^{32}P]) approximately equal to the activity of cells in the inoculum. These results indicated random dilution of labelled membrane between mother and bud cells, and further showed that there was no difference in the rate of membrane growth in the two populations.

A refinement of this experiment has been made possible with the unusual budding yeast *Sterigmatomyces halophilus*. This yeast (Fig. 5.8) forms buds remote from the mother cell on the ends of fine projections called sterigma (Fell, 1966), which can be disrupted by

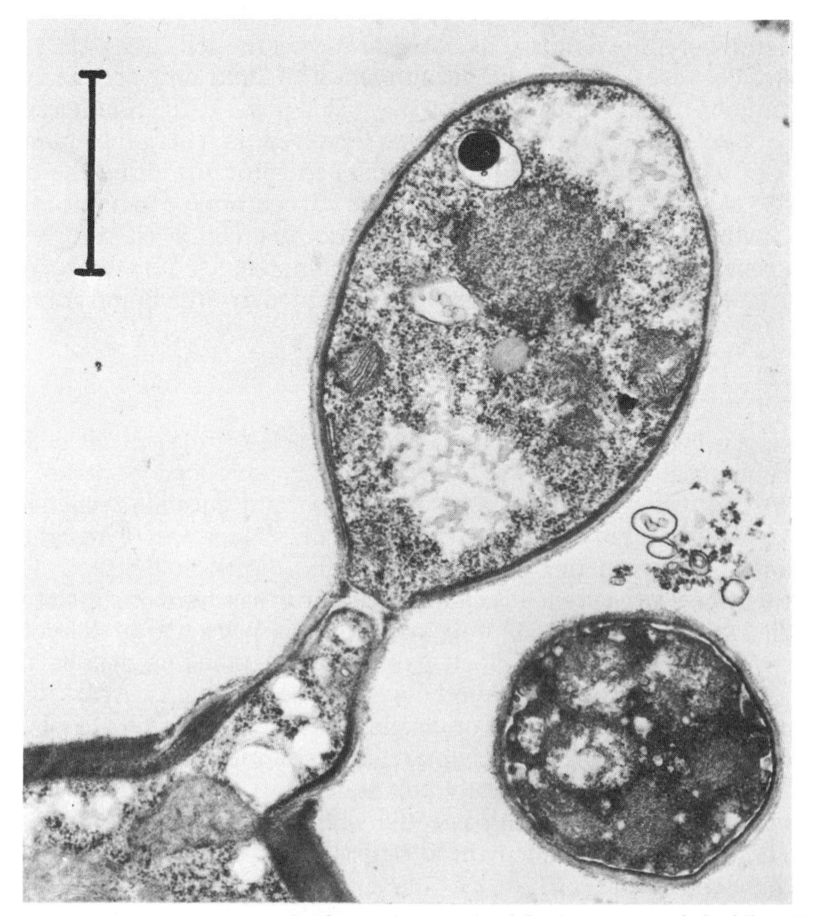

Fig. 5.8 Electron micrograph of a sectioned cell of *Sterigmatomyces halophilus*. The scale marker represents 1 μm. (Preparation of I. Salmon and A. Collinge.)

relatively gentle sonication (I. Salmon, A. Ajmeri and R. K. Poole, unpublished results), thus dissociating the mother and bud and allowing biochemical analysis to be carried out on the separated cell types. In recent experiments (Salmon, 1980; Salmon and Poole, 1980a) cells from exponentially-growing cultures were separated according to age by large-scale isopycnic centrifugation in gradients of Urografin. Cells from successive fractions were sonicated and the resultant buds and mothers separated on the basis of size by further density gradient centrifugation. The assay, of two enzymes that are considered to be "marker enzymes" for the inner mitochondrial membrane (Lloyd and Poole, 1979) (in French Press extracts of the fractions) gave the results shown in Fig. 5.9. In the unseparated mother–daughter pairs (Salmon

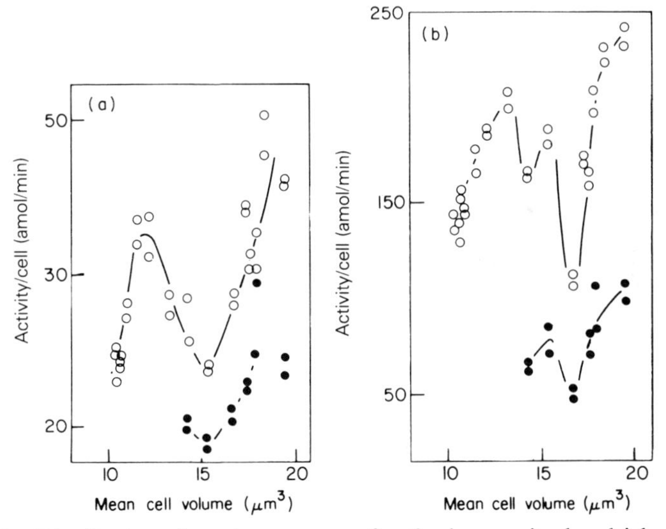

Fig. 5.9 Distribution of marker enzymes for the inner mitochondrial membrane between mother and bud cells during the cell cycle of *Sterigmatomyces halophilus*. The experimental method is described in the text. Plotted as a function of mean cell volume (and thus age) in the cell cycle are the activities per cell of succinate dehydrogenase (a) and cytochrome *c* oxidase (b) in the mother and bud fraction (open symbols) or, in those fractions where the developing buds could be removed and separated, the bud fraction (closed symbols). (Experiment of I. Salmon.)

and Poole, 1980b), and in those buds that could be recovered and studied, both enzyme activities oscillated, with each enzyme behaving similarly in bud and (by difference) in mother cells. The results indicate free exchange of cytoplasmic contents between the two cell types or perhaps a very precise control over the development of mitochondrial structure in the bud so that it reflects exactly that in the mother. Although mitochondria have been observed in the sterigma in random cell sections, the detailed sequence of the mitochondrial duplication cycle remains to be described.

C. VACUOLES

1. *Micro-organisms and Plant Cells*

In *S. cerevisiae*, morphological changes of the vacuoles represent one of the most conspicuous features of the cell cycle. During bud initiation, the large vacuoles shrink and fragment into small vacuoles (Wiemken *et al.*, 1970a). These subsequently segregate between mother and daughter cell and then fuse and expand. Vacuolar fragmentation (Fig. 5.10) also occurs following synchronization by a feed-starve regime (Severs *et al.*, 1976). The changes in buoyant density of the cells that accompany this vacuolar cycle have been exploited by Wiemken *et al.* (1970b), Hartwell (1970) and Sierra *et al.* (1973) as the basis for isopycnic separations of cell ages (Chapter 2). Vacuole permeability appears to be highly regulated; release of sequestered vacuolar constituents, in particular nitrogen sources, together with protein turnover, may be regulated by the G_1 arrest signal (Sumrada and Cooper, 1978). Such a mechanism could provide the cell with a means of maintaining intracellular homeostasis and thus the capacity for completion of a cell cycle during external nutrient deprivation.

In *Chlorella*, vacuole volume increases during the first 10 h of the cycle and is followed by a period of decline (Atkinson *et al.*, 1974). However, surface area continues to increase because of an increase in vacuole number, thus achieving an overall increase in tonoplast area throughout growth.

Similar structural changes in the vacuole occur during interphase of some higher plant cells. Two well-defined stages were described by Sacristán-Gárate *et al.* (1974c) on the basis of a stereological study of *Allium* meristematic cells. The number of vacuoles per cell increased between G_1 and S, while mean volume decreased, suggestive of vacuole division. Between mid-G_2 of one cycle and mid-G_1 of the following cycle, vacuoles again increased in volume. It is perhaps significant that the phase of vacuolar enlargement coincided with the previously-observed phase of mitochondrial lysis (Sacristán-Gárate *et al.*, 1974b) but positive identification of the vacuolar inclusions as mitochondria could not be made.

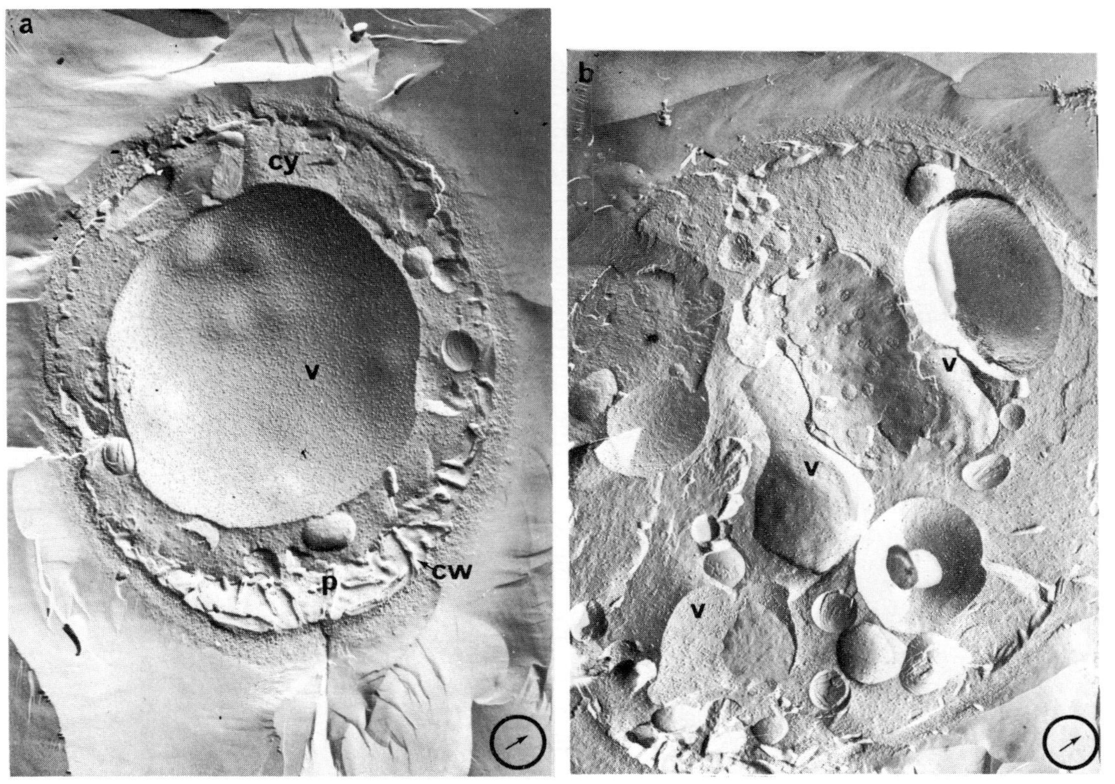

Fig. 5.10 Morphological dynamics of vacuoles in *Saccharomyces cerevisiae* synchronized by a regime of repeated feeding and starving. (a) Shows a cell in resting stage (G_0) sampled immediately after inoculation. The cleavage plane of the freeze-fractured preparation has followed a concave course through the membrane of the large vacuole (V). Also shown are the narrow zone of cytoplasm (cy) and the plasmalemma B-face (p) surrounded by the cell wall (cw). In (b) is a cell sampled 60 min after inoculation (end of "lag phase"). Many vacuoles (V) are distributed through the cell, none being as large as in (a). Two vacuole membrane faces show close association with the nuclear membranes (n). In (a) and (b) the magnification is $\times$ 17 000 and the direction of shadow is shown by the encircled arrow. (Reproduced with permission from Severs *et al.*, 1976.)

The caution that should be applied to the determination of sizes, shapes and numbers of organelles from examination of thin sections, and which was illustrated by consideration of mitochondrial studies (Section II.B), is equally applicable to studies of vacuoles. Davison and Garland (1977) have clearly shown that in two yeasts, *Candida utilis* and *Schiz. pombe*, vacuoles form lobed clusters whose morphology cannot be adequately described if only random thin sections are examined.

In *Tetrahymena*, the fact that dividing cells do not form food vacuoles (Chapman-Andresen and Nilsson, 1968; Nachtwey and Dickinson, 1967) has been exploited by Wolfe (1973) for selection synchrony (see Chapter 2).

2. *Animal Cells*

Quintart *et al.* (1979) have studied the buoyant density distributions of lysosomes and other organelles through the cell cycle by subjecting hepatoma tissue-culture cells (synchronized by selective detachment at mitosis combined with a colcemid metaphase block) to the classical procedures of analytical subcellular fractionation. Although the specific activity of acid hydrolases (marker enzymes for lysosomes) varied by less than 10% over the cycle, the distribution of the enzymes following isopycnic centrifugation of homogenates in sucrose gradients varied dramatically (Fig. 5.11). In contrast, no significant changes occurred in the distribution of mitochondrial or plasma membrane markers. In early G_1, the median equilibrium density of lysosomal markers (cathepsin B, acid phosphatase and N-acetyl-β-glucosaminidase) decreased appreciably after 5 h and then increased progressively. In addition, increased solubility of the last enzyme was noted at 5 h after mitosis, presumably as a result of increased size and thus fragility of the lysosomes. The observed changes in buoyant density perhaps arose by the modifications of endocytosis that are known to occur in the cell cycle (Quintart and Baudhuin, 1976; Riley and Dean, 1978; Berlin *et al.*, 1978; Quintart *et al.*, 1979; Berlin and Oliver, 1980). Morphometrical studies have also shown modifications of both average volume and fractional volume of lysosomes (Okamura *et al.*, 1977; Baudhuin *et al.*, 1979; Quintart *et al.*, 1980).

D. RIBOSOMES AND POLYSOMES

Although protein synthesis in eukaryotic cells is controlled at the translational and transcriptional levels, relatively little effort has been

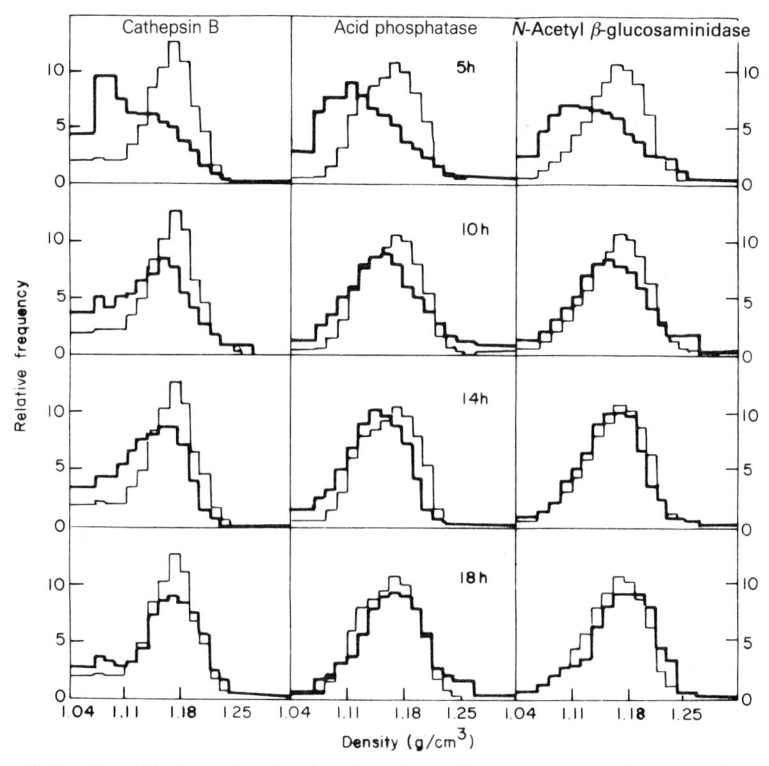

Fig. 5.11 Equilibrium density fractionation of homogenates from synchronized hepatoma cells: distribution of lysosomal marker enzymes. The abscissa is a density scale that is divided into 15 fractions of equal density increments from 1·07 to 1·27 g cm^{-3}. Outside this range, fractions were pooled. The ordinate represents the relative frequency (activity/unit density increment); the area of each histogram equals unity after normalization. Thin lines give the distribution found in mitotic cells for comparison. Times refer to the homogenization of cells after mitosis. Recoveries from homogenates were 95·7 ± 17·2% for cathepsin B, 102·6 ± 11·2% for acid phosphatase and 100·4 ± 6·7% for N-acetyl β-glucosaminidase. (Reproduced with permission from Quintart *et al.*, 1979.)

directed towards a study of the changes in the cell's translation machinery during the cell cycle.

In budding yeast, the proteins of the small and large ribosomal subunits are synthesized continuously (Shulman *et al.*, 1973), and in fission yeast ribosomal numbers per cell increase throughout most of the cell cycle (Maclean, 1965), but these results may not truly reflect the complexity of ribosomal and polysomal dynamics. As early as 1961, Plesner showed that in *Tetrahymena*, synchronized by heat shocks, a class of 70 S ribosomes appeared shortly before division that were distinguishable by their stability at both low and high Mg^{2+} concentrations. Evidence was presented that the stability was due to their association with newly synthesized protein. More recent evidence for a cycle of synthesis and degradation of ribosomes in this organism has been presented by Jakoi *et al.* (1976). Cytoplasmic ribosomes are synthesized at the start of G_1 in *Euglena gracilis* (Ledoigt and Calvayrac, 1979).

The appearance and disappearance of specific classes of polysomes that has been observed in several organisms is of special interest because it may shed some light on the cycle-dependent synthesis of specific proteins. In *Chlorella*, synchronized by a light–dark regime, the numbers of both free and membrane-bound polysomes increase in a stepwise fashion in the dark period (Galling, 1970). During a short portion of the preceding light period, there is an increase in the rate of appearance of membrane-bound ribosomes, relative to free ones. Unlike synchronized *Tetrahymena* (Whitson *et al.*, 1966), in which only monosomes are found just prior to division, the minimum level of polysomes (expressed as a percentage of the total ribosomes) occurs early in the cycle.

In HeLa cells (Scharff and Robbins, 1966; Erlandson and De Harven, 1971), complete disaggregation of polysomes occurs during mitosis, whilst in *Physarum* (Mittermayer *et al.*, 1966), the disaggregation is less complete and occurs shortly after metaphase. In Chinese hamster ovary cells, the decline in protein synthesis during mitosis was shown by Fan and Penman (1970a,b) to result from decreased attachment of ribosomes to messenger RNA. In a more detailed study of HeLa cells, Eremenko and Volpe (1975) have shown dramatic oscillations in the proportions of ribosomes, their subunits and polysomes (Fig. 5.12) during the cell cycle. New populations of "heavy" polysomes appear in G_2 and disappear again by G_1 of the next cycle. The 40 S and 60 S subunits are completely absent in G_2 and M, whereas the 80 S monomers do not fully disappear. The ribosomal subunits in G_2 and M may be largely used to form ribosomes, most of which in turn are incorporated into polysomes. Analysis of the incorporation of [^{3}H]-leucine into nascent polypeptide chains during the cell cycle showed that the trisome peak (II in Fig. 5.13) is not labelled during G_1 and S. The inverse relationship between the low absorbance profiles of polysomes and their relatively high level of label in G_1 and S indicate that their specific radioactivity is about twice as

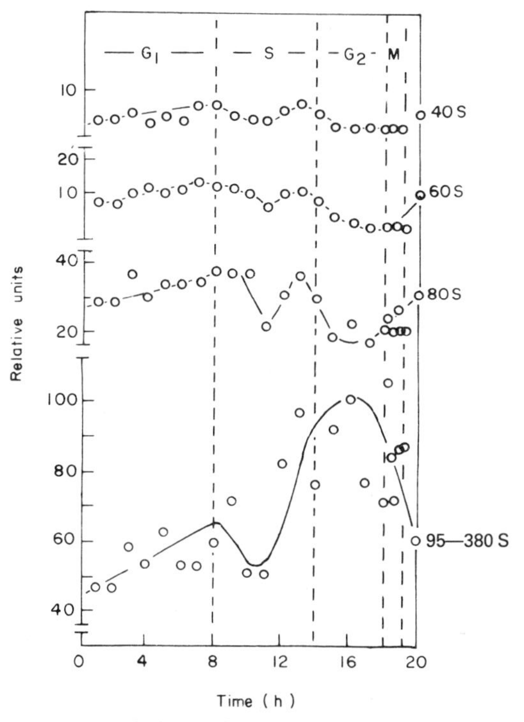

Fig. 5.12 Quantitative variations of ribosomal subunits, monomers and polyribosomes during the HeLa cell cycle. Absorbance profiles of ribosomal preparations made at hourly intervals during interphase and every 15 min during mitosis were quantitatively analysed to yield the plots shown. (Reproduced with permission from Eremenko and Volpe, 1975.)

high in young cells than in the older cells. This strongly suggests that differences in the extent of translation occur during G_1 and G_2.

The proportions of membrane-bound and free polysomes change during the cell cycle of mouse plastocytoma cells, with maximal amount of membrane-bound polysomes occurring in G_1, concomitant with higher levels of immunoglobulin synthesis (Abraham *et al.*, 1973).

E. CHLOROPLASTS

A cycle of growth and division of chloroplasts has been observed in numerous algae. In *Chlamydomonas*, an organism widely used for studies of chloroplast biogenesis, there is disagreement as to whether

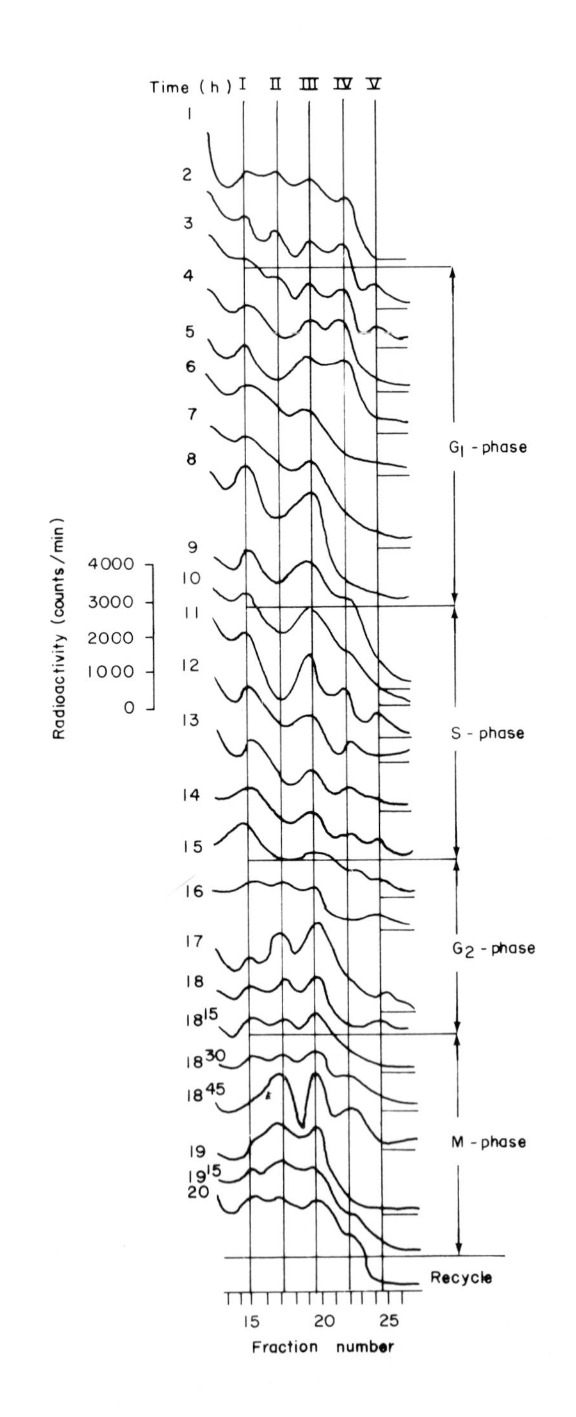

Time (h)
I
II
III
IV
V
1
2
3
4
5
6
7
8
9
10
11
12
13
14
15
16
17
18
18^15
18^30
18^45
19
19^15
20
Radioactivity (counts/min)
4000
3000
2000
1000
0
G1 - phase
S - phase
G2 - phase
M - phase
Recycle
15
20
25
Fraction number

the chloroplast divides after (Ladygin *et al.*, 1974) or before (Ettl, 1976) the nucleus. Osafune *et al.* (1972b) have reported a temporary association of the chloroplast with mitochondria, but the significance of this relationship is not understood. The major membrane polypeptides of the chloroplast are synthesized in an orderly, sequential fashion (Beck and Levine, 1974). In light–dark synchronized cultures, group I polypeptides (mol. wt 50 000 to 55 000 and associated with photosystem I) are synthesized predominantly during the early light period, before chlorophyll synthesis. The initiation of synthesis of the various group II polypeptides, which are associated with a membrane fraction enriched for photosystem II activity, occurs later. The synthesis of all major polypeptide groups ceases in the dark. In contrast, Bourguignon and Palade (1976) showed that incorporation of newly-synthesized polypeptides into thylakoid membranes occurred continuously but differentially throughout the cycle. Maximal rates of incorporation for the majority of polypeptides were detected shortly after division in G_1 (in the dark phase). The rates of radioactive labelling decreased gradually to a low level at the end of the dark period and then rose slightly at the beginning of the light period. Bourguignon and Palade suggest that the discrepancies between these results and those of Beck and Levine (1974) are due in part to incomplete separation of polypeptides in Beck and Levine's SDS gels. Chloroplast lipids are also synthesized in the light phase in a sequential or multistep process in synchronized *Chlamydomonas* (Beck and Levine, 1973), although no conclusion can be drawn yet regarding obligatory associations between lipids and proteins during membrane assembly.

The synthesis of membrane proteins in the light phase of the synchronized cell cycle is temporally, but not necessarily causally, related to variations in the ratio of free ribosomes to thylakoid-bound ribosomes (Chua *et al.*, 1973, 1976). Thus, in the light, about 20 to 30% of the total chloroplast ribosomes are membrane-bound, whilst in the dark only a few or no membrane-bound ribosomes are present. The data of Schor (1971) and Iwanij *et al.* (1975) on the rates of protein synthesis in the presence of spectinomycin, an inhibitor of chloroplast ribosomes, also suggest that the chloroplast protein-syn-

---

Fig. 5.13 Profiles of labelled nascent polypeptides on polyribosomes isolated at short intervals throughout the HeLa cell cycle. The patterns relate to the total incorporation of [^{3}H]-leucine into polyribosomes. The base line for each curve is shown as a short segment on the right side. Roman numerals on the top refer to the labelled polysome peaks corresponding to 120 S (I), 150 S (II), 200 S (III), 270 S (IV) and 350 S (V) respectively. The numbers shown on the left indicate the hours of the mitotic cycle. (Reproduced with permission from Eremenko and Volpe, 1975.)

thesizing apparatus may be quiescent in the dark. Approximately 90% of the chloroplast (and cytoplasmic) rRNA is transcribed during G_1 in the light period (Wilson and Chiang, 1977). A sequential synthesis of protein–chlorophyll complexes during the light phase of synchrony also occurs in *Chlorella pyrenoidosa* (Adler, 1976). Of particular interest to the question of chloroplast development in this organism is the large proportion of total cell volume that the chloroplast occupies. The single chloroplast enlarges rapidly at the start of the cycle to occupy 40% of the cell volume, after which it declines to occupy 33% of the cell. The rapid growth is accompanied by an extension of the organelle from its lateral position to a form in which it encloses most of the cell. Its plasticity allows it to surround other organelles. There is some evidence that microtubules play a part in its division. There is no evidence for autonomous pre-mitotic cleavage of the chloroplast.

Chloroplast replication occurs shortly before, or at, cell division in *Euglena gracilis* (Gojdics, 1953; Leedale, 1967; Orcival-Lafont *et al.*, 1972; Ledoigt and Calvayrac, 1979; Pellegrini, 1980), the marine alga *Hymenomonas carterae* (Stacey and Pienaar, 1979), and in cells of the meristematic tissue of *Isoetes lacustris* (Whatley, 1974). In *Euglena gracilis*, chloroplasts remain relatively compact with closely packed lamellae during the light portion of a light–dark synchronizing cycle and then become distended during the dark (division) period. This change persists for at least one cycle even when the cells are left in continuous light, suggesting that the cycle is a true consequence of cell age (Cook *et al.*, 1976). Pellegrini (1980) detected no plastidial reticular phase in the *Euglena* cell cycle, in contrast to earlier claims of its existence (Orcival-Lafont *et al.*, 1972; Calvayrac and Lefort-Tran, 1976). In the alga *Olisthodiscus luteus*, which lacks cell walls, the many small, discoidal chloroplasts divide synchronously, slightly before cell division (Gibbs *et al.*, 1976).

Vanden Driessche (1966) found that chloroplasts of the giant alga *Acetabularia mediterranea* grown under a 13 h light–12 h dark regime underwent a circadian rhythm of swelling and contraction. Greatest volume was noted during the light periods, in contrast to the findings of Cook *et al.*(1976) with *Euglena*.

F. SURFACE MEMBRANES

1. *Morphological Changes in the Plasma Membranes of Mammalian Cells*

The dramatic changes in surface morphology that occur during the cell cycles of mammalian cells have been well documented (Abraham *et al.*, 1973; Porter *et al.*, 1973; Hale *et al.*, 1975; Haars and Hampel, 1975; Ching *et al.*, 1976; Sanger and Sanger, 1980). In the fairly typical case of Chinese hamster ovary cells, those at mitosis bear microvilli and many longer outgrowths of variable diameter called filopodia, which attach to the substrate. After division of cells selected by mitotic detachment (Terasima and Tolmach, 1963a), early G_1 cells begin to flatten, most of the filopodia having been lost. Subsequently, blebs or "zeiotic knobs" (Godman *et al.*, 1975) appear, only to disappear during the transition from G_1 to S, so that the surface of the S-phase cell is almost smooth and featureless. During G_2, microvilli increase in number as the cells assume a more convex or rounded-up shape. Just prior to mitosis, long filopodia appear that persist through the division process. Chick and mouse embryo cells exhibit a similar pattern of change (Hale *et al.*, 1975; Lehtonen, 1980). The flattened, blebbed form of cells found in G_1 sometimes persists throughout the cycle and appears to be a function of culture density and cell-to-cell contact (Rubin and Everhart, 1973). It is clear that there are temporal correlations between the normal sequence of morphological changes and nuclear events, but a causal connection has not been established. Our understanding of the events is hindered by ignorance of the precise functions of the various surface protuberances. Filopodia have been considered to be sensory organelles (Albrecht-Beuhler, 1976) and are also involved in the transport of particles towards the cells (Albrecht-Beuhler and Goldman, 1976). Likewise the relevance of the blebbed morphology is unclear, although it may be involved in membrane assembly (Pasternak *et al.*, 1974; see below).

As noted by Allred and Porter (1977), the spherical shape of mitotic cells is presumably of importance for cytokinesis, simply because all cells appear to assume this shape. The change in shape is partly due to the action of cytoplasmic microtubules.

Freeze-cleavage of the cell membranes of L cells reveals intramembranous 70 Å particles, thought to be formed from the association of glycoproteins and lipids. The surface density of these particles oscillates during the cell cycle (Scott *et al.*, 1971), dropping dramatically during mitosis and increasing again during G_1. In contrast, Torpier *et al.* (1975) were unable to find any cycle-dependent variation in the densities of membrane particles in BHK21 cells, and Knutton (1976) made similar observations with P815Y cells.

The surface changes which occur during the growth cycle of tissue culture cells described above and by others (Fox *et al.*, 1971; Noonan *et al.*, 1973; Furcht and Scott, 1974) have involved cells grown in monolayer culture. Pasternak and his co-workers (see Pasternak, 1976a,b) have studied structural changes in the plasma membrane of synchronized P815Y mastocytoma cells grown in suspension culture, with a view to eliminating those changes resulting from attachment to a substrate. The volume of these spherical cells doubles during G_1 and G_2 and the apparent surface area thus increases 1·6-fold (Fig. 5.14). Knutton *et al.* (1975) have shown that there is actually a two-fold increase in surface area and that the extra surface area required

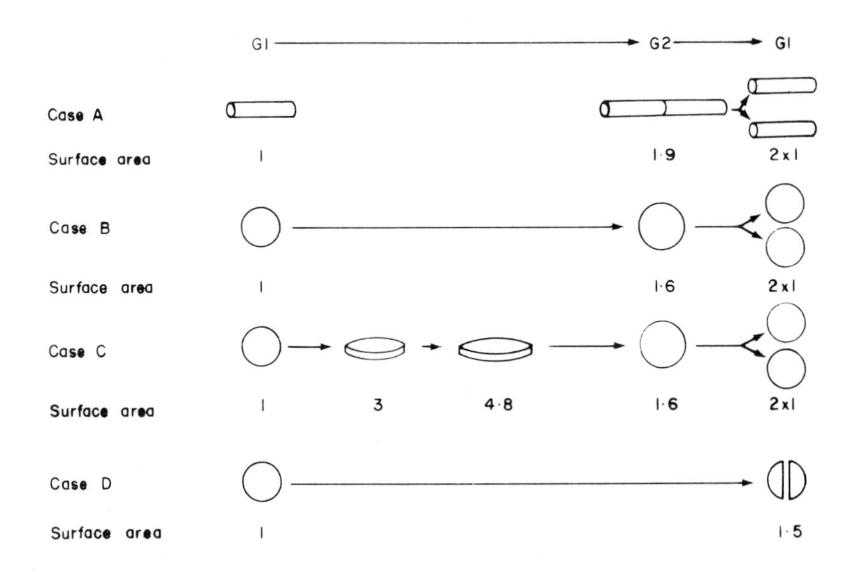

Fig. 5.14 Relationship between surface area and the cell cycle. (A) Bacterial rod; a cylindrical shape, with radius = 0·1 of length, is assumed; (B) spherical cell, e.g. mammalian cell in suspension culture; (C) flattened cell, e.g. mammalian cell in monolayer culture; a disc-like shape during interphase, with radius = 10 × height, is assumed; (D) egg in first cleavage, e.g. sea-urchin egg. A doubling in volume is assumed to occur between G_1 and G_2 in cases (A), (B) and (C). Note that this is a highly diagrammatic approximation only; monolayer cells do not round-up completely during mitosis (case (C)), nor do sea-urchin eggs cleave to form two perfect hemispheres (case (D)). (Reproduced with permission from Pasternak, 1976b.)

during mitosis is provided by an "unfolding" of microvilli (Fig. 5.15). The extent to which the cell surface is folded increases from 3% (approx. 4 microvilli μm^{-2}) in early G_1 to 6% (approx. 7·5 microvilli μm^{-2}) in G_2 (Knutton, 1976). That membrane thickness and, in particular, the density of intramembranous particles are invariant throughout the cell cycle (Knutton, 1976; cf. Scott *et al.*, 1971) suggests that plasma membrane components are inserted continuously and in concert, confirming earlier biochemical studies (Graham *et al.*, 1973).

2. *Changes in the Biochemical Composition and Function of the Plasma Membrane*

Bulk measurements of the major chemical components of plasma membranes during the cell cycle have yielded somewhat conflicting results. Gerner *et al.* (1970) found that incorporation of precursors into the protein, carbohydrate and lipid components of the plasma membrane was elevated soon after division of synchronized KB cells. Bosmann and Winston (1970) showed that protein and glycoprotein were synthesized throughout the cell cycle of L5178Y cells but at somewhat higher rates during S. In HeLa cells also, incorporation of radioactive fucose into plasma membranes was found to take place primarily in the S-phase (Nowakowski *et al.*, 1972). Phospholipid synthesis occurred throughout the cell cycle of synchronized mast cells, but at elevated levels in S phase (Bergeron *et al.*, 1970). In contrast, Bosmann and Winston (1970) claimed that the synthesis of glycolipid and lipid in L5178Y cells is restricted to the G_2 and mitotic periods. In a comprehensive study, Graham *et al.* (1973) showed that, after separating cells by size according to their position in the cycle, the bulk of plasma membrane components (proteins, phospholipids and individual sugars) increased continuously during interphase so as to double between G_1 and G_2. Johnsen *et al.* (1975) detected about 35 protein bands (five of which appeared to be glycoproteins) following gel electrophoresis of solubilized plasma membranes from synchronized HeLa cells (Fig. 5.16). A few minor bands appeared only in cells at mitosis. Incorporation of [^{3}H]-fucose into glycoproteins occurred throughout the cycle, with only minor changes in the rate of incorporation into individual glycoproteins (Johnsen and Stokke, 1977). In contrast, modification of membrane proteins by phosphorylation may be restricted to G_1 (Mastro, 1979).

The observation that expression of antigens and other surface markers decreases to a minimum in interphase and is restored to a maximal value in mitosis was the first result to be interpreted as a cell cycle-dependent change in plasma membrane architecture. The amount of anti-H-2 antibody approximately doubles in G_1 and then remains constant (Sumner *et al.*, 1973), whilst the expression of H-2, as measured by immune cytolysis in the presence of complement, is minimal in late interphase (Pasternak *et al.*, 1971). In fact, *non*-immune

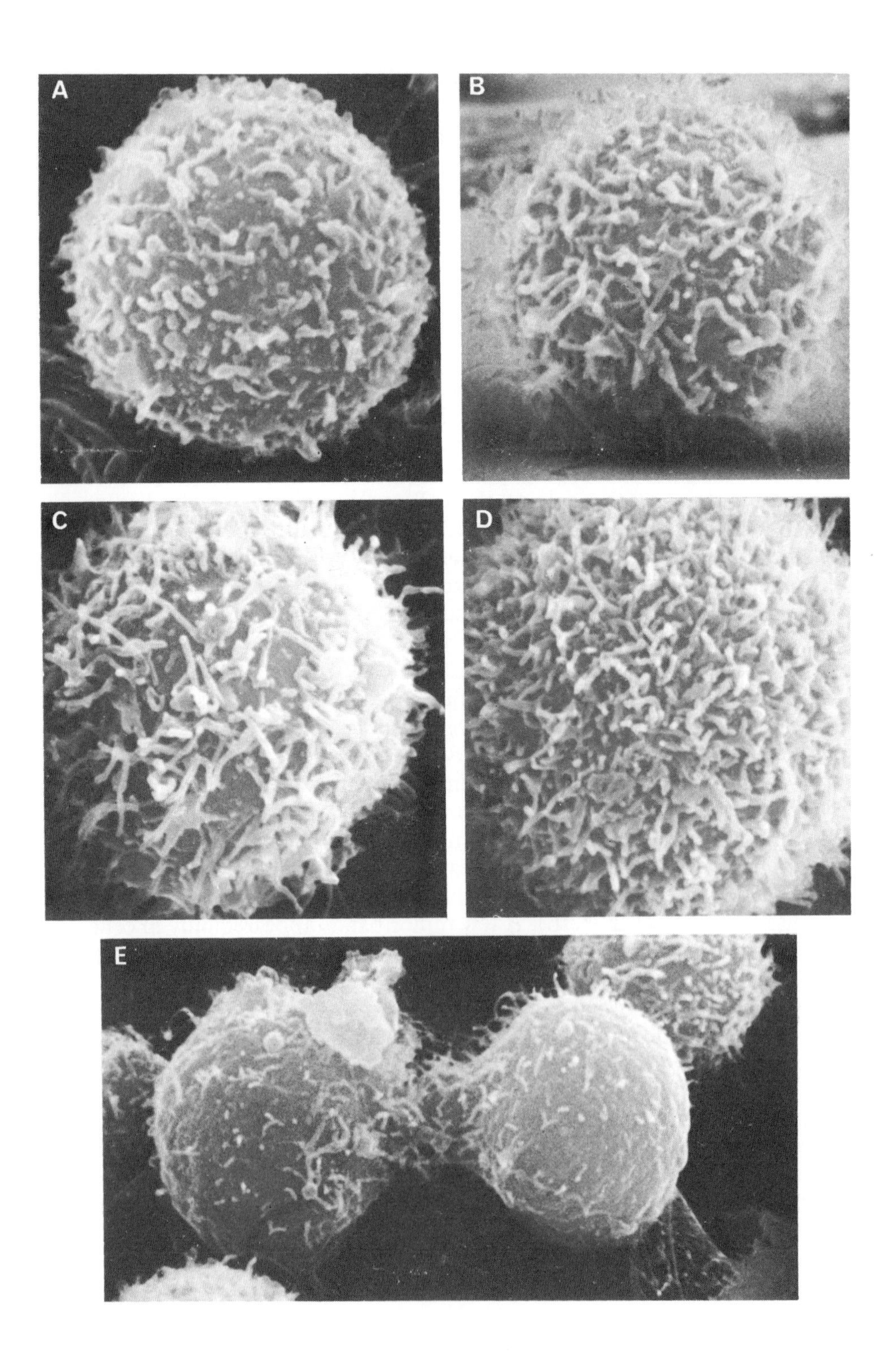
A
B
C
D
E

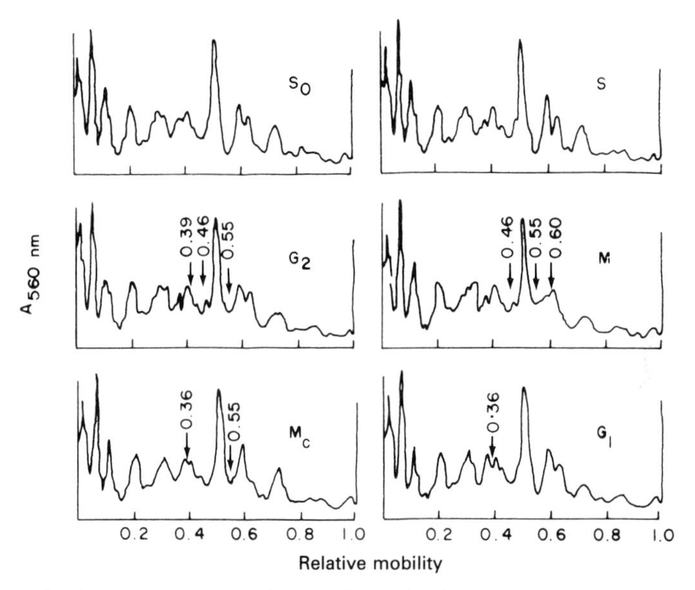

Fig. 5.16 Sodium dodecyl sarcosinate-polyacrylamide gel electrophoresis of plasma membrane proteins prepared from HeLa cells in various phases of the cell cycle. The gels were stained with Coomassie Brilliant Blue and scanned at 560 nm. Changing bands are indicated by arrows and relative mobilities. (Reproduced with permission from Johnsen *et al.*, 1975.) Cells harvested at the moment of release from a thymidine block are labelled S_0; other symbols have the usual meaning, except for M_c, which refers to cells harvested 5 h after Colcemid addition.

cytolysis achieved by exposure to hypotonic milieu, detergents or temperature shock, is also minimal in late interphase (Sumner *et al.*, 1973), perhaps as a result of the increased presence of microvilli.

Further evidence for the periodic exposure of particular proteins at the cell surface comes from observations of a high molecular weight protein in synchronous populations of mouse L cells (Hunt *et al.*,

---

Fig. 5.15 Scanning electron micrographs of synchronized P815Y mastocytoma cells. The scale marker in all micrographs equals 2 μm. Micrograph A shows an early G_1 cell. The majority of such cells have up to 3% of their surface elaborated into numerous ~0·1 μm diameter microvilli of varying lengths (×7 000). B shows a typical late G_1 cell in which much of the surface is obscured by microvilli. Branching of some microvilli is apparent and may seem to emerge from small flat protuberances of the cell surface (×6 200). C is a typical S cell. The often-branched microvilli range from 0·1 to 0·2 μm in length, stand erect from the surface and frequently emerge from large flat protuberances to the surface (×6 700). D shows a G_2 cell with its surface obscured by microvilli. E is a cell in the latter stages of cytokinesis. Most microvilli are in the region around the division plane, while the remainder of the surface is smooth and typical of early G_1 cells (×4 600). (Reproduced with permission from Knutton *et al.*, 1975.)

1975) and of a similar structure in NIL8 cells (Hynes and Bye, 1974). These proteins are preferentially accessible to the non-penetrant lactoperoxidase iodination probe during G_1. Furthermore, in P815Y cells, Pasternak *et al.* (1974) observed a gradual increase in the [^{125}I]-labelling of surface proteins during progression from G_1 to G_2. In HeLa cells, the externally-exposed proteins, which all appear to be glycoproteins, become generally more accessible to iodination in mitosis (Gudjonsson and Johnson, 1978).

An ultrastructural study of synchronized BHK 21/c 13 cells revealed a dramatic redistribution of some glycopeptides at mitosis and an increase in the availability of some basic protein groups that suggested their re-orientation in the membrane (Blanquet *et al.*, 1977). In HeLa cells, the exposure of an arginine-rich protein at the cell surface decreases as the cells go from mitosis into G_1 (Stein and Berestecky, 1975). The significance of these changes in membrane organization is not understood at present. The enhanced adhesiveness (which is abolished by trypsination) of CHO cells in G_1 has been tentatively attributed to changes in the presence or exposure of surface components (Hellerqvist, 1979).

Rabito and Tchao (1980) have used ouabain, which binds to the Na^+-K^+ATPase, to estimate the quantity of the enzyme during thymidine-synchronized growth of dog kidney cells; [^{3}H]-ouabain binding decreased during S and G_2 to reach a minimum at mitosis, when the membrane organization was disrupted. The transepithelial electrical resistance behaved similarly. These results are contrary to those obtained in fibroblasts (a non-epithelial cell line) by Graham *et al.* (1973) in which maximal "pump" activity was reached during mitosis. Membrane potential also varies during the cell cycle of Chinese hamster lung cells (Sachs *et al.*, 1974). An integrated study of membrane conductivity, pump activities and potential changes is called for.

Perhaps the most widely-studied example of chemical changes in the plasma membrane is that of its transient ability to bind plant lectins (for a review, see Prescott, 1976a). The assumed correlation between binding of lectins, which is maximal in mitotic cells (Fox *et al.*, 1971; Noonan and Burger, 1973; Mannino and Burger, 1975), and agglutination (Burger, 1969) has been questioned by Pasternak (1976a). Cell cycle-dependent changes in binding of Concanavalin A also occur at the cell membrane of *Amoeba* (Chatterjee *et al.*, 1979).

The importance of the lipid matrix in regulating the dynamics of cell membranes led to the study of membrane microviscosity as a function of the division cycle of neuroblastoma cells by de Laat *et al.*

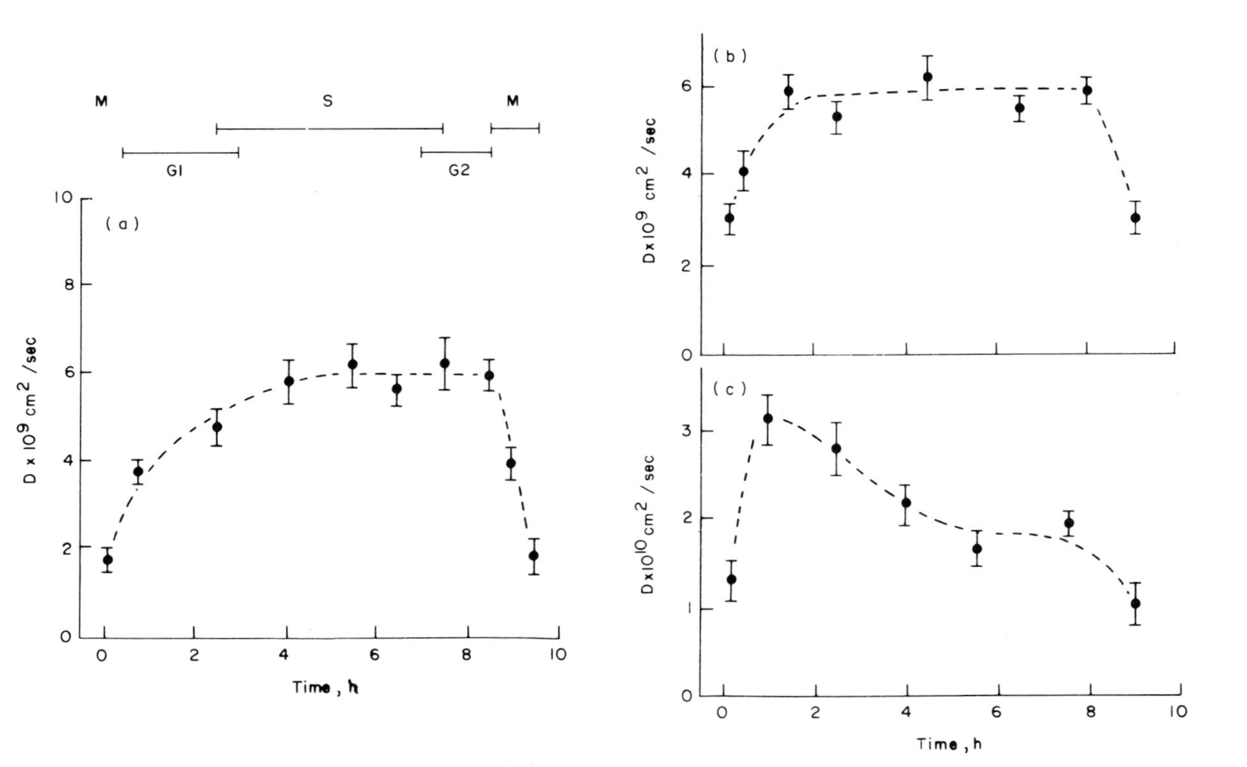

Fig. 5.17 Changes in the apparent diffusion coefficients for membrane probes during the cell cycle of neuroblastoma cells. The results shown are for two probes of lipid mobility, 3,3′-dioctadecylindocarbocyanine iodide; (a) and a fluorescein-labelled analogue ganglioside GM_1, (b) Membrane proteins were labelled with rhodamine conjugated rabbit antibodies against mouse lymphoid cells; (c) Fluorescent photobleaching recovery curves were measured at the times shown at two different locations per cell with at least two successive bleaches per location. The radius of the focused laser beam was approx. 1 μm in all experiments. (Reproduced with permission from De Laat *et al.*, 1980.)

(1977). Microviscosity was maximal at mitosis and dropped during G_1 to reach a minimum in S-phase. These and other results suggest that the cell membrane of mitotic cells is more rigid than that of interphase cells (Furcht and Scott, 1974; Garrido, 1975). The lateral mobility of membrane lipids is modulated in a fashion consistent with this, with diffusion coefficients reaching a minimum at mitosis (de Laat *et al.*, 1980). The mobility of proteins appears to be regulated by other constraints, because their diffusion coefficients decrease throughout G_2 and M (Fig. 5.17).

Other studies, however, have revealed maximal fluidity at mitosis (Collard *et al.*, 1977; Beiderman *et al.*, 1979; Lai *et al.*, 1980a, b) or no significant cell cycle dependent changes (Obrénovitch *et al.*, 1978). Interpretation of these data is frustrated by the possibility that the fluorescent probes used may be located in several membranes. Changes in fluorescein fluorescence and excitation polarization spectra were observed in synchronous cultures of S3 fibroblasts (Cercek *et al.*, 1978), but the significance of these fluctuations remains obscure.

G. CELL WALLS

1. *Yeast Cell Walls*

The fission yeast *Schiz. pombe* has long been regarded as a model organism for the study of binary fission and wall growth. May (1962) used fluorescent antibodies to show that this cylindrical cell extends exclusively at the poles; this result was confirmed in the experiments of Biely *et al.* (1973), though there were quantitative differences in the incorporation of wall precursors at the two poles (Johnson, 1965). Johnson (1965) found that only 20% of cells extend at both ends and Mitchison (1957) reported an even lower value. In contrast, no fewer than 80% of cells extend at both poles according to Streiblová (1977). To facilitate the study of polarized extension, Streiblová and Wolf (1972) exploited the finding that the division scars persist and are recognizable for several generations (Calleja *et al.*, 1980). Cell growth was found always to begin at the older "primary" end (Johnson, 1965) and persist virtually throughout the stage of extension growth (Streiblová and Wolf, 1972). The secondary (or younger) poles elongate only during the latter half of the cycle. After termination of pole growth, at the beginning of the constant volume stage (Mitchison, 1970), the cell plate appears, eventually dividing the cell into siblings of equal volume but unequal length (Johnson *et al.*, 1979). Formation of the cell plate or septum, (of unknown chemical nature; Johnson *et al.*, 1974) is initiated by the alignment of endoplasmic reticulum, followed by the appearance of dense granules and the progressive growth and thickening of an electron-transparent torus towards the cell centre (Oulevey *et al.*, 1970; Johnson *et al.*, 1973).

The wall is thought to extend (Johnson, 1968) by having glucan synthetase activity so closely coordinated with endoglucanase activity (Fleet and Phaff, 1974) that the cellular turgor pressure can cause only limited stretching in the weakened region of the broken glucan molecules. Indeed, endoglucanase activity has been found in vesicles believed to be near the cell poles (Cortat *et al.*, 1972). The broken glucan molecules are rejoined by covalent addition of glucose or oligoglucan (Kjosbakken and Colvin, 1975) across their breaks before complete rupture of the wall. This feature is also common to bacterial cell wall growth; peptidoglycan is inserted under strain-free conditions and only after the covalent links are formed are the intervening stressed peptide bonds cleaved (see Section III.B1; A. L. Koch, M. L. Higgins and R. J. Doyle, personal communication; Higgins and Shockman, 1976).

Bud formation in *S. cerevisiae* is initiated by the formation of a small surface bulge. Chung *et al.* (1965) demonstrated that the surface of the developing bud wall was almost entirely newly-synthesized, containing very little "old" wall material. Incorporation of 6-[^{3}H]-glucose into developing buds showed that wall growth occurs primarily at the cell tip (Johnson and Gibson, 1966a; Biely *et al.*, 1973) as it does in *Schiz. pombe* (Johnson, 1965) and *Pichia farinosa* (Johnson and Gibson, 1966b). Newly-synthesized mannan, labelled with Concanavalin A (Tkacz *et al.*, 1971) is deposited in the wall surrounding the bud (Tkacz and Lampen, 1972).

A recent affinity cytochemical study has shown that biotinylated sites on the yeast surface are stationary and not transferred to the newly synthesized wall of the bud (Skutelchsky and Bayer, 1979). The localized secretion of acid phosphatase follows the pattern of cell surface growth and is localized in the bud; this spatial organization requires the *cdc* 24 (see Chapter 7) gene product (Field and Schekman, 1980).

Despite extensive knowledge of the chemistry and organization of cell wall polymers in budding yeast, the pattern of their increase during the cell cycle remains equivocal. Early work indicated synthesis of glucan, but not of mannan, to be periodic (Phaff, 1971; Hayashibe *et al.*, 1970). Sierra *et al.* (1973) subsequently showed that both polymers were synthesized continuously during the cell cycle of *S. cerevisiae* (Fig. 5.18a), perhaps as a result of the stability of glucan

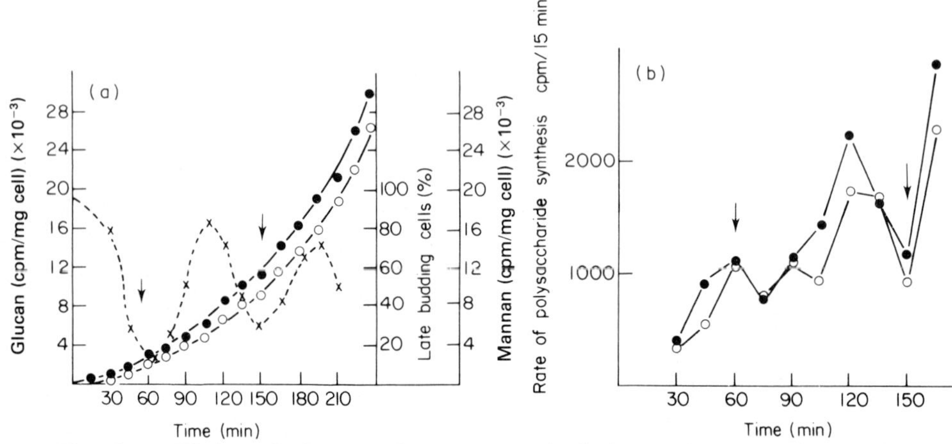

Fig. 5.18 Timing of glucan and mannan synthesis in synchronous cultures of *Saccharomyces cerevisiae*. (a) Cells (5 mg) collected from a discontinuous Ficoll gradient were suspended in a medium (100 ml) containing 5 μCi of [^{14}C] glucose at 28°C. Samples (5 ml) were withdrawn at time intervals and radioactivity incorporated into mannan (○) and alkali- and acid-insoluble glucan (●) determined. (b) shows the average *rates* of polysaccharide synthesized over 15 min intervals calculated from the data plotted in (a). In both graphs, arrows indicate initiation of budding in (a) X, the percentage of late budding cells. Reproduced with modification and permission from (a) Sierra *et al.*, 1973; and (b) from Biely, 1978.)

synthetases and of mRNA for wall mannan peptides (Elorza *et al.*, 1976). Others, however, have reported a reduction in the rate of synthesis of mannan and total glucan at the time of division (Biely *et al.*, 1973; Hayashibe *et al.*, 1977), and an analysis of these and other data by Biely (1978) indicates that synthesis may virtually cease at cell division and initiation of budding (Fig. 5.18).

The finding that the chitin of yeast cell walls is concentrated in bud scars (Bacon *et al.*, 1969; Cabib and Bowers, 1971) has provoked a detailed investigation of its role and synthesis. The chitin content of the wall increases once per cell cycle, with synthesis beginning shortly after the appearance of a new bud and ceasing before division is completed (Cabib and Farkăs, 1971). Subsequently, using fluorescent dyes presumed to bind to chitin, Hayashibe and Katohda (1973) concluded that the chitin of bud scars is deposited before bud emergence. In contrast, using a specific inhibitor of chitin synthesis, polyoxin D, Cabib and Bowers (1975) showed that chitin synthesis begins at, or shortly after, bud emergence. Exploiting the ability of an *Arthrobacter* β-glucanase to dissolve the non-chitinous wall fractions from *S. cerevisiae*, Vršanská *et al.* (1979) have shown that the initiation of chitin deposition in the walls of "virgin cells" is correlated with the formation of an annular structure that cannot be isolated at cell cycle stages preceding bud emergence or in the early stages of bud development. Thus, the annular chitin structure does appear to be formed progressively during bud growth and is not preformed before bud emergence.

Kopecká (1977) and Kopecká *et al.* (1977) have exploited Hartwell's temperature-sensitive *cdc* mutants of *S. cerevisiae*, blocked at various stages (Chapter 7), to demonstrate that the enzymes involved in wall biosynthesis are synthesized in the previous cycle. Protoplasts of *cdc* mutants blocked in G_0, G_1 or S were able to form an apparently-normal fibrillar wall fraction in a liquid medium, independently of DNA synthesis or its initiation. Synthesis of the wall β-1,3-glucan is triggered before initiation of DNA synthesis.

In certain filamentous fungi such as *Basidiobolus ranarum* and *Schizophyllum commune* there are close temporal, spatial and quantitative relations between mitosis and septation (Trinci, 1978). Thus, there is a constant interval between the initiation of mitosis and the initiation of septation in the apical compartments, only one septum is laid down per mitosis and this is formed at a site in the cytoplasm previously occupied by the metaphase plates of dividing nuclei. Such considerations have led Trinci (1979) to propose a close analogy between the cell cycles of yeasts and the so-called duplication cycles of fungal mycelia.

2. *Algal Cell Walls*

The cell walls of the alga *Chlorella* possess a remarkably resistant polymer-sporopollenin. The development of the wall in light–dark synchronized cultures of this organism has been described by Atkinson *et al.* (1972). Wall formation around the naked autospores, which are formed within the mother cell, commences with the appearance of small trilaminar plaques that fuse to form a sheath containing sporopollenin. Two phases of sporopollenin biosynthesis were detected during the cell cycle, one coinciding with the formation of the trilaminar component and the other 6 to 8 h earlier in karyokinesis, probably representing formation of a precursor. The dry weight and glucosamine content of the wall increase at the time of autospore formation (Richard and Broda, 1976; Takeda and Hirokawa, 1979).

The formation of the unique siliceous walls of diatoms has been studied in light–dark synchronized cultures by Volcani and his collaborators. The mature wall of *Navicula pelliculosa* is composed of two valves (an epitheca and a hypotheca), surrounded by a series of girdle bands, 3 bands on the epitheca and 2 on the hypotheca. The mother cell hypotheca will become the epitheca of the daughter and so, in preparation for division, the former requires a third girdle band. Following cytokinesis, the four distinguishable phases of valve formation are initiated (Chiappino and Volcani, 1977) by augmentation of an original, single, central band. Details of the process vary between diatoms (see references in Chiappino and Volcani, 1977; Pickett-Heaps *et al.*, 1979; Schmid and Schulz, 1979).

H. ORAL MORPHOGENESIS IN *TETRAHYMENA*

The oral apparatus of *Tetrahymena* is an organellar complex composed of ciliated and non-ciliated basal bodies, interconnected by a network of microtubules and filaments. A new mouth is differentiated at a site at a mid-ventral surface of the cell, posterior to the existing oral apparatus. At cleavage, the oral apparatus is retained by the proter (anterior cell), whilst the opisthe (posterior cell) receives the new structure. Oral development thus provides an example of extensive microtubule assembly that occurs regularly in the cell cycle and which can be easily studied microscopically (for references, see Nelsen, 1970). The stages and timing of this development in synchronous cultures were described by Frankel (1962). The duration of the process appears to be independent of the generation time (Antipa, 1980) and occurs in the last 90 min of the cell cycle (Suhr-Jessen *et al.*, 1977). The structure appears first as a field of argentophilic bodies described as the "anarchic field of kinetosomes", later to become organized into the oral apparatus. Heat shocks block development of the organelle in the opisthe, so that, by the end of seven heat shocks, most cells are in the "anarchic field" stage (Frankel, 1964) in which the kinetosomes are in various stages of ciliation (Buhse *et al.*, 1973). By the beginning of cytokinesis, oral regions of both daughter cells are in the same stage of development and remain synchronized until the cell divides.

The structure behaves, in its response to repeated heat shocks, in precisely the manner predicted for an organelle composed of "division proteins" (see Chapter 1). Treatment of a synchronous culture with heat shocks, cold shocks or metabolic inhibitors, when applied after the "transition point" are without effect on the development of the oral apparatus and cell division, but when applied earlier inhibit most cells from dividing and cause complete regression of the oral apparatus. An age-dependent division delay, similar to those caused by temperature change and anti-metabolites has been observed following colchicine treatment (Nelsen, 1970), suggesting that the assembly of microtubules prior to the transition point, as well as protein synthesis, is required for division. Treatment with colchicine after the transition point blocked oral development but not division, suggesting that microtubule formation is required for oral morphogenesis but not cell cleavage.

Whether the structural changes that occur in the oral apparatus are direct effects of the heat shock procedure and causally related to cell division, or are only indicators of the behaviour of structures (proteins?) more directly relevant to division is an open question (Zeuthen and Rasmussen, 1972). However, further persuasive evidence for the importance of the oral apparatus comes from the observation that a temperature-sensitive mutant of *T. thermophila* cannot be synchronized by heat shocks under conditions that are restrictive for oral development (Suhr-Jessen, 1978).

III. Surface and Membrane Growth in Bacteria

The relative, if only apparent, structural simplicity of prokaryotic cells makes them attractive models for the study of wall and membrane growth. Indeed, the early studies of Cole and Hahn (1962) provided the first convincing demonstration of the mode of synthesis of the cell surface in any micro-organism.

A. THE TEMPORAL PATTERN OF ACCUMULATION OF BACTERIAL CELL SURFACE COMPONENTS

There is overwhelming evidence that the synthesis of RNA and bulk proteins (Dennis, 1971; Ecker and Kokaisl, 1969), the main constituents of the dry weight of bacterial cells, as well as many specific proteins (Mitchison, 1971) takes place throughout the cell cycle. With regard to the protein components of membranes, however, there is less agreement.

1. *Proteins*

Synthesis of membrane proteins has been studied in synchronous cultures of *Bacillus subtilis* obtained by filtration (Sargent, 1973a, 1975a) and of *Escherichia coli* by the membrane elution technique (Churchward and Holland, 1976a). In *B. subtilis*, firmly-bound membrane proteins were found to be labelled discontinuously with no net synthesis occurring early in the cycle, followed by a period of rapid synthesis over the latter part of the cycle. Using pulse-labelling, it was shown that synthesis occurred at a constant rate that doubled at a fixed time in the cycle. On the basis of these experiments, it was suggested that proteins are added to the membrane from the cytoplasm and that during the period of zero net synthesis there is an efflux of proteins from the membrane. The increased rate of membrane synthesis was coincident with nuclear division (Table 5.1; Sargent, 1975a) and perhaps termination of chromosome replication (Sargent, 1975b).

Table 5.1. *Timing of cell cycle events in synchronous cultures of* Bacillus subtilis *168S*

	Time (min)	
	Succinate medium	Glucose medium
GENERATION TIME	115	60
1st cycle		
Nuclear division	30	30
Doubling in rate of membrane protein synthesis	40	25
DNA initiation	90	40
Cell separation I	80	40
2nd cycle		
Nuclear division	160	100
Doubling in rate of membrane protein synthesis	150	95
DNA initiation	200	100
Cell separation II	200	100

Each time represents the point in the cycle at which the rate changed fastest. In each case, the period of most rapid change occurs over one third of a cell cycle, so that the error in these figures is ±20 and 10 min for succinate and glucose, respectively. (Reproduced with permission from Sargent, 1975a)

Similarly, in *E. coli*, total envelope protein was synthesized throughout the cycle, with a stepwise doubling in the rate of synthesis in the first half of the cycle (Churchward and Holland, 1976). Analysis of the pattern of synthesis of about 29 individual envelope polypeptides revealed that all but one followed the pattern of bulk measurements. One envelope polypeptide (mol. wt 76 000) was, however, synthesized only near the time of division, and the bulk of it was lost from the envelope during the succeeding generation. Subsequently, Boyd and Holland (1979) demonstrated that the linear accumulation of envelope protein is restricted to the outer membrane (Fig. 5.19), the rate of accumulation doubling 10 to 15 min before division. The doubling is not controlled by a gene duplication mechanism.

Other individual membrane proteins, however, have been reported to be synthesized only periodically during the bacterial cell cycle. These include succinate dehydrogenase (Sargent, 1973a), cytochrome "b_1" (Ohki, 1972), potassium transport proteins (Kubitschek *et al.*, 1971), the bacteriophage λ receptor protein (Ryter *et al.*, 1975) and many periplasmic proteins (Shen and Boos, 1973). Unfortunately, diverse methods have been used for cell cycle analysis and some of these results are equivocal. For example, in contrast to Ohki's (1972) findings, all membrane-bound cytochromes in aerobically-grown *E. coli* have been found to accumulate continuously after analysis of the cell cycle by age fractionation (Scott *et al.*, 1982; Scott and Poole, 1981), in concert with other membrane-bound components of the respiratory chain (Poole *et al.*, 1980a). Changes in the activities of enzymes associated with bacterial walls and membranes are included in Table 4.3.

The delay between synthesis and incorporation into the membranes of proteins appears to be different for proteins destined for the inner and outer membranes respectively (Ito *et al.*, 1977), but this has not been studied as a function of cell age.

2. *Lipids*

There is conflicting information in the literature on the rates of lipid synthesis during the cell cycle and this has been summarized in Chapter 4 (Table 4.9). The experimental approach has in general been to measure incorporation of selected phospholipid precursors into cell material and in relatively few cases has the validity of extrapolating to synthesis of overall membrane lipid been established. In the experiments of Pierucci (1979), increases in the rate of [^{14}C] acetate incorporation were approximately coincident with increases in [^{14}C] thymidine incorporation (Fig. 5.20). Since the composition of newly-synthesized phospholipid has been shown to be independent of the age of cells (Zuchowski and Pierucci, 1978), the observed increases

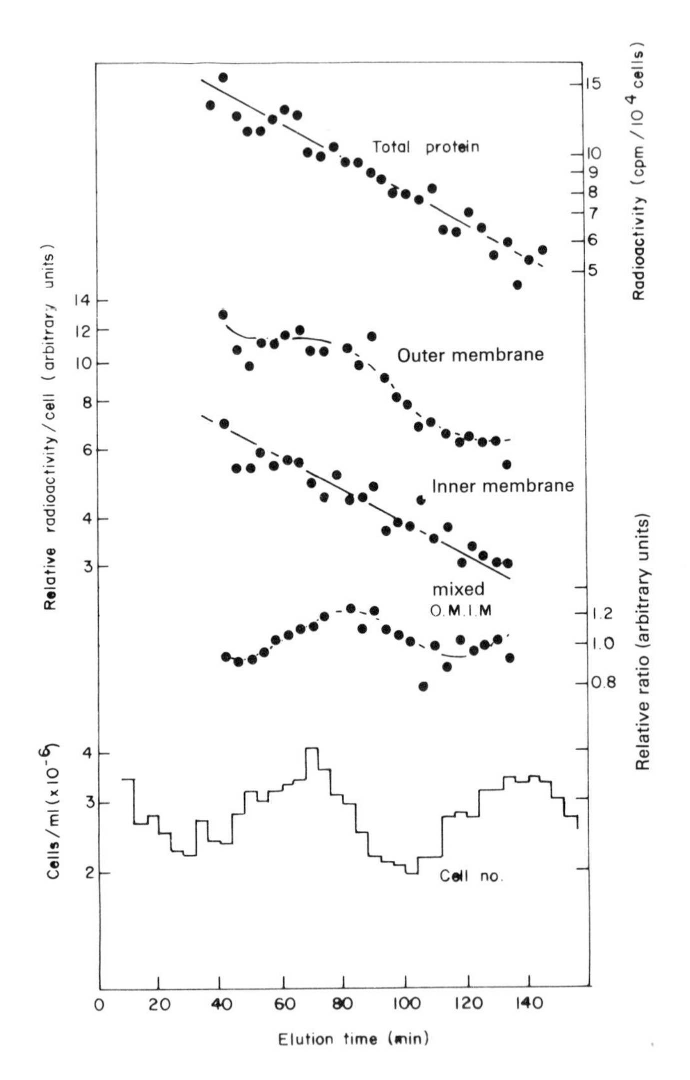

Fig. 5.19 Membrane protein synthesis in the cell cycle of *Escherichia coli*. *E. coli* LEB 16 growing exponentially in proline/alanine medium (+20 μg/ml thymine) was pulse-labelled with [^{14}C]-leucine for 5 min and fractionated on an age basis by the filter elution technique. Cell number and acid-precipitable radioactivity were determined in various cell fractions in each eluted sample. For membrane fractions the data were presented as a ratio of [^{14}C/^{3}H], where [^{3}H] is derived from the [^{3}H]-leucine-labelled culture, added as internal standard before the preparation of the membranes. Cell age increases from right to left in the figure. The bacterial number profile deviates considerably from the theoretical age distribution curve (Powell, 1956), presumably because of the variation in individual cell doubling times (Schaechter *et al.*, 1962). (Reproduced with permission from Boyd and Holland, 1979.)

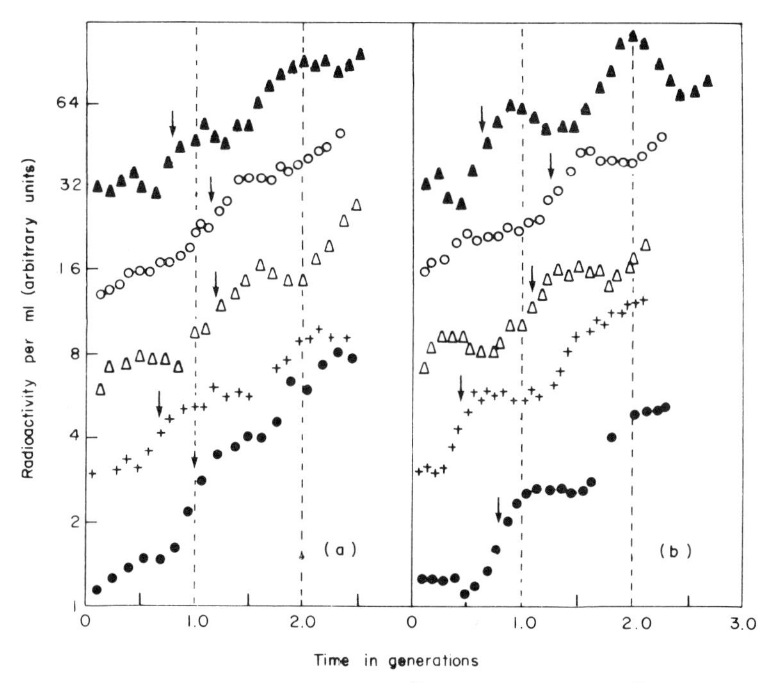

Fig. 5.20 Rate of incorporation of (a) [^{14}C]-acetate and (b) [^{14}C]-thymidine in synchronously-growing populations of *E. coli* b/r A. (a) Cultures were pulse-labelled with [^{14}C]acetate at a variety of cell ages. The cells were washed and the phospholipid was extracted. The growth rate and the medium composition of each culture were as follows: (●) 1 doubling per h, glycerol; (+) 1·2 doublings per h, glucose; (△) 1·76 doublings per h, glucose supplemented with methionine, histidine, and arginine; (○) 1·0 doublings per h, glucose supplemented with methionine, histidine, arginine, leucine, threonine, and proline; (▲) 2·1 doublings per h, glucose supplemented with Casamino Acids and tryptophan; (b) cultures were pulse-labelled with [^{14}C]-thymidine at a variety of cell ages, and the radioactivity in trichloroacetic acid-insoluble material was determined. The media compositions were the same as in (a). The growth rates that they supported were: (●) 0·97 doublings per h; (+) 1·3 doublings per h; (△) 1·86 doublings per h; (○) 2·0 doublings per h; (▲) 2·2 doublings per h. Arrows indicate the midpoints of the increases in rate; the dashed vertical lines are drawn at the midpoints of the increase in cell numbers. (Reproduced with permission from Pierucci, 1979.)

in the rate of synthesis of phosphatidylethanolamine must be accompanied by increases in the rate of synthesis of both phosphatidylglycerol and cardiolipin. Thus, an overall increase in phospholipid synthesis occurs, probably coincidently with the initiation of chromosome replication. The age in the cycle at which these events occur is a function of growth rate (Fig. 5.20).

3. *Two Case Studies: Membrane Synthesis in* Rhodopseudomonas *and* E. coli

One of the most detailed studies of membrane synthesis comes from the work of Kaplan and his collaborators. When synchronously-dividing, photosynthetic cultures of *Rhodopseudomonas sphaeroides* undergo a shift from D_2O- to H_2O-containing medium, the density of the intracytoplasmic photosynthetic membrane (ICM) undergoes a discontinuous decrease (Fig. 5.21; Lueking *et al.*, 1978). This decrease in density is coincident with cell division and the magnitude of the change reflects the dilution of "old" macromolecules with "new" macromolecules. Fraley *et al.* (1978) and Wraight *et al.* (1978) have shown that, during the cell cycle, the reaction centre and light-

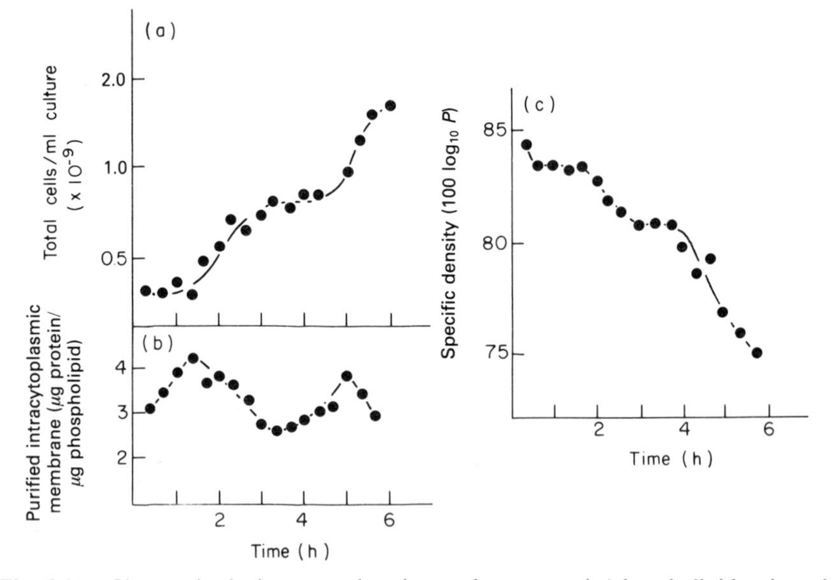

Fig. 5.21 Changes in the intracytoplasmic membrane protein/phospholipid ratio and density during the synchronous growth of *Rhodopseudomonas sphaeroides*. Division synchrony was obtained by a modification of the technique of Cutler and Evans (1966). Following transfer of the cells to H_2O-based medium, cell samples were collected every 5 min and purified intracytoplasmic membrane fractions were obtained employing discontinuous sucrose gradients. The lipid phosphorus values were multiplied by a factor of 25 to obtain the total quantity of phospholipid. Membrane densities were determined employing linear CsCl gradients; (a) total cells per ml culture; (b) micrograms of protein to micrograms of phospholipid in purified intracytoplasmic membranes; (c) intracytoplasmic membrane specific density. (Reproduced with permission from Lueking *et al.*, 1978.)

harvesting bacteriochlorophyll protein complexes and carotenoids undergo continuous synthesis and incorporation into the membrane. In addition, greater than 90% of the ICM protein (including cytochromes *b* and c_2) is continuously synthesized and at constant relative rates. Succinate dehydrogenase and NADH oxidase activities, however, increase discontinuously. Phospholipid incorporation into the ICM is also discontinuous, and increases coincident with cell division; phospholipid turnover is negligible. In addition, the protein to phospholipid ratio of the ICM as well as the buoyant density of the ICM changes in a cell cycle-specific fashion (Fraley *et al.*, 1979). Through the use of the fluorescent probe, α-parinaric acid (Fraley *et al.*, 1978), it has been shown that the cell cycle-specific changes in the ICM described above are accompanied by cyclic alterations in the microviscosity of the membranes (Fraley *et al.*, 1979).

More recent work has shown that each phospholipid species studied is accumulated discontinuously into the ICM. Pulse label studies suggest that this phenomenon arises from regulation of the cell cycle-specific transfer of phospholipid from its site of synthesis to the ICM, rather than from regulation of phospholipid synthesis (B. Cain, C. Deal, R. T. Fraley and S. Kaplan, unpublished, cited by Kaplan *et al.*, 1980).

Thus, despite the bilayer continuity of the ICM (Prince *et al.*, 1975), a barrier exists to the lateral movement of phospholipids between the two membrane domains. It has been suggested (Kaplan *et al.*, 1980) that at cell division, pre-existing photosynthetic membrane invaginations undergo division concomitant with the insertion of newly synthesized phospholipid into the ICM.

As in *R. sphaeroides*, many individual, membrane-bound, energy transducing components of *E. coli* appear to accumulate continuously and at constant relative rates during the cell cycle. These components include all cytochromes whose absorption maxima in the α-region of their visible spectra may be resolved by fourth-order finite difference analysis of low temperature (77 K) spectra (Scott and Poole, 1981) and the major terminal oxidase, cytochrome *o* (Fig. 5.22) when assayed with unequivocal specificity by flash photolysis of the CO-liganded enzyme under anoxic conditions (Scott *et al.*, 1981). Electron paramagnetic resonance (e.p.r.) spectroscopy of intact cells after zonal sizing reveals that two potentiometrically-distinct ferredoxin-type FeS clusters as well as signals from high-spin haemoproteins all increase continuously through the cell cycle (Poole *et al.*, 1981). The results of these direct spectral measurements of individual membrane proteins are consistent with the data of Lutkenhaus *et al.* (1979), who found

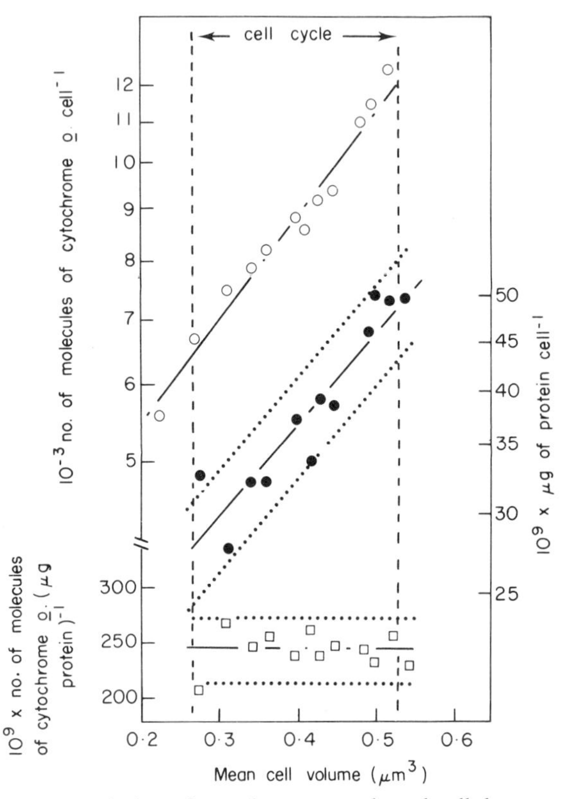

Fig. 5.22 The accumulation of cytochrome *o* and total cellular protein during the cell cycle of *Escherichia coli* K12. Cell cycle analysis was performed by size (=age) fractionation on sucrose gradients in a zonal rotor. Cytochrome *o* amounts were calculated from measurements of the Soret absorption bands in photodissociation spectra recorded at −100°C. The lines fitted to the data for amounts of cytochrome *o* per cell (○) and total protein per cell (●) are those of best fit, the correlation coefficients, r, being 0·98 and 0·92 respectively. The line fitted to the data for amount of cytochrome *o*:(mg protein)$^{-1}$ (□) represents the mean value; the coefficient of variation (100 × standard deviation.mean) is 7·0%. The vertical dashed lines define a doubling in mean cell volume and indicate the average cell volume at birth and division, respectively, assuming that growth in cell volume is linear through the cell cycle. Where shown, pairs of dotted lines indicate the extent of two standard deviations from the fitted curves. (Reproduced with permission from Scott *et al.*, 1981.)

no protein (of about 750 indentified on gels) that accumulated discontinuously.

The spatial organization of some of these carriers has been investigated by e.p.r. spectroscopy of membrane preparations ("oriented multilayers"), in which the membrane planes lie more-or-less parallel to each other. Several cytochromes (Poole *et al.*, 1980a) and FeS clusters (Blum *et al.*, 1980) have been shown to be oriented with respect to the membrane plane. The very high degree of orientation of a low-spin cytochrome (perhaps cytochrome *o*) in membranes derived from an asynchronous culture (Poole *et al.*, 1980a) and the direct demonstration (Poole *et al.*, 1981) that the orientation of ferredoxin-type clusters is the same in membranes prepared from (a) cell cycle fractions or (b) an asynchronous culture (Fig. 5.23), lead to the conclusion that there is no substantial cell cycle-dependent modulation of the orientation of these components.

As described in Chapter 4, the activity of the Mg^{2+}-ATPase in cell-free extracts, and its sensitivity to Ruthenium Red oscillate during the cell cycle (Scott *et al.*, 1980). In spite of this, however, the amount of the catalytic F_1 component of the ATPase, as measured immunochemically in whole cell digests, accumulates continuously during the cell cycle (N. Smith, D. Boxer, R. I. Scott and R. K. Poole, unpublished results). It remains to be established whether the enzyme is membrane-associated throughout the cell cycle, or whether its binding is cell cycle dependent as has been suggested for the ATPase of *Alcaligenes eutrophus* by Edwards *et al.* (1978).

4. *Temporal Incorporation of Transport Proteins*

The well-established fact that the cell envelope of Gram-negative bacteria grows throughout the cell cycle (see above, also Bayne-Jones and Adolph, 1933; Schaechter *et al.*, 1962; Mitchison, 1971) does not necessarily imply that any component other than the peptidoglycan layer grows throughout the cycle, since the inner and outer membranes are somewhat flexible. Kubitschek (1968a) has suggested that the envelope grows linearly throughout most of the cell cycle, but that the membrane-bound transport machinery is formed just before cell division. Cell growth is thus assumed to be limited by the availability of functional transport systems. Measurements of transport capacity in *E. coli* through the cell cycle have been used to support this view (Kubitschek, 1968b; Kubitschek *et al.*, 1971). In contradiction to these proposals and results, Koch (1975) has shown that cells from an asynchronous culture are capable of growth almost immediately when switched to a new carbon and energy source whose metabolism requires new membrane function, suggesting that the carrier concerned (the y gene product of the *lac* operon) is synthesized, incorporated and can start functioning in transport at any time in the cell cycle. This conclusion is reinforced by the demonstration (cited by Koch,

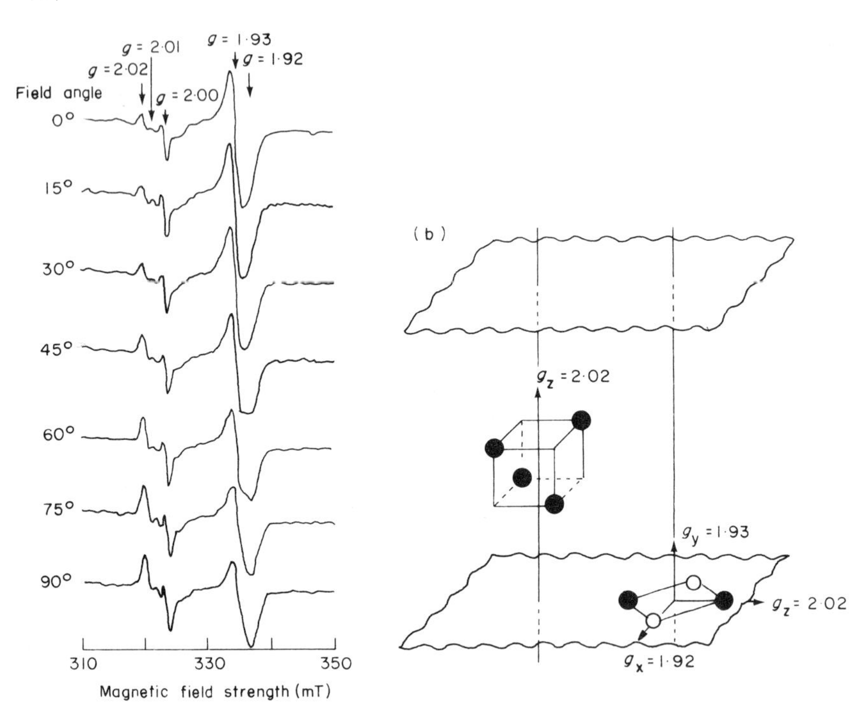

Fig. 5.23 The orientation of Fe–S clusters in the cytoplasmic membrane of *Escherichia coli* K12. Multiple parallel layers of cytoplasmic membrane from an asynchronous culture of *E. coli* were prepared by a centrifugation technique and rotated in the magnetic field of an e.p.r. spectrometer (Poole *et al.*, 1980a). (a) shows the angular dependence of the e.p.r. spectra of such multilayers after reduction with sodium succinate. The field angle is the angle that the applied field makes with the normal to the multilayer plane. Prominent features of the spectra are labelled. The sample temperature was 40 K and the incident microwave power 10 mW at 9·06 GHz. Similar orientation-dependent effects have been observed in layers prepared from cells at 0·2, 0·48 and 0·92 of the cycle after age fractionation in sucrose gradients in a zonal rotor (Poole *et al.*, 1981). (b) illustrates the orientation of the ferredoxin-type cluster seen in (a) and of a HiPIP signal that is proposed to occur throughout most of the cell cycle. The plane of the membrane is horizontal (almost perpendicular to the page). The upper and lower faces of the membrane are shown. The structure on the left is a tetranuclear cluster (bacterial ferredoxin and HiPIP types) containing four iron atoms (shaded) and four acid-labile sulphur atoms. The $g_z = 2·02$ signal lies parallel to the membrane normal. Since no model exists which relates the magnetic anisotropy to the structure of tetranuclear iron–sulphur clusters, the illustrated orientation of this component is only assumed. The structure on the right is a binuclear cluster (plant ferredoxin type) containing two iron atoms (shaded) and two acid-labile sulphur atoms. The structure shown is that of Gibson *et al.* (1966) in which the $g_z = 2·02$ axis corresponds to the Fe–Fe axis and is shown as lying in the membrane plane. The $g_x = 1·92$ axis also lies in this plane, whereas the $g_y = 1·93$ axis lies parallel to the membrane normal. (Reproduced with permission from Blum *et al.*, 1980.)

1975) that a pulse of inducer for one-tenth of a cell cycle gives rise to permease activity in cells of all sizes in an asynchronous culture. Other experiments by Cohn and Horibata (1959) and Novick and Weiner (1957, 1959) have been cited by Koch in support of his findings.

B. TOPOGRAPHY OF GROWTH OF BACTERIAL SURFACES

1. *Cocci*

Studies of cocci in which the walls of chain-forming cells were uniformly labelled with immunofluorescent (Cole and Hahn, 1962; Cole, 1965), autoradiographic (Briles and Tomasz, 1970) or morphological markers (Higgins *et al.*, 1971) have indicated that new unlabelled segments of peripheral wall are issued symmetrically from the septal regions of these cells. In *Streptococcus faecalis*, septal synthesis forms discrete zones of new envelope, which are separated from the old by raised bands of wall material (Higgins *et al.*, 1971).

Based on observations of thin sections, in which the thickness and shape of the cell wall was studied, a model of surface growth has been proposed (Higgins and Shockman, 1970, 1971) in which active wall precursors are incorporated along the leading edge of the cross wall. This incorporation would cause the leading edge to extend centripetally into the cell (like the closing of an iris diaphragm), finally producing a complete septum. Higgins and Shockman (1976) have used computer-assisted, three-dimensional reconstruction of growth zones from data provided by median sections to propose a model in which precursors are channelled into cross wall assembly, which proceeds at a constant rate until the cross wall closes. Separation of the bilayered cross wall is presumed to occur by a second channel of precursors which thicken the separating layers of the cross wall (Fig. 5.24). Autolytic enzymes, whose activities fluctuate during the cycle (Hinks *et al.*, 1978) are involved in regulating both the rate at which the cross wall layers are split and the amount of remodelling that takes place. Decreases in flux through this second channel during the cell cycle coupled with internal hydrostatic pressure, would result in increasing curvature of the peripheral walls and the promotion of cell division. An important assumption of this model, namely that the septal material is under tension before it is externalized to form peripheral walls, has been developed by A. L. Koch, M. L. Higgins and R. J. Doyle (personal communication); who propose that the shape of the central growth zone is determined by forces similar to surface tension forces.

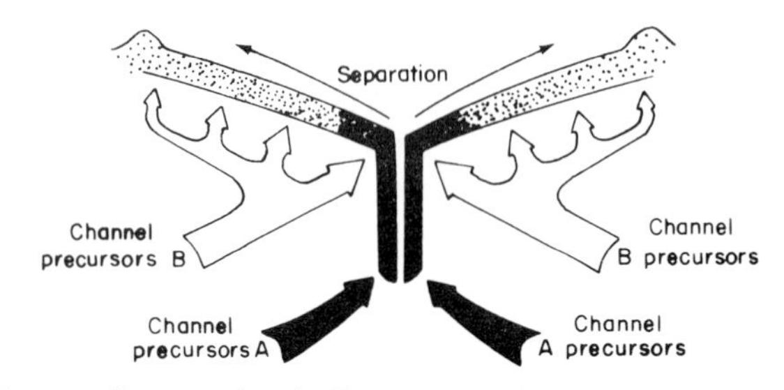

Fig. 5.24 Cross wall separation in *Streptococcus faecalis*. It is proposed that the cell wall of *S. faecalis* grows by the regulated flow, and integration, of two channels (A and B) of cell wall precursors into a growth zone. Channel A is involved in the synthesis of new cross-wall material, and the flow of precursors through this channel is apparently constant until septation. The location for the integration of these precursors into new cross wall surface is not presently known. Channel B brings the cross wall to a constant thickness at its base before septation and then intercalates into and increases the thickness of the cross-wall as it separates into two layers of peripheral wall. The actual role played in this process by stress or the stretching of the separated cross wall layers cannot be assessed at this time. As the peripheral wall moves from the septal region of the cell by new peripheral wall growth, channel B precursors would be used in ever-decreasing amounts to thicken and expand the surface area of the peripheral wall with time. Also, the flow of precursors through channel B would be reduced through the cycle correlating with an increase in the curvature of the peripheral wall. This supposedly would reduce the total amount of surface that must be assembled for septation to occur. Thus, the reduction of channel B precursors stimulates cell division. (Reproduced with permission from Higgins and Shockman, 1976.)

It is thus possible to blow soap bubbles of similar shape to the central growth zone of *Streptococcus*.

2. *Rod-shaped Bacteria*

(a) *Dispersive vs. conservative membrane growth.* The mode of growth of the surfaces of rod-shaped bacteria ia not known with any certainty, despite a voluminous literature. This is largely due to (i) the high rate of peptidoglycan turnover, which would obscure any pattern of growth, and (ii) the weak bonding and thus lateral mobility of experimentally useful antigens.

The bulk of data reviewed by Sargent (1979) relating to growth of the peptidoglycan layer suggests that synthesis occurs by intercalation of new material into old (Fig. 5.25), but several recent claims have contradicted this pattern (e.g. Sturman and Archibald, 1978; Koppes *et al.*, 1978a, b; Pooley *et al.*, 1978). The age-dependent degree of lysis of *Bacillus subtilis* by lysozyme may reflect discontinuities (spatial or temporal) in the synthesis of lytic sites, wall synthesizing enzymes or remodelling autolytic enzymes (Edwards, 1980a).

With respect to the outer membrane, there is some convincing evidence for conservation of growth zones (e.g. Begg and Donachie, 1973, 1977) but, again, interpretation of results has been questioned (see Smit and Nikaido, 1978). Matrix protein, which represents about 35% of the total outer membrane protein, is not spatially conserved, and is not even localized for short periods at the time of incorporation (Begg, 1978).

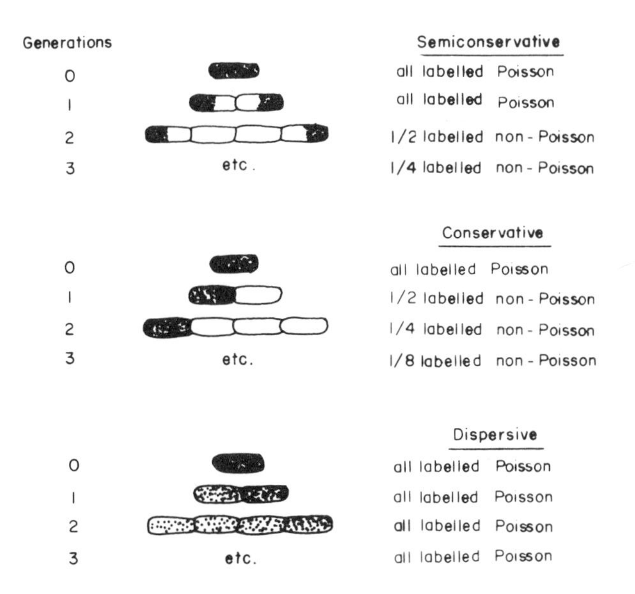

Fig. 5.25 Patterns of "old" and "new" membrane material predicted by the semi-conservative, conservative, and dispersive growth models. The dark patches shown at zero time in this figure represent labelled membrane material. The drawings show the hypothetical distribution of label in the three models. The right side of the figure shows the fraction of labelled cells and the expected fit of the observed distribution to the Poisson distribution. (Reproduced with permission from Green and Schaechter, 1972.)

Particular interest has been focused on the mode of growth of the cytoplasmic (inner) membrane because of the suggestion (discussed in Section III.B2b) that DNA is membrane-bound and that its replicated strands are separated as a result of localized membrane growth.

With only one exception (Morrison and Morowitz, 1970), experimental data do not support the idea that membrane phospholipids are synthesized and conserved in specific zones on the cell surface (Lin *et al.*, 1971; Tsukagoshi *et al.*, 1971; Green and Schaechter, 1972; Mindich and Dales, 1972; Tsukagoshi and Fox, 1979). In view of the rapid mobility of lipids within the planes of membranes (Finean *et al.*, 1978), this conclusion does not appear surprising. Although, as pointed out by Sargent (1979), bacterial membranes do have high protein : lipid ratios, and the mobility of membrane components may be further decreased by association with the adjacent rigid wall, measurements of protein mobilities in the cytoplasmic membrane of *E. coli* (Haest *et al.*, 1974; M. T. Davison and P. B. Garland, personal communication) indicate rates that appear to be incompatible with the lateral asymmetry required for semi-conservative segregation of newly synthesized membrane proteins. Wilson and Fox (1971) and Green and Schaechter (1972) concluded that segregation of proteins was dispersive or that only small fragments of membrane were conserved at division. By contrast, Kepes and his co-workers have examined the segregation of a number of markers (Fig. 5.26) and concluded that segregation is semi-conservative with a central growing zone (Autissier and Kepes, 1971, 1972; Autissier *et al.*, 1971). Results on cytochrome segregation that are consistent with this idea have been reported by Poole *et al.* (1978). Cadenas and Garland (1979) have used the same penicillin selection method to study the segregation of membrane-bound respiratory nitrate reductase in *E. coli* for the three generations after cessation of enzyme synthesis caused by removal of nitrate from the medium. On the basis of these experiments (Fig. 5.27) and the direct assay of enzyme activity in the fractions produced by the penicillin selection method, they concluded that segregation of this enzyme is dispersive and not semi-conservative. Similar conclusions as to the intercalation of new permeases have been drawn by Koch and Boniface (1971).

(b) *Membranes and chromosome replication.* The most widely discussed model of genome segregation is that of Jacob *et al.* (1963) who suggested that chromosomes are attached to the cell membrane by the chromosome origin and are then separated and packaged into daughter cells by extension of the cell surface between these attachment points. The evidence for and against this model has been critically discussed by Sargent (1979). There seems little doubt that the chromosome *is* attached to the membrane at the origin in rod-shaped bacteria as well as in cocci (Yamaguchi and Yoshikawa, 1973, 1977; Ryder *et al.*,

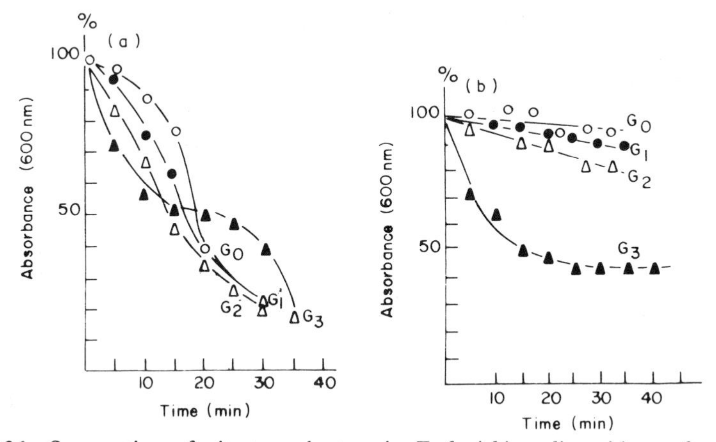

Fig. 5.26 Segregation of nitrate reductase in *Escherichia coli*—evidence for conservative membrane growth. *E. coli* K12 was grown anaerobically overnight on mineral medium with nitrate as nitrogen source and glucose as carbon source to induce the synthesis of nitrate reductase. The cells were de-induced by anaerobic growth on ammonium-glucose medium. Cells de-induced for 0, 1, 2, and 3 generation times were submitted to a penicillin lysis test for viable cells (a) in the same medium used for induction (with nitrate) or (b) in the medium used for de-induction (with ammonium), supplemented with 10 mM $KClO_3$. In (a), cells containing nitrate reductase were lysed quickly after 0, 1 and 2 generations. After 3 generations, a fraction of the population resistant to lysis arose and were taken to be those cells lacking the enzyme. In (b) the cells were killed by chlorate and not lysed by penicillin. (Reproduced with permission from Kepes and Autissier 1972.)

1975; Beeson and Sueoka, 1979; see also Sargent, 1980) but confirmation of Jacob's model will require further investigations, in particular of the mode of growth of the membrane and the demonstration of a conservative growth pattern.

3. Caulobacter *and Budding Bacteria*

Those events related to morphogenesis and differentiation in prokaryotes are best studied in organisms with complex fine structures and morphologically distinctive cell cycles, such as the endospore-forming *Bacillus* species, *Streptomyces*, *Arthrobacter*, (e.g. Kolenbra and Hohman, 1977), *Myxobacter*, the cyanobacteria and the prosthecate bacteria that include *Caulobacter* and the budding bacteria (Whittenbury and Dow, 1977). In budding bacteria, in contrast to rod-shaped bacteria, wall growth is always polar, allowing morphogenetic development (Fig. 5.28). The basic cell cycle events of the much-studied *Caulobacter crescentus* are described in Chapter 7.

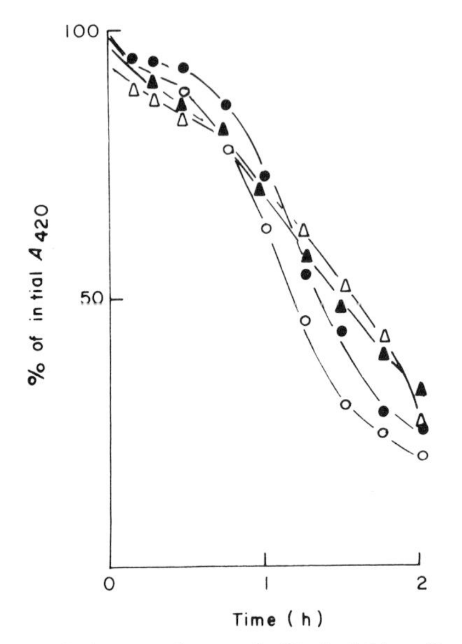

Fig. 5.27 Segregation of nitrate reductase in *Escherichia coli*—evidence for dispersive membrane growth. The principle of the experiment is the same as that explained in the legend to Fig. 5.26. This figure shows penicillin lysis of cells of various generations after transfer from inducing to non-inducing growth medium. Cells at the time of transfer are referred to as zero generation (▲), and thereafter cells were sampled from the non-inducing growth medium after one (△), two (○) and three (●) generations. All these samples were exposed to the EDTA/penicillin lysis procedure, and lysis was followed spectrophotometrically at 420 nm in cuvettes of 1 cm light path. (Reproduced with permission from Cadenas and Garland 1979.)

In synchronous cultures of *C. crescentus* started with flagellated cells, lipid and protein incorporation into membranes occurred during most of the cell cycle, the rate slightly declining before division (Galdiero, 1973b). Parental membrane appeared to be equally partitioned between the daughter cells at division. In contrast, Agabian *et al.* (1979) have shown that the association of membrane proteins with a specific cell type results not only from their periodicity of synthesis, but also from the asymmetric pattern of segregation between the progeny stalked and swarmer cells. The asymmetry is generated primarily by the preferential association of specific soluble proteins with the membrane of only one daughter cell. The majority of proteins that exhibit this segregation behaviour are synthesized during the entire cell cycle and exhibit relatively long functional messenger RNA half-lives. Swarmer and stalked cells also differ with respect to their pattern of membrane phospholipid biosynthesis (Mansour *et al.*, 1980).

The single flagellum of *C. crescentus* is an example of a polar structure whose expression, as assayed by using a flagellotropic phage (Fukuda *et al.*, 1976), occurs at a specific time in the cell cycle. Flagellar synthesis is also periodic in the cell cycle (Shapiro and Maizel, 1973). The temperature-sensitive flagella mutants isolated by Marino *et al.* (1976), and which have been shown to be defective in synthesis of the structural flagellin monomer, open new avenues of investigation into flagellar assembly during the cell cycle.

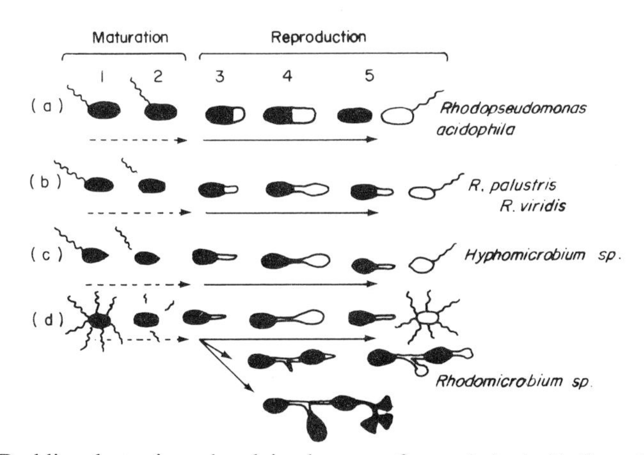

Fig. 5.28 Budding bacteria ordered in degrees of morphological/cell cycle complexity. In all cases, motility, flagella shedding, maturation of the cell, daughter cell synthesis by obligate polar growth, and asymmetric cell division give rise to an immature daughter cell and a mature mother cell: (a) *Rhodopseudomonas acidophila*. Simple dimorphic life cycle not involving tube/filament synthesis; (b) *Rhodopseudomonas palustris* and *R. viridis*. Involvement of tube/filament synthesis prior to daughter cell formation; (c) *Hyphomicrobium* species. Reduction in width coupled with an increase in length of the tube/filament compared with *R. palustris*; (d) *Rhodomicrobium* species. Enviromentally induced life cycle variations ranging from a simple dimorphic cell cycle through complex matrix formation to exospore production. (Reproduced with permission from Whittenbury and Dow, 1977.)

The proteins of the envelope-associated nucleoid are derived primarily from membranes and are sequentially associated with the nucleoid during the cell cycle (Evinger and Agabian, 1979). One protein has been shown to segregate preferentially with the swarmer cell nucleoid at division.

Summary

Cycles of organelle growth and division that are tightly co-ordinated with progress through the cell cycle, have been observed in higher and lower eukaryotic cells and described in considerable detail in the cases of nuclei, mitochondria, vacuoles and chloroplasts. The close spatial connections that occur at specific points in the cycle between different classes of organelles remain to be explained as do the mechanism(s) by which organelle division is timed and the progeny equally distributed between the daughter cells. Ribosomes and polysomes exhibit complex dynamics; their cycle of rapid synthesis and degradation are of special interest because of possible correlations with the discontinuous synthesis of certain proteins. The cell surface of mammalian cells changes dramatically during the cell cycle in morphology, composition, fluidity and the accessibilty of membrane components at the cell surfaces. In lower eukaryotes, particularly yeast, cell wall growth occurs by localized, intricate interplay between lytic and synthetic activities.

There is much evidence for the continuous synthesis and accumulation of membrane proteins in bacteria. In *E. coli*, membrane growth occurs without discernible effect on the spatial organization of membrane proteins. Patterns of lipid synthesis are more equivocal and perhaps continuous. Oscillatory and stepwise changes in the rate of lipid synthesis have also been described. In *Rhodopseudomonas sphaeroides*, marked variations in membrane density and viscosity are the result of cell cycle-specific transfer of phospholipids from the site of synthesis to the intracytoplasmic membrane. The topography of surface growth, especially of membranes, in bacteria remains controversial; the relatively few claims that new material is inserted in a localized fashion and that growth zones are conserved deserve attention because of the potential importance of such a mechanism in effecting genome segregation.

6. Bioenergetics and Transport

"[Lifes'] dance to the music of time."

Anthony Powell

There have been relatively few reports concerning studies of energy metabolism during the cell cycle and, where such experiments have been carried out, the underlying mechanisms responsible for the complex patterns that have been observed are often poorly understood. If the undisturbed cell cycle is to be studied, then selection procedures are the method of choice to produce synchronous cultures, and where possible, adequate control experiments should be performed to ensure that the experimental procedures employed do not induce metabolic perturbations. As we shall see, these two considerations have not always been employed in many of the published reports. This section is concerned with the patterns of respiratory activity that have been observed during the cell cycle, and discusses some of the control mechanisms that may be involved, including regulation by adenine nucleotides or substrate availability.

I. Changes in Respiratory Activities during the Cell Cycles of Eukaryotes

A. MAMMALIAN CELLS IN CULTURE

Information on changes in oxygen uptake rates during the cell cycles of mammalian cells is very incomplete. In HeLa cells synchronized either by amethopterin and thymidine (Gerschenson *et al.*, 1965), or thymidine treatment followed by mitotic detachment (Robbins and Morrill, 1969), oxygen uptake rates are low in mitosis. In the latter case, rates of respiration may be correlated with rates of macromolecular synthesis. Respiration rates are also low during mitosis in mouse cells synchronized by mitotic selection (Ishiguro *et al.*, 1978), although another minimum is also observed at the end of G_1 and a maximum at the end of S phase.

Exposure of Ehrlich ascites tumour cells to anaerobiosis or the respiratory inhibitors rotenone or antimycin A, results in an increased rate of glycolysis; after anaerobiosis cells become blocked in G_1, after exposure to rotenone, they are blocked in the G_2 + M boundary after exposure to antimycin A they do not leave the S-phase (Loffler *et al.*, 1980).

B. YEASTS

In cultures of *Saccharomyces cerevisiae*, synchronized by a regime involving starvation and feeding, respiration rates increased in a stepwise pattern, remaining constant between successive increases (Scopes and Williamson, 1964). However, the steps were less than doublings; volume and mass also failed to double over the initial cycles and the intervals between steps were less than the generation time. Similar results were obtained by Greksák *et al.* (1971) but a continuous, exponential increase in oxygen uptake rates was reported by Cottrell and Avers (1970). Oscillatory patterns of respiratory activity (one peak per cycle) were observed in cultures of *S. cerevisiae*, prepared by a size-selection technique, grown in defined (Küenzi and Fiechter, 1969; Nosoh and Takamiya, 1962) or complex media (von Meyenburg, 1969; Wiemken *et al.*, 1970a). Such oscillatory expression was observed when the carbon source was glucose, but not maltose in the experiment of Dharmalingham and Jayaraman (1973).

Synchronous cultures of the fission yeast, *Schiz. pombe*, prepared by a nutritional induction method in a complex growth medium, showed a single stepwise increase in respiration rate per cell cycle (Osumi and Sando, 1969; Osumi *et al.*, 1971). A more complex oscillatory pattern (Fig. 6.1) was found in synchronous cultures prepared by a sedimentation-velocity selection method of growing with 1% glucose as the carbon source, (Poole *et al.*, 1973). Total cell protein and respiration rates doubled overall during the cycle, but the latter rose to maxima twice per cell cycle. The first maximum occurred at about 0·5 of a cycle and the other during cell division. Respiratory maxima and minima showed marked differences in responses to the stimulatory effect of 16 μM-carbonyl cyanide m-chlorophenyl-hydrazone (CCCP); oxygen uptake at respiratory maxima was stimulated to a far greater extent than at minima. Antimycin A or CN^- did not attenuate or alter the periodicity of these oscillations; heat production increased uniformly through the cell cycle. These observations led the

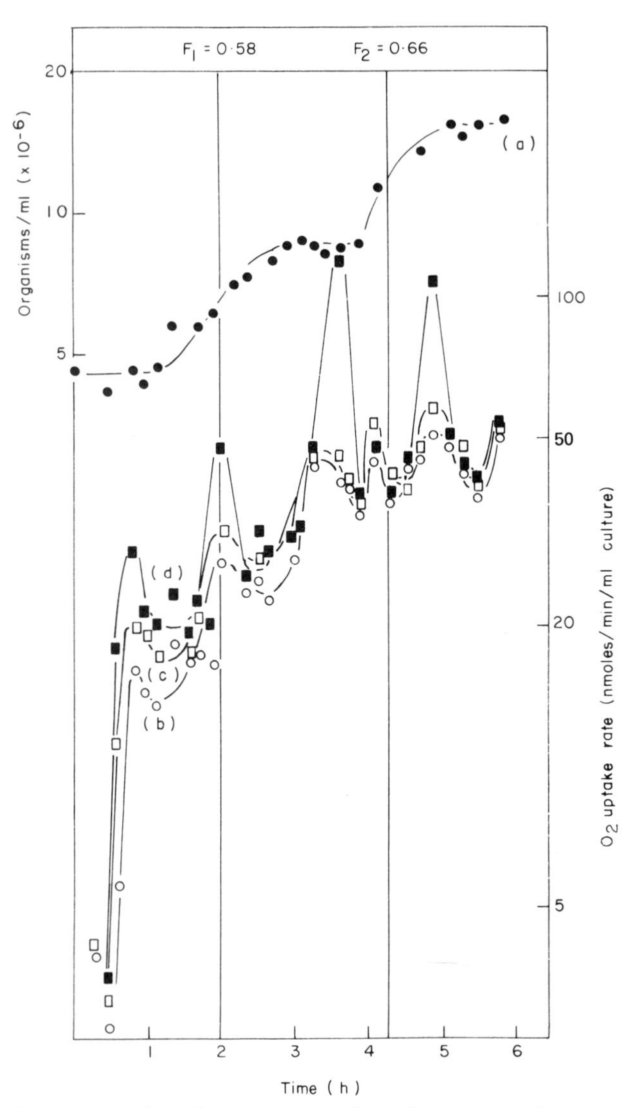

Fig. 6.1 Oxygen uptake of yeast suspensions from a synchronous culture of *Schizosaccharomyces pombe* and the effect of CCCP. F_1 and F_2 denote synchrony indices of the first and second doublings in cell numbers (a) respectively. Oxygen uptake measurements on culture samples removed at frequent intervals from a synchronous culture were made in the absence (b) or the presence of 8·1 μM-CCCP (c) or 16·2 μM-CCCP (d). (Reproduced with permission from Poole *et al.*, 1973.)

authors to the conclusion that the respiration of such synchronous cultures has two components, one of which increases exponentially at all times in the cycle, cannot easily be uncoupled and is preferentially blocked by inhibitors of electron transport. The second component consists of easily uncoupled electron transport chains, which are only inhibited by high concentrations of antimycin A or CN^-. The activity of this component oscillates with a periodicity of about 0·5 of a cell cycle, presumably because of the periodic synthesis of a rate-limiting entity. When glucose is replaced by 1% glycerol as the carbon source, oxygen uptake rates show two step per cycle (Fig. 6.2) with timings of the abrupt increases being similar to those found in glucose-grown cells (Poole and Lloyd, 1974): this confirms that respiratory oscillations observed with glucose arise from interactions between the respiratory system and glycolysis.

The respiration rates of cultures of *Schiz. pombe* synchronized by

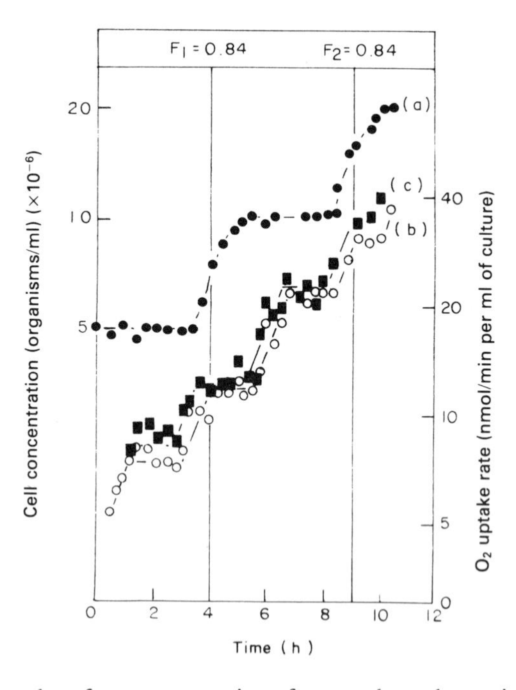

Fig. 6.2 O_2 uptake of yeast suspensions from a glycerol-containing synchronous culture of *Schiz. pombe* and effect of CCCP. F_1 and F_2 denote the synchrony indices of the first and second doublings in cell numbers (●) respectively. O_2-uptake measurements on culture samples removed at frequent intervals from a synchronous culture were made in the absence (○) or the presence (■) of 0·3 μM-CCCP. (Reproduced with permission from Poole and Lloyd, 1974.)

the addition and subsequent removal of 2 mM-2′deoxyadenosine (an inhibitor of DNA synthesis) also show an oscillatory pattern (Poole, 1977b). Oscillatory respiration was observed 1·5 h before the removal of the inhibitor and continued through the subsequent period of synchronous growth (Fig. 6.3). However, three maxima were found during this period and, in contrast with the findings from synchronous cultures prepared by selection, respiration was preferentially stimulated by CCCP at minima rather than maxima; heat dissipation increased continuously either in the presence or absence of this uncoupler. Thus the underlying mechanisms responsible for the expression of respiratory activity in this organism are quite different in selection- and induction-synchronized cultures (see later).

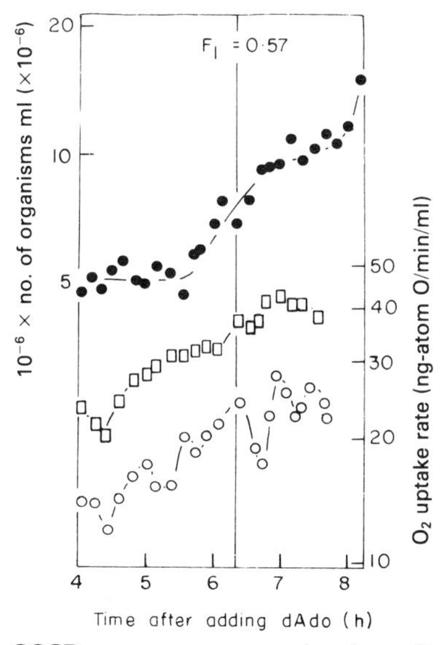

Fig. 6.3 Effect of CCCP on oxygen uptake in cell suspensions from a deoxyadenosine-synchronized culture of *Schiz. pombe*. The inhibitor was added to an exponential culture at zero time and, after 4 h incubation, removed by centrifugation; cells were resuspended in fresh medium lacking the inhibitor. F_1 is the synchrony index of the first synchronous doubling (mid-point indicated by vertical line) in cell numbers (●). Oxygen uptake measurements on culture samples removed at approximately 10 min intervals were made in the absence (○) or in the presence of 9·8 μM-CCCP (□). (Reproduced with permission from Poole, 1977b.)

Glycolytic activity measured in a gradient diver by evolved CO_2 during growth of *Schiz. pombe* in a complex medium, increased linearly in cloned cells or small synchronous cultures prepared either by a size-selection technique or heat treatment (Hamburger *et al.*, 1977).

Similar linear increases in glycolytic activity were observed in heat-shocked synchronous cultures growing in defined medium (Kramhøft *et al.*, 1978). The rate of CO_2 evolution measured manometrically on samples withdrawn from a synchronous culture prepared by size-selection, remained constant until just before cell division (0·75 of the cell cycle) and then doubled; addition of 2′-deoxyadenosine or mitomycin C (an inhibitor of nuclear division) to these synchronous cultures did not affect this pattern of CO_2 evolution (Creanor 1978a). When grown synchronously in yeast extract, the pattern was a linear increase with a rate change, and when synchronized by the addition and subsequent removal of 2′-deoxyadenosine there was a continuous increase in the rate of CO_2 evolution. In these cultures, O_2 uptake rates increased periodically as two steps per cell cycle irrespective of

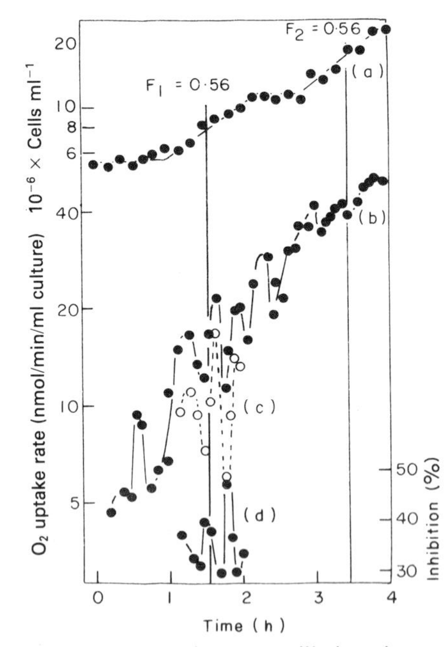

Fig. 6.4 Effects of cyanide on respiratory oscillations in a synchronously dividing culture of *C. utilis* with glucose as carbon source. The synchronous culture initially contained 10% of the original population of the exponentially growing culture and was concentrated fourfold by centrifugation at 200 *g* for 2 min followed by resuspension in conditioned growth medium. At intervals, samples were withdrawn from the culture and rates of respiration were determined in the presence or absence of KCN. (a) Cell numbers; (b) respiration rate in the absence of KCN; (c) respiration rate in the presence of 10 μM-KCN: (d) inhibition of respiration (%) resulting from addition of KCN. (Reproduced with permission from Kader and Lloyd, 1979.)

whether 2′-deoxyadenosine or hydroxyurea was added to the cultures (Creanor, 1978b). Therefore, it is suggested that increases in O_2 uptake rates or glycolytic activity under these conditions are independent of DNA synthesis, nuclear division or cell division.

Synchronous cultures of the budding yeast, *Candida utilis*, prepared by continuous flow size selection, showed respiratory oscillations when the carbon source was either glucose (Fig. 6.4), acetate or glycerol (Kader and Lloyd, 1979). Whereas the periodicity of the oscillations (about one-third of the cell cycle time i.e. 0·5 h) was unaltered by the nature of the carbon source, respiratory maxima and minima showed different uncoupler and inhibitor sensitivities when grown on different substrates. When either acetate or glycerol was the carbon source, the inhibitory effects of CN^- and DCCD [an inhibitor of ATP synthetase (Kováč *et al.*, 1968)] were greater at respiratory maxima, while maximal uncoupling of respiration by CCCP was achieved at respiratory minima (Table 6.1). When cells were grown synchronously in the presence of glucose, the effects of these compounds were different. Heat evolution in these cultures increased smoothly. The respiratory oscillations

Table 6.1. *Cell cycle times and the effects of KCN, DCCD and CCCP on the respiration of synchronous cultures of* Candida utilis

	Cell cycle time (min)		Effect of		
Carbon source	First cycle	Second cycle	KCN	DCCD	CCCP
Glucose	94·6 ± 17·4(10)	102·8 ± 12·1(9)	−	−	−
Acetate	110·8 ± 16·4(12)	118·1 ± 12·1(11)	+	+	−
Glycerol	91·4 ± 11·3(9)	98·2 ± 9·5(7)	+	+	−

Cell cycle times (mean values ± standard deviations, for the number of experiments shown in parentheses) were measured from the time of selection to the first mid-point of increase in cell numbers (first cycle), and from the first to second mid-points of increases in cell numbers (second cycle).

+ indicates maximum effect at respiratory maxima; − indicates maximum effect at respiratory minima. (Reproduced with permission from Kader and Lloyd, 1979.)

observed in acetate and glycerol-grown synchronous cultures can therefore be explained as reflecting *in vivo* mitochondrial respiratory control (Chance and Williams, 1956). It was suggested that the oscillations in glucose-grown cells reflect a more complicated control mechanism, in that secondary interactions between the products of mitochondrial energy conservation and the initial reactions of glucose catabolism *via* feedback loops and allosteric control sites are also involved.

Variations in the participation of the Embden-Meyerhof-Parnas pathway and hexosemonophosphate pathway in glucose metabolism were reported following radiorespirometric measurements of phased cultures of *C. utilis* (Dawson and Westlake, 1975; Dawson, 1977; Dawson and Steinhauer, 1980).

C. PROTOZOA AND ALGAE

Studies of synchronously-dividing cultures of trypanosomatids offer attractions in that these organisms possess a single mitochondrion (Simpson, 1972). Thus, these systems may yield information concerning mitochondrial biogenesis which may not be possible from studies with other organisms. However, few studies have been performed with unperturbed synchronously-dividing cultures.

Large-scale synchronous cultures of *Crithidia fasciculata* have been produced by a sedimentation-velocity size-selection technique in an iso-osmotic gradient (Edwards *et al.*, 1975). The cultures prepared in this way showed continuous increases in dry weight, protein and RNA, which doubled over one cell cycle. Similarly, oxygen uptake rates per ml doubled overall during one cycle (Fig. 6.5), but oscillated to give five maxima per cell cycle. Addition of KCN to samples removed from the culture at all times in the cycle produced approximately the same degree of inhibition of respiration.

Early work with *Tetrahymena pyriformis* indicated that the respiration rate of single cells increased linearly through most of the cell cycle but slowed or decreased just before division (Zeuthen, 1953). Subsequent experiments showed that the pattern was dependent on the growth conditions (Lövlie, 1963). In cultures synchronized by temperature shocks, the respiration rate per cell fell during cell division (Padilla *et al.*, 1966). However, in synchronously-dividing cultures of this organism prepared by continuous flow size-selection, a quite different pattern of respiratory activity was observed in that oxygen uptake rates rose to three maxima in a cell cycle time of about 3·5 h (Fig. 6.6), doubling overall during one cycle (Lloyd *et al.*, 1978b). Heat production also showed fluctuations in these cultures.

Measurements of rates of oxygen uptake of the soil amoeba, *Acanthamoeba* have been measured as a function of cell age using the gradient diver technique (Hamburger, 1975). The rate of uptake

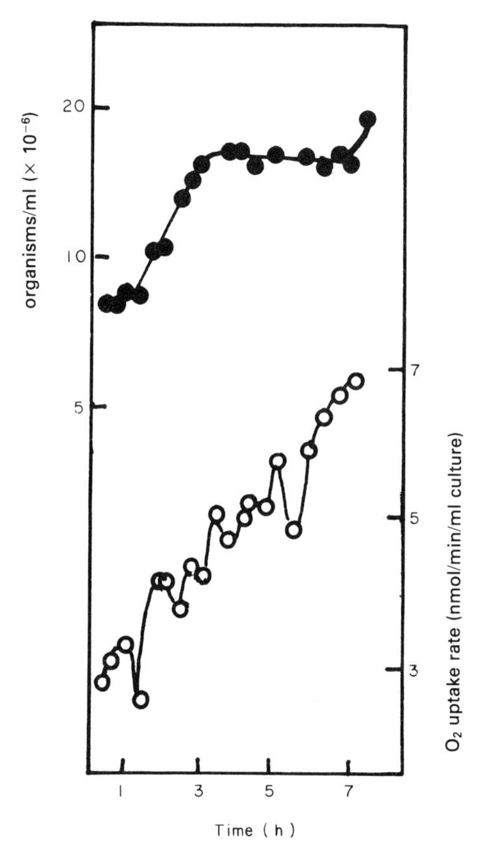

Fig. 6.5 Changes in oxygen uptake rates during synchronous growth of *C. fasciculata*. (●) Cell numbers, (○) respiration rates. (Reproduced with permission from Edwards *et al.*, 1975.)

increased linearly throughout the cycle, or in some cases remained constant for a period just before cell division. If a large cell was taken from the stationary phase, it quickly underwent several cell divisions and oxygen uptake rates oscillated but did not increase. Synchronously-dividing cultures of *A. castellanii*, prepared by a size-selection technique that introduced minimal metabolic perturbations (Chagla and Griffiths, 1978), showed complex respiratory oscillations (Edwards and Lloyd, 1978). In a cell cycle time of about 8 h, seven distinct maxima in oxygen uptake rates were found (Fig. 6.7); these respiratory oscillations showed differential sensitivities to KCN and FCCP. Respiratory maxima were more sensitive to inhibition by CN^- than were minima, while FCCP produced a greater stimulatory

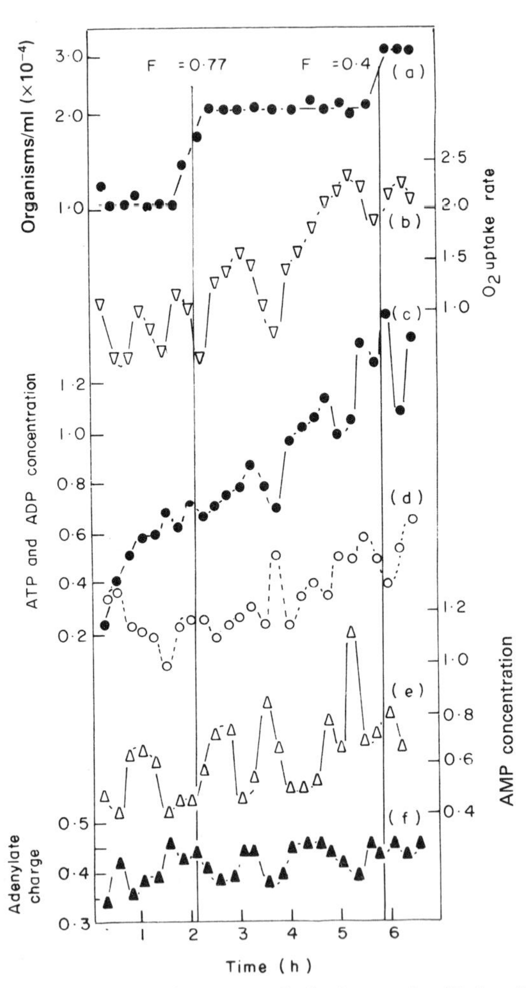

Fig. 6.6 Changes in oxygen uptake rates and adenine nucleotide levels during growth of a synchronous culture of *T. pyriformis*. Samples were removed at frequent intervals for oxygen uptake measurements and 1 ml portions were mixed with 0·3 ml chloroform for adenine nucleotide estimations. Percentage selection from the exponentially growing culture used for the preparation of the synchronous culture by continuous-flow size selection was 10%: inlet flow rate, $1{\cdot}5\,\mathrm{l\,min^{-1}}$; rotor speed, 1750 r/min. (a) Cell numbers; (b) oxygen uptake rates, expressed as nmol O_2 $\mathrm{min^{-1}}$ (ml culture)$^{-1}$; (c) ATP; (d) ADP; (e) AMP; (f) adenylate charge. All adenine nucleotides are expressed as nmol (ml culture)$^{-1}$. (Reproduced with permission from Lloyd *et al.*, 1978b.)

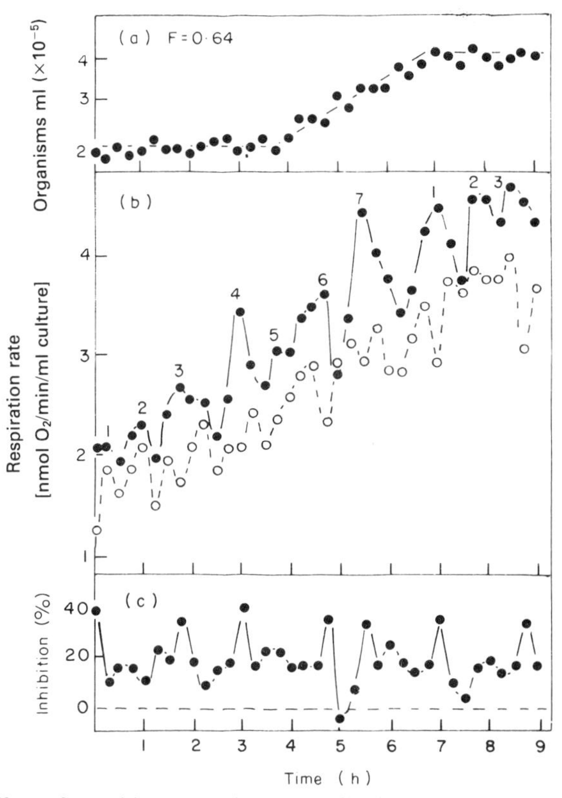

Fig. 6.7 Effect of cyanide on respiratory oscillations in a synchronously-dividing culture of *A. castellanii*. The synchronous culture contained 15% of the exponentially growing culture. At 15 min intervals, samples were withdrawn from the culture and rates of oxygen uptake were measured in the presence or absence of 0·5 mM-cyanide. (a) Cell numbers and synchrony index, *F*; (b) respiration rate in the presence (○) or absence (●) of 0·5 mM-cyanide; (c) inhibition of respiration (%) resulting from the addition of cyanide. (Reproduced with permission from Edwards and Lloyd, 1978.)

effect on samples withdrawn from the culture when respiration was low, suggesting that the underlying mechanism responsible for these changes is *in vivo* mitochondrial respiratory control (see later). In control experiments where exponentially-growing cultures were subjected to identical experimental procedures, smooth increases in respiration rates and cell numbers were observed (Fig. 6.8). Thus the changes in oxygen uptake rates in these synchronously-dividing cultures are genuinely associated with the undisturbed cell cycle.

A similar method of synchrony (although a different sub population was selected because the smallest cells were found to be non-viable) has been successfully applied to cultures of the cellular slime mould *Dictyostelium discoideum* (C. Woffendin and A. J. Griffiths, unpublished data). Oxygen uptake rates double overall during one cell cycle of about 6 h, but oscillate with a period of about 1 h (Fig. 6.9). Control experiments show smooth increases in cell numbers and respiration rates.

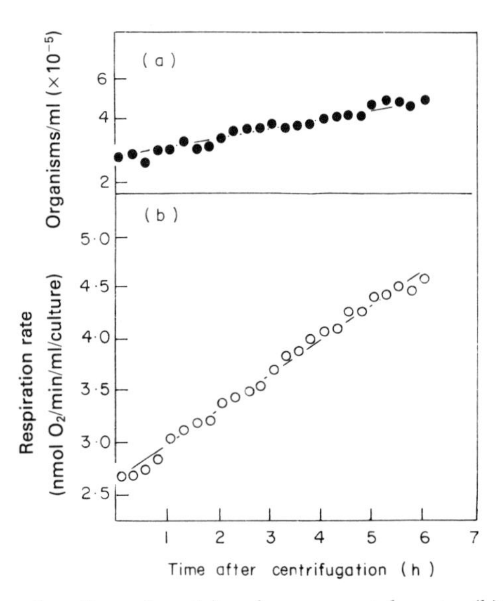

Fig. 6.8 Changes in cell numbers (a) and oxygen uptake rates (b) after centrifugation and resuspension of an exponentially growing culture of *A. castellanii*. The exponentially growing culture was centrifuged at 300 r/min (10 *g*; $r_{av.}$ 10 cm) for 2 min, and then remixed and returned to the growth vessel. (Reproduced with permission from Edwards and Lloyd, 1978.)

The acellular slime mould *Physarum polycephalum* is an attractive system for studies of the mitotic cycle since nuclear division is naturally synchronous. The rate of oxygen uptake by single macroplasmodia increased in two steps during a synchronous mitotic cycle (Forde and Sachsenmaier, 1979). Mid-points of these increases (Fig. 6.10) were at 0·4 (mid-interphase) and 0·94 (just prior to mitosis) and these respiratory changes were closely paralleled by changes in cytochrome oxidase activity but not by fumarase, malate dehydrogenase or succinate dehydrogenase. It was suggested by these authors that lowered

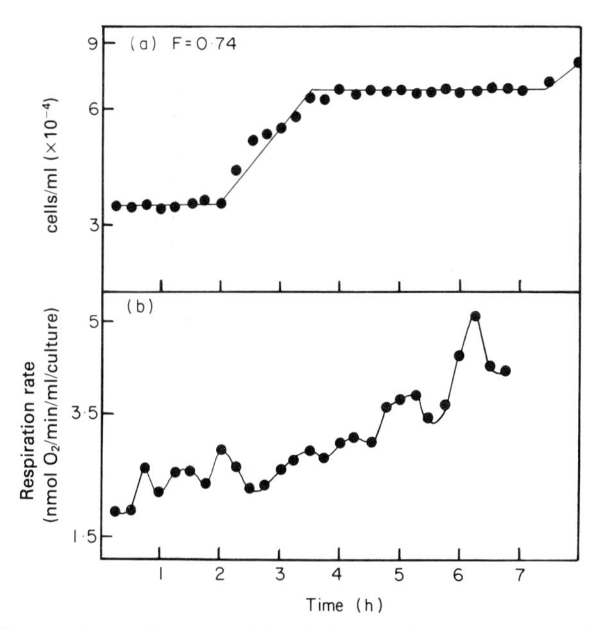

Fig. 6.9 Changes in respiratory activity during synchronous growth of *D. discoideum*: (a) Cell numbers; (b) Respiration rate. (Unpublished results of C. Woffendin and A. J. Griffiths.)

ATP levels during mitosis (Chin and Berstein, 1968) possibly correlating with changes in the rate of cytoplasmic streaming (Sachsenmaier *et al.*, 1973) may result in increased respiratory activities.

Cultures of *Astasia longa* synchronized by heat shocks have respiration rates (per cell) which decrease during cell division (Wilson and James, 1966). More complex changes in oxygen uptake rates are

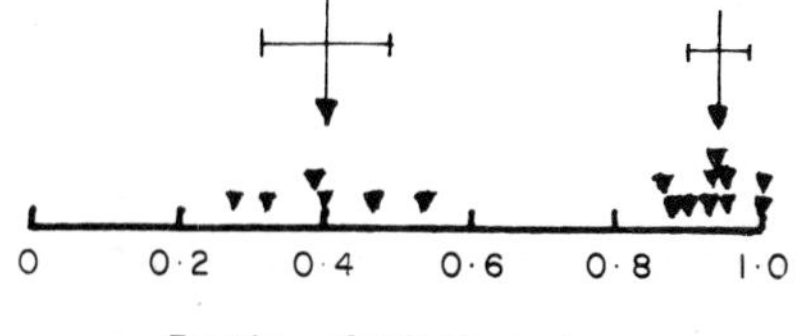

Fig. 6.10 Timings of plateaux in oxygen uptake rates during the mitotic cycle of *P. polycephalum*. The mitotic cycle is represented linearly from 0–1·0 and each triangle represents the mid point of the period when the oxygen uptake rate of a single microplasmodium increases. The mean and standard deviations are indicated by an arrow and crossbar. (Reproduced with permission from Forde and Sachsenmaier, 1979.)

observed during the cell cycle of *Chlamydomonas reinhardii* in which the reversible formation of giant mitochondria by the fusion of many small mitochondria coincides with a decreased oxygen uptake rate (Osafune *et al.*, 1972a, b). Variations in endogenous rates of respiration and those in the presence of added respiratory substrates are observed during the growth cycle of *Chlorella* (Talbert and Sorokin, 1971). While the observed pattern of respiratory activity depends on the method of synchronization employed, small cells (separated by a centrifugation technique) possess a higher respiratory activity than large cells. In temperature-induced synchronous cultures of *Polytomella agilis*, a sharp doubling of oxygen-uptake per cell occurs once in the cell cycle (Cantor and Klotz, 1971).

II. Changes in Nucleotide Pools during the Cell Cycles of Eukaryotes

The importance of ATP in cellular energy metabolism has long been appreciated (Lipmann, 1941). More recent studies involving enzyme kinetics *in vitro* have indicated that all metabolic pathways are linked through the adenine nucleotide system and thus the concentrations of ATP, ADP and AMP should all be considered when assessing the energy status of the cell (see Chapman and Atkinson, 1977). For this reason, the adenylate energy charge concept was formulated (Atkinson and Walton, 1967) and is defined as:

$$\text{Adenylate energy charge} = \frac{[\text{ATP}] + \frac{1}{2}[\text{ADP}]}{[\text{ATP}] + [\text{ADP}] + [\text{AMP}]}$$

Thermodynamic and kinetic considerations of ATP-utilizing and ATP-generating reactions *in vitro* predict that the energy charge *in vivo* must be stabilized in the range 0·8–0·95 during growth and normal metabolism (Chapman and Atkinson, 1977); indeed, in the majority of organisms and tissues examined such values are generally found (Chapman, A. G. *et al.*, 1971). There are instances, however, where low adenylate charge values are found e.g. during carbon starvation in *S. cerevisiae* (Ball and Atkinson, 1975) and *Prototheca zopfii* (Lloyd *et al.*, 1978a), and also in growing cultures of two protozoa, *Acanthamoeba castellanii* (Edwards and Lloyd, 1977b) and *Tetrahymena pyriformis* (Lloyd *et al.*, 1978b). In the last case, values of adenylate charge of mid-exponential phase cultures increased from less than 0·2 to greater than 0·4 as growth progressed. In *A. castellanii*, intracellular adenylate charge values obtained during the initial 30 h of growth were less than 0·2, and large quantities of AMP were found extracellularly.

Interpretation of adenylate charges in eukaryotes is complicated by the localization of nucleotides within discrete subcellular compartments. Also, the adenine nucleotide translocase responsible for the exchange of ATP and ADP in mitochondria does not transport AMP, and thus intra- and extra-mitochondrial pools of AMP are separated (Knowles, 1977). For this reason in eukaryotic micro-organisms, measurements of ATP: ADP ratios are more informative than adenylate charge values. It has also been shown that the phosphate potential (ATP)/(ADP) (Pi) is an important regulatory mechanism (Erecińska *et al.*, 1977; Wilson *et al.*, 1977).

A. MAMMALIAN CELLS

A few reports suggest that periodic variations in nucleotides occur during the cell cycles of mammalian cells, indicating that either energy supplies or energy requirements are not constant. In HeLa cells, levels of nucleotide triphosphates are at a minimum at mitosis and maximal 4 h after the addition of thymidine (Gerschenson *et al.*, 1965). Pyrimidine nucleoside triphosphates increase sharply at the onset of S-phase and it is suggested that the synthesis of these components may be triggered at or just before the initiation of DNA replication (Bray and Brent, 1972). In Chinese hamster cells synchronized by mitotic detachment, ATP levels per cell more than double and then decline at the end of G_2 closely paralleling changes in cell volume (Chapman, J. D. *et al.*, 1971). ATP levels per cell show cyclic changes in mouse L cells that have been synchronized by mitotic detachment (Fig. 6.11); they declined in G_1, reaching a maximum at the end of S, and then fell steadily until the end of mitosis (Ishiguro *et al.*, 1978).

B. YEASTS

In synchronously-dividing cultures of *Schiz. pombe* prepared by continuous flow size-selection and grown at 32, 30, 27 or 25°C, pool sizes of ATP, ADP and AMP, and ATP/ADP ratios and adenylate energy charge all showed complex fluctuations (El-Khayat, 1980): adenylate

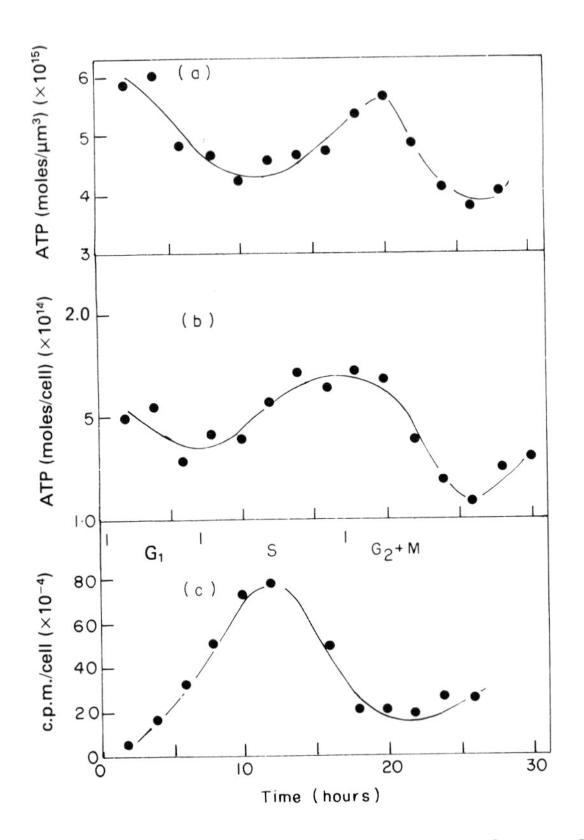

Fig. 6.11 Changes in ATP content during the cell cycle of mouse L cells. ATP content was assayed by the luciferin–luciferase system: (a) and (b) ATP levels per unit volume and per cell respectively, and (c) incorporation of [^{3}H]–Tdr into perchloric acid-insoluble material. (Reproduced with permission from Ishigura *et al.*, 1978.)

Fig. 6.12 Changes in individual adenine nucleotide pools, total adenylates, energy charge and the ATP/ADP ratio in dAdo-synchronized cultures of *Schiz. pombe.* Culture samples were rapidly mixed and extracted with trichloroacetic acid; adenylates were assayed by the luciferase method. (a) Cell numbers (●), total cell volume (○) and cell plate index (□); (b) Total adenylates (●) and the individual pool sizes of ATP (□), ADP (■) and AMP (○); all expressed as nmol (ml culture)$^{-1}$. (c) Energy charge (○) and ATP/ADP ratio (●). dAdo was added at the first arrow (and solid vertical line) and removed by centrifugation and resuspension in fresh medium at the second arrow (and solid vertical line). The vertical broken line indicates the mid-point of the first synchronous doubling in yeast numbers. (Reproduced with permission from Poole and Salmon, 1978.)

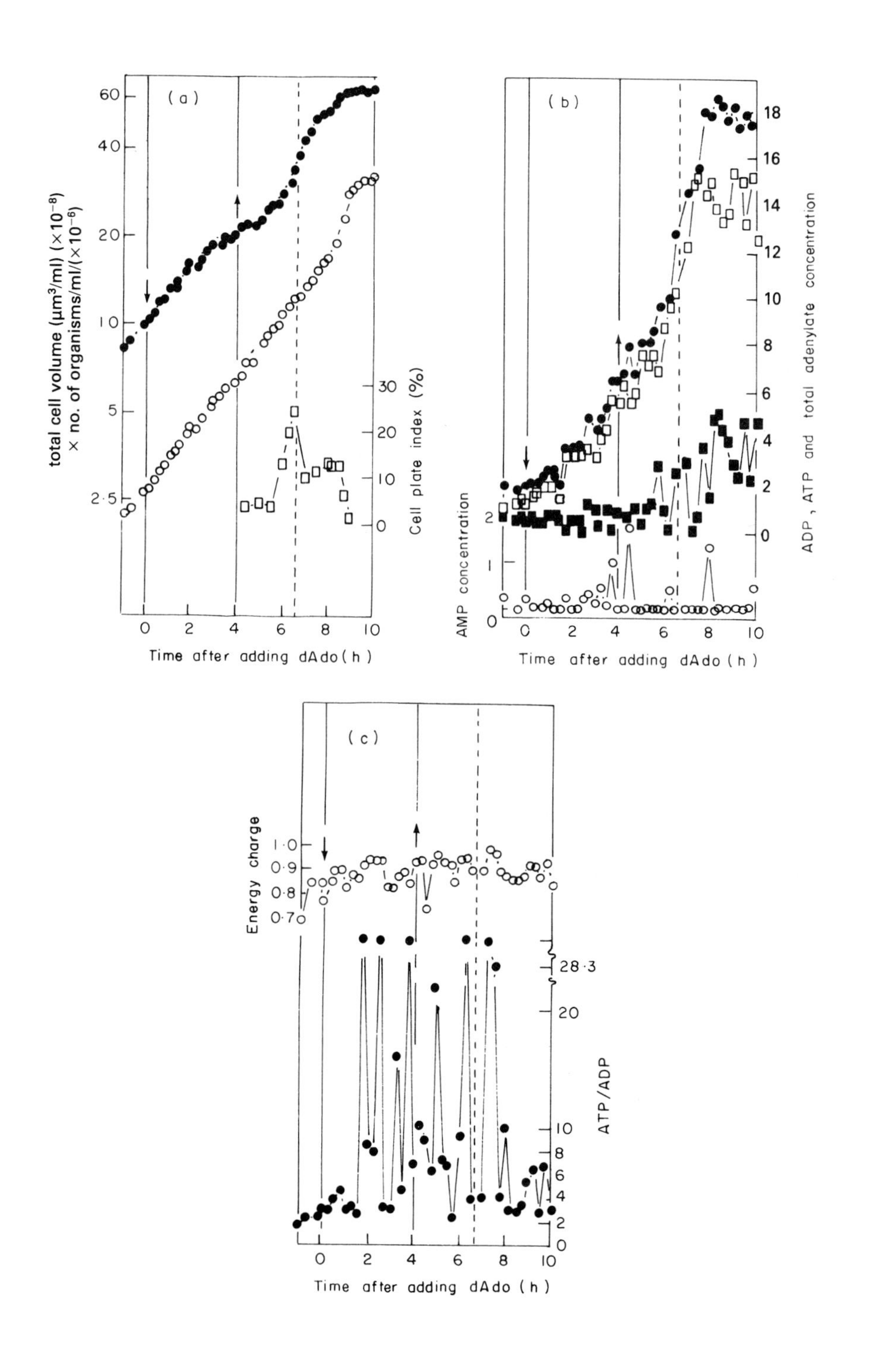
(a)
total cell volume (μm³/ml) (×10⁻⁸)
× no. of organisms/ml/(×10⁻⁶)
60
40
20
10
5
2·5
30
20
10
0
Cell plate index (%)
0
2
4
6
8
10
Time after adding dAdo (h)
(b)
18
16
14
12
10
8
6
4
2
0
ADP, ATP and total adenylate concentration
AMP concentration
2
1
0
0
2
4
6
8
10
Time after adding dAdo (h)
(c)
Energy charge
1·0
0·9
0·8
0·7
28·3
20
10
8
6
4
2
0
ATP/ADP
0
2
4
6
8
10
Time after adding dAdo (h)

charge values were in the range 0·73–0·9 and ATP/ADP ratios 1·2–4·9. In cultures of *Schiz. pombe* synchronized by a 4 h exposure to 2′-deoxyadenosine followed by its removal (Fig. 6.12), pool sizes of adenine nucleotides oscillated (Poole and Salmon, 1978). A good correlation was obtained between maxima in respiration rates (which were relatively insensitive to the uncoupling effects of CCCP compared with respiratory minima) and minimum values of ATP/ADP ratios and adenylate energy charge. Similar variations were also observed during the last 2·5 h before the removal of the inhibitor.

Adenylate charge values vary during phased growth in a chemostat of *Candida utilis* (Thomas and Dawson, 1977). In nitrogen- or sulphate-limited growth, adenylate charge values varied between 0·78 and 0·94, while in iron-limited cultures (which it was suggested were energy-limited) values varied between 0·44 and 0·78. Adenylate-charge values were lowest in the middle of the cycle when respiration was highest.

Analysis of the principle phosphorus-containing metabolites of *S. cerevisiae* by ^{31}P-NMR after glucose starvation and re-feeding has shown that a change in ATP level is *not* the intracellular "start" signal; an increase in intracellular pH is a possible alternative candidate in this role (Gillies and Shulman, 1980).

C. PROTOZOA AND ALGAE

Adenylate energy charge values and individual levels of ATP, ADP and AMP (Fig. 6.13) increased periodically during the cell cycle of the trypanosomatid, *Crithidia fasciculata*, prepared by a size-selection technique (Edwards *et al.*, 1975). The periodicity of these maxima was also about 1·0 h (generation time of about 5 h) and values of adenylate charge varied between 0·47 and 0·66. In these cultures, respiration rates also oscillated, doubling over one cell cycle and a cell cycle map (Fig. 6.14) indicates that maxima in respiration rates closely coincide with maximum levels of pools and ATP and AMP. No changes were unique to nuclear or kinetoplast S-phases.

In heat-shocked synchronous cultures of *T. pyriformis*, nucleoside phosphates show cyclic variations, although these results must be interpreted with caution as nucleoside triphosphate levels are extremely sensitive to heat treatment (Scherbaum *et al.*, 1962). Exposure of cells to 34°C decreases nucleoside triphosphate levels but after returning to 29°C, ATP levels increase in the first cycle and then decrease at cell division. A 35% increase in intracellular ATP, with equivalent

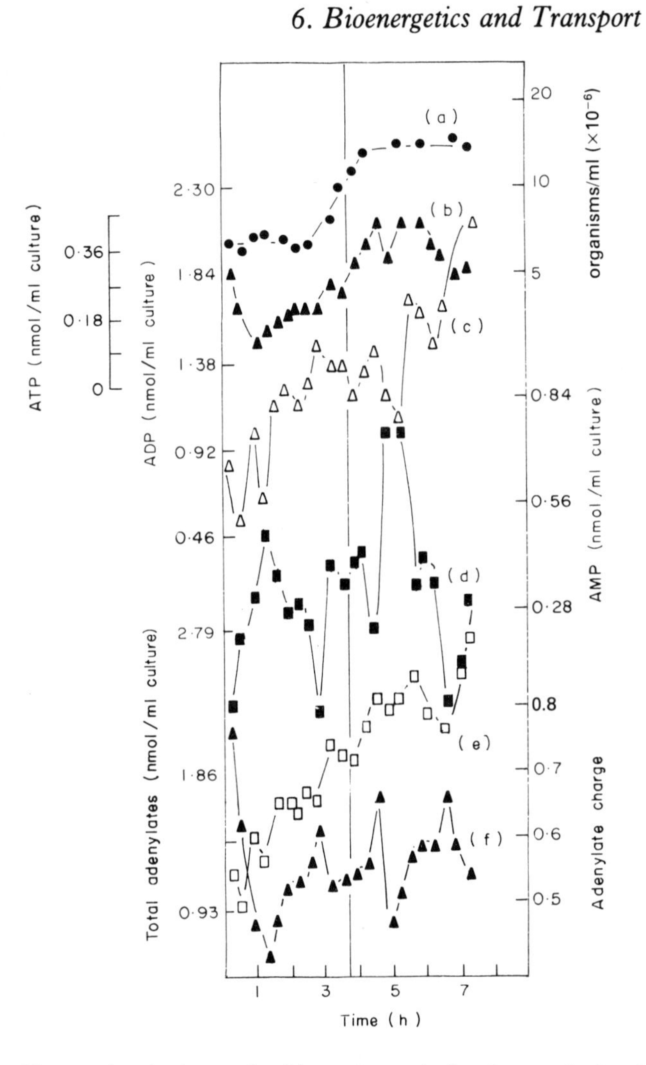

Fig. 6.13 Changes in adenine nucleotide pools, total adenylates and adenylate charge during synchronous growth of *C. fasciculata*. Samples (1 ml) were removed at frequent intervals from the culture and rapidly mixed with 1 ml n-butanol. (a) Cell numbers, (b) ATP, (c) ADP, (d) AMP, (e) total adenylates and (f) adenylate charge. (Reproduced with permission from Edwards *et al.*, 1975.)

decreases in ADP and AMP levels, was observed 20 minutes after the end of heat treatment (Stocco and Zimmerman, 1975) and ATP, GTP, CTP and UTP levels reach maximum values just before synchronous division (Plesner, 1964; Echetebu and Plesner, 1977). Elevation of

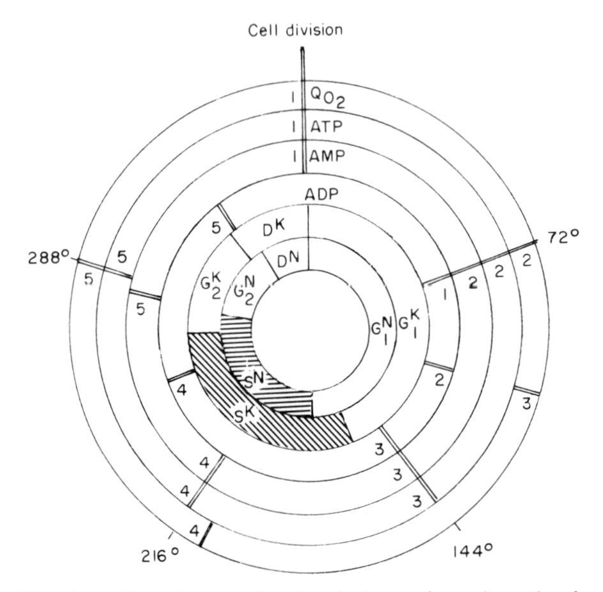

Fig. 6.14 Circular cell-cycle map for the timings of consistently observed maxima of oxygen uptake rates and adenine nucleotide pools in *C. fasciculata*. The phase angles of the oscillatory components are plotted with respect to cell division (at a phase angle of 0°). The phases of nuclear and kinetoplast events are replotted from the data of Cosgrove and Skeen (1970). (Reproduced with permission from Edwards *et al.*, 1975.)

temperature to 34°C also impairs RNA synthesis and it may be that the preformed nucleotide pool may be utilized for RNA synthesis at this higher temperature (De Balros *et al.*, 1973).

Quite different results were obtained with synchronous cultures of *T. pyriformis* prepared by a size-selection method (Lloyd *et al.*, 1978b). Levels of ATP oscillated in phase with respiration rates, while levels of ADP and AMP oscillated in phase with one another but out of phase with the other two parameters (Fig. 6.6). Adenylate charge values varied over the range 0·34–0·4. The phase relationships of timings of maxima of adenine nucleotides and oxygen uptake rates are shown in Fig. 6.15.

Oscillatory accumulation of adenine nucleotides was also observed in synchronously-dividing cultures of *A. castellanii* prepared by a minimally-perturbing size-selection procedure (Edwards and Lloyd, 1978). The periodicity of the oscillations was similar to that observed in respiration rates (about 1 h) and maximum amplitudes (peak to trough, % of minimum) for ATP, ADP and AMP were 108, 194 and

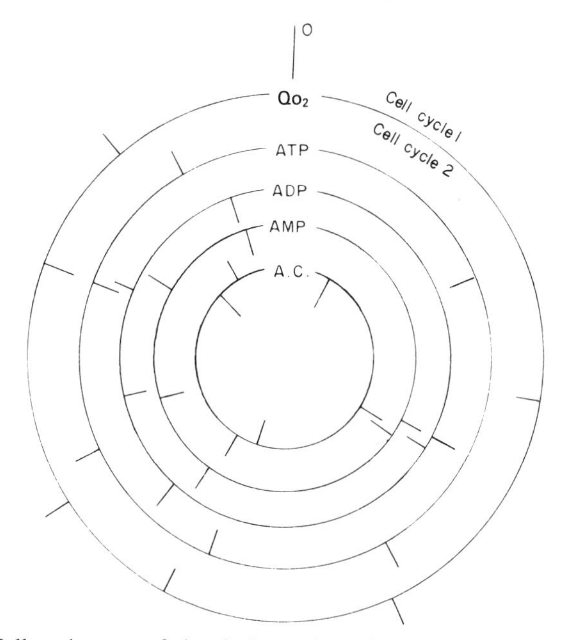

Fig. 6.15 Cell-cycle map of the timings of maxima of oxygen uptake rates Qo_2), ATP, ADP and AMP pool sizes, and adenylate charge (A.C.) in *T. pyriformis*. The phase angles of these components are plotted with respect to the mid-points of cell division (at a phase angle of 0°). On each circle, the outer markers indicate maxima in the first cell cycle, and the inner markers, those in the second cell cycle. Results are expressed as the means of seven experiments for respiration rates and five experiments for adenine nucleotides similar to the one shown in Fig. 6.6. (Reproduced with permission from Lloyd *et al.*, 1978b.)

520, respectively (Fig. 6.16). Adenylate charge values varied between 0·63 and 0·88 and ATP/ADP ratios varied between 0·9 and 10·9. ADP levels oscillated in phase with respiratory maxima but out of phase with ATP/ADP ratios, which, together with the differential sensitivities of respiratory maxima and minima to an inhibitor and an uncoupler of oxidative phosphorylation (see earlier), suggest that the overall changes in respiratory activity represent *in vivo* mitochondrial respiratory control (Chance and Williams, 1956) i.e. control of respiration by the availability of ADP. In the presence of adequate supplies of substrate and phosphate, the respiration of isolated mitochondria is low (State 4) and is stimulated by the addition of ADP (State 3). State 3 is the "active" state of respiration during which ATP is generated. Stimulation of State 4 can also be achieved by the addition of uncouplers and a state similar to State 3 is achieved (State 3u) although no

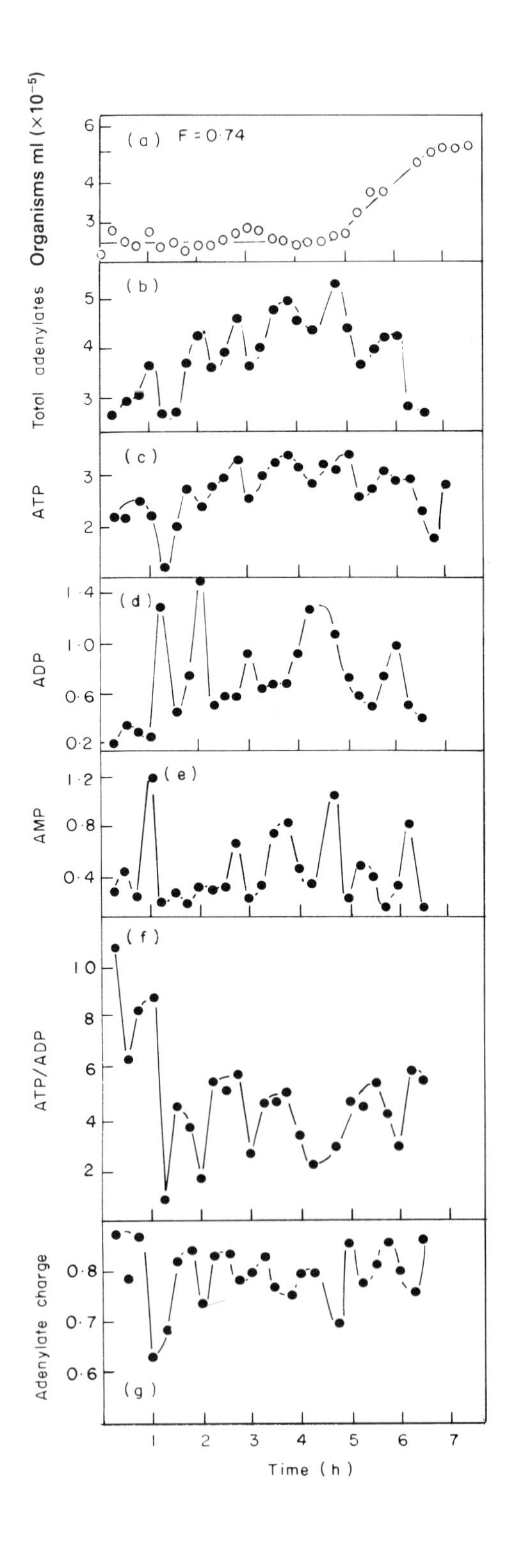
Organisms ml (×10⁻⁵)
(a) F = 0·74
Total adenylates
(b)
ATP
(c)
ADP
(d)
AMP
(e)
ATP/ADP
(f)
Adenylate charge
(g)
Time (h)

ATP is generated (Chance and Hollunger, 1963; Chance *et al.*, 1963). The inhibitory effect of CN^- is greater on State 3 respiration than State 4 respiration (Ikuma and Bonner, 1967). Thus, during its cell cycle, *A. castellanii*, is in State 3 at respiratory maxima and State 4 at respiratory minima. The mitochondrial redox state measured by the fluorescence of mitochondrial flavoprotein or NADH also shows periodic changes of the order of about 1 h (Fig. 6.17) in synchronously-dividing cultures (Bashford *et al.*, 1980). How this discontinuous accumulation of energy is coupled to macromolecular synthesis is discussed in Chapter 4. The phase relationships between

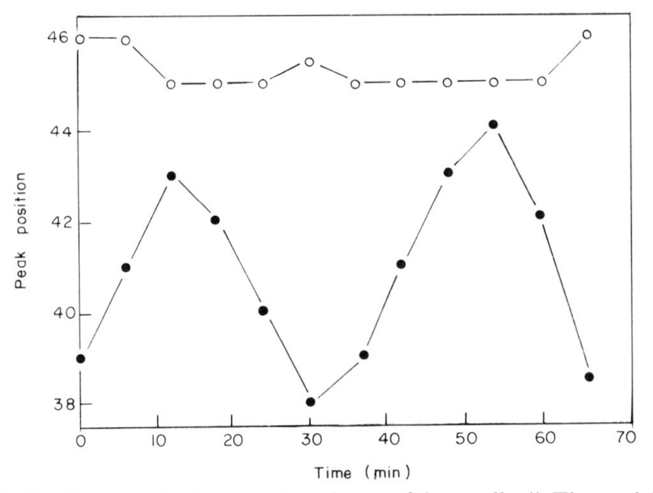

Fig. 6.17 Oscillations of redox state in cultures of *A. castellanii*. The modal intensity in arbitrary units of the flavoprotein fluorescence for both synchronous (●) and asynchronous (○) cultures of *A. castellanii* is shown. The exponential culture contained $4{\cdot}6 \times 10^6$ cells/ml at time zero. The synchronous culture was obtained by size selection of 10% of the original population and then concentrated tenfold. The selection procedure was completed 10 min before scanning was commenced. (Reproduced with permission from Bashford *et al.*, 1980.)

Fig. 6.16 Changes in adenine nucleotide pool levels and adenylate charge values in a synchronously-dividing culture of *A. castellanii*. The synchronous culture contained 10% of the exponentially growing population; 1 ml samples were withdrawn at 15 min intervals for measurements of adenine nucleotides after rapid quenching and extraction in chloroform. (a) Cell numbers and synchrony index, *F*; (b) total adenylates; (c) ATP; (d) ADP; (e) AMP; (f) ATP/ADP ratio; (g) adenylate charge. Adenylate concentrations are expressed as nmol (ml culture)$^{-1}$. (Reproduced with permission from Edwards and Lloyd, 1978.)

respiration rate, ADP and cellular protein content clearly indicate that in this system the machinery of energy generation is enslaved by the energy requirements of protein synthesis.

The naturally-synchronous mitotic cycle of the acellular slime mould *Physarum polycephalum* shows cyclic variations in levels of nucleotides. Levels of ATP, ADP (and adenosine), but not AMP, fall to low values at prophase, remain low through mitosis and then increase (Chin and Berstein, 1968). A later study showed that all nucleotide triphosphates increased prior to mitosis and then decreased suggesting that accumulation of triphosphates results from a diminished demand for RNA precursors due to decreasing RNA synthesis in late telophase (Sachsenmaier *et al.*, 1969). Deoxyribose nucleotide triphosphates (dNTP) were found to be higher in S than G_2 (Bersier and Braun, 1974) whereas dNTP and rNTP increased before and after the initiation of DNA synthesis and also in mid G_2, about 5 h after mitosis (Fink, 1975).

III. Changes in Photosynthetic Activity

There have been a number of reports showing that the photosynthetic capacity of eukaryotic algae varies at different stages of the cell cycle. In *Cylindrotheca fusiformis* maximum rates of photosynthesis occurred at 6 h in the light (generation time of 16 h), whereas dark respiration decreased from about 2 h until cell division (Paul and Volcani, 1976). In the xanthophycean alga *Bumilleriopsis filiformis* synchronized by light-dark regimes, the cell cycle may be divided into two stages (Hesse *et al.*, 1977). Stage 1 is the period of cell enlargement, and during this stage photosynthetic activity (measured as O_2 evolution) fluctuated. Stage 2 is the phase of nuclear division followed by cell division and minimum photosynthetic activity occurred at the beginning of cell division. Maximum photosynthetic activity, accompanied by corresponding changes in Hill reaction and re-reduction of cytochrome *f* by photosystem II, occurred after the completion of cell division. In *Scenedesmus obliquus*, synchronized by a light–dark regime (generation time of 24 h), photosynthetic activity was maximal at 8 h and lowest at 16 h, just prior to the beginning of cell division (Bishop and Senger, 1971). Photosystem I capacity remained the same throughout the cycle but photosynthetic electron flow at 16 h was 50% lower than that at 8 h (Senger and Bishop 1977): the quantum yield and Hill reaction are also lowest at 16 h (Senger and Bishop, 1979). Chlorophyll synthesis begins after 2–8 h in this organism and, whereas the photosynthetic rate per oxygen flash remains constant, the light-saturated photosynthetic rate per unit chlorophyll is maximal after 3 h and then steadily declined (Myers and Graham, 1975).

In *Chlorella*, the rate of photosynthesis is greater in young cells (2–4 h) than in older cells (Sorokin and Krauss, 1961). In *C. reinhardii*, the rates of photosynthesis as well as the amounts of individual chloroplast components (e.g. chlorophyll, cytochromes, ferredoxin etc.) increase after about 4 h growth and then increase throughout the light period (Armstrong *et al.*, 1971). Photoreductive activity and the activities of photo-systems I and II are maximal in mid-light phase and then decline at the end of the dark period (Schor *et al.*, 1970); chloroplast cytochrome changes are different to those of chlorophyll and thus the thylakoid membrane in *C. reinhardii* may undergo a series of alterations during the cycle by insertion of chlorophyll and cytochromes at different stages (Schor *et al.*, 1970).

IV. Energy Metabolism During the Cell Cycles of Prokaryotes

The mode of development of respiratory activity during the cell cycles of bacteria has received little attention (for a review see Poole, 1980). During synchronous growth of *E. coli* after continuous flow centrifugation in a minimal medium containing alanine as the carbon source, respiration rates oscillated with a periodicity of about one third of the cell cycle (Evans, 1975). However, growth with glucose and casein hydrolysate, gave respiratory oscillations with two maxima peaks at 0·45 and 0·95 of the cycle (Fig. 6.18) in a cell cycle time of 44 min (Poole, 1977d). An ATPase-deficient mutant of this organism also showed a similar pattern of oxygen uptake during the cell cycle while growth in the presence of glycerol resulted in two steps of respiration rates (Fig. 6.19). These results suggest that the respiratory oscillations in the cell cycle of *E. coli* do not represent respiratory control. More recent work with synchronous cultures of *E. coli* prepared by a sucrose density-gradient selection procedure has shown that respiration rates increase smoothly throughout the cell cycle when succinate is the sole carbon and energy source (Poole *et al.*, 1980b). Variations in the levels of ribonucleoside triphosphates are also observed during synchronous growth of *E. coli* (Huzyk and Clark, 1971). Levels of ATP and GTP increase by only 50% during one cycle whereas UTP and CTP show more complex changes in amounts (Huzyk and Clark, 1971).

In *Photobacterium phosphoreum*, the intensity of bioluminescence increases continuously by a factor of 1·5 every doubling time (Watanabe

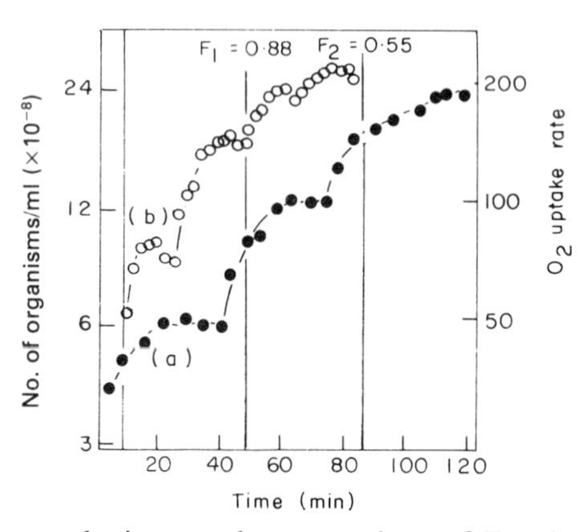

Fig. 6.18 Oxygen uptake in a synchronous culture of *E. coli* strain A1002 grown with glucose and Casamino acids. The population at the time of selection was $4{\cdot}4 \times 10^8$ cells ml^{-1}. Flow rate of the culture through the continuous action rotor was 0.16 min^{-1} and the rotor speed 15 900 r min^{-1} ($2{\cdot}2 \times 10^4$ g-min at half-maximal radius). After discarding the first 300 ml of effluent, a portion (75 ml), which contained 6·9% of the original culture, was collected and concentrated 13-fold. Samples (2 ml) were cultured in open O_2 electrode vessels. ●, Cell numbers: ○, O_2 uptake rates calculated at intervals corresponding to 2·5 min from the polarographic trace and expressed as ng-atom O/min/ml. F_1 and F_2 denote the synchrony indices of the first and second doublings of the cell numbers, respectively. (Reproduced with permission from Poole, 1977d.)

and Nakamura, 1980). Rates of oxygen uptake per ml culture show a discontinuous increase during synchronous growth of *Alcaligenes eutrophus* (Edwards and Jones, 1977). With lactate as the carbon source, respiration rates per ml increase as two steps per cell cycle (at 0·4 and 0·9) in a cycle time of 76 min (Fig. 6.20). Addition of CCCP to samples withdrawn from the culture at different times during synchronous growth (Fig. 6.21) produces a greater stimulation when respiration is low (38–50% stimulation). This observation suggests that *in vivo* respiratory control may be the mechanism responsible for the expression of respiratory activity in this organism. The respiration of synchronously-dividing cultures of *Bacillus subtilis* also increases as two steps per cell cycle (Edwards, 1980b; Edwards and McCann, 1981).

Insufficient data has accumulated on respiratory activity during the cell cycle of bacteria to enable general conclusions to be drawn. With one exception (*Alcaligenes eutrophus*), respiration rates do not appear to be regulated by respiratory control as they are in some protozoa and yeasts. Instead, aerobic growth may be limited by the rate of

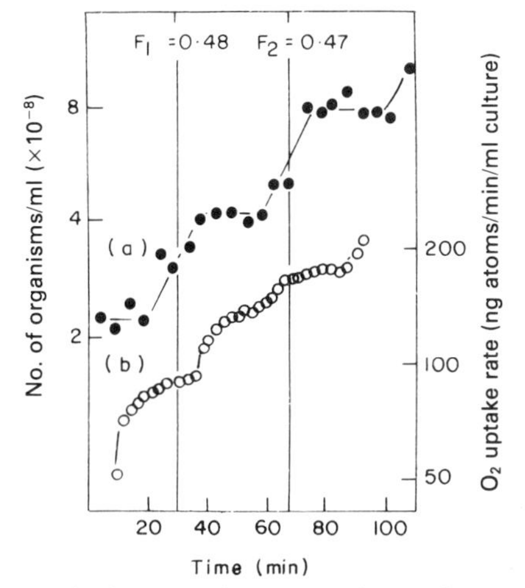

Fig. 6.19 Oxygen uptake in a synchronous culture of *E. coli* strain A1002 grown with glycerol and Casamino acids. The population at the time of selection was $3{\cdot}3 \times 10^8$ cells min^{-1}. Flow rate of the culture through the continuous action rotor was 170 ml min^{-1} and the rotor speed 15 000 r min^{-1} ($1{\cdot}85 \times 10^4$ g-min at half-maximal radius). After discarding the first 300 ml of effluent, a portion (200 ml), which contained 8·6% of the initial population, was collected and concentrated 10-fold. Samples (2 ml) were cultured in open O_2 electrode vessels. ●, Cell numbers; ○, O_2 uptake rates calculated at intervals corresponding to 2·4 min from the polarographic trace. F_1 and F_2 denote the synchrony indices of the first and second doublings of the cell numbers, respectively. (Reproduced with permission from Poole, 1977d.)

respiration and the concomitant rate of ATP generation by oxidative phosphorylation (Andersen and von Meyenburg, 1980).

V. Transport

The control of energy metabolism within cells (as well as the control of the rate of synthesis of macromolecules) may depend on the rate of entry of solutes into the cell (Poole, 1980). However, this direct control mechanism is unlikely to occur if molecules are initially transported into a soluble pool. In many cases pools may fluctuate widely during the cycle e.g. in *Navicula pelliculosa* silicic acid enters two compartments one of which is a soluble pool and may expand 2·5-fold during the cycle (Sullivan, 1979). There are only a few examples where changes in rates of uptake of precursors or related compounds coincide with similar variations in associated metabolic phenomena in some

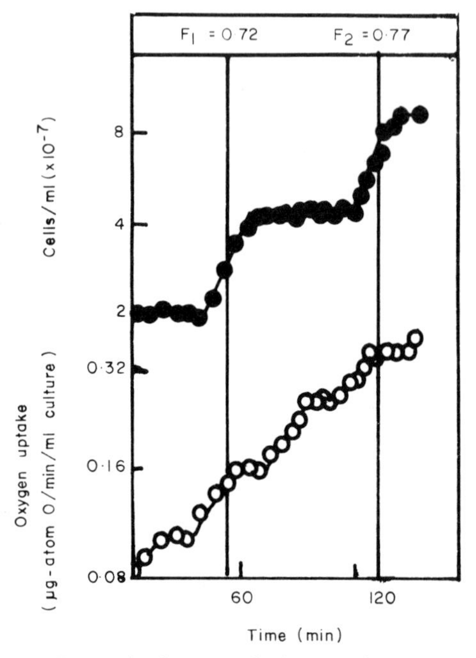

Fig. 6.20 Changes in respiration rate during synchronous growth of *A. eutrophus*. (●) = cell numbers and (○) = oxygen uptake rates [μg-atom O $min^{-1} ml^{-1}$ culture]. F_1 and F_2 denote synchrony indices of the first and second doublings in cell numbers, respectively. (Reproduced with permission from Edwards and Jones, 1977.)

cases e.g. in HeLa cells rates of oxygen uptake closely parallel changes in rates of uptake of thymidine (Robbins and Morrill, 1969) and variations in rates of uptake of uridine and phosphate accompany changes in the rate of RNA synthesis after release of 3T3 cells from contact-inhibited-growth (Cunningham and Pardee, 1969). However, it is often difficult to ascertain whether such uptake variations control or merely accompany these changes in metabolic activity.

In photosynthetic organisms, where synchronization is often induced by light–dark transitions, uptake of phosphorus in *Platymonas striata* (Ricketts, 1977a, b) and $Si(OH)_4$ in *N. pelliculosa* (Sullivan, 1977) is maximum just before cell division, which usually coincides with the time when illumination of cultures is terminated. However, similar results for $Si(OH)_4$ uptake in *N. pelliculosa* are obtained when cells are grown in continuous light (Darley *et al.*, 1976), although NO_3^- uptake in *Chlorella sorokiniana* is critically dependent on illumination conditions (Tischner and Lorenzen, 1979).

Increased rates of uptake of glucose occurred just after budding and at the mid point of the cycle of *S. cerevisiae* (Golombek and Wintersberger, 1974), while rates of adenine and serine uptake were constant for the initial part of the cycle and then doubled in the S-phase (Kubitschek and Edvenson, 1977). In *Schiz. pombe*, rates of accumulation of radioactively-labelled glucose and glycine were constant

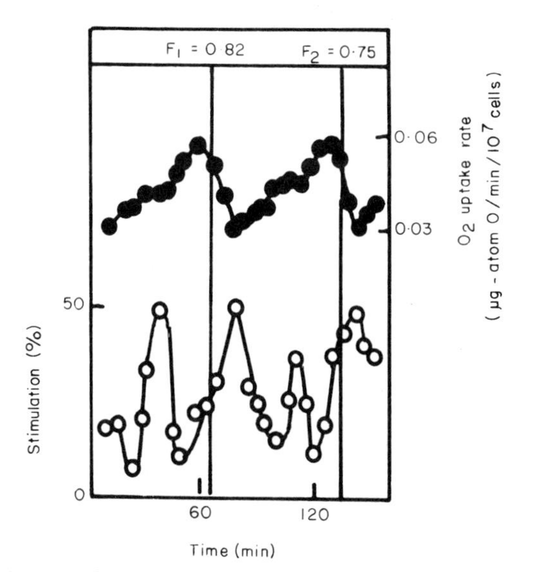

Fig. 6.21 Respiration of *A. eutrophus* during synchronous growth. Oxygen uptake (●) and (○) stimulation of oxygen uptake rate (%) following the addition of 16 μM-CCCP. (Reproduced with permission from Edwards and Jones, 1977.)

for most of the cycle, implying a constant rate of cell mass increase (Kubitschek and Claymen, 1976). In Chinese hamster ovary or L-cells, transport of α-amino-isobutyric acid is reduced at mitosis and early G_1, and then doubles in rate in late G_1 (Sander and Pardee, 1972). Whilst there were only slight variations in the rate of transport of 2′-deoxyglucose during the cell cycle of murine erythroleukemia cells, rates of uptake of α-amino-isobutyric acid were low in S-phase (Gazitt, 1979). It is proposed that transport of α-amino-isobutyric acid in L–M and KB cells is related to the activity of the Na^+/K^+ ATPase, which varies at different stages of the cycle (Soprano and Kuchler, 1978). More recent experiments have shown that the rate of transport of α-amino-isobutyric acid in 3T3 cells is dependent on cell density (Moya and Glaser, 1980).

In *E. coli*, transport of maltose increases at, or just before, cell division (Dietzel *et al.*, 1978) but uptake of glucose, acetate, phosphate, sulphate, leucine, glycine and thymidine are constant for much of the cycle (Kubitschek, 1968a, b). The rate of transport of L-α-glycerophosphate in the cell cycle of this organism increases in a stepwise fashion, coinciding with a stepped increase in cytochrome *b* (Ohki, 1972). Methionine uptake in *Alcaligenes faecalis* is constant for much of the cycle, doubling at the time of division (Lark and Lark, 1960). Thus, in bacteria, at least, it appears that fluctuations in rates of uptake cannot entirely account for the observed discontinuities in energy metabolism.

Summary

Energy metabolism during the cell cycle is an area which has received little attention. Various patterns of oxygen uptake rates have been reported including (i) continuous increase with inflections at particular cell cycle stages, (ii) stepwise or (iii) oscillatory patterns, depending on the cell type, the growth medium and the method of synchrony. Perhaps more than any other cellular process, energy metabolism appears to be most easily perturbed and hence selection procedures are the method of choice for preparing synchronous cultures; adequate control experiments must be performed to test for the possible effects of metabolic perturbation.

Inhibitor studies and measurements of adenine nucleotide pool levels have suggested that a number of different mechanisms may operate in the control of overall respiratory activity. These include, for example, respiratory control, i.e. the control of respiration by the availability of ADP, and the Pasteur effect. The rate of transport of particular substrates into cells is probably not the rate-limiting step for respiratory activity sizes although intracellular pool of metabolites may vary.

During the cell cycle of *Acanthamoeba castellanii*, it is not the energy state of the cell that drives the direction of metabolism towards biosynthesis or energy-generation reactions; instead, changes in the ATP/ADP ratios reflect the changes which result from the varying biosynthetic requirements of the cell. Further work is necessary to determine if a similar phenomenon occurs in other cell types. In bacteria, for example, the converse may be true, the cell's capacity for growth being limited by rates of respiration and oxidative phosphorylation.

7. Genetics of the Cell Division Cycle

"... the course of nature ... seems delighted with transmutations."

Sir Isaac Newton Query 30, Opticks

I. Introduction

Progression through the cell cycle depends both on the consecutive and simultaneous functioning of many genes. Some of these genes have now been identified by mutation studies, although little is yet known of the control mechanisms that underlie the normally precise temporal order of activation of these genes. This chapter describes recent studies, mostly of the last decade, that involve mutant cells defective in specific stages of the cell cycle. Particular emphasis is placed on the genetic analysis of the cell cycle in the yeast *Saccharomyces cerevisiae*, because it is in this system that the greatest advances have been made in identifying large numbers of cell cycle genes and the molecular bases of their functions.

II. Terminology

Since progression through the cell cycle is, by definition, an indispensable function of proliferating cells, mutations that unconditionally inactivate gene products essential for cycle progress are lethal. Therefore, cell cycle mutations must be of a conditional nature. Most work has been done with heat sensitive mutants, i.e. those for which a low temperature is permissive and a high temperature is nonpermissive, or restrictive. Some cold-sensitive mutations have also been isolated and studied and, in principle, additional types of conditional mutations could be isolated, particularly suppressible mutations where the suppressor is temperature-sensitive (Rasse-Messenguy and Fink, 1973). An important feature of the use of conditional lethal mutants is that, theoretically, a large fraction of the genome is susceptible to such lesions and so a large number of functions involved in the cell cycle should be amenable to analysis by this approach.

The definition of a cell cycle mutation adopted here is that of Hartwell (1978) and describes those mutations that lead to defects in a stage-specific event (or landmark) that occurs once per cycle, such as DNA replication or nuclear division. As Hartwell (1978) and Pringle (1978) point out, this definition excludes mutations in genes whose products control the continuous processes of growth, and a broader definition may be desirable (Chu, 1978).

Under restrictive conditions (most frequently, the higher temperature), the mutant cells in a previously-asynchronous population will first exhibit defective behaviour at the same "diagnostic landmark" in the cell cycle (Hartwell, 1974, 1978). Previously, Hartwell *et al.* (1974) had used the term "initial defect" to describe this stage. The

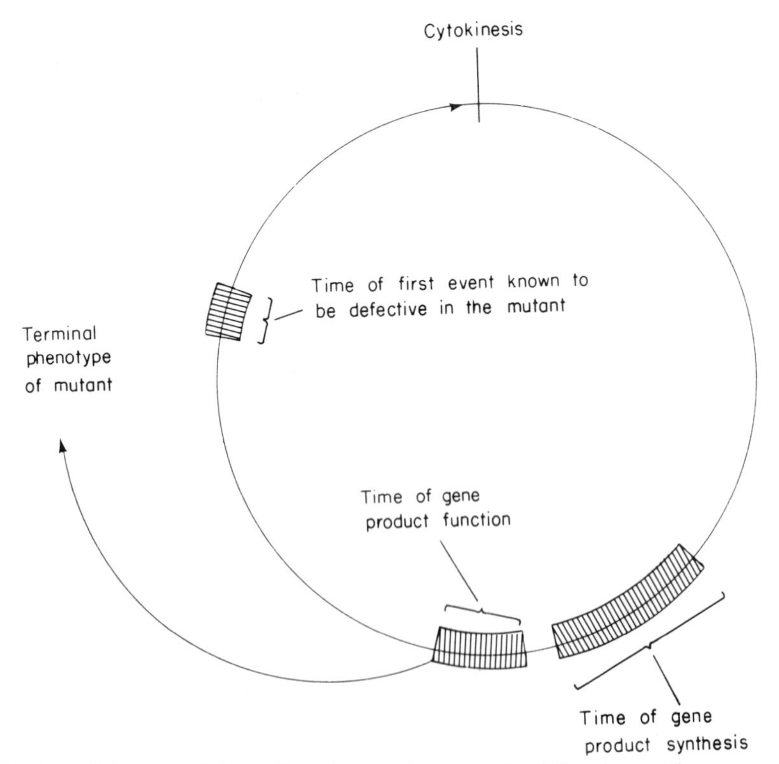

Fig. 7.1 Diagram of the cell cycle showing several of the times relevant to consideration of a particular hypothetical conditional lethal cell cycle mutant. (Reproduced with permission from Pringle, 1978.)

mutant cells then assume the same, sometimes aberrant, and characteristic morphology that is termed the terminal phenotype (Hartwell, 1974; Simchen, 1978) or termination point (Hartwell *et al.*, 1970; Fig. 7.1).

The "execution point" of such mutants was originally defined as the cell cycle stage when the defective gene product would normally have functioned in the wild-type cell (Hartwell *et al.*, 1970) or completed its function (Hartwell *et al.*, 1970; Hartwell, 1978; Nurse *et al.*, 1976). The alternative terms "transition point" (Nurse *et al.*, 1976), "block point" (Howell, 1974) and "shift up point" (Ashihara *et al.*, 1978) have been advocated. Cells shifted to the restrictive condition before this point are incapable of further division, but after this point the restrictive condition is without effect on the succeeding division. The terms used here are in general those given by the original authors.

Similarly, we use the gene symbols given in the research literature, even though they do not always conform with the proposals made elsewhere for bacteria (Demerec *et al.*, 1966), yeasts (Sherman and Lawrence, 1974) or other fungi (Clutterbuck, 1973).

III. Mutants of Eukaryotic Cells

A. *SACCHAROMYCES CEREVISIAE*

1. *Isolation and Genetic Characterization*

S. cerevisiae can grow vegetatively in either the haploid or diploid state, the same sequence of cell cycle events being traversed in both modes of growth. This enables recessive cell cycle mutants to be isolated in the haploid state and the ability of two such mutants to exhibit complementation to be studied in the diploid state.

More than 2 000 temperature-sensitive mutant clones (obtained by mutagenesis with nitrosoguanidine and ethyl methane sulphonate) have been examined morphologically by Hartwell and his collaborators: they isolated 148 temperature-sensitive cell division cycle (*cdc*) mutants initially on the basis of their uniform, and sometimes unusual, phenotypes after incubation at the restrictive temperature 36°C (Hartwell *et al.*, 1970). Complementation studies placed these recessive mutations into 32 groups and tetrad analysis revealed that each of these groups defines a single nuclear gene, many of which are represented by more than one allele (Hartwell *et al.*, 1973). Fourteen of these genes have been located on the yeast genetic map; functionally related genes were found not to be tightly clustered. Time-lapse photomicroscopy revealed that, for most mutants, those cells that exist before the execution point arrest development in the existing cycle at 36°C; those cells that are after this point divide and then arrest in the next cycle. Such mutants are said to exhibit first cycle arrest (Fig. 7.2). Many mutants complete several cell cycles after the shift to the

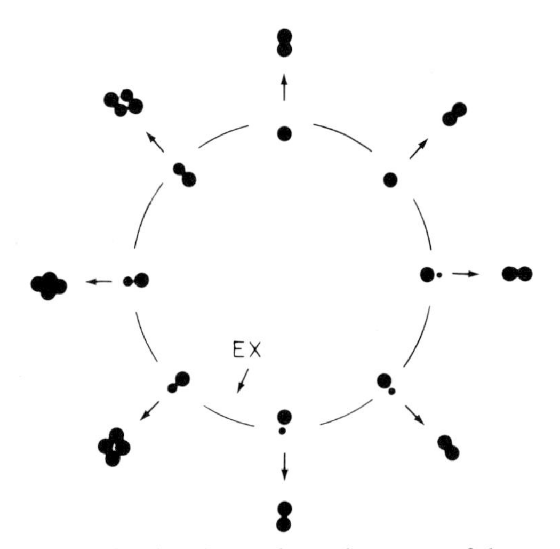

Fig. 7.2 Determination by time-lapse photomicroscopy of the execution point in a temperature-sensitive *cdc* mutant of the budding yeast, *Saccharomyces cerevisiae*. Cells growing at the permissive temperature were shifted to the restrictive temperature and photographed at the time of the shift (cells on the inner circle); the same cells were photographed after 6 h at the restrictive temperature (cells on the outer circle). Cells were cut from photographs taken at the time of the shift and arranged on the inner circle in order of bud size, a measure of position in the cell cycle. Each cell on the outer circle developed from the corresponding cell on the inner circle and is composed of a parent cell with a large bud. Cells early in the cycle (before the execution point, EX) arrested development in the first cell cycle at the restrictive temperature; each cell later in the cell cycle (after EX) finished the first cycle and arrested in the second cycle. (Redrawn from Hartwell, 1978.)

restrictive temperature (multicycle arrest). In several *cdc* genes (e.g. *cdc* 16; Hartwell *et al.*, 1973), both types of responses are exhibited by different alleles after a shift to the restrictive temperature. In such cases, it is likely that alleles conferring first cycle arrest are temperature-sensitive for function, whilst alleles that result in multicycle arrest are temperature-sensitive for synthesis, the gene product normally being present in excess and not constituting a constraint on the course of division.

The *cdc* mutations described in *S. cerevisiae* have been tabulated by Simchen (1978). To this list may be added mutants defective in *cdc* 36 (14 alleles), *cdc* 37 (1 allele) and *cdc* 38 (1 allele), all of which, like *cdc* 28, appear to be defective in G_1 at "start". Nine alleles have now been identified for the latter gene (L. H. Hartwell, personal communication). The terminal phenotypes of strains carrying mutations in each of the complementation groups have been given by Hartwell *et al.* (1974) and Simchen (1978).

Additional cell cycle mutations have been found among mutants selected for other properties:

(a) Mutant *tra* 3 was isolated as a strain fully derepressed for the enzymes of several pathways of amino acid biosynthesis, but was subsequently found to be temperature-sensitive for growth and to arrest as single unbudded cells in G_1 (Wolfner *et al.*, 1975).
(b) Mutations in *tmp* 1 were shown to be allelic with *cdc* 21 (Game, 1976).
(c) A new cell cycle mutation, *cdc* 40, was induced by u.v. irradiation of a diploid heterozygous for *cdc* 5 (Kassir and Simchen, 1978).
(d) Mutations in six new complementation groups, as well as in seven new alleles of known *cdc* genes, were obtained amongst temperature-sensitive mutants that showed depressed DNA synthesis at the restrictive temperature (Johnston and Game, 1978). These workers also found ten other mutants with a characteristic terminal phenotype, some of which may reflect new *cdc* genes.

Sudbery *et al.* (1980) have isolated mutants exhibiting small cell size, by a procedure involving the combined use of the mating hormone α-factor and cell size separation (Carter and Sudbery, 1980). The mutants define two genes *whi*-1 and *whi*-2 and are altered with respect to the conditions under which "start" occurs (see Section III.A6c).

2. *Mutants Defective in Bud Emergence*

The *cdc* 24 gene product is probably intimately involved in bud emergence, because mutations in this gene block budding (but not DNA synthesis or nuclear division) and because the execution point is precisely at the time of bud emergence (Hartwell *et al.*, 1974). At the restrictive temperature, the mutant produces large, round multinucleate cells. The chitin that normally forms a ring at the point of bud emergence on the mother cell wall continues to be synthesized but is not organized into discrete rings (Sloat and Pringle, 1977, 1978).

Mutants defective in *cdc* 1 exhibit first cycle arrest and bear no buds or very tiny ones (Fig. 7.3), at the restrictive temperature (Hartwell *et al.*, 1970; Hartwell, 1971b). Again, the execution point is coincident with bud emergence at 0·2 of a cycle. Mutations in *cdc* genes 19, 22, 25, 29 and 32 also prevent bud emergence (Hartwell *et al.*, 1973) but have not been as well characterized.

From the behaviour of mutant strains carrying lesions in *cdc* 4 (initiation of DNA synthesis), *cdc* 8 (DNA synthesis) or in both genes, it has been proposed that a cellular clock controls bud initiation and that the running of the clock is independent of DNA synthesis, nuclear division, and cell separation. Premature bud initiation is normally prevented as a consequence of the successful initiation of DNA replication (Hartwell, 1971a). Strains defective in *cdc* 4 accumulate with a prominent double spindle plaque (Byers and Goetsch, 1973) similar to that associated with bud emergence in cultures growing normally.

3. *Mutations affecting Cytokinesis and Cell Separation*

Strains carrying mutations in *cdc* genes 3, 10, 11 or 12 do not complete the process of cytokinesis or cell membrane separation, based upon an empirical test (Hartwell, 1971b) so that multiple elongated buds develop, which do not separate from the parent cell (Fig. 7.4). Several rounds of DNA replication and nuclear division persist at the restrictive temperature. In those mutants that exhibit first cycle arrest (*cdc* 3, 10 and 11), the execution point occurs within the first 0·6 of the cycle, considerably before cytokinesis itself at 0·9 of a cycle. Byers and Goetsch (1976) studied 24 strains defective in *cdc* genes, which arrested in the cycle with one or more buds present. The highly ordered ring of filaments that normally lies within the neck between bud and mother cells persisted in 20 of these strains. In the remaining 4 strains, the filamentous ring was missing and all proved to be specifically defective in cytokinesis. The involvement of the ring in cytokinesis was further suggested by the similar kinetics of (a) the loss of potential for cytokinesis and (b) the rate at which the ring was lost after a shift in the restrictive temperature. Clarke and Carbon (1980) have cloned the DNA segment carrying the *cdc* 10^+ gene.

4. *Mutations affecting Nuclear Division*

Several mutants, belonging to 11 complementation groups (*cdc* 2, 6, 7, 9, 13, 14, 15, 16, 17, 20, 23) have been described that appear to

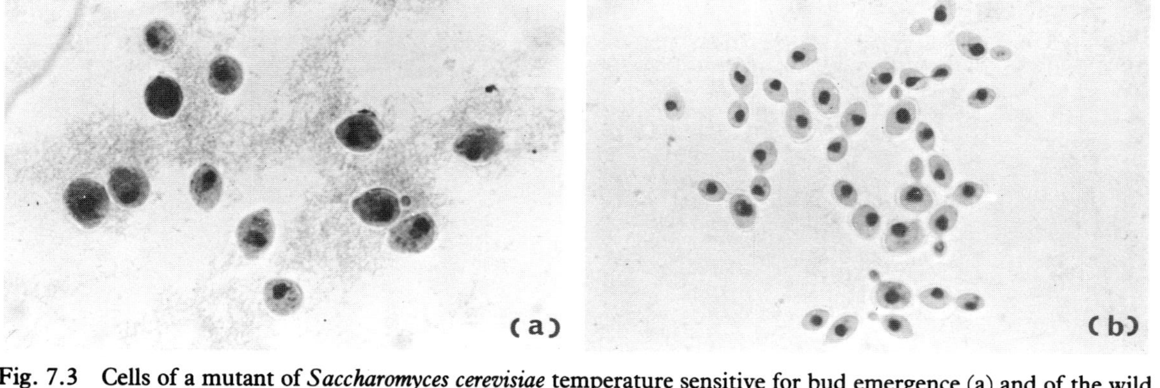

Fig. 7.3 Cells of a mutant of *Saccharomyces cerevisiae* temperature sensitive for bud emergence (a) and of the wild type strain (b). Both cultures were incubated at the temperature restrictive for the mutant and stained to reveal nuclei. The mutant is 369 D1 (*cdc* 1–1/*cdc* 1–1) and the wild type is the haploid A 364 AD4. Note that cells in the mutant culture remain mononucleate. (Reproduced with permission from Hartwell, 1971b.)

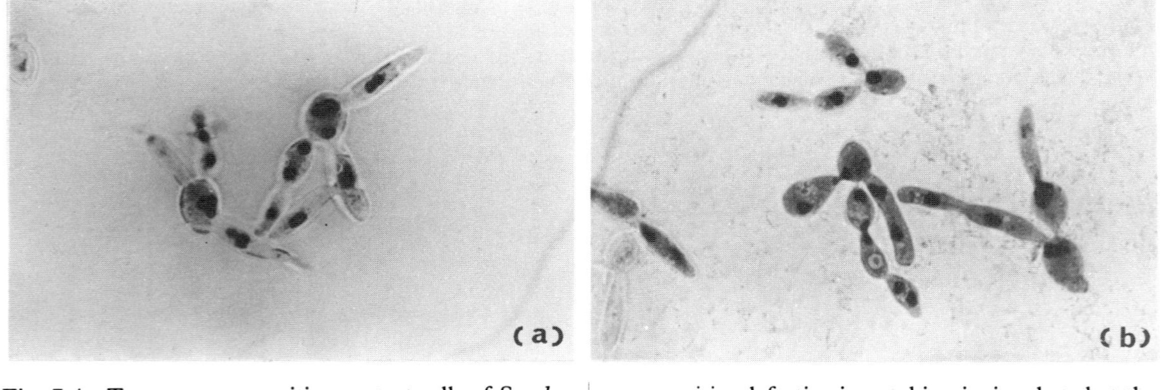

Fig. 7.4 Temperature-sensitive mutant cells of *Saccharomyces cerevisiae* defective in cytokinesis, incubated at the restrictive temperature. The strains are (a) 104 D7 (*cdc* 3–1/*cdc* 3–1) and (b) 471 D3 (*cdc* 12–1/*cdc* 12–1). (Reproduced with permission from Hartwell, 1971b.)

be defective in nuclear division (Hartwell, 1971b; Hartwell *et al.*, 1973). The mutations affect the early, medial and later stages of nuclear division, although the execution points in all mutants are clustered much earlier in the cycle, between 0·02 and 0·45. Further analysis showed that the gene with the earliest execution point (*cdc* 7) was actually defective in DNA synthesis, an early cycle event (Hartwell, 1973a). Thus, a temporal programme of gene activity appears to be initiated early in the cycle, in preparation for nuclear division.

Byers and Goetsch (1973) have studied microscopically the terminal phenotypes of many *cdc* mutants that exhibit first cycle arrest. In those blocked in medial nuclear division in which the nucleus becomes elongated as it is squeezed into the neck (*cdc* 2, 6, 9, 13, 16, 17, 20, 21, 23) as well as in DNA synthesis and its initiation (see below), final spindle elongation is blocked. In strains arrested in later nuclear division, however, the very elongated nucleus contains a spindle reaching to its extremities, suggesting that the spindle plays a role only in this final elongation.

5. *Mutations Affecting DNA Replication and its Initiation*

(a) *Nuclear DNA*. The bulk of cellular DNA in yeast normally replicates during S-phase, a short interval of the cycle that begins at about the time of bud emergence (Chapter 3; Ogur *et al.*, 1953; Williamson, 1964a, b; Williamson and Scopes, 1960).

Lesions in the *cdc* 28, *cdc* 4 and *cdc* 7 genes (Hartwell, 1971a, 1973a: Hereford and Hartwell, 1973) block a precondition for DNA synthesis, since cells carrying these lesions cannot start new rounds of DNA replication after a shift to the restrictive temperature, but can finish rounds that are in progress. Inhibition of DNA replication by mutation allows one cycle of budding and migration of the nucleus into the neck; these effects are mimicked when DNA synthesis is inhibited by hydroxyurea (Slater 1973). The temperature sensitivity of initiation of nuclear DNA synthesis has recently been demonstrated in ether-permeabilized ρ^0 cell derived from a *cdc* 7 mutant (Oertel and Goulian, 1979). Cell-free extracts from *cdc* mutants 4, 7, 28 and 8 (see below) grown at the permissive temperature are capable of stimulating DNA replication in a system consisting of nuclei from spleen cells of *Xenopus* (Jazwinski and Edelman, 1976). After incubation of cells for one generation at 36°C, the extracts showed little or no activity. In a control experiment, an extract from a *cdc* 10 mutant, defective in cytokinesis, showed little or no loss of stimulatory activity at the restrictive temperature.

The order of steps mediated by genes involved in the initiation of DNA synthesis have been shown by Hereford and Hartwell (1974) to be

$$\xrightarrow{cdc\ 28} \xrightarrow{cdc\ 4} \xrightarrow{cdc\ 7} \text{Initiation of DNA synthesis.}$$

Protein synthesis is also required for initiation of DNA synthesis Hereford and Hartwell, 1973). Mutants in *cdc* 7, but not *cdc* 4, on release from the permissive temperature, are able to complete a round of DNA synthesis even in the presence of cycloheximide, indicating that an essential programme of protein synthesis occurs between *cdc* 4 and *cdc* 7 gene products but that the initiation of DNA synthesis occurs without protein synthesis. Protein synthesis is also required to pass the *cdc* 28 block, and one of the two alleles, *cdc* 28.1, may be defective in the initiation of protein synthesis (Edwards *et al.*, 1978).

DNA molecules isolated from strains carrying either the *cdc* 4 or *cdc* 7 mutation are simple linear structures, containing neither replication "bubbles" ("eye forms") nor Y-shaped replication forks (Petes and Newlon, 1974), suggesting that these are true initiation mutants and not mutants with a leaky block in replication (Hartwell, 1974).

Two gene functions, *cdc* 8 and *cdc* 21, are required throughout the period of DNA replication (Hartwell, 1971b, 1973a). Mutation in either gene results in an immediate cessation of DNA replication upon a shift to the restrictive temperature (Fig. 7.5) and the arrest of cells with a terminal phenotype identical to that produced by established inhibitors of DNA synthesis. Both mutants have execution points near the end of S (Hartwell, 1971a; Hartwell *et al.*, 1973), close to the time of completion of the diagnostic landmark. One of these mutants, *cdc* 8, has been found to be temperature-sensitive for DNA synthesis *in vitro* using permeabilized cell systems (Hereford and Hartwell, 1971; Oertel and Goulian, 1979). This mutant is not defective in synthesis of deoxyribonucleotide precursors, since no DNA synthesis is observed in these permeabilized cells at the restrictive temperature when provided with deoxyribonucleotide triphosphates.

Purified DNA polymerases I and II from *cdc* 8 are no more labile than those from the wild type, suggesting the *cdc* 8^+ is not a structural gene for either enzyme (Prakash *et al.*, 1979). Instead, it seems more likely that the gene specifies an as yet unknown component of the DNA replication complex that may influence the latter's ability to read past damage on the template strand.

The defective gene product has been identified in only two *cdc* mutants. The *cdc* 21 mutant is defective in thymidylate synthetase

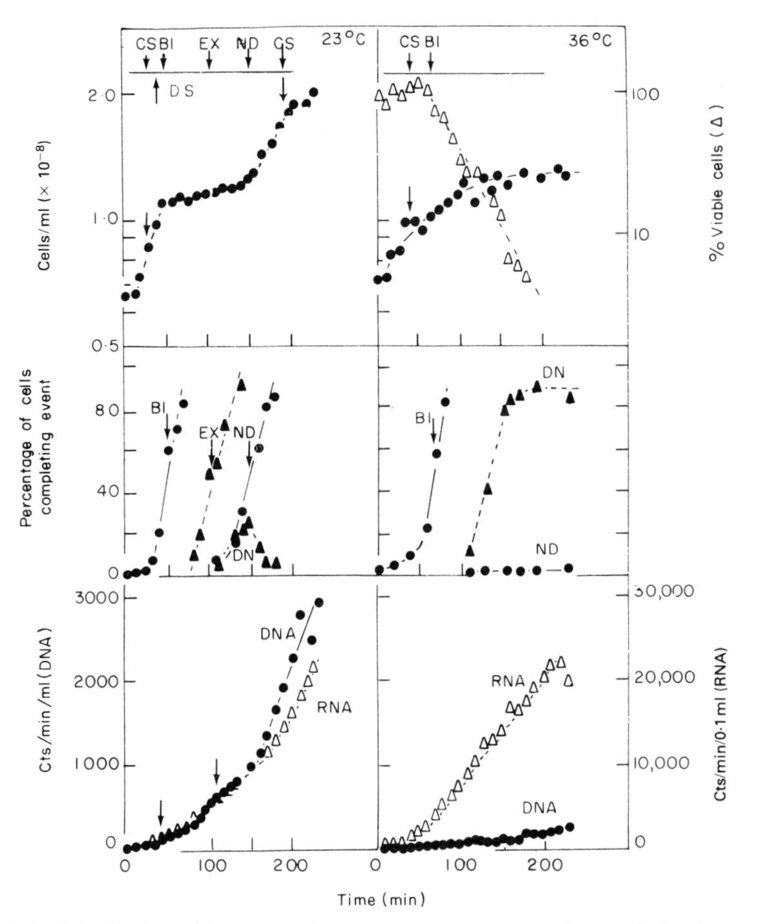

Fig. 7.5 Monitoring cell cycle events in a synchronous culture of *Saccharomyces cerevisiae* strain 198D1 (*cdc* 8) at the permissive (23°C) and restrictive (36°C) temperature. Approximately 5×10^9 cells growing in synthetic medium were collected by centrifugation and banded isopycnically in a Renografin density-gradient. Approximately 1×10^8 of the lightest cells were resuspended in 100 ml of synthetic medium containing 1 μCi [2-^{14}C]uracil/ml (final specific activity 39 mCi/m-mole) and the culture was divided into two portions, one of which remained at 23°C (left panel) while the other was placed at 36°C (right panel). Samples were removed at various times and monitored for the increase in cell number to determine the time of cell separation (CS), for the percentage of the cells that had completed bud initiation (BI), and nuclear division (ND), and for the percentage of the cells with dividing nuclei (DN). DNA synthesis was monitored by the incorporation of radioactivity into DNA, and RNA synthesis by the incorporation of radioactivity into RNA. Samples from the 23°C culture were transferred to agar plates at 36°C, photographed immediately and again several hours later to determine the time of execution (EX). Samples from the 36°C culture were removed, diluted and plated onto agar plates at 23°c to determine the number of viable cells; percentage viable cells is calculated relative to a value of 100% at the beginning of the experiment. Arrows designate the time at which 50% of the cells complete a particular event. (Reproduced with permission from Hartwell, 1971a.)

(Bisson and Thorner, 1977); at the restrictive temperature, it arrests with a morphology identical to that of a temperature-sensitive thymidine 5′-monophosphate auxotroph. The other product so far identified in a *cdc* mutant defective in DNA replication, is a DNA ligase, involved in both replication and repair of DNA in *cdc* 9 (Johnston and Nasmyth, 1978).

In contrast to mutants *cdc* 4 and *cdc* 7, approximately half of the DNA molecules isolated from mutant *cdc* 8 incubated at the restrictive temperature contain one or more small replication bubbles (Fig. 7.6), thus confirming its classification as a DNA propagation mutant. Mutant *cdc* 21, classified by Hartwell (1973a) as a possible propagation mutant, also contains DNA with replication bubbles at the restrictive temperature (T. D. Petes, unpublished, cited by Petes and Newlon, 1974).

(b) *Mitochondrial DNA*. The *cdc* mutants have been used to explore the coordination of the synthesis of the bulk of cellular DNA in the nucleus with that of mt DNA, a subject that has generated some controversy (Chapter 3). In 1973, Cottrell *et al.* and Cryer *et al.* reported that in mutant strain 314, mt DNA synthesis was sustained at the restrictive temperature (36°C) whilst synthesis of total cellular DNA was greatly diminished. This mutant was originally described (Hartwell, 1971a) as harbouring a mutation in gene *cdc* 8 but was later

---

Fig. 7.6 Small DNA replication bubble in DNA isolated from a temperature-sensitive DNA propagation mutant of *Saccharomyces cerevisiae* (*cdc* 8) at the restrictive temperature. A culture of *cdc* 8 (strain 198–1) was grown at 23°C in 10 ml of minimal media supplemented with 0·2 g glucose, 5 mg yeast extract, 200 μg uracil, 200 μg adenine, 500 μg tyrosine, 500 μg histidine and 500 μg lysine. The doubling time of the culture under these conditions is approximately 3 h. At a cell density of 2×10^6 cells ml^{-1}, the cultures were shifted to 36°C for 5 h. Cells were then collected and spheroplasted and 0·2 M hydroxyurea was added to the spheroplasting solutions. Nuclear DNA was separated from other cellular components by centrifugation in sucrose gradients spread for electron microscopy. Each branch of the bubble in the photographed molecule is 0·4 μm in length; arrows show the forks. Nuclear DNA was prepared from *cdc* 4 (strain 135.1.1) and *cdc* 7 (strain 4008) as described above. For both initiation mutants all the DNA molecules observed in the experiments should come from cells arrested in the cell cycle at the beginning of the S period. For *cdc* 8, however, although most of the molecules of DNA should come from cells blocked near the beginning of the S period, some DNA molecules should come from cells within the S period. Since only 3–5% of the DNA molecules in an asynchronous culture are replicating at any given time (Petes *et al.*, 1973), most of the DNA molecules observed in the *cdc* 8 experiment should contain no replication structures before the shift to the restrictive temperature. (Reproduced with permission from Petes and Newlon, 1974.)

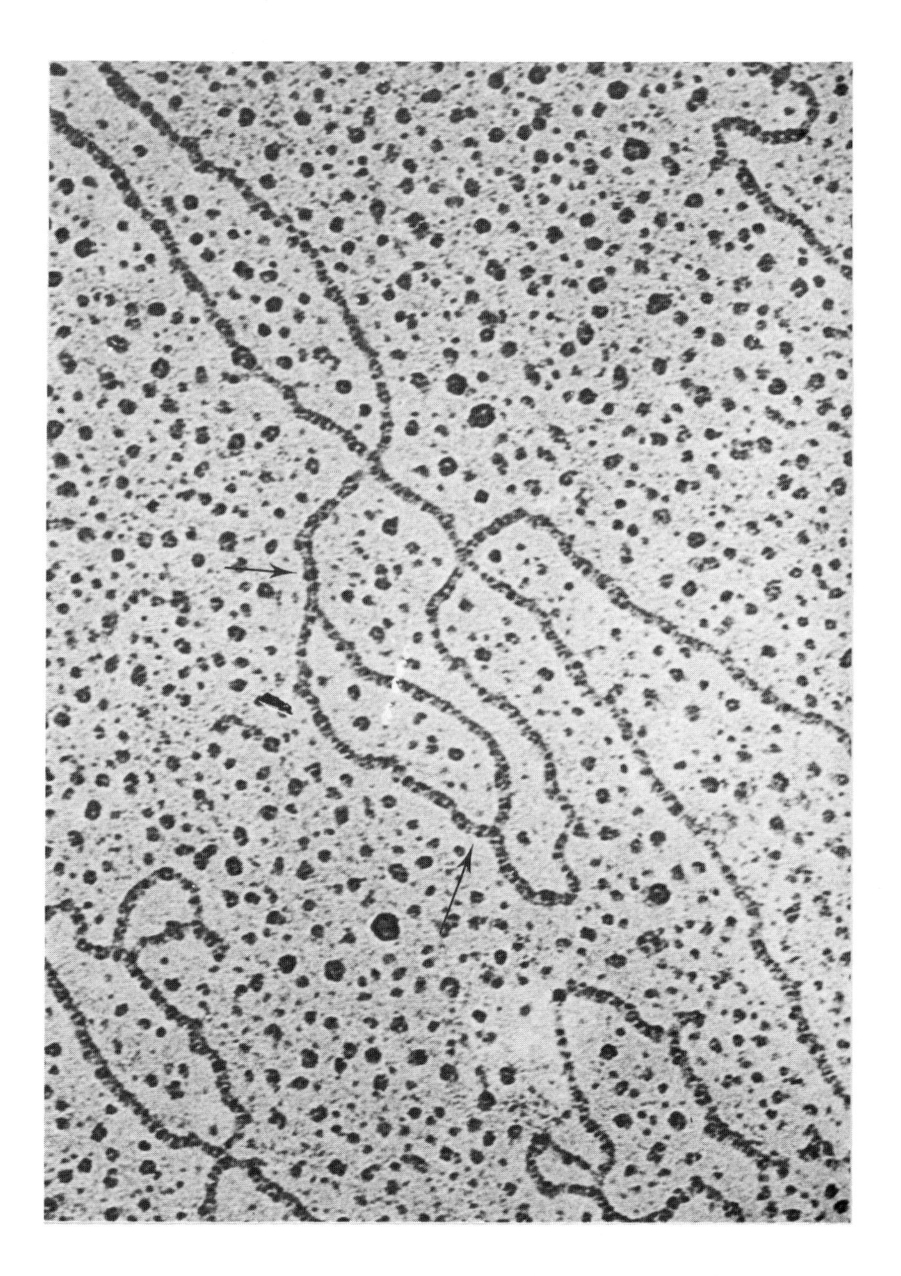

correctly described as a *cdc* 4 mutant (Hartwell, 1972). This error has caused some confusion in interpreting data from various laboratories. Newlon and Fangman (1975) subsequently showed that mt DNA replication continued for a long time at the restrictive temperature in five *cdc* mutants in which nuclear DNA synthesis ceases within one cycle: *cdc* 4, 7 and 28 (defective in initiation of nuclear DNA synthesis) and *cdc* 14 and 23 (defective in completion of nuclear division). The products of these genes, therefore, are apparently not required for the initiation of mt DNA replication. In contrast, in mutants *cdc* 8 and 21, defective in continued replication of nuclear DNA, mt DNA replication virtually ceased at the non-permissive temperature (Fig. 7.7). Cultures of these mutants also contain petites at an unusually high frequency; this phenotype is caused by the defect in DNA synthesis (Newlon *et al.*, 1979). However, both Wintersberger *et al.* (1974) and Cryer *et al.* (1973) reported a less clear-cut result, observing only preferential synthesis of mt DNA at the non-permissive temperature in a mutant that contained a different allele of *cdc* 8 (Hartwell *et al.*, 1973).

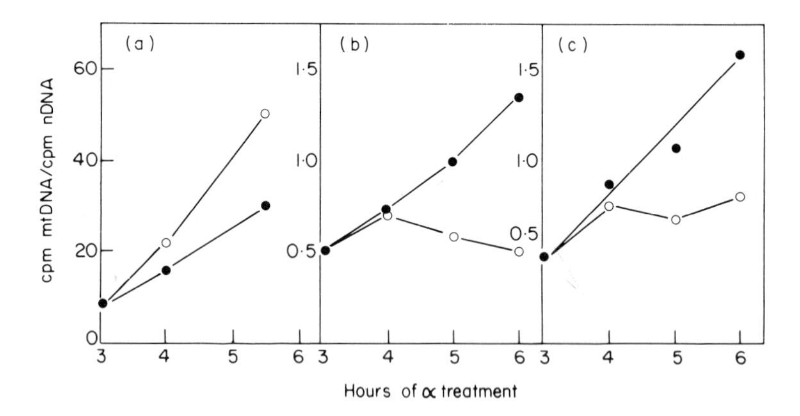

Fig. 7.7 Mt DNA synthesis in *Saccharomyces cerevisiae* mutants *cdc* 8 and *cdc* 21 arrested with *α* factor. *α* Factor was added to cultures of wild type, *cdc* 8 and *cdc* 21 growing at 23°C. Thirty minutes later 50 μCi/ml of [6-^{3}H]uracil was added to each culture and incubation at 23°C was continued. Three hours after the addition of *α* factor, half of each culture was shifted to 36°C. At the time of the shift and at hourly intervals, samples were withdrawn from all cultures, the cells were spheroplasted, and lysates were prepared. The ratio of mt DNA to nuclear DNA in each sample was determined from the DNA density profile on a CsCl gradient. (●) 23°C; (○) 36°C. (a) wild type; (b) *cdc* 8; (c) *cdc* 21. (Reproduced with permission from Newlon and Fangman, 1975.)

Richmond (1976) has isolated a temperature-sensitive mutant (*ts* 47) which appears to be blocked at an early stage of nuclear division. In contrast to nuclear DNA, the synthesis of which stops at about the same time as cell division, mt DNA is synthesized for over 6 h at 37°C accumulates to represent about 50% of total cellular DNA. The mutant phenotype is due to two recessive mutations; both are necessary for temperature sensitivity and no phenotype has been found for either mutation alone.

A number of *cdc* mutants (*cdc* 8–13, 21–1, 4–2 and 28) have been used by Mahler *et al.* (1975) to demonstrate that glucose derepression in yeast is independent of both the initiation and elongation of progeny DNA in the nucleus. Using mutants that are deficient in the initiation of nuclear DNA synthesis, and which overproduce mt DNA (see above), it was concluded that mitochondrial gene dosage is without effect on the regulation of derepression.

(c) *Other DNA*. In addition to mt DNA and nuclear DNA, every known strain of *S. cerevisiae* contains a class of extranuclear, circular, double-stranded DNA molecules, with a contour length of 2 μm, which constitutes about 3% of the total cellular DNA (Clark-Walker and Miklos, 1974). Petes and Williamson (1975a, b) have shown that, compared with a wild-type parental strain, *cdc* 8 mutants contain a greatly enhanced proportion of these molecules and that they exhibit a double-branched appearance, indicative of replication. The 2 μm circles were shown to require a functional *cdc* 8 gene product for a normal rate of replication. Double-stranded circles similar to those found in *cdc* 8 were observed also in *cdc* 21 (a DNA propagation mutant) but not in *cdc* 7 (a DNA initiation mutant). In an extension of this study, Livingston and Kupfer (1977) reported that the 2 μm DNA does not replicate at the restrictive temperature in cells bearing the *cdc* 28, *cdc* 4 and *cdc* 7 mutations (which prevent the passage of cells from G_1 to S) and in *cdc* 8 mutants. Replication of this plasmid-like structure also continued at the restrictive temperature for about one generation time in a *cdc* 13 mutant defective in nuclear division. Cell-free extracts from mutants *cdc* 7 or *cdc* 8 show diminished ability to stimulate DNA synthesis directed by supercoiled 2 μm DNA *in vitro* (Jazwinski and Edelman, 1979). Thus, the synthesis of all three DNA species (nuclear, mitochondrial and 2 μm plasmid) is dependent on at least two common gene products, the *cdc* 8 product (probably involved in DNA elongation) and the *cdc* 21 product (involved in thymidine metabolism).

Many strains of *S. cerevisiae* produce an extracellular substance that kills sensitive cells; these strains also carry two double-stranded (ds) RNA species encapsulated in intracellular virus-like particles (VLPs) (Wickner, R. B., 1976). Overproduction of ds RNA and VLPs occurs at the restrictive temperature in mutants defective in either the initiation of DNA synthesis or in medial or late nuclear division, but not in mutants *cdc* 8 and 21 defective in DNA chain elongation (Fischer and Shalitin, 1977; Shalitin and Weiser, 1977; Weiser and Shalitin, 1978).

These observations remain to be explained, but recent results that implicate an unidentified gene product in killer ds RNA replication (Wickner, R. B., 1976) suggest a correlation between the synthesis of mitochondrial DNA and VLPs (Weiser and Shalitin, 1978).

6. *Integration of Cell Cycle Events*

One of the major long-term objectives in the study of mutants defective in specific stages of the cell cycle is an understanding of the temporal and causal order of cycle events, sequential gene expression and the coordination of cycle events.

(a) *Causal order.* Information on the causal order of cycle steps may be obtained by cataloguing those cycle events which do or do not occur after arrest at the diagnostic landmark when cultures are shifted to the restrictive temperature. When the occurrence of the diagnostic landmark is specifically inhibited by expression of the mutation, some of the events that normally follow at the permissive temperature are also prevented and are, therefore, deemed to be dependent on completion of the inhibited landmark. In contrast, those that proceed unaffected by expression of the mutation may be assumed to be independent of the inhibited landmark.

Figure 7.8 shows the order and interdependence of landmark events in *S. cerevisiae*, based on consideration of the terminal phenotypes of *cdc* mutants (Table 7.1) and is an elaboration of an earlier scheme (Hartwell *et al.*, 1974). Although mutations in 32 *cdc* genes had been discovered at that time, many were not included because strains carrying these lesions progress through several cycles at the restrictive temperature before development is arrested. Mutant *cdc* 1 (Hartwell, 1971b) was excluded because macromolecule synthesis, as well as bud emergence, was rapidly arrested at the restrictive temperature. This inhibition of growth was suspected to prevent the occurrence of some events which are not normally dependent upon bud emergence but which are dependent on growth. An examination of the mutant collection showed that six clearly-definable landmarks of the cell cycle

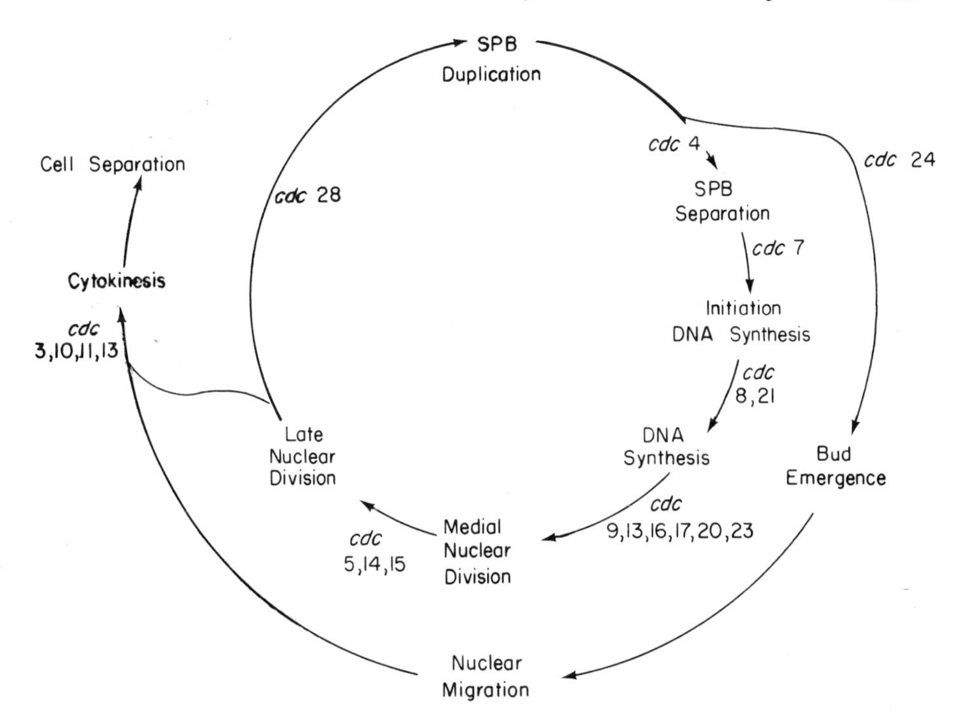

Fig. 7.8 Dependent pathway of landmarks derived from mutant phenotypes in the cell cycle of *Saccharomyces cerevisiae*. *Cdc* gene designations are placed immediately preceding their diagnostic landmark. The diagram relates to mutant phenotypes as follows: upon a shift to the restrictive temperature, mutant cells arrest asynchronously at the position designated by the *cdc* number; all events flowing from this point do not occur while all other events do. (Reproduced with permission from Hartwell, 1978.)

could be placed in the following sequence: initiation of DNA synthesis; DNA synthesis; medial nuclear division; late nuclear division; cytokinesis; cell separation.

Any mutant with an initial defect in one process failed to complete other, later events in the sequence. For example, mutants defective in the initiation of DNA synthesis (*cdc* 4 and *cdc* 7) are unable to complete *any* subsequent events in this sequence (but *can* undergo bud emergence and nuclear migration). Similarly, mutants defective in late nuclear division (*cdc* 14, 15) can complete all events prior to this in the postulated sequence (*and* undergo bud emergence and nuclear migration) but cannot complete cytokinesis or cell separation. The simplest explanation of these results is that these six events constitute a dependent pathway in which completion of each event is

Table 7.1. *Summary of mutant phenotypes of the* cdc *mutants of* Saccharomyces cerevisiae

cdc★	Initial defect	Events completed at restrictive temperatures							
		BE	iDS	DS	NM	mND	lND	CK	CS
28	Start	–	–	–	?	–	–	–	–
24	BE	–	++	++	?	++	++	–	–
4	iDS	++	–	–	★	–	–	–	–
7	iDS	+	–	–	+	–	–	–	–
8	DS	+	+	–	+	–	–	–	–
21	DS	+	+	–	+	–	–	–	–
2	mND	+	+	+	+	–	–	–	–
6	mND	★	★	★	★	–	–	–	–
9	mND	+	+	+	+	–	–	–	–
13	mND	+	+	+	+	–	–	–	–
16	mND	+	+	+	+	–	–	–	–
17	mND	+	+	+	+	–	–	–	–
20	mND	+	+	+	+	–	–	–	–
23	mND	+	+	+	+	–	–	–	–
14	lND	+	+	+	+	+	–	–	–
15	lND	+	+	+	+	+	–	–	–
3	CK	++	++	++	++	++	++	–	–
10	CK	++	++	++	++	++	++	–	–
11	CK	++	++	++	++	++	++	–	–
	CS	++	++	++	++	++	++	++	–

Cells were shifted from 23 to 36°C at the time of cell separation. A minus sign indicates that an event does not occur, a plus indicates that the event occurs once, and a double plus indicates that the event occurs more than once.

★Although mutations in 32 *cdc* genes have been discovered, only 19 of these genes are included here for consideration in developing a model of the cell cycle. Most of those not included were left out because they progress through several cycles at the restrictive temperature before development is arrested and this prevents an analysis of DNA synthesis during their terminal cycle. The mutant *cdc* 1 was excluded because macromolecule synthesis, as well as bud emergence, is rapidly arrested at the restrictive temperature, and this inhibition of growth probably prevents the occurrence of some events which are not normally dependent upon bud emergence, but which are dependent on growth.

Abbreviations are bud emergence (BE), initiation of DNA synthesis (iDS), DNA synthesis (DS), nuclear migration (NM), medial nuclear division (mND), late nuclear division (lND), cytokinesis (CK) and cell separation (CS). (Reproduced with permission from Hartwell *et al.*, 1974.)

a necessary prerequisite for the occurrence of the immediate succeeding event (Fig. 7.8).

In contrast, bud emergence and nuclear migration can occur in all mutants defective in DNA synthesis and nuclear division. Since the nucleus normally migrates into the neck between the bud and parent cell, it seems reasonable to suppose that nuclear migration is dependent on, and subsequent to, bud emergence (Fig. 7.8).

The two separate pathways are considered to converge at cytokinesis (Fig. 7.8), based on the finding that nuclear division is completed in mutant *cdc* 24 (defective in bud emergence) but neither cytokinesis nor cell separation occurs. Thus, cytokinesis and cell separation are dependent on bud emergence as well as on nuclear division. The pathway of causally-related landmarks for *S. cerevisiae* shown in Fig. 7.8 has certain features in common with the pathways derived for *Schiz. pombe* to be described in Section B3.

(b) *Sequential gene product expression.* The use of the *cdc* mutants has shown that some of the events of the yeast cell cycle are dependent one upon another. One likely possibility for this dependence of "late" events on "early" events is that the biochemical steps that comprise each event are themselves organized into dependent sequences. Two gene-mediated events may be related to each other in one of four independent ways (Fig. 7.9).

The strategies for determining such sequences have been described in detail by Jarvick and Botstein (1973) and Hereford and Hartwell (1974). One method (Hartwell *et al.*, 1974; Hereford and Hartwell, 1974) requires a comparison of the behaviour of strains carrying each mutation alone with that of a double mutant (defective in both genes).

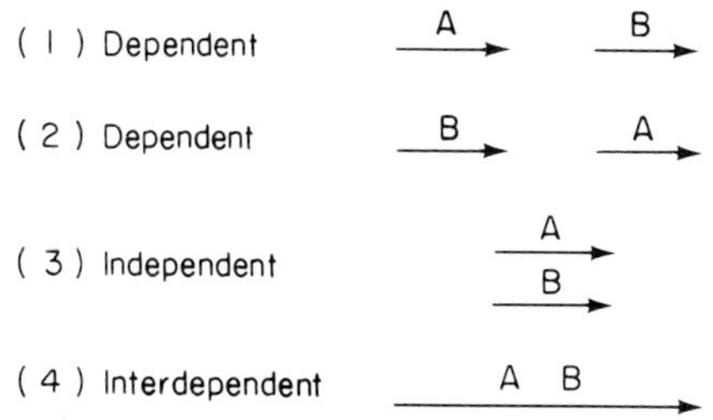

(1) Dependent A → B →
(2) Dependent B → A →
(3) Independent A → / B →
(4) Interdependent A B →

Fig. 7.9 Four possible relationships between 2 gene-mediated steps. Two steps may be related in a dependent sequence (1 or 2), in which case the first step must occur before the second. They may be independent (3), in which case either step can occur in the absence of the other, or they may be interdependent (4), where both steps must occur concomitantly. (Reproduced with permission from Hartwell, 1978.)

The four alternatives can be unambiguously resolved using a reciprocal shift method that involves shifting cells from one condition that produces a stage specific block (e.g. the restrictive temperature for a *cdc* mutant) to another (e.g. the presence of an inhibitor such as hydroxyurea), and determining whether or not cell division occurs. A pathway deduced from a combination of the two methods is shown in Fig. 7.10 (Hartwell, 1976, 1978).

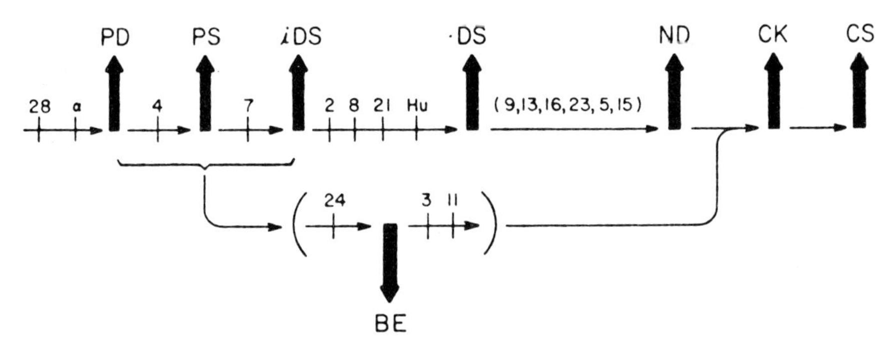

Fig. 7.10 Summary of dependent relationships in the *Saccharomyces cerevisiae* cell cycle. Succeeding steps are related in a dependent sequence of given order. The relationships between arrows or symbols enclosed within parentheses are undefined. Completion of steps *cdc* 24, 3 and 11 are known to be dependent on the α-mating factor-sensitive steps and independent of the hydroxyurea-sensitive step; however, their relationship to steps mediated by *cdc* 4 and 7 is unknown. Designation: numbers = *cdc* genes; α = mating factor; HU = hydroxyurea; PD = nuclear plaque duplication; PS = nuclear plaque separation; iDS = initiation of DNA synthesis; DS = DNA replication; ND = nuclear division; CK = cytokinesis; CS = cell wall separation; BE = bud emergence. The pathway is a summary of data presented by Bücking-Throm *et al.* (1973), Byers and Goetsch (1973), Hereford and Hartwell (1974) and Hartwell (1976). (Reproduced with permission from Hartwell, 1976.)

(c) *Identification of the "start" event.* One mutant that has proved to be of particular interest and significance in the elucidation of the initiation of dependent pathways is that carrying a lesion in gene *cdc* 28. The product of this gene mediates an event called "start" (Hartwell *et al.*, 1970). This event initiates the cycle, which then bifurcates into two independent pathways (Fig. 7.8). This step precedes, and is a prerequisite for, the *cdc* 4 and *cdc* 7 mediated steps leading to DNA synthesis and for the *cdc* 24 mediated step that leads to bud emergence.

The terminal phenotype of the *cdc* 28 mutant is an unbudded cell arrested after cell separation but before the initiation of DNA synthesis. A similar phenotype is evident when cycle progress is arrested in the stationary phase of growth (Williamson and Scopes, 1960; Unger and Hartwell, 1976; J. Pringle and R. Maddox, unpublished, cited by Reid and Hartwell, 1977). Arrested *cdc* 28 mutants contain a single modified spindle plaque (Byers and Goetsch, 1973) reminiscent of that observed in cells treated with α-hormone (Fig. 7.11). These authors suggest that the *cdc* 28 gene product and mating hormone are required for the formation of a double plaque, which in turn integrates the events of the normal cell cycle by triggering both the initiation of DNA synthesis and bud emergence. The *cdc* 28^{+} gene product has not been identified, but the methodology developed by Nasmyth and Reed (1980) for the molecular cloning of the gene is a promising start in characterizing the product of this and other *cdc* genes.

Two genes controlling the "start" event have been described by Sudbery *et al.* (1980). The *whi*-1 mutation results in bud initiation when the parent cell is only half the size at which bud initiation occurs in the wild type. Such mutants are apparently analogous to the *wee* mutants of *Schiz. pombe* (see Section III.B2). A second type of mutation in *whi*-2 is involved in the mechanism whereby cells arrest in G_1 in stationary phase.

Further information on "start" and other G_1 events has come from examination of the sedimentation behaviour of folded chromosomes from *cdc* mutants arrested in G_1 (Piñon, 1979). The results indicate that by the time of execution of the *cdc* 7 step, the folded genome has assumed the form characteristic of G_1 nuclei. The folded chromosomes from *cdc* 28 and *cdc* 25 mutants arrested in G_1 at the restrictive temperature, however, are unstable, suggesting that these gene products are required for the functional integrity of the folded chromosome.

Information on the state of amino acid biosynthesis also appears to be part of the signal for "start", since a dual role of the *tra* 3 gene product in both control of amino acid biosynthesis and cell division has been demonstrated (Wolfner *et al.*, 1975).

Although the progression to cytokinesis and cell separation of the normal cell cycle is characterized by the continual traverse of two independent developmental sequences, it appears that a defect in cytokinesis, cell separation or bud emergence does not prevent a cell from embarking on a new sequence initiated by the "start" event. First, cells can go through an indefinite number of cell cycles despite a failure to separate. Mutants defective in cell separation have been isolated by filtration from mutagenized cultures (L. H. Hartwell, unpublished, cited by Hartwell *et al.*, 1974). Secondly, mutants defective in cytokinesis (see Section III.A3) undergo multiple rounds of bud emergence, DNA synthesis and nuclear division, sometimes

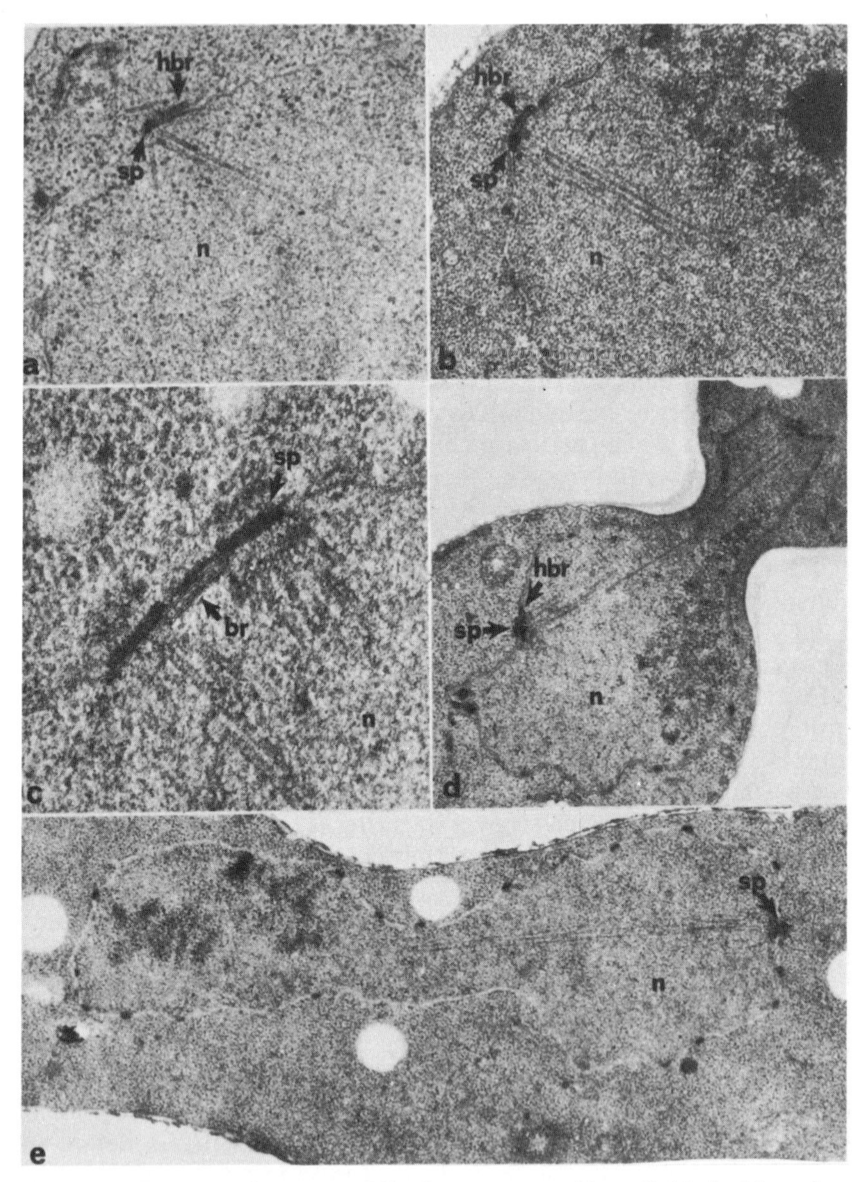

Fig. 7.11 Electron microscopy of *Saccharomyces cerevisiae* cells blocked by α-hormone or by *cdc* temperature sensitivity. (a) A small single plaque with a prominent half-bridge (hbr) in a cell arrested by α-hormone (×35 000); (b) A similar single plaque in a cell defective in *cdc* 28 (×35 000); (c) A double plaque in the multiple budded uninucleate strain defective in *cdc* 4 (×105 000); (d) A complete spindle of a strain defective in *cdc* 7. The nucleus (n) resides in the narrow neck between mother cell and bud, as in other medial nuclear division strains (×35 000). (e) One pole of an elongate, complete spindle in a late nuclear division strain defective in *cdc* 14 (×26 000). (Reproduced with permission from Byers and Goetsch, 1973.)

attaining an octanucleate stage. Finally, mutants defective in bud emergence (e.g. *cdc* 24) frequently undergo additional cycles of nuclear division to become tetranucleate at the restrictive temperature.

(d) *Coordination of growth with division.* Progress of a cell through the cell cycle, culminating in division is normally accompanied by an increase in cell size (Chapter 8). The isolation of cell cycle mutants has permitted a new experimental approach to the mechanisms underlying this coordination to be made. As a test of the degree to which the constituent processes of the "growth cycle" and the "DNA-division cycle" (Mitchison, 1971; see Chapter 1) are linked, *cdc* mutants that arrest at various points in the DNA division cycle were shifted to the restrictive temperature and cell growth was then monitored (Johnston *et al.*, 1977). Mutants blocked in the initiation of DNA synthesis (*cdc* 7), in spindle plaque duplication (*cdc* 28) and separation (*cdc* 4), nuclear DNA synthesis (*cdc* 8), bud emergence (*cdc* 24) and nuclear division (*cdc* 2, 13, 14) all attained a two to fourfold increase in growth (dry weight, volume and protein content) after the arrest of cell division at the restrictive temperature. Other *cdc* mutants continue to synthesize stable RNA at an uninhibited rate for at least 1 or 2 cycles after arrest (Culotti and Hartwell, 1971; Hartwell, 1971a,b; Hartwell, 1973a). *Cdc* mutants were also used to demonstrate that the abnormally small cells arrested in G_1 do not initiate new cycles (as defined by the earliest known events of the DNA-division cycle) until growth to a critical size has been achieved (see Chapter 8). Coordination of growth and division in each cycle is thus assured because the doubling of cell mass (rather than completion of the stage-specific programme of cycle events) is rate-limiting and this programme pauses at the *cdc* 28 step until growth catches up.

7. *Genetic Approaches to the Relationship between Mitotic and Meiotic Cycles*

The meiotic cycle in yeast is initiated by conjugation of G_1 haploid cells of opposite mating type to yield a diploid G_1 cell which then undergoes DNA replication. DNA synthesis is followed by extensive recombination and two meiotic nuclear divisions. Cytokinesis and cell separation are accomplished by the growth of new wall material within the cytoplasm of the mother cell. In the ascomycetes (including *S. cerevisiae*), an ascus is formed in which the four haploid products are held together (Moens and Rapport, 1971). For a recent review of meiosis, the reader is referred to Baker *et al.* (1976). In view of the similarity, at least superficially, of the landmarks in the mitotic and meiotic cycles it is of interest to enquire whether mitotic gene functions, specified by *cdc* genes, are required for meiosis.

Meiosis has been examined at the non-permissive temperature in diploids homozygous for one of several *cdc* mutations. Nearly all of the genes involved in the control of nuclear events during mitosis (spindle pole body duplication and segregation, DNA replication, chromosome segregation and the "start" mutations *cdc* 28 and *tra* 3) were found to be essential for meiosis also (Simchen, 1974; Byers and Goetsch, 1975a; Zamb and Roth, 1977; Simchen and Hirschberg, 1977; Schild and Byers, 1978; Kassir and Simchen, 1978; Shilo, V. *et al.*, 1978). Genes specifying bud emergence (*cdc* 24) and three genes for cytokinesis (*cdc* 3, 10 and 11), were found not to be essential for meiosis (Simchen, 1974). Two other "start" mutations (*cdc* 25 or 35) did not affect initiation of meiosis (Shilo, V. *et al.*, 1978). The meiotic programme, therefore, is achieved by modification of the mitotic cell cycle.

Genes that function uniquely in meiosis have also been identified (Esposito and Esposito, 1969; Roth, 1975; Baker *et al.*, 1976).

In addition to the use of the *cdc* mutants in the genetic dissection of meiosis (which is outside the scope of this book), a study of meiosis may be used to further our understanding of the primary defects in the *cdc* mutants (Schild and Byers, 1978; Simchen and Hirschberg, 1977). Distinction between the primary lesion of a mutant and its diverse secondary effects may be inferred from the behaviour of the mutant in meiosis because this process includes *two* divisions that share certain features but differ in others. An illustrative case is the analysis by Simchen and Hirschberg (1977) of the *cdc* 4 mutation, reported to block the initiation of DNA synthesis (Hartwell, 1971a; Petes and Newlon, 1974) and spindle plaque separation (Byers and Goetsch, 1973). One effect of the expression of the *cdc* 4 mutation on meiosis is the occurrence of three distinct terminal phenotypes at the restrictive temperature, suggesting that the function of the gene is required at three stages in the process. Thus, the primary defect of the *cdc* 4 mutation cannot be the failure to initiate DNA replication, since only one round of replication occurs and this is prior to nuclear division. Simchen and Hirschberg (1977) have suggested that a defect in separation of spindle plaques is a more likely explanation or that the initiation of DNA synthesis and spindle plaque separation are both dependent on a third, unidentified process.

The *cdc* mutants have also proved valuable in determining the cell's commitment to complete a mitotic cycle in a poor (sporulation) medium in which meiosis represents an alternative developmental pathway. It was found (Hirschberg and Simchen, 1977) that commitment to mitosis occurred early in the cell cycle (in G_1) and was dependent upon the function of *cdc* 4^+.

8. *Genetic Approaches to the Regulation of Mating*

The transition between the mitotic cell cycle and either the dormant stationary phase or the events of mating and zygote formation occur early in G_1 (Hartwell, 1973b; Sena *et al.*, 1973). The mitotic cycles of the two participant cells are synchronized by the production of small polypeptide pheromones. Each cell produces a diffusible factor (α or **a**) that arrests its partner, of opposite mating type, at the same step in G_1 as that mediated by the product of the *cdc* 28 gene (Bücking-Throm *et al.*, 1973; Hereford and Hartwell, 1974; Wilkinson and Pringle, 1974). Populations of strains carrying other *cdc* mutations, when synchronized at the *cdc* block do not mate well, illustrating that mating is restricted to a step in the cycle close to the normal functioning of the *cdc* 28 gene (Reid and Hartwell, 1977). Four genes (*cdc* 1, 4, 24 and 33) have been identified that are essential both for the mitotic cell cycle and for mating (Reid and Hartwell, 1977).

B. *SCHIZOSACCHAROMYCES POMBE*

1. *General Features of the Organism and Mutants*

Like *S. cerevisiae*, the fission yeast *Schiz. pombe* has a number of features which make it an attractive experimental organism for the isolation and study of cell cycle mutants. The essential morphological features of the cell cycle have been described in Chapter 1.

Mutant strains of two major classes have been studied: (1) those in which the normal coordination between cellular growth and division is disrupted, resulting in the division of cells of abnormal (small) size and (2) temperature sensitive mutants blocked at specific cell cycle stages, superficially analogous to the *cdc* mutants of *S. cerevisiae*.

In addition, Yamamoto (1980) has described mutants resistant to antimitotic benzimidazole compounds. Mutants *ben* 2 and *ben* 3, defining two of the three linkage groups, exhibit temperature-dependent resistance. By analogy with similar *Aspergillus* mutants (see Section III.H), it is suggested that one of these *ben* genes codes for tubulin.

2. *The* wee *Mutants*

From about 71 heat sensitive mutants isolated, Bonatti *et al.* (1972) selected 11 as having possible defects in the cell cycle. As the result of a more extensive search of about 500 heat sensitive mutants, Nurse *et al.* (1976) selected 28 that formed abnormally sized cells at the restrictive temperature. The mutants isolated by Nurse and his collaborators have been more extensively studied. The first such strain to be described, *cdc* 9–40 (subsequently renamed *wee*-1; Nurse *et al.*, 1976) was obtained by nitrosoguanidine mutagenesis of the wild-type strain 972h⁻ (Nurse, 1975). The mutant is heat-sensitive and, at the restrictive temperature, divides at about half the volume of wild-type cells and with a reduced macromolecular content. In this mutant, and an independently-isolated mutant *wee* 2–1, the timing of DNA synthesis is delayed by 0·23 to 0·29 of a cycle, compared with the wild-type (Fig. 7.12; Nurse, 1975; Nurse and Thuriaux, 1977). The altered

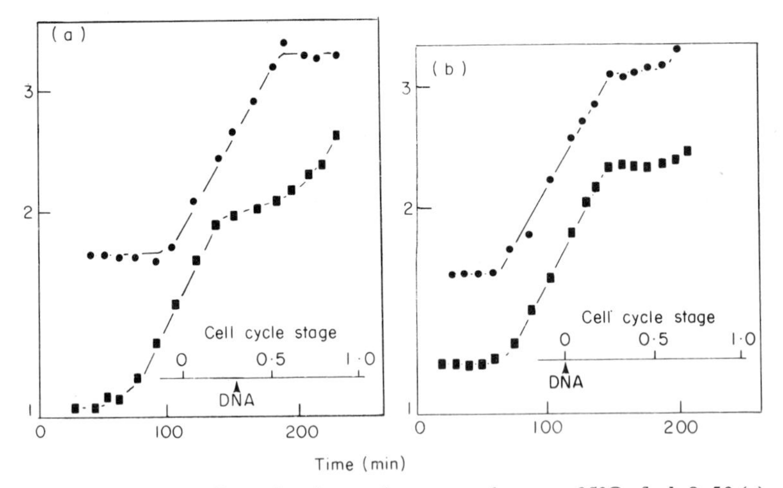

Fig. 7.12 DNA and cell number in synchronous cultures at 35°C of *cdc* 9–50 (a) and wild-type 972h⁻ (b) strains of *Schizosaccharomyces pombe*. Strains were grown at 35°C. Small cells were selected from these cultures by rate-sedimentation centrifugation through a lactose gradient in a zonal rotor. The small cells were used as an inoculum for synchronous cultures and DNA and cell number determined. A cell cycle map is given for both cultures starting at cell separation and showing the time of the midpoint of the rise in DNA. The actual values of the experimental variables per ml, equivalent to one unit on the arbitrary log scale, is given in brackets in the symbol key. (a) ●, DNA ml^{-1} (8·63 ng); ■, cell number ml^{-1} (4×10^6). (b) ●, DNA ml^{-1} (7·47 ng); ■, cell number ml^{-1} (3×10^6). (Reproduced with permission from Nurse, 1975.)

gene product in *wee* 1–50 has no direct effect on the timing of DNA replication. This, taken with the observation that DNA replication occurs in cells of the same size (=protein content), whether in the mutant growing at the restrictive temperature or in small wild-type cells produced by environmental manipulation, led to the proposal that there is a minimum cell mass below which initiation of DNA replication cannot occur. This control determines the timing of S during the cell cycle of the *wee* mutant, but would be cryptic in fast growing, wild-type cells of normal size. It was further suggested that the initiation of S-phase is also controlled by a dependency upon completion of the previous mitosis. This control would be operative if cells divided at a large mass (Nurse, 1975; Nurse *et al.*, 1976; Nurse and Thuriaux, 1977). Four mutant strains (*wee* 1–1, *wee* 1–50, *wee* 2–1 and *cdc* 2) together with the wild-type, which vary in cell mass at division over a three-fold range, were subsequently used to support the hypothesis and to demonstrate a transition from one control to the other at a particular cell mass at division (Nasmyth *et al.*, 1979).

The *wee* mutants also offer a means of studying the consequences of altered gene concentration (Fraser and Nurse, 1978, 1979). Although the mean gene concentration (i.e. DNA:protein ratio) in a *wee* 1–50/*wee* 1–50 diploid is approx. 1·4-fold higher than in a wild-type diploid, the mean concentrations of mRNA and rRNA are the same as the wild type. This compensation is achieved by differences in the patterns of messenger and ribosomal RNA synthesis during the cell cycle. In cells of different sizes, there is relatively little variation in the timing of DNA replication but, the smaller the cell at division, the later are the steps in rates of RNA synthesis. It is proposed that after DNA replication, there is a delay before a doubling in the rate of transcription occurs (Fraser and Nurse, 1978, 1979) and that the mass-related delay can maintain the mean content of poly-adenylated RNA as a constant proportion of cell mass, irrespective of cell size (Barnes *et al.*, 1979).

In synchronous cultures of the *wee* 1–50 diploid strain, a fraction of the cells loses viability in each cell cycle. It is tempting to propose (Fraser and Nurse, 1979) that because of the heterogeneity of cell sizes in a population just prior to division, some cells are too small to activate the doubling in rate of RNA synthesis before the division and become non-viable. It is perhaps noteworthy that mutants smaller than *wee* 1–50 have not yet been found.

A further 25 *wee* mutants were isolated by Thuriaux *et al.* (1978). Mutants carrying the two genes *wee*-1 and *wee*-2 divided at about half the length of wild-type cells. The transition point is just before the

Table 7.2. *Mutants of* Schiz. pombe *affected in the cell division cycle*

Gene	Number of alleles described	Transition point from temperature shifts of exponential cultures (Howell, 1974)	Defect	Notes	Reference
cdc 1	2	0·69 to 0·74	ND		Nurse *et al.* (1976)
cdc 2	3	0·69 to 0·78	ND		Nurse *et al.* (1976)
cdc 3	2	0·73	late cell plate formation		Nurse *et al.* (1976)
cdc 4	2	0·78 to 0·80	late cell plate formation	intragenic complementation	Nurse *et al.* (1976)
cdc 5	1	0·79	ND	shows some residual division at 35°C	Nurse *et al.* (1976)
cdc 6	2	0·38 to 0·44	ND		Nurse *et al.* (1976)
cdc 7	1	0·50	early cell plate formation	significance of early TP clouded by single allele	Nurse *et al.* (1976)
cdc 8	3	0·73 to 0·87	late cell plate formation		Nurse *et al.* (1976)
cdc 9				Renamed *wee*-1 (see below)	Nurse *et al.* (1976)
cdc 10	2	−0·10 to −0·15	DS	TP is at 0·90 in previous cycles. Mutants in further genes isolated by K.	

				Nasmyth (unpublished)	
cdc 11	3	0·8 to 0·81	early cell plate formation		Nurse *et al*. (1976)
cdc 12	1	0·82	late cell plate formation		Nurse *et al*. (1976)
cdc 13	1	0·64	ND	forms multiple cell plates	Nurse *et al*. (1976)
cdc 14	1	0·48	early cell plate formation	shows some residual division at 35°C	Nurse *et al*. (1976)
cdc 15	2	0·72 to 0·81	early cell plate formation		Nurse *et al*. (1976)
—	1	0·88	ND	sterile; unique phenotype, very elongated nucleus	Nurse *et al*. (1976)
cdc 16	1	0·82	control over septation		Minet *et al*. (1979)
cdc 17	3	execution point 0·05	DNA ligase		Nasmyth (1977, 1979b)
wee-1	probably 24	0·70 to 0·74	size control over mitosis	only 3 mutants are t.s.	Thuriaux *et al*. (1978)
wee-2	1	0·68	size control over mitosis	slightly t.s.	Thuriaux *et al*. (1978)

A mutant of *Schiz. pombe* resistant to Colcemid (N-di-acetyl-N-methylcolchicine) tended to form clumps of cells that often contained elongated cells with multiple cell plates (Stetten and Lederberg, 1973). This mutant has not been further studied.

Abbreviations: DS = DNA synthesis; ND = nuclear division; TP = transition point; t.s. = temperature-sensitive.

point where mitosis normally occurs and is similar to the transition point in most *cdc* mutants (below; Nurse *et al.*, 1976). Temperature shift experiments show that mutant cells at 0·45 to 0·70 of the cycle (i.e. in the latter part of G_2) can be accelerated into mitosis; presumably a mitosis size control operates in the wild-type which prevents premature mitosis and this is altered in both *wee*-1 and *wee*-2.

3. *The* cdc *Mutants*

Nurse *et al.* (1976) described the isolation and properties of 27 temperature-sensitive mutants which were unable to complete the cell cycle at the restrictive temperature. These mutants define 14 unlinked recessive genes that are involved in the occurrence of various cell cycle landmarks (Table 7.2). Mutants defective in *early* cell plate formation become elongated with 4 to 8 nuclei per cell after 7 h incubation at the restrictive temperature (35°C); on prolonged incubation, up to 16 nuclei may accumulate. No cell plates or cell plate material are formed. Cells defective in *late* cell plate formation, however, accumulate as enlarged "dumbell" shapes with 2 to 4 nuclei per cell after 7 h at 35°C and disorganized cell plate material is observed in all cells. In *cdc* 3 and *cdc* 12 mutants, axial cytoplasmic microtubules and microfilaments are seen at the restrictive temperature (Streiblová and Girbardt, 1980). Those mutants defective in DNA synthesis and nuclear division become elongated with a single nucleus at 35°C. With the exception of *cdc* 13–117, no cell plates or cell plate material are formed. One of these mutants, *cdc* 2–23, has been used to demonstrate homeostatic control over cell length and cell cycle duration (Fantes, 1977; Chapter 8).

Unlike the situation in *S. cerevisiae*, the transition points (or execution points) in most *cdc* mutants of *Schiz. pombe* occur only shortly before the landmark event in which they are involved. This implies that the thermolabile gene product is synthesized or functions at the time of occurrence of the diagnostic landmark and that the gene product plays a role intrinsic to that event (Hartwell, 1978).

Based on the behaviour of a range of *cdc* mutants, Nurse *et al.* (1976) proposed a scheme to illustrate the interdependence of cell cycle events in *Schiz. pombe* (Fig. 7.13). Here, as in *S. cerevisiae*, cell division is dependent on nuclear division which in turn is dependent on DNA synthesis. Furthermore, the initiation of a second nuclear cycle, consisting of DNA replication and nuclear division, does not require cytokinesis of the preceding cycle but does require completion of some stages of nuclear division in the preceding cycle.

Further evidence for the dependency of separation (cell plate formation) on nuclear division is afforded by a study of a double mutant, *cdc* 16–116/*cdc* 2–33 (Minet *et al.*, 1979). Strains carrying the monogenic, recessive mutation *cdc* 16–116 undergo uncontrolled septation during the cell cycle at the restrictive temperature, whilst *cdc* 2–33 mutants are defective in nuclear division. When a population of the

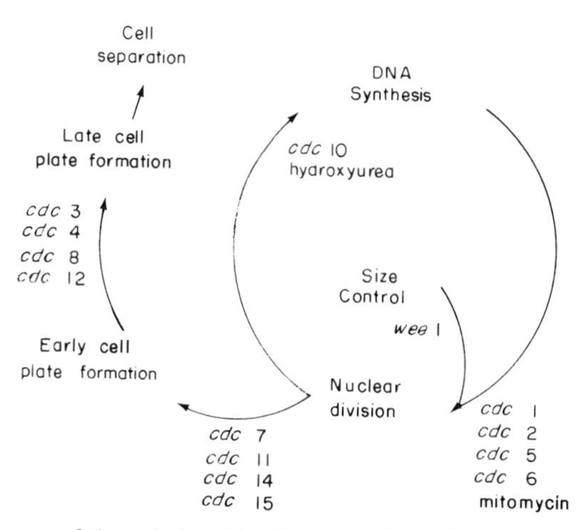

Fig. 7.13 Summary of the relationships between the various events of the cell cycle and the genes that control them in *Schizosaccharomyces pombe*. The *cdc* genes and the cell cycle inhibitors are placed just before the event in which they are involved. The connecting arrows are not analogous to a biochemical pathway but only formally represent the interdependent relationships of the various cell cycle events. (Reproduced with permission from Nurse *et al.*, 1976.)

double mutant is shifted from 25 to 35°C, cells which are before the transition point for *cdc* 2–33 at 0·78 of a cycle (Table 7.1; Nurse *et al.*, 1976) complete neither nuclear division nor cell separation and accumulate as elongated and mononucleate cells typical of *cdc* 2–33 alone (Fig. 7.14). The single mutant, *cdc* 16–116, accumulates two types of cell after 3 h growth at the restrictive temperature. Cells shifted before the transition point at 0·82 of a cycle complete mitosis and then cease elongation and undergo several rounds of septation in the absence of cytokinesis to form enucleate compartments. Cells shifted later in the cycle separate, but *one* daughter then produces an extra asymmetrically located septum. It is interesting that in *S. cerevisae* also, successive rounds of landmark development (bud emergence) occur in a temperature-sensitive mutant; in *Schiz. pombe*, the frequency of the repeated events is greater than the usual cycle time, whilst in *S. cerevisiae* it is equal (Hartwell, 1974; Hartwell *et al.*, 1974). Models to explain these observations, which involve two antagonistic controls, one initiating septation as a consequence of mitosis and one preventing initiation of further septa, are described by Minet *et al.* (1979).

Three independent conditional-lethal mutants of *Schiz. pombe* which appear to be altered in the structural gene for DNA ligase have been described by Nasmyth (1977). Cells of the most extensively studied

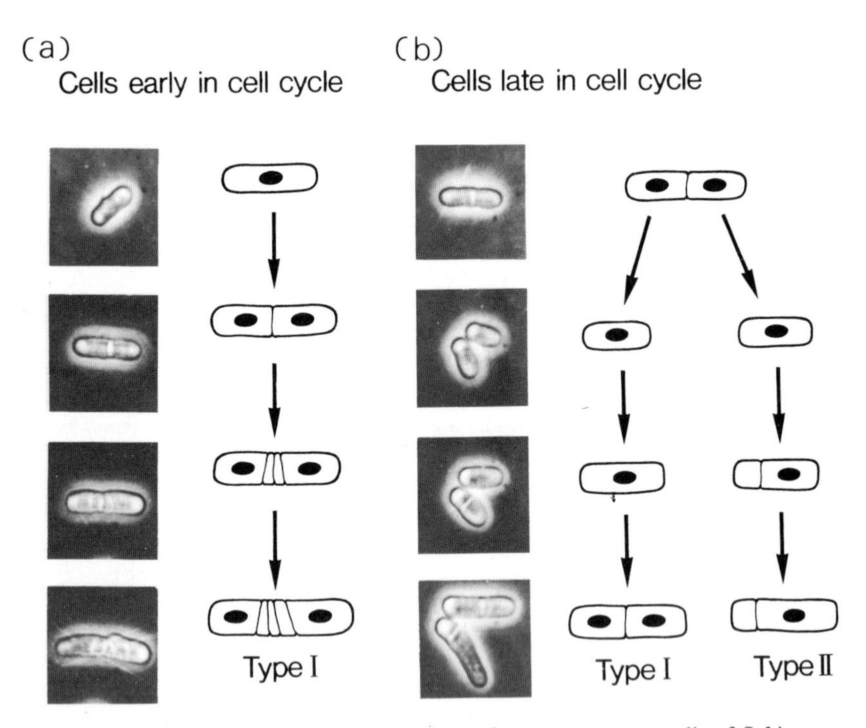

Fig. 7.14 Cellular morphologies displayed by *cdc* 16–116 mutant cells of *Schizosaccharomyces pombe* within 4 h at 35°C. (a) Cells shifted when in an early phase of their cycle (i.e. until completion of nuclear division), failing to divide, produce a type I cell. (b) Cells shifted at a later stage of the cycle, undergoing cell separation and produce one type I and one type II daughter cell. The type I cell reproduced here has only one septum, but multiseptate cells were observed upon further incubation (data not shown). Septa show up on the phase-contrast pictures as a strongly contrasting structure and were redrawn for more clarity in the accompanying sketches. Nuclei cannot be detected by phase-contrast microscopy under the conditions used, but were incorporated in the drawings on the basis of a Giemsa-staining analysis. (Reproduced with permission from Minet *et al.*, 1979.)

mutant, *cdc* 17–K42, enter S-phase and undergo a complete round of DNA synthesis but mitosis does not follow. The fact that the nascent DNA that accumulates at the restrictive temperature is composed of short Okazaki fragments, together with evidence from direct assays of DNA ligase activity, imply that the *cdc* 17 locus is the structural gene for DNA ligase.

C. *TETRAHYMENA PYRIFORMIS*

The genetic system of *T. pyriformis* is characterized by the existence of two nuclei (a macronucleus and a micronucleus) and the phenomenon known as macronuclear or phenotypic assortment that limits expression to only one allele at a given locus. Either allele at a genetic locus may be expressed in the macronucleus, whilst the micronucleus is unaffected by reassortment and transmits alleles regardless of which allele is actually expressed. In effect a cell becomes functionally homozygous with respect to expression but heterozygous with respect to transmission (for references see Frankel *et al.*, 1975). The rather complex screening procedure used by Frankel *et al.* (1975, 1976a) exploited this system to detect not only temperature-sensitive dominant mutations but also recessive mutations that had come to expression as a result of macronuclear assortment.

Amongst a large collection of heat- or cold-sensitive mutants of *T. thermophila* (formerly *T. pyriformis* syngen 1), generated by combining techniques to manipulate large numbers of clones with a method for the selection of self-fertilized cells, Bruns and Sanford (1978) have reported several isolates manifesting cell cycle blocks. Martindale and Pearlman (1979) have also reported an enrichment technique that gives considerable increases in the yields of temperature-sensitive mutants defective in growth and division. Mutants isolated by these techniques have not been gainfully employed in cell cycle studies so far.

Fourteen single gene recessive mutations that fall into six complementation groups have been isolated as phenotypic "assortants." In such clones, "the continuation of growth and of periodic cycles of oral development in cells that cannot even begin to divide brings about the eventual formation of large misshapen monsters" (Frankel *et al.*, 1976b; Fig. 7.15). The clones were thus given the designations *mo* 1, *mo* 2, *mo* 3, *mo* 6, *mo* 8 and *mo* 12. Between 1 and 5 (in *mo* 3) alleles were found at each locus, with some differences in expression (penetrance) being found among different alleles at a given locus. Mutant alleles at these "cell division arrest" loci (*cda*) have been renamed (Frankel *et al.*, 1980a); examples are *cdaA*1 (formerly *mo* 1[a]), *cdaA*4

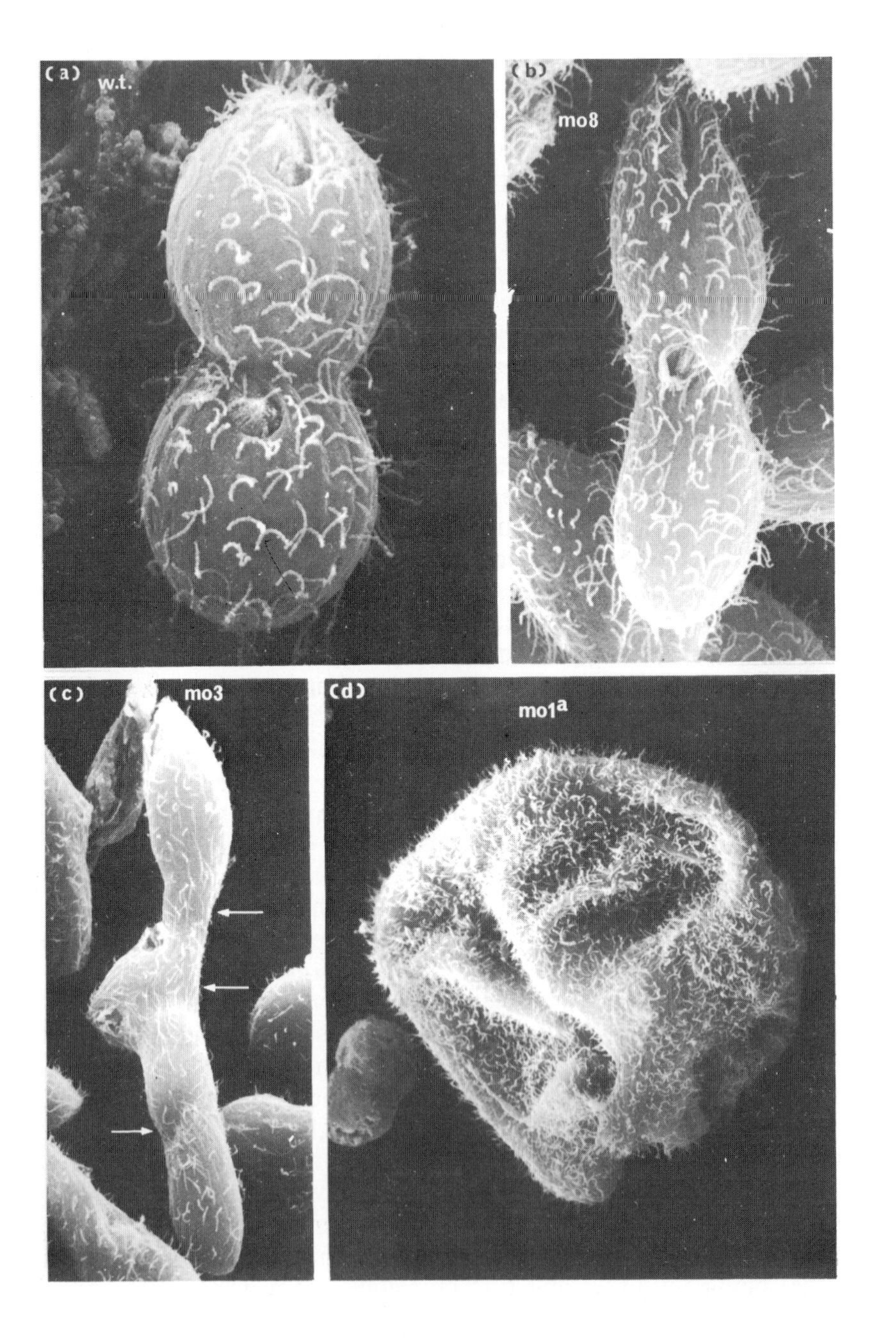
(a)
w.t.
(b)
mo8
(c)
mo3
(d)
mo1a

(formerly *mo* 1[d]), *cdaC2* (formerly *mo* 3[b]) and *cdaH1*. Two further recessive mutations in *mo* 1 have been isolated that differ only in size, retaining a regular shape. Clones made up of "tiny" cells were analyzed (Jenkins, unpublished, cited in Frankel *et al.*, 1976a), but were not studied further. It is possible that such mutants could prove useful in studies of the existence and function of size control mechanisms (cf. the *wee* mutants of *Schiz. pombe*). In a "fat" clone, arrest of cell division at 39·5°C was shown only in some cells, perhaps as a result of generalized, severe reduction in the rate of development at all stages.

Mutations at the five *mo* loci studied further have distinct phenotypic effects on cell division at the restrictive temperature (Frankel *et al.*, 1976a; Frankel *et al.*, 1977). Mutants with *mo* 1 alleles fail in the normal formation of the equatorial fission zone, a zone of discontinuities in the longitudinal ciliary meridians. Mutant alleles at *mo* 8 and *mo* 12 cause incomplete and spatially abnormal fission zones, whilst alleles at *mo* 3 and *mo* 6 allow the formation of complete initially-normal fission zones but interfere with cytokinesis itself. The four alleles at *mo* 1 exhibit either "weak" (*mo* 1[b] and *mo* 1[c]) or "strong" (*mo* 1[a] and *mo* 1[d]) expression.* In the latter type, division arrest is accompanied by formation of cytoplasmic projections where cytoplasmic constriction would normally have taken place, contributing to the bizarre appearance of the monster forms (Fig. 7.15).

The defects in *mo* 1, *mo* 3 and *mo* 12 are expressed in the first division after a shift to the restrictive temperature (Fig. 7.16). In *mo* 6, however, fission is blocked after one or more divisions at the restrictive temperature. Preliminary reports (J. Frankel, unpublished, cited by Hartwell, 1978; Frankel and Mahler, unpublished, cited by

* Superscript letters denote alleles.

Fig. 7.15 Scanning electron micrographs of wild-type and cell-division mutants of *Tetrahymena pyriformis*, syngen 1. (a) A wild-type (strain B) cell dividing at 28°C in rich medium. The division furrow is situated just anterior to the new oral apparatus (arrow) (×2 550); (b) A mo8[a] cell, 4 h at 39·5°C in rich medium following 2 h at 39·5°C in non-nutrient medium. The new oral apparatus (arrow) is wedged within the division furrow; (c) A mo3[a] cell, 2·5 h at 39·5°C in rich medium. This is a tandem chain of four cell units, with three arrested furrows visible (arrows). Due to some twisting around the long axis, not all of the oral structures are visible. As the time of fixation was only one generation after onset of the restrictive temperature, this cell had presumably been arrested in division at the time of the temperature shift and was attempting a second division at the time of fixation (×840); (d) A mo1[a] cell, 24 h at 39·5°C. Oral structures are probably hidden in crevices (×740). (Reproduced with permission from Frankel *et al.*, 1977.)

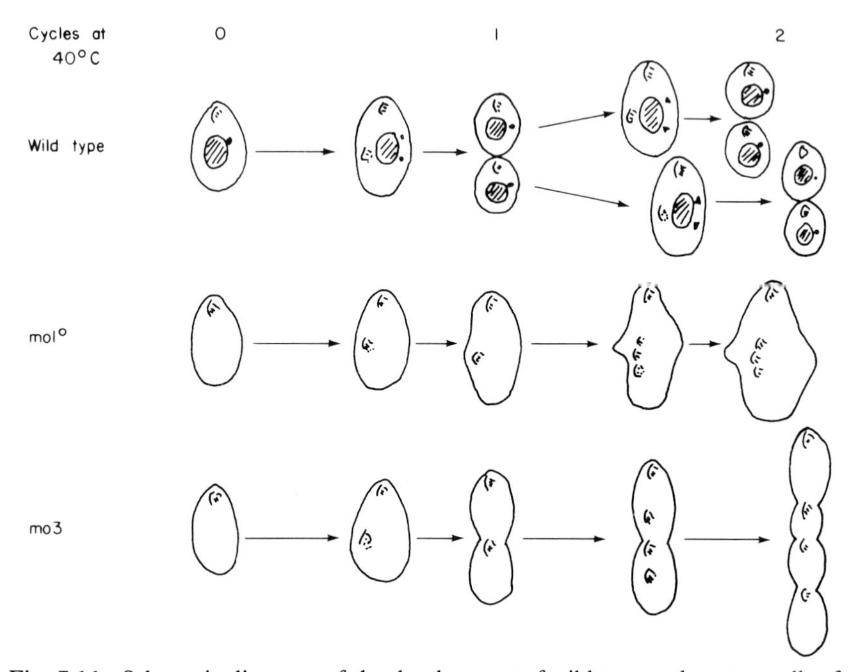

Fig. 7.16 Schematic diagrams of the development of wild-type and mutant cells of *Tetrahymena pyriformis* through two cell cycles at 40°C. Cell outlines, oral structures, and developing oral primordia are shown in all three series of diagrams. In addition, the course of division of micronuclei and macronuclei is shown in wild-type cells; nuclei are omitted from diagrams of the mutants. (Reproduced with permission from Frankel *et al.*, 1976b.)

Frankel *et al.*, 1977) indicate that, in some of the mutants, the execution point and the diagnostic landmark are coincident.

As in yeast (Hereford and Hartwell, 1974), the use of double mutants affords a powerful means of deducing the sequence and dependence of gene-mediated steps in the cycle. In each of various double homozygotes, the characteristic phenotypes of *both* mutations were expressed simultaneously, suggesting that the five loci studied mediate independent processes required for cytokinesis (Fig. 7.17). For example, in the *mo* 1^a–*mo* 3^a and *mo* 1^a–*mo* 8^a double homozygotes, cells became arrested before the onset of constriction (characteristic of *mo* 1^a) but also underwent the specific phenotypic changes associated with the *mo* 3 and *mo* 8 mutations respectively. This indicates that *mo* 1 is preventing division through a pathway distinct from those mediated by *mo* 3 and *mo* 8. Similarly, a *mo* 3–*mo* 8 double homozygote was

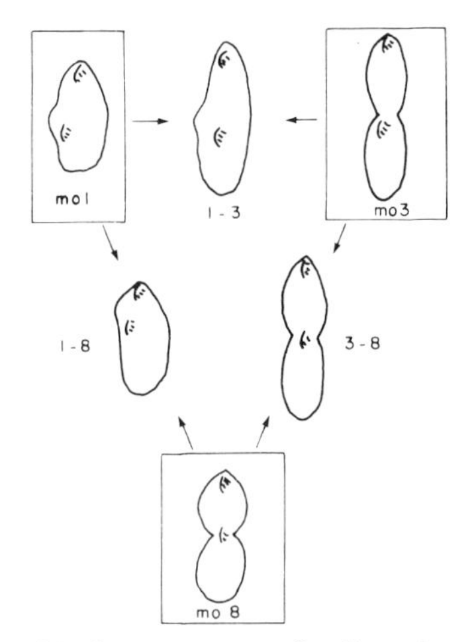

Fig. 7.17 Single and double homozygotes of cell cycle mutants of *Tetrahymena pyriformis*. (Reproduced with permission from Frankel *et al.*, 1977.)

used to demonstrate that both these genes contribute independently to cell division and that the functions of gene products do not lie on a single dependent pathway.

Further similarities between the cell cycle of *T. pyriformis* and *S. cerevisiae* are afforded by a study of causal relationships between cycle events (Hartwell, 1978; Frankel *et al.*, 1976b). The *mo* mutants have revealed that DNA replication in the macronucleus and division of the micronucleus are not dependent on cell division in the previous cycle (Cleffmann and Frankel, 1978). However, macronuclear division is dependent on prior formation of the fission zone (Frankel *et al.*, 1976b). In a quite different class of mutant, the "pseudomacrostome" (*psm*) mutant, switch to an alternative mode of development involves repeated replacement of the oral apparatus at the restrictive temperature; no nuclei divide but macronuclear DNA accumulation continues.

Figure 7.18 shows a model for the relationships between clusters of sequential events in the cell cycle of *T. pyriformis*. The model is considered by Frankel *et al.* (1976b) to be the simplest consistent with the experimental data. A single linear pathway coordinates the clusters comprising (i) micronuclear division and oral development and (ii) macronuclear division and cytokinesis, respectively, consistent with the mutant-derived data showing that events in the former cluster can continue repeatedly while those in the latter are blocked. When events in both clusters do occur, however, they follow each other in a relatively

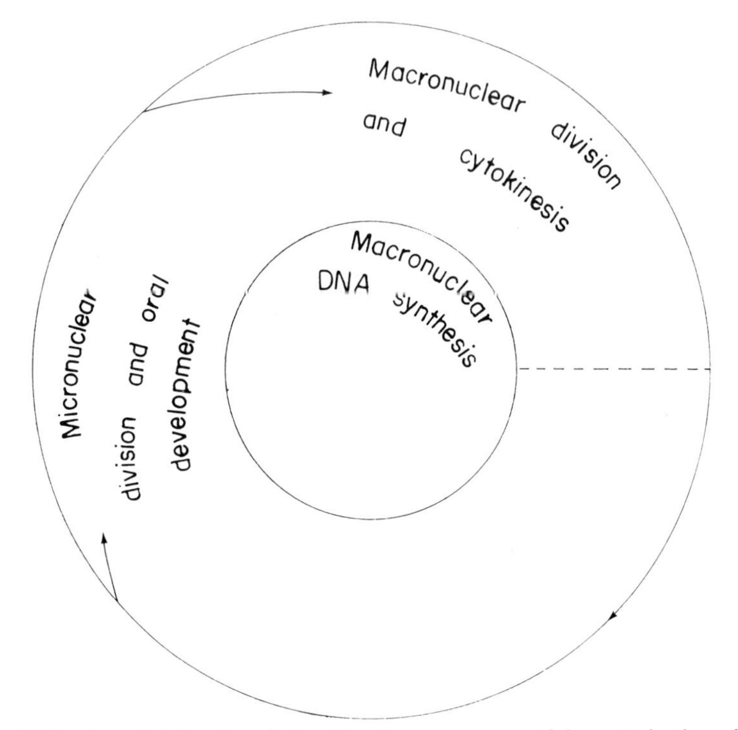

Fig. 7.18 A model for the relationships between sequential events in the cell cycle of *Tetrahymena pyriformis* based on experiments with temperature-sensitive mutants. The model is the simplest consistent with the available data and consists of two parallel independent timer sequences with a "checkpoint" (dashed line) affecting phase relations. (Redrawn from Frankel *et al.*, 1976b.)

close sequence. A separate linear pathway is shown to control macronuclear DNA synthesis and is related to the parallel sequence by a "check point".

The temperature-sensitive periods in three *cda* mutants have now been identified (Frankel *et al.*, 1980a) and found to coincide with or immediately precede the onset of phenotypic abnormality. In yeasts also, the ends of the temperature sensitive periods and the diagnostic landmark are frequently juxtaposed. The phenomenon of excess delay (Chapter 1) has been shown by Frankel *et al.* (1980b) to be the result of a setback in the sequence of gene expression leading to cell division. This has important implications for the accurate measurement of execution points at temperatures at which excess delays are elicited.

D. *PARAMECIUM TETRAURELIA*

A start has been made on the isolation and characterization of temperature-sensitive cell cycle mutants of *P. tetraurelia*. Peterson and Berger (1976) have isolated 56 "*ts*-0" mutants that arrest cell division within one cycle at the restrictive temperature. A quarter of the mutants show substantial reduction of DNA synthesis under these conditions; two allelic mutants continue to synthesize protein after arrest of DNA synthesis and may be directly involved in the replication processes.

E. *CHLAMYDOMONAS*

Mutants of *C. moewusii* and *C. reinhardii* with abnormal cell division were isolated by Lewin (1952) and Warr (1968) respectively. Both groups of mutants were characterized by the formation of multinucleate cells. Cultures of Warr's (1968) mutant, *cyt*, exhibited a partial blockage in cytokinesis, so that cultures consisted of a mixture of uninucleate, binucleate and multinucleate cells. A further increase in the mean number of nuclei per cell was elicited by treatment with benzimidazole or cobalt (Warr and Durber, 1971) and other substances that raised the intracellular levels of free cysteine or cystine. Thus, *cyt* 1 cells are in some way hypersensitive to such changes (Warr and Gibbons, 1973; Warr and Quinn, 1977).

Mutants which are temperature-sensitive for cell cycle progress were first isolated from *C. reinhardii* by Howell and Naliboff (1973). These workers screened microscopically a large number of heat sensitive mutants induced in a haploid culture by u.v., and selected clones in which single cells completed no more than one round of division, i.e. produced 8 or less daughter cells at 33°C. For such *cb* or "cycle blocked" mutants, the "block point" (equivalent to the transition point or execution point) was calculated from measurement of the residual division observed when an exponentially-growing culture of the mutant was shifted to the restrictive temperature (Fig. 7.19). Confirmation of the block points determined in this way was obtained by shifting to the restrictive temperature cells from light–dark induced synchronous cultures. Ten *cb* mutants chosen for further study were shown to be blocked at distinct stages in the cell cycle (Fig. 7.20) (Howell, 1974; Howell and Naliboff, 1973).

Like the *cdc* mutants of *S. cerevisiae*, the *cb* mutants do not in general show full reversal of inhibition of cell division after a shift to

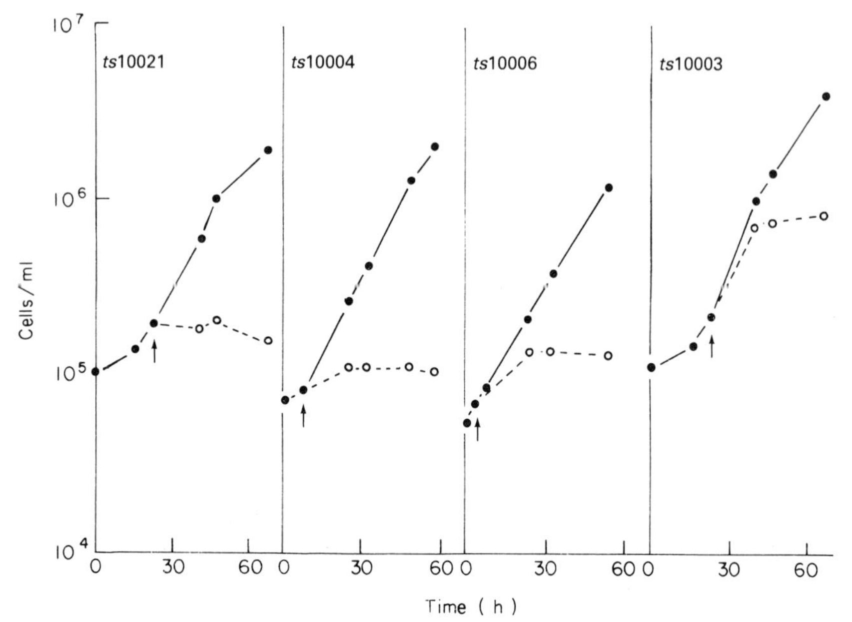

Fig. 7.19 Temperature shift-up of cell cycle mutants of *Chlamydomonas reinhardii* during asynchronous growth. Cultures were grown autotrophically at 21°C. At the times indicated by the arrow (↑) each culture was split into equal portions. Half of the culture was retained at 21°C (—●—) and the remainder shifted to 33°C (– –○– –). The block points calculated from the extent of residual division after the temperature shift are: for strain *ts* 10021, 0·96; for strain *ts* 10004, 0·79; for strain *ts* 10006, 0·42; for strain *ts* 10003, 0·01. (Reproduced with permission from Howell and Naliboff, 1973.)

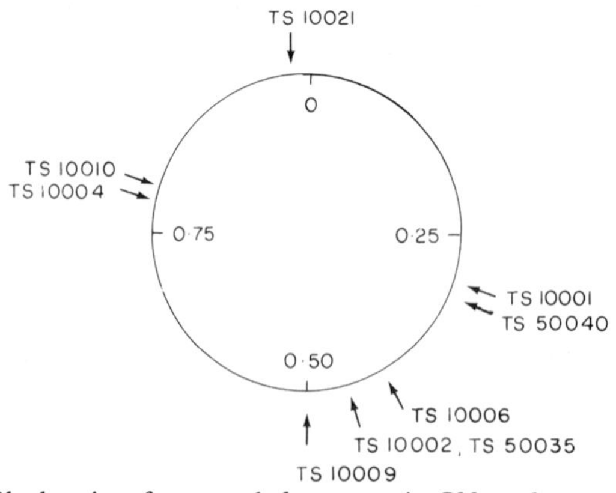

Fig. 7.20 Block points for several *cb* mutants in *Chlamydomonas reinhardii*. Note the pronounced clustering in the 2nd and 4th quarters of the cycle. (Reproduced with permission from Howell, 1974.)

the permissive temperature. Furthermore, again like *cdc* mutants, most *cb* mutants continue to increase in size after inhibition of cell division at the restrictive temperature, sometimes attained a tenfold increase in mean cell volume. Several mutants completed more than one cycle at the restrictive temperature after first encountering the block point.

A start has been made in establishing interrelationships between cell cycle events by using asynchronous cultures of the *cb* mutants (Howell, 1974). For one mutant, *ts* 10009, both DNA synthesis and cell division are individually dependent on completion of function of the mutated gene, whilst the stepwise accumulation of two autoregulated enzymes, aspartate transcarbamylase and glutamate dehydrogenase, are independent events.

Sato (1976) has also isolated temperature-sensitive mutants of *C. reinhardii* that show various levels of growth inhibition at the restrictive (higher) temperature. In one mutant, *ts*-60, which was also sensitive to low temperature (15°C), thermolability and colchicine resistance were shown to be conferred by mutation in a single gene.

F. *USTILAGO VIOLACEA* AND *U. MAYDIS*

Among a collection of over 400 temperature-sensitive mutants of the smut fungus, *Ustilago maydis*, five mutants partly blocked in DNA synthesis at the restrictive temperature (32°C) were described (Unrau and Holliday, 1970). One of these, *tsd*-1, formed long uninucleate filaments at the non-permissive temperature that died after 4 h. It is likely that such mutants are blocked in G_2, but the question of whether the block is directly on DNA synthesis or whether the primary block is in nuclear division can only be answered by further characterization of the mutant.

Another of these mutants, *pol* 1–1, previously designated *ts* 92A, carries a single recessive mutation in a structural gene for DNA polymerase (Jeggo *et al.*, 1973). Instead of normal growth by yeast-like budding, this mutant forms filamentous, normally uninucleate cells at the restrictive temperature. It has been suggested that the mutant has an altered form of the polymerase required for chromosome replication.

A quite different class of *U. violaceus* mutants, with altered expression of the genes involved in morphogenesis, has been studied by Day and Cummins (1975). These authors presented evidence that the period of cell cycle inducibility of a locus governing a morphogenetic pathway (mating) is regulated by a separate control gene, the *cc* locus, with two known alleles cc^{str} (a stringent or restricted period of inducibility) and cc^{rel} (a relaxed or non-restricted period of inducibility).

Cell division cycle (*cdc*) mutants have also been reported in this organism, identified by their uniform morphology at the restrictive temperature (Cummins and Day, 1975).

G. *PHYSARUM POLYCEPHALUM*

Several groups of workers have reported the isolation of temperature-sensitive growth mutants in the slime mould *Physarum polycephalum* (see Table 7.3). The naturally synchronous mitoses (Howard, 1932) and DNA synthesis (Braun *et al.*, 1965) of the plasmodial phase offer an attractive experimental system for investigating the control of the nuclear division cycle. The alternative growth phase of the life cycle, the haploid uninucleate amoebae, however, are much more amenable to traditional microbiological and genetic manipulations. Haugli and Dove (1972) first isolated temperature-sensitive mutants in amoebal populations and constructed plasmodia homozygous for the mutation, but these failed to express the mutant characteristics. Using appropriate CL strains, however, it is possible to study the expression of the mutations in haploid plasmodia derived from mutant amoebae (Wheals, 1970), although most investigators have found that only 10 to 20% of growth temperature-sensitive mutants isolated in amoebae or plasmodia show temperature sensitivity in the other phase (Gingold *et al.*, 1976; Wheals *et al.*, 1976; Sudbery *et al.*, 1978; Laffler *et al.*, 1979). Del Castillo *et al.* (1978), however, reported that 73% of mutants isolated as temperature-sensitive in amoeba were also mutant in the plasmodial phase. Burland and Dee (1979, 1980) have subsequently shown that the proportion of phase-specific mutants is influenced by the choice of culture media for amoebae and plasmodia and that many genes are common to the growth of both amoeboid and plasmodial phases (c.f. Wheals *et al.*, 1976).

Further work is now urgently required to examine large numbers of temperature-sensitive growth mutants for those with the required characteristics.

H. *ASPERGILLUS NIDULANS*

Several features of the mode of growth of the filamentous fungus *Aspergillus nidulans* have attracted workers to attempt isolation of

Table 7.3. *Temperature-sensitive mutants of* Physarum polycephalum *defective in the mitotic cycle*

Mutant	Defect(s)	Affects on plasmodial (p) or amoebal (a) phase	Reference
CT31	Division rate but not growth drastically reduced	a	Wheals *et al.* (1976)
CT54	Extended prometaphase without effect on division rate	a	Wheals *et al.* (1976)
CT97	Aberrant nucleolar reconstruction	a	Wheals *et al.* (1976)
E7	DNA synthesis? Nucleolar reconstruction	p	Gingold *et al.* (1976)
18 mutants, 6 of which are possible cell cycle mutants	Arrest in one or more of: increase in optical density, RNA, DNA, or protein at the restrictive temperature. Some show abnormal size and morphology of nuclei	both	Del Castillo *et al.* (1978) Wright *et al.* (1976, 1980)
MA67	Affected in step in nuclear replication cycle in late G_2	p	Laffler *et al.* (1979)
ATS20, ATS22	Block in nuclear division	both	Burland and Dee (1980)

temperature-sensitive mutants defective in mitosis, nuclear division or other events of the "duplication cycle". The analogies between this cycle and the cell cycles of mononucleate organisms that multiply by budding or fission have been surveyed by Trinci (1978, 1979). First, mitosis in *A. nidulans* (as in higher eukaryotes) is accompanied by the formation of a microtubular spindle and by chromosome condensation.

Secondly, the nuclear membrane remains intact during mitosis, thus enabling, in principle at least, the process to be investigated in intact nuclei. Thirdly, mutants defective in nuclear division may be detected by looking directly for hyphae with less than the normal number of nuclei.

Table 7.4. *Mitotic mutants of* Aspergillus nidulans

Mutant	Phenotype at restrictive temperature
nim A–W	Blocked in interphase
bim C	Blocked in early nuclear division; microtubules fail to interact and nuclear spindle does not form
bim A	Blocked in medial nuclear division; short nuclear spindle does not elongate
bim B (III), *bim* F (II)	Blocked in late nuclear division, stage I; nuclear spindle elongates partially
bim D (IV)	Blocked in late nuclear division, stage II. Forms very long spindle composed of central spindle microtubules and lacking kinetochore microtubules
bim E (VI)	Slow nuclear division; no clearly defined termination phenotype (Morris, 1976b)
tub A (VIII)	Altered α-tubulin
ben A (VIII)	Altered β-tubulin
sep A–D	Unable to form septa
nud A–E	Blocked in nuclear movement

Roman numerals in parentheses designate linkage groups.
Adapted from Morris (1980).

The largest and most extensively-studied collection of cycle mutants is that of Morris and his colleagues who identified these mitotic mutants by screening microscopically about 1 000 temperature-sensitive strains blocked at the restrictive temperature (42°C) (Morris, 1976a). They found 45 mitotic mutants and assigned them to 38 complementation groups (Table 7.4). Nine mutants were blocked in various stages of mitosis (Fig. 7.21) and designated *bim* (*b*locked *in m*itosis). A tenth, similar mutant has been isolated by Orr and Rosenberger (1976a), and a cold sensitive *bim* mutant has also been isolated (C. F. Roberts, cited by Morris, 1980). Several of the β-tubulin mutants also exhibit a *bim*-like phenotype at the restrictive temperature (Morris *et al.*, 1979).

Table 7.5. *Fraction of nuclei able to reach mitosis at the nonpermissive temperature in replication cycle mutants of* Aspergillus nidulans

Mutant	Hyphae with mitotic nuclei (%)[a]	Age of nuclei at execution point [b]
12	83	0·17
316	63	0·37
161	50	0·5
35	49	0·51
296	16	0·84
50	2	0·98
59	1	0·99
182	1	0·99

[a] Corrected for fraction of hyphae in parent strain which do not reach mitosis under the same conditions

[b] The point where benomyl blocks mitosis is taken as age 0 and 1·0

To determine the execution points of the mutations, conidia were germinated at 30°C and transferred to a medium containing benomyl (to trap nuclei in mitosis) at 42°C. The table shows the proportion of hyphae with mitotic nuclei scored after 3 h at 42°C. Note the clustering of points at the beginning, middle and end of the cycle. (Reproduced with permission from Orr and Rosenberger, 1976b.)

Mutants in the largest group (*nim*; *n*ever *i*n *m*itosis) were blocked before mitosis, while smaller numbers failed to septate (*sep*) (Fig. 7.21) or exhibited an abnormal distribution of nuclei along the mycelium (*nud*). All mutations were recessive, located on all eight chromosomes, and were not clustered on any one chromosome according to their phenotypic expression. In none of the temperature-sensitive mitotic mutants has the gene product yet been identified.

Orr and Rosenberger (1976a) found temperature-sensitive mutants of *A. nidulans* blocked in nuclear division, representing 11 complementation groups. In addition to one mutant blocked in mitosis (see above), five were presumed to have mutations in steps related to chromosome replication and others were defective in nuclear division. Execution points for nine mutants were measured by transferring asynchronous cultures to the restrictive temperature and determining the fraction of nuclei able to reach mitosis (Orr and Rosenberger, 1976b). The execution points fell in clusters, either early in the cycle, near its midpoint, or close to mitosis (Table 7.5). Mutants with execution points in the middle or at the end of the cycle were those that continued to synthesize DNA at the restrictive temperature.

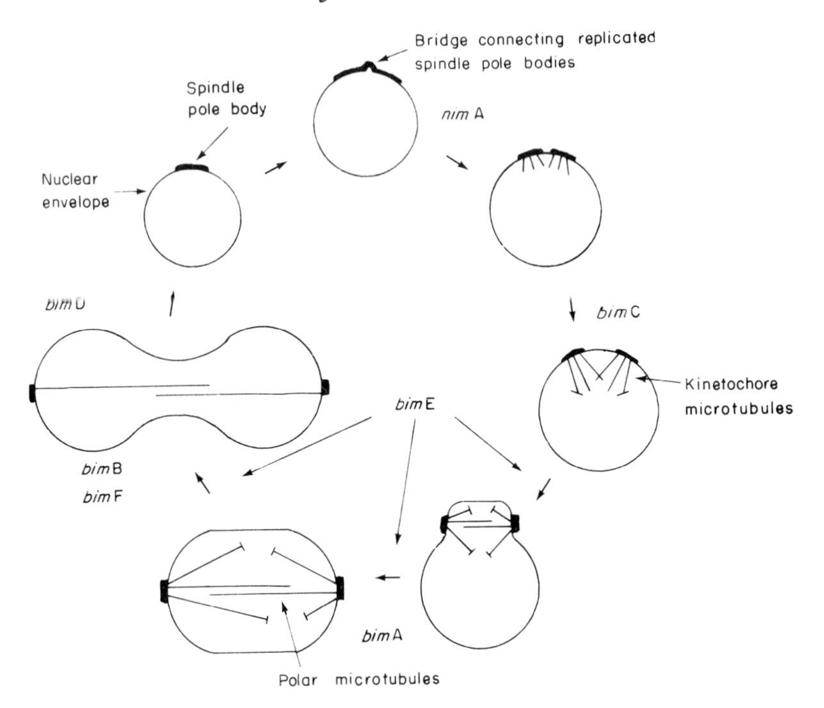

Fig. 7.21 Terminal phenotypes of *bim*A–F and *nim*A mutants of *Aspergillus nidulans* at restrictive temperature as determined from light and electron microscopy. The steps at which various mutations block the nuclear division cycle are indicated. The terminal phenotypes displayed behind the *bim*C block (short non-interacting microtubules) and the *bim*D block (an extremely elongated nucleus without kinetochore microtubules) have not been observed in wild-type nuclei undergoing mitosis. (Reproduced with permission from Morris, 1980.)

I. MAMMALIAN CELLS

1. *General Features*

The isolation and characterization of cell cycle mutants of mammalian somatic cells is currently in a state of rapid development. Although, at first, the prospect of obtaining a large number of temperature-sensitive mutants was gloomy, because of the diploid nature of the eukaryotic genome, the success in isolating numerous potentially-useful mutants has been encouraging (Table 7.6).

The selection of a large number of recessive mutations, in particular in the Chinese hamster cell lines (CHO) might be attributed to extensive functional hemizygosity with one set of genes or chromosomal segments, being inactive or missing. Such a condition may have arisen during the evolution of the permanent cell line in culture (Siminovitch, 1976). Double mutation or mutation plus gene inactivation have not been ruled out as alternative explanations in this or other cell lines. There is, however, some persuasive evidence, reviewed by Siminovitch and Thompson (1978) that a number of genes in CHO cells are present and expressed in two copies. Recessive mutations in such cases could arise from a combination of two non-independent events—mutation and gene conversion or mitotic crossing over (Williams, 1976). Further tests are required to determine the extent of functional hemizygosity in permanent cell lines and to develop information about mutation frequencies and the genetic events leading to useful (cell cycle) mutant phenotypes.

The procedures devised for the isolation of cell cycle specific mutants have been reviewed by Basilico and Meiss (1974) and Baserga (1978). In general, enrichment has been achieved by exposing mutagenized cultures to the restrictive temperature, in the presence of agents that kill normally-dividing cells. The cells are then shifted back to the permissive temperature to allow growth of the temperature-sensitive mutants.

Two important conclusions may be drawn from the survey of mutants presented in Table 7.6. First, many different cell lines can each give rise to a variety of useful mutants. Secondly, the mutants isolated to date are affected at a variety of steps in the cell cycle, although the great majority are G_1-like (Hirschberg *et al.*, 1980) and include those that appear to enter a resting G_0-like state. The basis for this bias is unclear; even when the selective procedures used should have favoured mutants in other stages of the cycle (Roscoe *et al.*, 1973a; Baserga, 1978), mutants in G_1 were preferentially obtained.

The availability of such mutants should now allow their exploitation in the dissection of the mammalian cell cycle. The experimental approaches that may be used to classify the mutants and map their temporal location in the cycle, together with examples of successful applications of these approaches, have been described by Levine (1978). The first approach includes the mapping of the mutant block relative to identifiable phases of the cycle (G_1, S, G_2, M). In principle, a second cell cycle mutation can serve as an appropriate point of reference. While this approach has been successfully used with mutants of yeast (see Section III.A) and viruses, it has not yet been used with mammalian cells. A second approach has been to classify mutants into groups that respond similarly to agents that suppress, bypass or

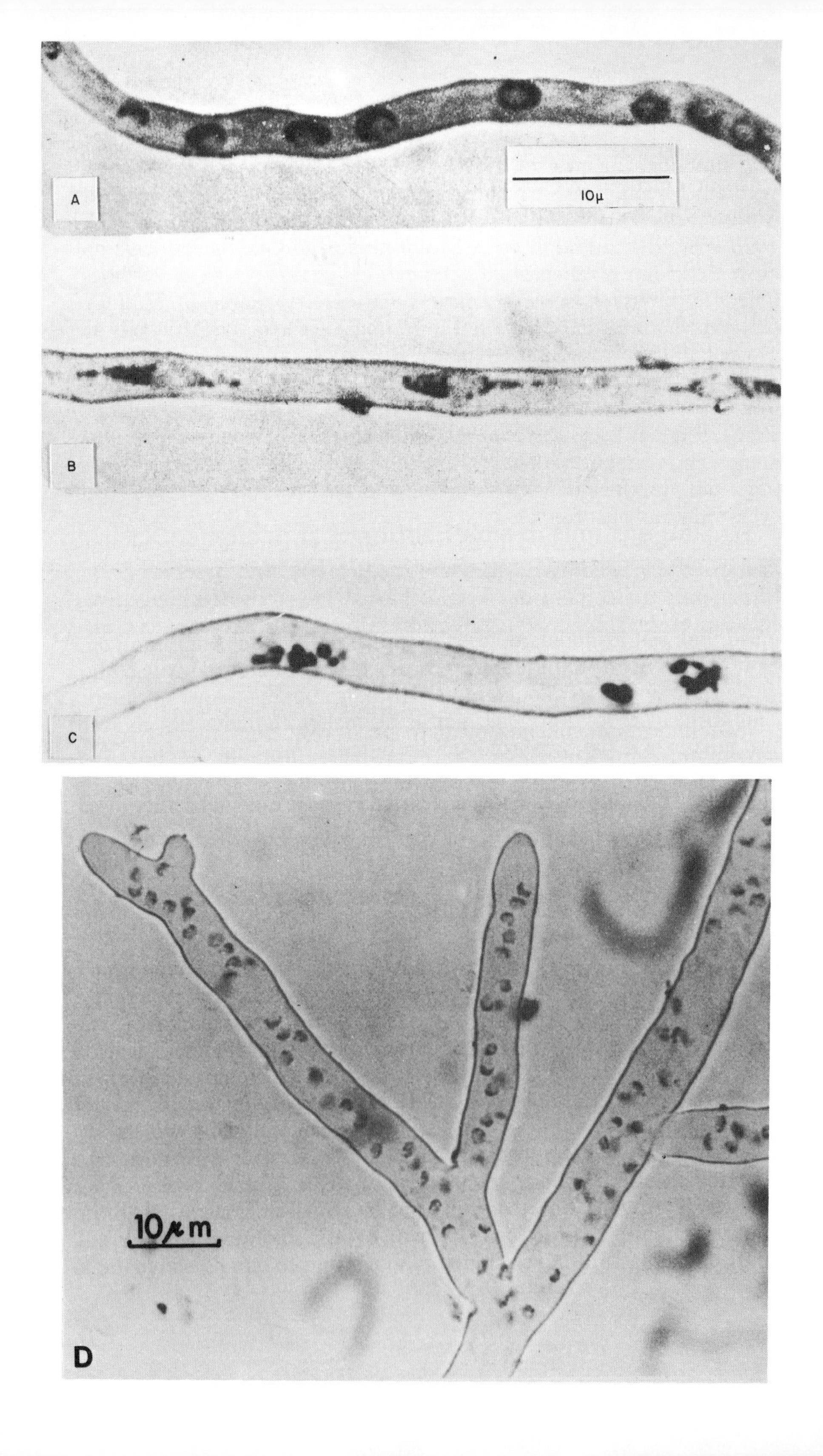
A
10μ
B
C
10μm
D

complement the mutational block, thus permitting the cells to grow under nonpermissive conditions. Here, second site (suppressor) mutations, external agents (e.g. viruses) or somatic cell hybrids have been useful tools. Thirdly, mutants may be tested at the permissive and nonpermissive temperatures for their effects upon the replication and expression of extrachromosomal elements, such as viruses. Finally, the mutants may be classified by the development of assays that permit the mutant defect to be complemented by the addition of the wild-type gene product.

The precise molecular defects in the mutants of Table 7.6 have remained elusive. In fact, in a provocative and pessimistic paper, Stanners (1978) has questioned the usefulness of attempts to characterize at a molecular level the temperature-sensitive cell cycle mutants. He advocates increased attention to the isolation and characterization of what he calls "specific mutants", i.e. those obtained by specific selection procedures and thus conditionally defective in defined biochemical functions. Examples of "non-specific" temperature-sensitive mutants that have been partly characterized (though these are not cell cycle mutants *sensu stricto*) are given by Kit (1980). Examples of temperature sensitive mammalian cells with identified defects in known gene products are those described by Ingles (1978), Nakamo *et al.* (1978) and Adair *et al.* (1978). The potential of these mutants for cell cycle analysis has yet to be realized.

2. *The K12 Mutant of Chinese Hamster Wg1A Cells*

The number of mutants isolated from various cell lines is now so great (Table 7.6) that it is no longer feasible or perhaps useful to describe the properties of each. The K12 mutant of Chinese hamster cells, however, serves as an appropriate representative to illustrate the approaches used and the successes achieved.

The rationale for a selective procedure was suggested by the relative ease with which mitotic cells may be detached from a monolayer culture at low Ca^{2+} concentrations (Terasima and Tolmach, 1963a). By shifting a culture of ethyl methane sulphonate-mutagenized Wg1A

Fig. 7.22 Morphology of temperature-sensitive mutants of *Aspergillus nidulans*. (A), (B) and (C) compare the nuclear morphology of the wild-type with *nim* mutants. (A) wild-type mycelium grown at 32°C for 12 h and stained with aceto-orcein. (B) UVts 136 grown at 32°C for 12 h, then at 42°C for 4 h and stained with aceto-orcein. (C) is the same as (B) but shows metaphase figures. (D) shows the phenotype of *sep* A2 grown at 32°C overnight, switched to 42°C for 4 h and then stained with aceto-orcein. (Reproduced with permission from Morris, 1980.)

Table 7.6. *Cell cycle mutants derived from mammalian cell lines*

Organism	Cell line	Mutant	Arrest point	Reference
Chinese hamster	Wg1A	K12	late G_1; entry into S	See text
	Wg1A	K18, K27, K33, 4/3, 4/2, 3/1	late G_1; entry into S	Melero (1979)
	Wg1A	*ts*K/34C	G_1	Tenner *et al*. (1977) Otsuka and Scheffler (1978) Landy-Otsukă and Scheffler (1978)
	CCL39	BF113	G_1 after one cycle	Scheffler and Buttin (1973)
	CHO	MS1–1	cytokinesis	Thompson and Lindl (1976)
	CHO	*ts*D, *ts*224 and others	various	Okinaka and Barnhart (1978)
	$^+$CHO	cs^4-D3	G_0/G_1 (reversible)	Crane and Thomas (1976) Crane *et al*. (1977) Berger *et al*. (1979) Lomniczi *et al*. (1977)
	$^+$CHO	CR^R E5	G_1	Ling (1977)
	GM75	TS111	cytokinesis	Hatzfeld and Buttin (1975)

	D_6(pseudo-diploid)	*ts*-1	G_1	Hori (1977)
	CCL61	1129 and 1132	reversibly in G_1 and less reversibly in G_2	Ohlsson-Wilhelm *et al*. (1976, 1979)
	CHO-K_1	11C3	late S/G_2	Marunouchi and Nakano (1980)
	CHL-V79	G_1^+-4 G_1^+-5	G_1	Liskay and Prescott (1978)
	CHO	*ts*AMA^R-1	G_1	Ingles (1978)
Syrian hamster	BHK	NW1	cytokinesis	Smith and Wigglesworth (1972)
	BHK21/13	*ts* Af8, *ts* 11, ts 13	early to mid G_1	Meiss and Basilico (1972) Burstin *et al*. (1974) Burstin and Basilico (1975) Kane *et al*. (1976) Ming *et al*. (1976) Talavera *et al*. (1976) Liskay and Meiss (1977) Talavera and Basilico (1977) Chang and Baserga (1977)

Table 7.6. *continued*

Organism	Cell line	Mutant	Arrest point	Reference
				Moser and Meiss (1975, 1977) Floros *et al.* (1978) Rossini and Baserga (1978) Jonak and Baserga (1979) Rossini *et al.* (1979, 1980)
	BHK211	*ts* HJ4	late G_1, early S function	Talavera and Basilico (1977)
	BHK21	*ts* BTN-1 *ts* BN-2 *ts* BN 75	S? G_1 and S S	Nishimoto and Basilico (1978) Yanagi *et al.* (1978) Nishimoto *et al.* (1980)
	BHK21	*ts* 422E	prior to division cells have 4n amount of DNA and very large nuclei	Toniolo *et al.* (1973) Mora *et al.* (1980)
Hamster	HM-1	*ts*-546	mitosis (metaphase)	Wang (1974) Wang and Yin (1976)
	HM-1	*ts*-655	mitosis (prophase)	Wang (1976)

	HM-1	*ts* 694, *ts* 559	G_1	Chen and Wang (1977)
	HM-1	*ts* 550c	G_1 and G_2	Chen and Wang (1977)
	HM-1	*ts* 687	mitosis	Wissinger and Wang (1978)
Mouse	CAK	B54	late G_1	Liskay and Meiss (1977) Liskay (1974) Farber and Liskay (1974)
	L	*ts* A 159	S	Thompson *et al.* (1970, 1971) Sheinin (1976a,b) Setterfield *et al.* (1978) Sheinin *et al.* (1977, 1978a)
	L	*ts* c1	late S or G_2	Setterfield *et al.* (1978) Sheinin and Guttman (1977) Thompson *et al.* (1971)
	Balb/3T3	A8, A83 and 5 others	G_1	Naha *et al.* (1975) Naha (1979a,b) Naha and Sorrentino (1980)

Table 7.6. *continued*

Organism	Cell line	Mutant	Arrest point	Reference
	Balb/c 3T3	*ts*-2	S	Slater and Ozer (1976)
	FM3A	*ts* 85	late S and G_2	Mita *et al.* (1980)
	FM3A	*ts* c1.B59	cytokinesis	Nakamo *et al.* (1978)
Murine leukemic	L51787	*ts* 2 *ts* 39	mitosis and cytokinesis	Shiomi and Sato (1976, 1978) Sata and Hama-Inaba (1978)
African green monkey	BSC-1	*ts* 3, *ts* 5, *ts* 9	S	Naha (1973)

[+] cold sensitive

Not shown are some mutants exhibiting random arrest in the cell cycle (e.g. Thompson *et al.*, 1977; Roufa and Reed, 1975; Wittes and Ozer, 1973). Examples of mutants not yet assigned to specific cell cycle blocks are those described by Nishimoto and Basilico (1978), Loomis *et al.* (1973) and Melero (1979).

cells from low to high temperature and allowing sufficient time for all cells to reach mitosis, it was possible to detach those cells blocked at this stage (Roscoe *et al.*, 1973a). The proportion of wild-type cells obtained was reduced by adding an inhibitor of DNA synthesis. K12 was one isolate (from 5×10^8 cells mutagenized) with an exceptionally low spontaneous reversion rate (<1 in 6×10^7). The K12 mutation is recessive, but is still expressed, presumably because there is only one copy of the chromosome that carries the altered gene, or perhaps because its allele is inactive (Smith and Wigglesworth, 1973).

At the restrictive temperature, DNA synthesis is markedly decreased but is restored on reversion to the permissive temperature (Roscoe *et al.*, 1973b). Temperature sensitivity is reversed following exposure of the mutant to SV40 virus (Smith and Wigglesworth, 1974). Confirmation that the mutant is blocked in the progression from G_1 into S was provided by Kit and Jorgensen (1976) who also showed that the transcription or post-transcriptional processing of thymidine kinase is affected by the K12 mutation. The point reached by the mutant at the permissive temperature, at which shifting to the nonpermissive temperature does not prevent entry into S (the "shift up time") is constant and occurs 1·8 h before the onset of S under diverse growth conditions (Ashihara *et al.*, 1978). The execution points for five other temperature-sensitive mutants isolated from Wg1A cells also lie in late G_1 (Melero, 1979; Fig. 7.23). The effect of the mutation in K12 on

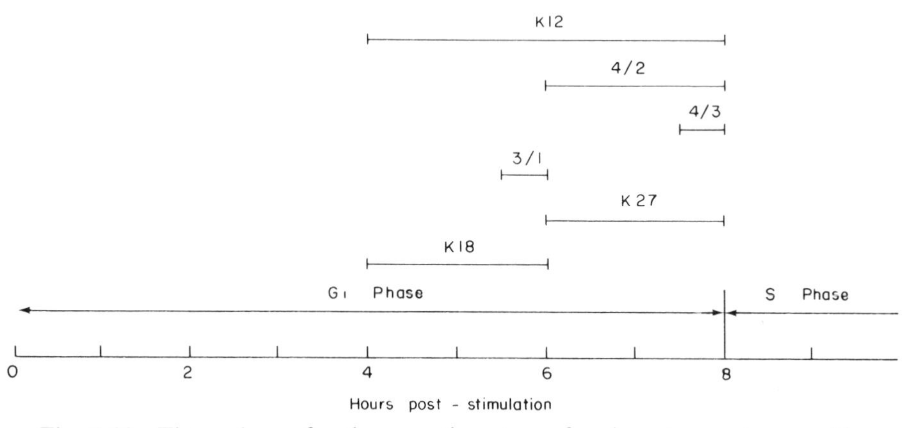

Fig. 7.23 Time scheme for the execution steps of various temperature-sensitive mutants of mammalian cells. Durations of the steps are represented by the horizontal lines. The 8-hour lag between stimulation of growth and initiation of DNA synthesis represents the minimal estimate for the duration of G_1. (Reproduced with permission from Melero, 1979.)

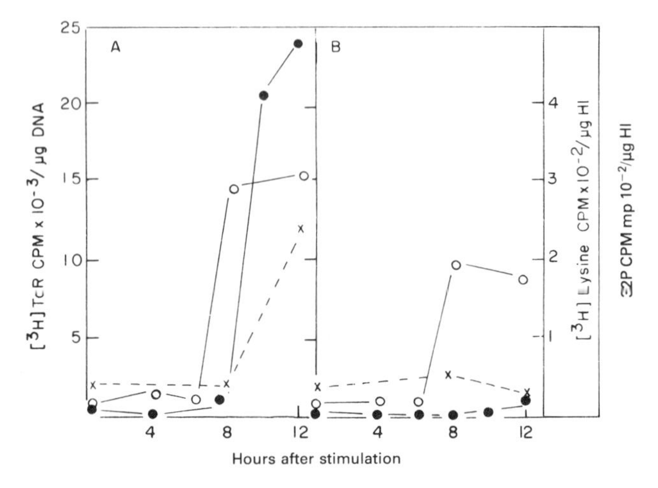

Fig. 7.24 Histone phosphorylation, DNA and histone synthesis in the K12 mutant of Chinese hamster cells at 34°C (a) and 40·6°C (b). Cells were stimulated by replacing growth medium with fresh medium containing 10% serum. ●—● [^{3}H] thymidine incorporation into DNA; ○—○ [^{32}P] incorporation into H1; ×- - -×, [^{3}H]-lysine incorporation into H1. (Reproduced with permission from Pochron and Baserga, 1979.)

another G_1 event, histone phosphorylation, has been studied by Pochron and Baserga (1979). Histone H1 phosphorylation occurred in late G_1, at either the permissive or restrictive temperature, in each of three mutants (*Af*8, *ts*13 and K12) that arrest at different points in G_1 (Fig. 7.24). The results were taken to suggest that each temperature-sensitive function was without effect on histone phosphorylation. Thus, G_1 events are not necessarily interdependent.

Attempts are now under way to characterize biochemically the defect in K12 cells. The use of heterokaryons, formed by fusing serum-starved K12 cells (blocked in G_1) with chick erythrocytes, showed that an unidentified cellular function, defined by the K12 mutation, is required for activation of chick erythrocyte DNA synthesis (Dubbs and Kit, 1976). Further steps to characterize the defect in K12 have been described by Melero and Fincham (1978) and Melero and Smith (1978). The synthesis of three polypepetides is markedly enhanced when K12 cells are incubated at 40·5°C. One (band B) serves as a useful biochemical marker for the expression of the mutation since its synthesis is not affected in either temperature-sensitive mutants or in hybrids in which the K12 mutation is complemented (Melero and

Fincham, 1978). In addition, the alteration in polypeptide B is irreversible and occurs during the same stage of the cell cycle at which the mutated function is expressed. The alterations observed in the synthesis of polypeptides A, B and C are dependent on the continuous production of RNA at the nonpermissive temperature, indicating that transcription is the level at which the K12 mutation regulates their synthesis. Further experiments *in vitro* indicated that the mRNAs which code for A, B and C accumulate in mutant cells incubated at 40.5°C. It was suggested that the genes coding for the three polypepetides form an operon-like unit which is regulated by the thermosensitive factor in K12 cells (Melero and Smith, 1978).

IV. Mutants of Prokaryotic Micro-organisms

A. INTRODUCTION

During the last 15 years or so, many mutants of prokaryotes (largely bacteria) have been described with abnormal morphology or which exhibit abnormal function in events that are generally considered to constitute the cell cycle (Tables 7.7 and 7.8). While many of these mutants may be regarded as relevant to studies of the cell cycle, in that they reveal the presence of control mechanisms acting on the component processes of the cycle, it is evident that they cannot be exhaustively covered here. Instead, attention will be focused on those mutants that resemble most closely the conditional mutants of eukaryotic cells described earlier in this chapter. However, as pointed out by Slater and Schaechter (1974) in a useful and thoughtful review, progress in the genetic analysis of eukaryotic cell cycles, and the yeast cell cycle in particular, has been remarkably rapid and straightforward. In bacteria, the study is frustrated by the extreme sensitivity of division to a wide range of growth inhibitors and mutations, such that it is difficult to define the degree of specificity with which these manipulations are involved in the regulation of division. In addition, the corollary to the morphological simplicity of bacterial growth and division is that "landmark" events are difficult to define.

B. MUTATIONS AFFECTING DNA SYNTHESIS AND REPAIR

Studies of chromosome replication in *Escherichia coli* have been facilitated by the isolation and characterization of temperature-sensitive

Table 7.7. *Mutants of* E. coli *defective in division and forming multinucleate filaments at the restrictive temperature*

Mutant or gene	Map position (min)	Reference
BUG6	3	Reeve *et al.* (1970)
fam-715 (ST 715)	74	Torti and Park (1976, 1980)
fcs A	86	Kudo *et al.* (1977)
fil_{ts}	88	Stone (1973)
fts A	2	Van de Putte *et al.* (1964) Ricard and Hirota (1973) Pages *et al.* (1975)
fts B	32–34	
fts C	4–9	
fts D	86	
fts E	73	
fts F	82	
fts G	29–30	
fts H(y–16)	69	Santos and de Almeida (1975)
fts H(ASH 124)	69	Holland and Darby (1976)
fts Q	2	Begg *et al.* (1980)
fts Z	2	Lutkenhaus *et al.* (1980)
JS10	72–75	Sturgeon and Ingram (1978)
sep	2	Allen *et al.* (1974) Walker *et al.* (1975) Fletcher *et al.* (1978)
ts 20	1	Nagai *et al.* (1971) Nagai and Tamura (1973)
ts 52	35	Zusman *et al.* (1972)
ts 612	75	Masamune (1975)

The map positions are from Bachmann and Low (1980) or extrapolated from the primary sources to the recalibrated linkage map.

mutants that are defective in DNA synthesis at nonpermissive temperatures. DNA replication is known to involve at least 14 dna^+ loci as well as *nal* A^+, cou^+ and lig^+. Mutations in these loci affect the initiation of DNA synthesis at the origin of replication, DNA elongation and other stages of replication (for references, see Henson *et al.*, 1979). In addition, the transfer of many conditional *dna* mutants to the nonpermissive temperature leads to filament formation (Hirota *et al.*, 1968a), indicating that for *E. coli*, but not *Bacillus subtilis* (Donachie *et al.*, 1971), cell division does not take place in the absence of DNA synthesis. In most *E. coli dna* A mutants, a shift to the restrictive condition allows rounds of replication to terminate until each completed genome segregates into a daughter cell. However, in one mutant in this gene, CRT 83, division stops immediately while rounds of replication continue until termination (Khachatourians and Clark, 1970).

The normal coupling between division and DNA replication is not absolute and may bc bypassed by specific mutations. For example, mutation in *div* A allows *E. coli* T46 (Hirota *et al.*, 1968b) to divide at the restrictive temperature, whereas *Salmonella typhimurium* 11G divides without any known "*div*" mutations (Spratt and Rowbury, 1971). The fact that such a bypass cannot be demonstrated when DNA synthesis is blocked by means other than the effect of a mutation (for references, see Slater and Schaechter, 1974) suggests that in these mutants the relationship between DNA synthesis and cell division operates by a different mechanism than in normal cells. This view is supported by studies on conditional DNA initiation mutants. In such mutants, cell division has a lag of about 1 h when cells are shifted to the restrictive temperature. This lag is related only to the time spent under restrictive conditions and is not related to the termination of rounds of replication or events triggered thereby (Hirota and Ricard, 1972; Shannon *et al.*, 1972).

Orr *et al.* (1979) have recently isolated a temperature-sensitive mutant of *E. coli*, resistant to chlorobiocin; the mutation is probably in the *cou* gene specifying a subunit of DNA gyrase. RNA synthesis and initiation of DNA replication are affected at the restrictive temperature, but unlike other mutants with defective metabolism of nucleic acids, it is not the frequency of divisions which is affected but the positioning of the septa, generating short anucleate cells.

There is now a substantial body of evidence which implicates DNA repair mechanisms in the control of cell division (Chapter 8). Some mutants that exhibit unusual division properties as a result of mutations in genes involved in such mechanisms are shown in Table 7.8. A fuller version of this Table is given by Helmstetter *et al.* (1979). Ultraviolet irradiation of *E. coli* initiates SOS or error-prone repair, an emergency response to extensive DNA damage. The system is inducible, dependent on the *rec A* gene and includes a variety of effects including inhibition of cell division and the synthesis of a new protein, termed X, that has been shown to be the *rec A* gene product (Emmerson and West, 1977; McEntee, 1977; Gudas and Mount, 1977). It has been suggested (Satta and Pardee, 1978; George *et al.*, 1975) that the *rec A* protein is directly responsible for filamentation but a recent kinetic analysis of cell division inhibition by Darby and Holland (1979) does not support this idea.

Table 7.8. *Mutants of* E. coli *with unusual division properties*

Mutant	Map position	Defect	Reference
MAC-1 (*div* A)	3 ± 1	Division independent of completion of chromosome replication.	de Pedro *et al.* (1975); de Pedro and Cánovas (1977)
min A *min* B	10, 29	Misplacement of structurally normal septa. Forms small chromosomeless cells. Minicell production shares common regulatory step with *cap R* and *BUG 6* (see Table 7.7).	Adler *et al.* (1967, 1969); Adler and Hardigree (1972); Frazer and Curtiss (1975); Teather *et al.* (1974); Khachatourians *et al.* (1973)
min A *min* B ts 52		Placement of septum plus thermosensitive cell division. Number of minicells reduced.	Zusman *et al.* (1978)
cap R (*lon*)	10	Filamentation after exposure to radiation and agents which damage DNA; suppressed by *sul A* or *sul B*.	Walker and Pardee (1967); Gayoa *et al.* (1976); Green *et al.* (1969); Howard-Flanders *et al.* (1964); Hua and Markovitz (1972); Leighton and Donachie (1970); Johnson and Greenberg (1975)

rec A	58	Recombination deficient and very sensitive to radiation and agents that damage DNA. Division in absence of chromosome replication.	Inouye (1971); Jenkins and Bennett (1976)
tif_{ts}	58	Filamentation at restrictive temperature. Thermoinducible repair and thermosensitive division suppressible by *rec A*, *zab* or *lex A*.	Castellazzi *et al*. (1972a, b); Castellazzi (1976); Witkin (1975); Kirby *et al*. (1967)
lex A	90	Sensitive to radiation and agents which damage DNA. Division in absence of chromosome replication. Enhancement of repair capacity with tsl_{ts} impairs division.	Howard-Flanders and Boyce (1966); Howe and Mount (1975, 1978); Mount *et al*. (1972, 1973)

The map positions are from Bachmann and Low (1980) or extrapolated from primary sources to the recalibrated linkage map.

Lex A mutants are sensitive to u.v. light and a variety of agents that damage DNA (Mount *et al.*, 1972). Like *rec A* mutants, they are effective in the coordinated expression of SOS functions, the *lex A* gene regulating the induction of protein X.

Two types of results suggest that the *lex A* product regulates cell division. First, one class of u.v.-resistant derivatives of *lex A* mutants are defective in cell division, forming filaments at 42°C, and carry a suppressor mutation designated *tsl* that is thought to affect the activity of the *lex A* product (Mount *et al.*, 1973). Secondly, cells lacking DNA arise following thymidine starvation of a *lex A* mutant because the mutant cells undergo additional cycles of division during abnormal growth (Howe and Mount, 1975, 1978). A similar effect of *rec A* mutations on cell division has been described (Capaldo and Barbour, 1975; Inouye, 1971).

In 1964, Goldthwait and Jacob isolated a thermosensitive mutant of *E. coli*, C600 T44 whose complex phenotype, which included filamentous growth at 41°C, was later shown by Castellazzi *et al.* (1972a) to be due to the *tif* mutation. All phenotypic traits were suppressed by three classes of mutations, *rec A*, *zab* and *lex* (Castellazzi *et al.*, 1972b) each of which also suppresses error-prone DNA repair. The *tif* mutation has several features in common with the *lon* mutation first described by Adler and Hardigree (1964). The *lon* mutation is not defective in DNA synthesis itself but is probably altered in the production of a factor linking DNA synthesis to division (for references, see Slater and Schaechter, 1974). The division defect is observed only after the action of DNA-damaging agents such as u.v. (Howard-Flanders *et al.*, 1964; Adler and Hardigree, 1964) and is reversed by *rec A* and *lex* mutations (Green *et al.*, 1969; Donch *et al.*, 1968).

Using a double mutant, George *et al.* (1975) showed that the cell filamentation promoted by *tif* is enhanced by the *lon* mutation. In addition, amongst "revertants" of the double mutant that were characterized by recovery of division ability without u.v. sensitivity, two mutations (which specifically *s*uppress *fi*lamentation) were mapped at two loci *sfi A* and *sfi B*.

George *et al.* (1975) have put forward a hypothesis (Fig. 7.25) to explain the intervention of DNA repair processes with the bacterial cell cycle, such that in response to DNA alterations the cell delays septation to permit extensive restoration of damaged DNA. A Repair Associated Division Inhibitor (RADI) is presumed to exert negative control over septation, as previously suggested by Inouye (1971) for the *rec A* gene product.

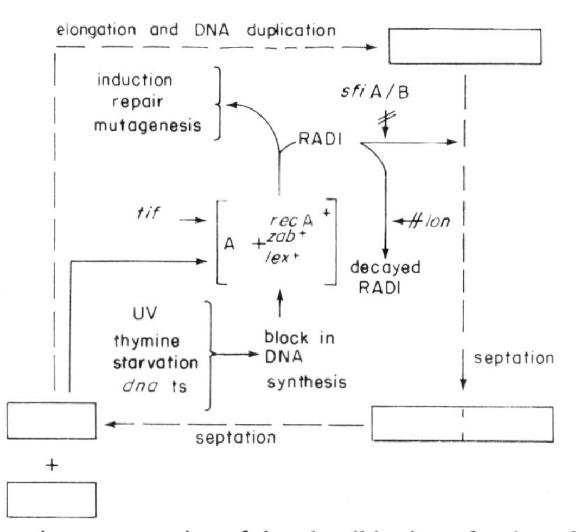

Fig. 7.25 Schematic representation of the plausible sites of action of *tif*, *sfi* and *lon* mutations in the RADI (Repair Associated Division Inhibitor) hypothesis. Thickened arrows indicate a stimulatory action or, if barred, an inhibitory action. "A" designates a hypothetical adenine derivative, and takes into account the stimulatory action of adenine on the expression of the tif^+ phenotype (Kirby *et al.*, 1967). The *sfi A* and *sfi B* mutations suppress the division defects caused by *tif* or *lon*. Suppression of the division defect takes place in cells which preserve their full repair capacity. The *sfi* mutations thus uncouple cell division from the control exerted by RADI. *lon* cells are unable to restore division normally after filamentation. Its action might be explained by the stabilization of RADI. The division block caused by *tif* requires the activity of the *rec* A^+, zab^+ and lex^+ genes. *tif* may be a regulatory mutation affecting either the activity or synthesis of repair enzymes. (Reproduced with permission from George *et al.*, 1975.)

C. MUTANTS AFFECTING SEPTUM FORMATION

1. *Mutants Unable to form Septa under Restrictive Conditions*

Slater and Schaechter (1974) have made several cautionary comments relevant to the study of bacterial cell cycle mutants and which are particularly pertinent to those mutants which form filaments. First, complex phenotypes may be due to pleiotropic effects of a single mutation or multiple mutations. Secondly, alleles of one gene regulating a cell cycle event may elicit different phenotypes, and until more complementation data is available to determine whether two mutants with the same apparent phenotype have a lesion in the same gene, results must be interpreted with care. Thirdly, as Mendelson and Cole have pointed out (1972), all conditional lethal mutants are, in some sense, division mutants and it is often difficult to determine the specificity of a mutation for division, especially since division is preferentially inhibited by so many interfering conditions.

Despite these problems, a large number of temperature-sensitive mutants, proposed to be defective in septum formation, and which thus form filaments at the restrictive temperature, have been described in *E. coli* (Table 7.7) and in other bacteria (Ahmed and Rowbury, 1971; van Alstyne and Simon, 1971; Nagai *et al.*, 1976). Most of these are heat-sensitive, but recently Kudo *et al.* (1977) and Sturgeon and Ingram (1978) have isolated cold-sensitive mutants. The mutants shown all continue to synthesize DNA, and often other macromolecules, after a shift to the restrictive temperature and are therefore quite distinct (Hirota *et al.*, 1968a) from the mutants described on Section IV.B. The mutants vary in the promptness with which cell division is arrested after a temperature shift; those in which division is arrested immediately are presumed to be temperature-sensitive for a factor required for division late in the cycle. They also vary in the capacity for reverse the filamentation and in the sensitivity of this process to inhibitors, high osmotic pressure and other factors, following a reciprocal shift.

The study of these mutants is still largely descriptive and little is known of the mechanism by which septum formation and its initiation are inhibited. Evidence for lesions in the cell wall of some *E. coli* mutants has been obtained (e.g. Stone, 1973; Reeve *et al.*, 1970; Ciesla *et al.*, 1972) but no cell wall defect was detected in the Gram positive bacterium *Bacillus* (Breakefield and Landman, 1973). Membrane proteins are affected in the *E. coli* mutants *ts*612 (Masamune, 1975).

In *E. coli*, there is a cluster of at least eight genes involved in cell envelope synthesis, cell division, or both (Fletcher *et al.*, 1978; Lutkenhaus and Donachie, 1979; Lutkenhaus *et al.*, 1980; Ricard and Hirota, 1973; Walker *et al.*, 1975; Wijsman, 1972). Two further genes in this cluster (*ftsQ* and *murG*) have been reported by Begg *et al.* (1980) and Salmond *et al.* (1980), respectively. However the *cs* gene (designated *fcsA*) (Kudo *et al.*, 1977) maps at 86 min and the *ftsH* gene or mutant Y16 (Santos and de Almeida, 1975) maps at 69 min (Bachmann and Low, 1980), both regions not previously described as being involved in cell division. In *S. typhimurium*, the *div C* locus also maps near *leu*.

Such mutants have already given some insight into the control mechanisms for septum formation (Slater and Schaechter, 1974). For example, the finding that shifting filaments to the restrictive temperature leads to rapid expression of the divisions missed at the restrictive temperature (Reeve and Clark, 1972; Ciésla *et al.*, 1972) indicates that the ability to measure normal cell lengths is not lost. There is some evidence that the placement of the division site and the completion or segregation of chromosomes are related (Reeve *et al.*, 1970; Inouye, 1969; Mendelson, 1972).

2. *Mutants that Construct Septa at Inappropriate Sites*

Cultures of certain bacteria contain two types of cells, normal rod-shaped cells and minicells (Table 7.8). In a fraction of the cells in the population, a nonmedial septation divides each affected cell into a large rod and a spherical minicell. Since minicells contain little or no chromosomal DNA (e.g. Adler *et al.*, 1967) they fail to divide further. Minicell production, at least in *E. coli*, does not interfere with normal median division, which may occur simultaneously, and so it is questionable whether minicell-producing strains should be considered as cell cycle mutants *sensu stricto*, although it is undeniable that they have contributed to understanding of the control of bacterial division (see Travis and Mendelson, 1977). The reader is referred to the reviews by Slater and Schaechter (1974) and Frazer and Curtiss (1975).

3. *Mutants that Fail to Separate into Individual Cells*

The chain-forming *ent A* mutants of *E. coli* K12 have been described by Normark (1971) and Normark *et al.* (1971).

4. *Mutants that Exhibit Diminished Cell Length*

Martinéz-Salas and Vicente (1980) have described an amber mutation in a newly found gene (*wee*; map position, 83·5 min) in an *E. coli* strain bearing a temperature-sensitive suppressor. At the restrictive temperature, mean cell length and volume are smaller than in the wild-type, even though mass and DNA content increase at the normal rate and cell division proceeds normally. The phenotype presumably results from faulty regulation of the elongation process, or abnormal coordination of the process with the cell cycle.

D. MUTANTS OF *E. COLI* DEFECTIVE IN MEMBRANE PROTEIN SYNTHESIS

A temperature-sensitive mutant of *E. coli* (*div* E42, previously called *ts*C42) has been shown by Ohki and his coworkers to be conditionally defective in the synthesis of a number of membrane proteins and mRNA for the *lac* operon at high temperature (Ohki and Mitsui, 1974; Ohki and Sato, 1975; Yamato *et al.*, 1979). Some of these proteins appear to be synthesized specifically at a particular phase in the cell cycle (Ohki, 1972) and growth of the mutant is arrested at a specific stage when transferred to high temperature, perhaps at the time when some of the membrane proteins are synthesized in the cell cycle. The mutation is located at about 22 min on the *E. coli* chromosome (Sato *et al.*, 1979).

E. MUTANTS OF *CAULOBACTER CRESCENTUS*

The *Caulobacter* cell cycle (for a review, see Shapiro, 1976) includes at least two cell types, and is characterized by the expression of temporally-distinct morphogenetic events (Fig. 7.26).

Conditional cell cycle mutants have been used to study the relationship between the cell cycle and the control of development. Approximately 400 temperature-sensitive strains were screened by Osley and Newton (1977) for mutants that accumulated with a char-

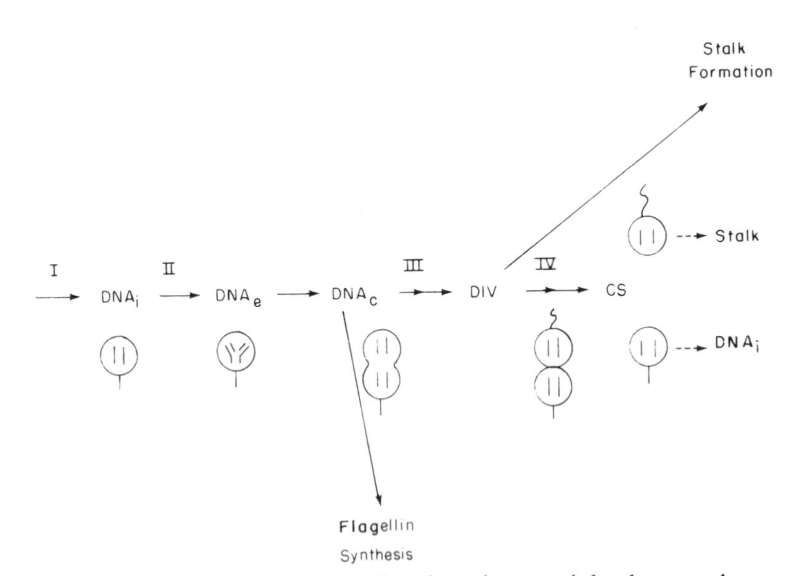

Fig. 7.26 Proposed organization of cell-cycle pathway and developmental events in *Caulobacter crescentus*. Steps in the *C. crescentus* cell cycle are arranged according to the order in which they occur in the cell cycle. The Roman numeral preceding a step refers to the group of mutations which blocks the appearance of that step and subsequent steps in the pathway. Thus, the sequence, DNA$_i$ (DNA initiation), DNA$_e$ (DNA elongation), DNA$_c$ (DNA completion), DIV (division), and CS (cell separation), corresponds to a *dependent* pathway in which each step must be completed for later steps to occur. (Reproduced with permission from Osley and Newton, 1977.)

acteristic phenotype at the restrictive temperature. The behaviour of ten such mutants showed that cell cycle events such as DNA replication and cell division are organized into a dependent pathway. The ability of these strains to develop normally under nonpermissive conditions suggested that flagellin synthesis and stalk formation are dependent on the completion of different cell cycle steps. The periodicity of flagellin A and flagellin B synthesis is achieved by coupling it to either DNA chain elongation or completion (Osley *et al.*, 1977). Stack formation is dependent on a later step in the cell division cycle (Fig. 7.26). Since neither flagellum formation nor stalk formation are themselves required for cell division, however, they appear to be controlled by independent pathways that branch from the cell cycle pathway (Osley and Newton, 1977).

F. MUTANTS OF *AGMERELLUM QUADRUPLICATUM*

Morphologically-aberrant mutants, presumably impaired in cell division, may be recovered at high frequency from mutagenized cultures of *Agmerellum quadruplicatum* and other blue–green bacteria (Ingram and Thurston, 1970; Ingram *et al.*, 1972; Ingram and Van Baalen, 1970; Kunisawa and Cohen-Bazire, 1970). Filamentous forms of a division-impaired mutant (SN29; Ingram and Fisher, 1973a) and a cold-sensitive conditional division mutant (SN12; Ingram and Fisher, 1973b) may be induced to divide into cells of almost normal size by an ethanol extract of spent-growth medium.

More recently, Ingram *et al.* (1975) have described a heat-sensitive conditional cell division mutant of *A. quadruplicatum* which produces abnormally small cells under conditions of nutrient limitation and forms multinucleoid filaments under normal growth conditions.

Summary

Conditional, temperature-sensitive mutants, defective in constituent events of the cell division cycle, are now in widespread use for investigating gene function through the cell cycle. Progress in the isolation and analysis of such mutants in *S. cerevisiae* has been rapid and relatively straightforward. Under restrictive conditions, the mutants isolated by Hartwell and others exhibit defective behaviour at an execution point (also termed "block point", "transition point" or shift-up point" in other systems) and eventually arrest development, often with a characteristic phenotype that can indicate the lesion involved. Such mutants in yeast have been used to investigate:

(i) the degree of interdependence of the syntheses of nuclear and extranuclear DNA;
(ii) the causal order and dependence of cell cycle events;
(iii) the coordination of growth processes with division control;
(iv) the similarities of the level of gene function in mitotic and meiotic cycles;
(v) to demonstrate and characterize a key "start" event in G_1. The success of these approaches has led to their wider application to other lower eukaryotes, notably fission yeast and *Tetrahymena thermophila*, and it is clear that many features of the *S. cerevisiae* cell cycle are manifest in these cycles also.

Despite the diploid nature of the genome of mammalian somatic cells, considerable success has been achieved in the isolation of recessive mutants that cause defects in all phases of the cell cycle, but especially in G_1. With few exceptions, progress has been slow in the analysis of mutants that both arrest at a unique point in the cycle and also have lesions identifiable at the molecular level.

The morphological simplicity of most prokaryotes and the extreme sensitivity of division to perturbing influences has frustrated progress in this area and countered the sophistication with which bacterial genetics may be studied. Mutants affected in DNA synthesis and repair, the appropriate location and timing of septum formation and the control of cell size offer exciting prospects for future work.

8. Control of Cellular Growth and Division

"Duration in change."
Goethe

The regulation of growth and division of cells is a central problem in biology. This chapter attempts to illustrate how answers have been found recently to some of the questions posed and describes avenues of research, particularly with micro-organisms, which are currently exciting or controversial, or seem to be promising for future investigations. Excellent, relevant reviews, which concentrate on animal cells and in particular on G_1 as that compartment of the cell cycle in which the key regulatory events of growth occur, are those by Baserga (1976), Prescott (1976a,b) and Pardee *et al.* (1978). The reviews by Mitchison (1971, 1977a,b) focus attention on micro-organisms, especially yeast.

Detailed analyses of G_1 will not be repeated here, although it will become clear that there is justification for probing this phase in micro-organisms as well as in animal cells. It is worth noting, however, that not all cells, whether they be of mammalian origin (Liskay, 1978; Liskay and Prescott, 1978; Rao and Sunkara, 1980; Liskay *et al.*, 1980) or microbes (Mitchison, 1971) have a G_1 phase. This fact has been used by Cooper (1979) to corroborate a unifying model that he proposes for the cell cycles of eukaryotes and prokaryotes. Cooper views G_1 merely as part of a longer period involved in the preparation for DNA synthesis (Fig. 8.1); its existence in some cells is due only to the fact that, in these cells, the cell cycle time is longer than the combined lengths of S, G_2 and M. Perhaps we may clarify this viewpoint by repeating an analogy used by Gould and Lewontin (1979) in a critique of the "adaptionist" view of evolution.

> "The embellished architectural features of fan vaulted ceilings that are seen where the fans intersect, and the richly painted tapering triangular spaces (spandrels) formed by the intersection of two rounded arches at right angles are both necessary byproducts of architectural constraints; their use as decorative features are secondary and in a sense misleading."

Similarly, the existence and variability of G_1 may be misleading. Hence, if key regulatory events occur in G_1, it may be because they must take place shortly before DNA synthesis; this *happens* to be in G_1 only when such a phase is allowed by a long cell cycle time, relative to the lengths of S, G_2 and M.

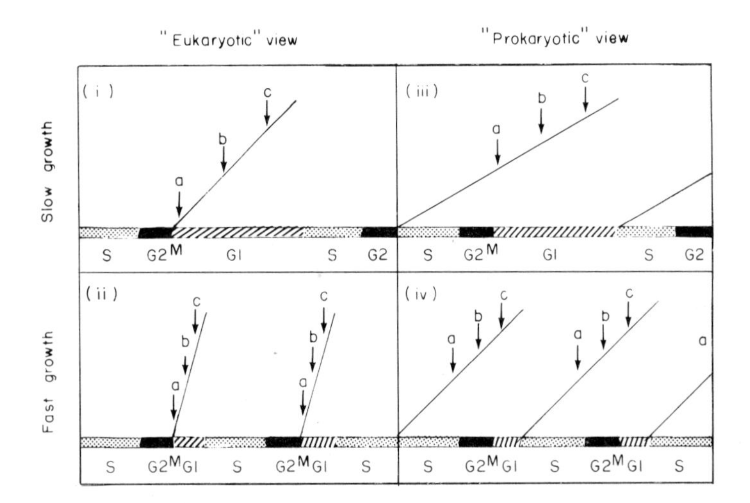

Fig. 8.1 A comparison of two views of the preparations for DNA synthesis. Three events that are observed to occur at various times in G_1 are indicated by a, b and c. (i) shows the current "eukaryotic" assumption, namely that G_1 is a period of preparation for S; (ii) shows the increased rate of sequential occurrence of G_1 functions when the length of the cell cycle is decreased. The model results in a paradox in that at very fast growth rates there is no G_1 period for execution of G_1 functions. Cooper's alternative model is shown in (iii) where a, b and c occur at the same time in G_1 as in (i) but are part of a longer preparation period. Increase in the growth rate (iv) results in a, b and c occurring earlier than G_1. (Modified and reproduced with permission from Cooper, 1979.)

Animal cells may exist in non-growing, non-dividing states for long periods. Whether such cells have withdrawn from the cell cycle into a distinct compartment, G_0, or are arrested in a prolonged G_1 is still debated. The stimulation to growth of these quiescent cells by growth factors and "the pleiotropic response" are not discussed in detail here; the reader is referred to reviews by Holley (1975), Baserga (1976) and Rudland and Jimenez de Asua (1979), and the book edited by Kimenez de Asua *et al.* (1980).

I. Deterministic vs. Probabilistic Controls

A. DETERMINISTIC MODELS

Experiments, mainly with microbial cells, have generated what may be called deterministic models of division control, which regard division as a result of a cyclic series of events, repeated with a high measure of reproducibility in successive generations. Deterministic models envisage division as the result of an orderly progression of growth processes, the rate of division being controlled by the rate of growth of the cell (e.g. Koch and Schaechter, 1962). Completion of each of the successive stages leading to division is dependent on the prior completion of earlier events. A requirement for further growth that has often been proposed is the attaintment of a critical cell size. Examples of cycle events whose occurrence appears to be size-dependent, and which will be treated in greater detail in this chapter, are the initiation of DNA replication in bacteria (Donachie, 1968), the traverse of "start" as marked by bud emergence in budding yeast (Johnston *et al.*, 1977) and cell division in fission yeast (Fantes, 1977).

How does a cell know how big it is? One possibility is that the rate of macromolecular biosynthesis is very precisely regulated, sensed and then used to "cue", with a certain random fluctuation, other much less precise processes of cell growth, including division (Koch, 1977). Zalkinder (1979) has proposed that such a cue in *Tetrahymena* could be a deficiency in the cell's pool of nutrients, itself caused by the attainment of a critical cell volume.

Clues to the requirements for biosynthetic functions prior to division have come from the measurements of "transition points". At the transition point (see Chapter 2) or "point of no return" (Mazia, 1974), the forthcoming division becomes insensitive to the addition of inhibitors of (for example) RNA and protein synthesis. One of the most intensively studied examples (described in detail by Mitchison, 1971) is in *Tetrahymena*. In this system, the delay of division (excess delay or set-back, see Chapter 1) caused by heat shocks follows the characteristic pattern, the extent of delay increasing as a function of age in the cycle when the shock is given, until a transition point is reached and division becomes insensitive to heat. Similar transition points and excess delay may be applied to a wide range of cell types, from bacteria to mammalian cells.

B. PROBABILISTIC MODELS

Perhaps the most radical departure from established hypotheses for division control was the formulation of a transition probability model by Smith and Martin (1973, 1974). In this hypothesis, the deterministic events of cell growth are restricted to the so-called B state of the cycle, which encompasses the S, G_2 and M phases of the conventional cell cycle terminology. G_1, or at a least part of it, constitutes the A state in which the cell is not committed to progress through the B state towards division. Thus, the hypothesis does not detract from the importance of G_1 (or A state) in regulation of the cell cycle (Prescott, 1976a,b) but instead serves to shift attention to the kinetics of departure from G_1. The hope is that such studies will reveal the biochemical basis for the variation in duration of G_1 that has to date remained tantalisingly obscure.

In the meantime, direct biochemical analyses of cells at various cell cycle stages have suggested that certain classes of cellular constituents may have particularly significant roles in progression through the cell cycle and the control of division. These include Ca^{2+} and Mg^{2+}, histones, and perhaps cyclic nucleotides, the subjects of later sections in this chapter.

C. TOWARDS A UNIFIED HYPOTHESIS OF CELL CYCLE CONTROL

The modified transition probability model that invokes two random events (Brooks *et al.*, 1980; Shields, 1980) satisfactorily explains many observations on the rates of division of mammalian cells, whilst the various deterministic models equally well explain most results for micro-organisms. Nurse (1980) considered the possibility that cell cycle controls are quite different for mammalian and microbial cells, but preferred the proposal that there is a size-related deterministic control *and* a probabilistic element in the division cycles of *all* cell types. In microbial cells, the probabilistic element might occur rapidly, making the deterministic size element more important in controlling division, with the balance reversed in mammalian cells. Thus, in budding yeast, Shilo *et al.* (1976) have integrated the two types of control by hypothesizing that the traverse of start requires growth of the cell to a critical size, followed by a random transition that starts the cell cycle (see, however, Nurse *et al.* (1977), Wheals (1977) and Shilo *et al.* (1977)). Evidence for a size-related feedback control of nuclear replication in ciliates (Frazier, 1973; Morton and Berger,

1975; Worthington *et al.*, 1976) taken together with the suggestion by Wolfe (1976) that the G_1 period of *Tetrahymena* is functionally divisible into A and B sectors, might in the future lead to a similarly integrated view of division control in protozoa.

II. The Control of Growth and Division in Budding Yeasts

Cell cycle regulation in budding yeast, as in most other eukaryotes, occurs in G_1, which seems to represent a particularly variable length of time in the cycle, depending on growth rate. Largely as a result of genetic studies (Chapter 7), a period within G_1 referred to as "start" has been defined, at which cells arrest after their growth has been inhibited by mating pheromones, nutritional deprivation or interruption of temperature sensitive gene function. Many studies suggest that once a critical cell size in G_1 has been achieved and the "start" event has been triggered, cell progress to division in a preprogrammed deterministic fashion and in a constant time that is relatively independent of continued growth in biomass.

A. SIZER CONTROLS IN BUDDING YEASTS

In some bacteria, it is well established that the dependence of cell size on growth rate is a consequence of the constancy of both mass per chromosome origin and of the lengths of the (C + D) period (see Chapter 3). In eukaryotes, two size controls (over initiation of DNA synthesis and nuclear division) have been demonstrated in *Schiz. pombe* (pp. 378–384). It is in budding yeast, however, that most progress has been made in defining sizer and timer controls from the genetic, physiological and biochemical points of view.

The results of Johnston *et al.* (1977) indicated that in *S. cerevisiae* a control mechanism operates where the completion of some specific event(s) in the DNA-division cycle requires growth beyond a minimum size. When nutrients are restored to starved yeast cells, the small unbudded cells of stationary phase populations require more time in fresh medium before budding than do larger cells. The close correlation between the time required by a cell and its size is shown in Fig. 8.2A. The duration of the subsequent cell cycle is normal and not correlated with bud size (Fig. 8.2B). A control mechanism of this type is sufficient to coordinate growth with division, if growth, rather than progress

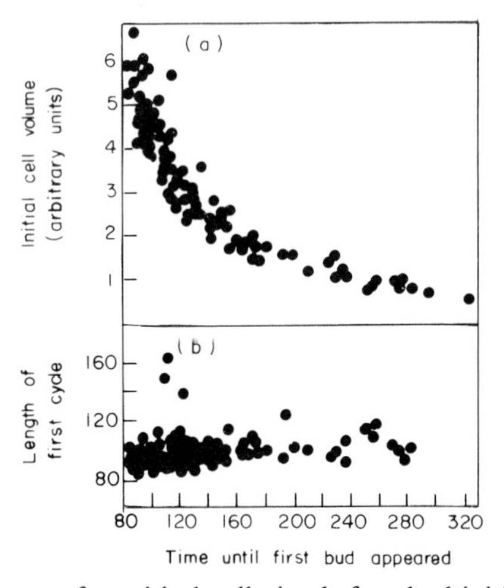

Fig. 8.2 Attainment of a critical cell size before bud-initiation in *Saccharomyces cerevisiae*. Strain C276 was inoculated at low cell density (less than 10^6 cells/ml) into rich liquid medium and allowed to grow into stationary phase at 23°C. At the time of the experiment, the culture contained 99·5% unbudded cells and had been in stationary phase less than 6 h (i.e. cell number increase had ceased less than 6 h earlier). A sample of cells was diluted with water and spotted on the surface of an agar plate at 23°C. Four fields, containing 104 cells, were photographed immediately after spotting and at intervals of approx. 8 min thereafter. The relative volume of each cell was calculated from the initial photographs, and the time of emergence of each cell's first bud was noted. For most of the cells the time of emergence of the cell's second bud was also determined. (a) Initial cell volume plotted against the length of the lag period in min (i.e. the time until emergence of the first bud); (b) length of the first cell cycle in min (i.e. interval between the time of emergence of the first bud and the time of emergence of the second bud) plotted against the length of the lag period. (Note that in this plot each cell simply retains the position along the abscissa that it had in (a). (Reproduced with permission from Johnston *et al.*, 1977.)

through the DNA-division cycle, were always rate-limiting for cell proliferation. This control would prevent cells from becoming too small, and the problem of becoming too large would not arise. The early event that requires the attainment of a critical size lies in G_1 at, or before, the step identified by the *cdc* 28 mutation (Johnston *et al.*, 1977). Even under conditions of acutely nutrient-limited growth, the attainment of a critical size is still mandatory for the initiation of budding (Johnston, 1977).

A similar model of cycle control has been proposed by Jagadish *et al.* (1977) and Carter and Jagadish (1978a). In synchronous cultures (Fig. 8.3) of slowly-growing cells, cell numbers increased from n to

2n at the first division, but from 2n to only 3n at the second division. This observation was interpreted as resulting from the division of the parents for a second time while the small daughters, the products of the first division, developed further before reaching the size required for initiation of a crucial cycle event. Recently, Johnston *et al.* (1979)

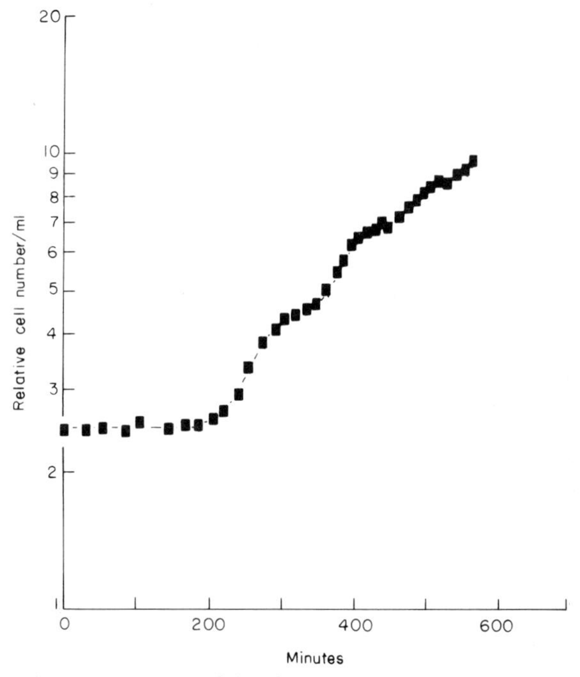

Fig. 8.3 A synchronous culture of *Saccharomyces cerevisiae* growing in rich liquid medium. An exponential culture was separated into fractions according to size. Fractions containing small unbudded cells were used to inoculate a synchronous culture growing in YEPG medium. Cell numbers in the culture were determined at intervals. Note the increase from 2·5 to 4·5 at the first division and from 4·5 to 6·5 at the second, equivalent to increases from n to 2n and 2n to 3n, respectively. (Reproduced with permission from Jagadish *et al.*, 1977.)

have shown that the size at which cells are able to initiate division varies with the growth rate. Adams (1977), in contrast, found that the mean cell volume of haploids at bud initiation was independent of the growth environment. These results have been attributed by Johnston *et al.* (1979) to errors in the measurement of cell size. Lorincz and Carter (1979) have also shown that the cell size at bud initiation varies inversely with generation time over the mass doubling time range 2·1 to 3·7 h. Since budding occurs very shortly after execution of the *cdc* 28^+-mediated start event, the results indicate that the critical time necessary for initiating the DNA division cycle varies with growth rate.

The "asymmetric" cell cycle of budding yeast manifest in the experiments of Jagadish *et al.* (1977) and seen in Fig. 8.4a, is an important feature of yeast growth (Hartwell and Unger, 1977; Slater

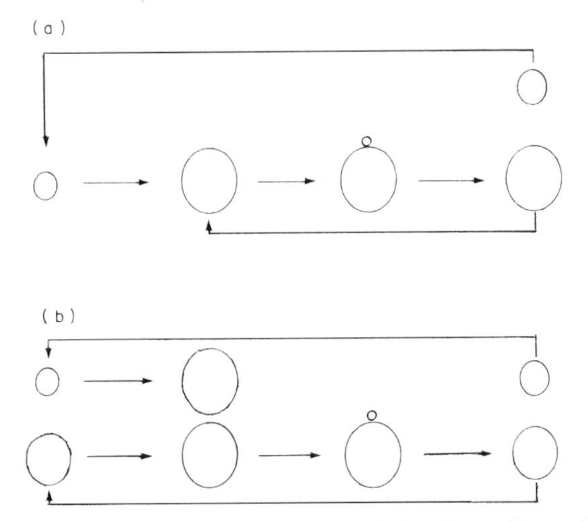

Fig. 8.4 Comparison of the growth pattern of *Candida utilis* under the phased method of cultivation with the model proposed by Hartwell and Unger (1977) for the growth of *Saccharomyces cerevisiae*. In *S. cerevisiae* (a) a newly formed daughter cell is smaller than the mother cell and therefore has a correspondingly longer generation time. Newly formed cells of *C. utilis* (b) are also unequal in size, but under the phased method of cultivation they grow at different rates so that both cells reach the maximum size at the same time. The model shows that despite the initial size difference, the generation times of the mother and daughter cells are equal. (Reproduced with permission from Thomas *et al.*, 1980a.)

et al., 1977; Yamada and Ito, 1979; Lord and Wheals, 1980). In other yeasts, e.g. *Sirobasidium magnum*, the mean generation time of daughter cells can exceed that of the mother cells by a factor of three and make synchronization of such organisms problematic (Flegel, 1978).

In *Candida utilis*, also, daughter cells at birth are smaller than the mother cells from which they are derived and might, therefore, be expected to take longer to reach division. Despite this, the phased method of cultivation gives prolonged synchrony (Müller and Dawson, 1968). This appears to be due to the fast rate of growth of the daughters (Fig. 8.4b), so that both mothers and daughters reach division size simultaneously (Thomas *et al.*, 1980a). A sizer control acting over initiation of the cell cycle (i.e. execution of "start") has been proposed for *Candida* growing in the phased cultivation system (Thomas *et al.*, 1980b). It is proposed that synchronization under these conditions is achieved by producing cells (both mother and daughter) that are below the critical size and then arresting them in G_1 until a fresh supply of medium is provided at the beginning of the phasing period. Buds start to appear at a constant time (2 h) after the start of the phasing period, suggesting that bud emergence is under the control of a timer.

In the dimorphic yeast *Candida albicans*, a threshold value for bud volume or mass is a prerequisite for nuclear division and migration during both synchronous mycelial formation (Soll *et al.*, 1978) and budding (Bedell *et al.*, 1980).

B. DURATION OF CELL CYCLE PHASES

A general conclusion that has been drawn (Mitchison, 1971; Prescott, 1976a,b) from examination of the lengths of G_1, S, G_2 and M phases in eukaryotic cells, including yeasts and other fungi (Martegani *et al.*, 1980), is that variation in G_1 accompanies changes in the overall duration of the cell cycle, whilst $(S + G_2 + M)$ remains relatively constant. Using three different media to vary the mass doubling time, Barford and Hall (1976) showed that the interval between initiation of DNA synthesis and division is almost constant (Table 8.1). Previously, von Meyenburg (1968) had shown that the unbudded portion of the cycle (equated with G_1; Williamson, 1965; Slater *et al.*, 1977; see, however, Rivin and Fangman, 1980a) selectively lengthened at slow growth rates. Hartwell and Unger (1977) made similar observations on cultures whose growth had been slowed by cycloheximide and other metabolic inhibitors. These workers also showed that the extensible portion of G_1 occurred between cell division and the point of sensitivity to mating factor.

Flow microfluorimetry of populations growing at different rates on different carbon sources (Slater *et al.*, 1977) and on different nitrogen sources (Johnston *et al.*, 1980) also revealed that G_1 variation is largely responsible for variation in cycle time, although small variations *do* occur in S and especially in G_2. Jagadish and Carter (1977) showed that for cells growing with generation times of less than 6·18 h, the

Table 8.1. *Estimated length of the cell cycle phases for* Saccharomyces cerevisiae

	Cell cycle phase lengths (min)				
Carbon source	G_1	S	G_2	M	G_1★
Glucose	5·35	45·47	22·29	15·67	3·64
Maltose	—	—	—	15·74	45·40
Galactose	29·89	45·96	29·35	15·58	64·02
Ethanol					
$\mu = 0{\cdot}07$	—	—	—	16·32	—
$\mu = 0{\cdot}096$	283·30[a]	47·81[a]	68·26[a]	16·32	19·31

[a]These values assume that the M phase length for cells grown on ethanol at $\mu = 0{\cdot}07\ h^{-1}$ and $\mu = 0{\cdot}096\ h^{-1}$ are equal.
μ = Specific growth rate (h^{-1}).
Lengths of the cell cycle phases were estimated using a method that is based on DNA labelling and nuclear staining of asynchronous cultures.
G_1 is the phase separating nuclear division from initiation of DNA synthesis. G_1★ is defined as that part of G_1 which separates nuclear division from cell division. This classification is introduced to categorize undivided cells that have entered G_1 following nuclear division. (Reproduced with permission from Barford and Hall, 1976).

cell cycle could be considered to comprise an expandable phase before "start" (the length of which is dependent on growth rate) and a constant phase from "start" to division, which occupies about 2 h. At longer generation times, the start-to-division period also increases somewhat (Table 8.2). Similar findings were obtained when the interval between initiation of DNA synthesis (defined by the execution point of the *cdc* 7 mutation) and cell division was studied (Carter and Jagadish, 1978b). Using hydroxyurea to determine the stage and time in the cycle when S-phase is completed, Jagadish and Carter (1978) have further shown that, when growth rate is varied by alteration of the dilution rate in chemostat culture, the expansion of cycle time is primarily a result of the increased time between division and the completion of S. Since the duration of S itself has been shown to be constant (Barford and Hall, 1976) it was concluded that the variability in cycle time was attributable soley to G_1. However, when growth rate was varied by altering the growth temperature, the cycle phases before *and* after completion of S-phase were extended. Although this result was not considered surprising by Jagadish and Carter (1978), it is

Table 8.2. *Effect of specific growth rate on the stage and time in the cell cycle of the α factor-sensitive step*

Specific growth rate (h^{-1})	Generation time (h)	Percentage increase in cell numbers in the presence of α factor	Stage in the cell cycle of the factor-sensitive step	Time before cell division of the factor-sensitive step
0·256	2·7	56·9	0·13	2·35
0·20	3·46	60·0	0·33	2·32
0·15	4·62	42·8	0·49	2·36
0·112	6·18	28·5	0·63	2·29
0·076	9·1	26·6	0·65	3·18

Haploid *Saccharomyces cerevisiae* cells of strain H185.3.4. carrying the *cdc* 28 mutation were grown at different steady-state specific growth rates in a glucose limited chemostat at 24°C. Samples at different specific growth rates were removed from the chemostat, and diluted in fresh medium at 24°C containing α factor. The α factor was purified and a concentration of α factor was chosen that produced greater than 95% unbudded cells and that inhibited budding for at least 8 h. Cell number was monitored at intervals until cell division in the presence of α factor ceased. The percentage of cells that divided in the presence of α factor was used to calculate the stage and time in the cycle of the α factor sensitive steps using the equation

$$X = 1 - \frac{\ln(N/N_0)}{0{\cdot}693}$$

where X is the stage in the cell cycle of the execution point (see Chapter 7), N_0 is the cell number at the time of temperature shift and N is the final cell number at the restrictive temperature. (Reproduced with permission from Jagadish and Carter, 1977.)

interesting that the "constant" $S + G_2 + M$ is not temperature-compensated in marked contrast to the clocks that underlie circadian rhythms (Chapter 1).

In direct contrast to these results, Rivin and Fangman (1980a) have reported recently that when growth rate is limited by the source of nitrogen, the length of S-phase varies but the portion of the total cycle

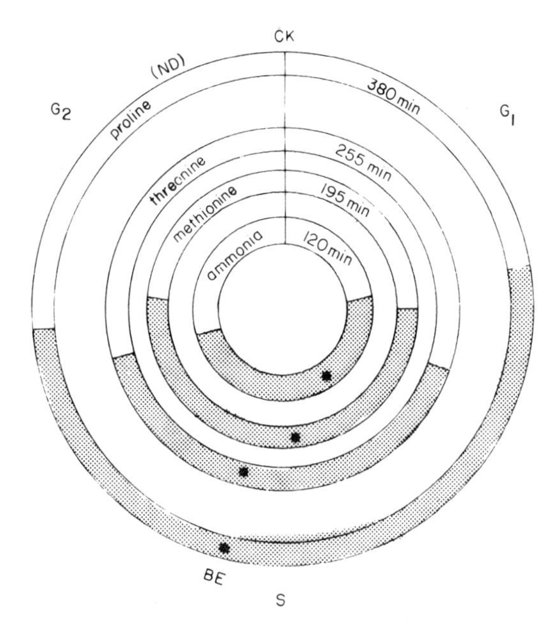

Fig. 8.5 Cell cycle expansion for 4 growth rates in *Saccharomyces cerevisiae*. Cytokinesis (CK) was used to align the different cycles. Bud emergence (BE) is indicated by * in each cycle. Nuclear division (ND) occurs at ~7% of the cycle before cytokinesis at all growth conditions. The diagram indicates the doubling time for each of 4 different nitrogen sources. (Reproduced with permission from Rivin and Fangman, 1980a.)

time that it occupies remains close to 50% (Fig. 8.5). An autoradiographic analysis has shown that the longer S-phases can be accounted for by a reduction in replication fork rates (Rivin and Fangman, 1980b). As suggested by Rivin and Fangman, the contrast between their results and previous ones may be due to strain differences; nevertheless, further work is needed to confirm and extend these important results.

C. NUTRITIONAL CONTROL OF CELL DIVISION

Starvation of *S. cerevisiae* for any one of a number of nutrients results in arrest in G_1 (Scopes and Williamson, 1962; Pringle, cited by Unger and Hartwell, 1976). Two events are required for such arrest: (a) passage through G_1 must be prevented by deficiency of a metabolic intermediate, and (b) sufficient protein synthetic capacity must be available for cells to finish the cycles that are in progress.

Unger and Hartwell (1976) hypothesized that the decrease in an unidentified "signal" compound could act as a sensor of the loss of sulphate in the medium. Thus, blocking sulphate metabolism at a point subsequent to the synthesis of the "signal" would not cause G_1 arrest. Methionine auxotrophs (Unger and Hartwell, 1976; Unger, 1977) were used to show that the signal for sulphate starvation is generated near the end of the sulphate assimilation pathway, at or beyond the formation of methionyl-tRNA.

G_1 arrest of nitrogen-starved cells can be prevented by addition of arginine (Cooper *et al.*, 1979), perhaps by starving them of other amino acids and thus of proteins that are required to complete the cycle or act as "signal" compounds.

An analysis of the effects of limiting concentrations of adenine, methionine or histidine on the rate of cycle initiation in appropriate auxotrophs has led Shilo, B. *et al.* (1978) to suggest that a common intermediate requiring these substances is involved in cell cycle regulation. A similar reduction in the rate of cycle initiation was achieved with concentrations of cycloheximide that slowed, but did not totally inhibit, protein synthesis. These results add support to the idea that a protein may be a key determinant whose concentration is "probed" by the cell before initiating a new cell cycle. This requirement may be additional to attaining a critical cell size.

Jain (1972) considered the G_1 phase to be a time for synthesis of polysaccharides, which are catabolized during subsequent stages of the cell cycle. This idea is reminiscent of the "energy reservoir" concept of Swann (1957). However, Jain's correlation of energy metabolism with cell cycle stages was made by monitoring growth of a batch culture in which G_1 cells were predominant at the beginning of the acceleration phase and in stationary phase. Thus, superimposed on any changes in energy metabolism that occur during the cell cycle are modulations imposed by nutrient concentration, gas tensions and cell density. It was reported that preliminary experiments with synchronous cultures confirmed a decreased utilization of glucose in S-phase, but clearly further studies are needed to establish the role of energy generation in control of the yeast cell cycle (see Chapter 6).

A role for labile proteins in G_1 traverse has been suggested for mammalian cells by Schneiderman *et al.* (1971) and Highfield and Dewey (1972). Shilo *et al.* (1979) have investigated the possible involvement of similar proteins in *S. cerevisiae*. Cells carrying the *cdc 25-2* mutation were released from "start" and pulsed with cycloheximide. Delays in budding resulted that were much longer than the pulse lengths and were not due to the decreased overall rate of protein synthesis. It was suggested that there is a requirement at "start" for a labile protein with a half-life of about 6 min. Thus, during the inhibitor pulse, the inability of cells to synthesize an essential protein (which is simultaneously being degraded) results in regression within G_1. The requirement for such a protein may allow the regulation of cycle initiation to adapt very rapidly to changes in the rate of protein synthesis and to modify or stop proliferation.

D. CONTROL IN G_1 AT THE TRANSCRIPTIONAL LEVEL

Certain compounds that decrease the rate of rRNA production, without accompanying effects on protein synthesis, cause cells to arrest in G_1 at "start". These include: nalidixic acid (Singer and Johnson, 1979); the zinc chelating agent, *o*-phenanthroline (Johnston and Singer, 1978); the methionine analogue, L-ethionine (Singer *et al.*, 1978); the phenylalanine analogue, β-2-DL-thienylalanine (Bedard *et al.*, 1980). The common macromolecular alterations accompanying cell cycle arrest by these agents are decreases in the rate of production of the 35S ribosomal precursor RNA (rpre RNA) as well as in the maturation of rpre RNA to mature RNA (Johnston and Singer, 1978; Singer *et al.*, 1978; Johnston and Singer, 1980).

Ornithine decarboxylase, although implicated in the regulation of RNA polymerase I activity, is not directly involved in cell cycle regulation (Kay *et al.*, 1980).

E. CONTROL BY MATING PHEROMONES

Conjungation between *a* and α mating types in *S. cerevisiae* is triggered by diffusible factors, specific to each mating type, which serve to synchronize cells prior to fusion (Hartwell, 1973b; Manney and Meade 1977). The substance secreted by cells of mating type α is called α factor (Duntze *et al.*, 1970) and arrests cells of type *a* at a defined point in G_1 (Chapter 7). The "*α* factor pheromone" is in fact a family of four linear oligopeptides; the demonstrated biological activity of the chemically synthesized molecules is an unequivocal demonstration of their mating function (for references, see Ciejek and Thorner, 1979). Likewise, cells of mating type *a* secrete into the medium *a* factor, which has now been purified and partially characterized (Betz and Duntze, 1979).

Throm and Duntze (1970) showed that the effects of *α* factor are reversible; that is, *a* cells eventually recover from growth arrest by the pheromone. Subsequent studies by Hicks and Herskowitz (1976) and Chan (1977) suggested that this is due to destruction of *α* factor activity by *a* cells. Indeed, chemical alteration of *α* factor has been demonstrated by Maness and Edelman (1978) and there is now good evidence (Ciejek and Thorner, 1979; Finkelstein and Strausberg, 1979) that the inactivation is due to proteolytic cleavage. Proteolysis is not, however, an obligatory event in the cell cycle arrest mediated by *α* factor (Finkelstein and Strausberg, 1979).

The mode of action of *α* factor is largely unknown, but it is expected that progressing research in this area will contribute to our understanding of this and other controls operating in G_1. Making use of the fact that the yeast cell wall becomes morphologically altered after interaction with *α* factor (Lipke *et al.*, 1976), Udden and Finkelstein (1978) have shown that saturation of a single site by *α* factor is sufficient to cause cell cycle arrest of a mating-type cells. This excludes the possibility that the *α* factor stoichiometrically inhibits an enzyme, a species of tRNA or any other substance that is present at more than copy per cell.

The recent isolation by Hartwell (1980) of temperature-sensitive mutants insensitive to developmental arrest by *α* factor opens the problem to genetic analysis. Extracts from one of Hartwell's mutants (*ste*5) exhibit adenylate cyclase activity that is resistant to *α*-factor at 34°C but fully sensitive at 23°C, regardless of the growth temperature (Liao and Thorner, 1980). Further evidence for implicating adenylate cyclase in the mode of action of *α*-factor comes from the observation that adding cAMP to a culture shortens the period of pheromone-induced G_1 arrest.

F. SUMMARY

Several authors have drawn analogies between the control of the cell cycle in bacteria and in *S. cerevisiae* (e.g. Johnston *et al.*, 1977; Adams,

1977; Carter and Jagadish, 1978b; Carter *et al.*, 1978). As in bacteria, faster-growing cells of this organism are larger than slow-growing cells (Mor and Fiechter, 1968; McMurrough and Rose, 1969; von Meyenburg, 1968). Furthermore, there is much evidence for the constancy of the $(S + G_2 + M)$ period, although this point is still debatable. These two features of the yeast cell cycle have their counterparts in bacteria and make such an analogy attractive. The data of Tyson, C. B. *et al.* (1979) on mean cell size and the proportion of budded cells during growth under various conditions, together with the response of the growth rate to a nutritional "shift up" (Carter *et al.*, 1978) are also consistent with a model in which initiation of the cell cycle occurs at "start" and is followed by a constant time period before attainment of a critical cell size. This critical size is dependent on growth rate. The duration of the budded phase $(S + G_2 + M)$ is constant and is given by Tyson, C. B. *et al.* (1979) in the empircal equation: budded phase $(B) = 0{\cdot}5\tau + 27{\cdot}0$ (all in minutes) where τ is the generation time. This compares with a value of $B = 0{\cdot}17\,\tau + 87{\cdot}4$ min obtained by Hartwell and Unger (1977).

Whether such a model can adequately account for the control of the cell cycle under all conditions remains to be established. A hierarchy of control mechanisms may be invoked to encompass the other control elements (e.g. nutritional status, transcription) presumed to play essential roles.

III. The Control of Growth and Division in Fission Yeast

In the fission yeast, *Schiz. pombe*, cell length at division is homeostatically controlled. Both James *et al.* (1975) and Fantes (1977) have shown that cells of size greater than the mean at birth have, on average, shorter cell cycle times. The main points to emerge from Fantes' careful study of *Schiz. pombe*, observed through several division cycles by time-lapse photomicrography, are as follows:

(1) There is no tendency for a cell that is abnormally long or short to produce daughter cells that divide at either larger or smaller volume than the average. Thus, deviations in length at division must be mainly compensated within one cell cycle. With abnormally long cells (mutant *cdc* 2-33, previously grown for 3 h at 35°C and then shifted to 25°C), the mean length decreased at successive divisions to approximate the steady state value after three divisions.

(2) In both wild type and mutant, there is a negative correlation

between the amount of growth during a cell cycle and cell length at birth. In other words, cells that are abnormally small at birth undergo greater than average growth during their cycle.

(3) There is a positive correlation between extension rate and size at cell division; large cells grow more rapidly than small ones. The effect of this is to increase, rather than decrease, the variation in cell length at division, thus working against homeostasis.

(4) There is a strong positive correlation between amount of growth and cycle time. From (2) and (4) it can be seen that:

(5) There is also an inverse relationship between birth size and cycle time, i.e. the larger the cell, the shorter is the cycle time. This conclusion was experimentally confirmed.

These relationships suggest a mechanism by which constancy of cycle time can be achieved. A cell that undergoes an abnormally long cycle will divide at a larger size than usual and will give rise to large daughters. Each of these (from (5)) will have shorter than average cycle times. Thus, a long interdivision time in one cycle results in the next cycle being shorter than average. This too was experimentally confirmed (Fig. 8.6).

Some aspect of Fantes' (1977) data have been subsequently confirmed and extended in a series of papers from Miyata's laboratory. Cell cycle length was shown to be negatively correlated with size (Miyata *et al.*, 1978a). Oversized cells (> 20 μm in length, produced in these experiments by a pulse of hydroxyurea) produced siblings with a constant, short cycle time of 70 min. This is substantially shorter than the minimum cycle time of 2·3 h observed by Fantes (1977) for cells longer than about 10·5 μm growing in either minimal or complete medium. Increasing periods of exposure to hydroxyurea, with consequent increase in septation delay, required progressively more cycles to restore normal cell lengths (Miyata *et al.*, 1978b).

Restoration of normal cell length in this organism may be demonstrated following pulses of inhibitors of (a) DNA synthesis e.g. 2′-deoxyadenosine (Mitchison and Creanor, 1971; Poole and Salmon, 1978) or (b) mitochondrial protein synthesis e.g. chloramphenicol (Quinton and Poole, 1977). In all these cases, the elongated cells that accumulate during the inhibitor treatment undergo abnormally-short division cycles on removal of the inhibitor (e.g. Fig. 8.7) in an attempt—in a teleological sense—to restore normal cell length.

Homeostatic controls over cell size at division are not restricted to yeast. A negative correlation between birth weight and generation time may be seen in *Amoeba* (Prescott, 1956) and *Tetrahymena* (Lövlie, 1963). Cell division is under homeostatic control in *Physarum poly-*

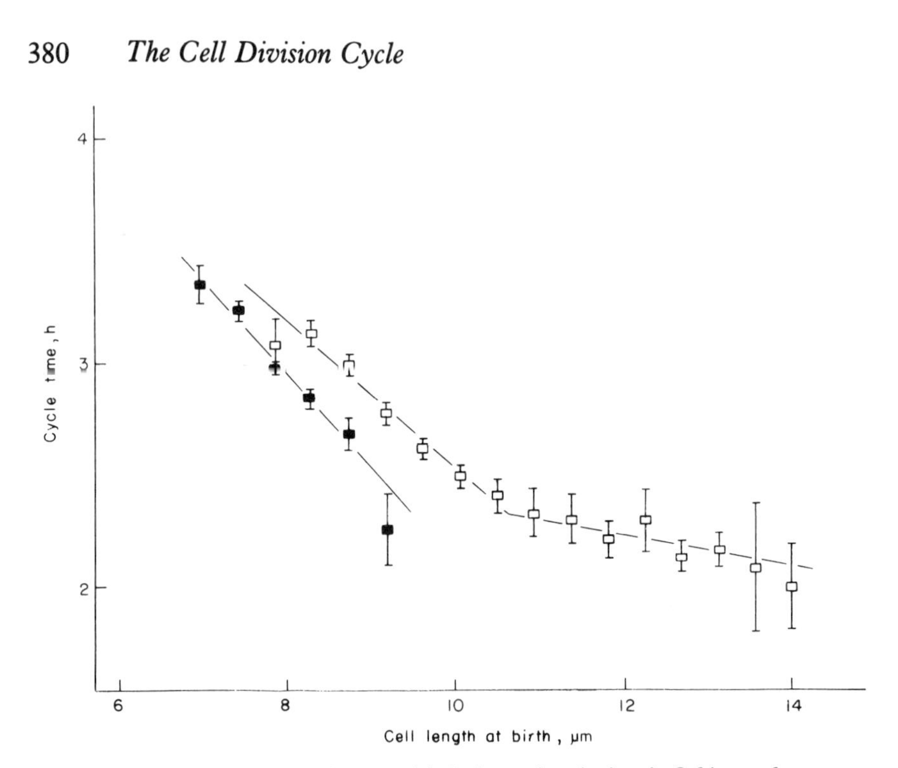

Fig. 8.6 The inverse relation between birth size and cycle time in *Schizosaccharomyces pombe*. Means and standard error bars are shown for each size class. The lines shown are calculated regression lines. The experimental points were obtained for strain 972 (■) in steady state at 25°C or for *cdc* 2–33 (□) at 25°C after a period of 3 h at 35°C. For cells shorter than 10·5 μm at birth, which includes the whole of the normal length range, there is an inverse relationship between cycle time and length at birth. For cells larger than this, the cycle time is about 2·3 h and varies little with cell size. (Reproduced with permission from Fantes, 1977.)

cephalum amoebae and deviations are corrected within a single cycle (Sudbery and Grant, 1975; Sachsenmaier *et al.*, 1970). Long delays to mitosis are followed by a minimum cycle, equal to about $\frac{3}{4}$ of the normal cycle on rich medium, the length of the "minimum cycle" being independent of growth rate (Sudbery and Grant, 1975).

As noted in Chapter 7, genetic analysis of the cell cycle of *Schiz. pombe* and, in particular, the isolation and characterization of the *wee* mutants has provided a new approach to the study of the cell cycle in this organism.

Mutant *wee* 1–50 divides at about the same cell size as the wild type at 25°C, but at half that size at 35°C (Nurse, 1975). A shift from 25 to 35°C accelerates G_2 cells into mitosis. This is evidence for control over mitosis; in other words, during the last part of G2, only the *wee* 1^+ gene product prevents cells from initiating mitosis. Confirmation of this suggestion was provided by the experiments of Fantes and Nurse (1977), which showed that transferring the wild-type from a poor medium (in which cells were small) to one able to support a faster growth rate ("nutritional shift-up") caused a rapid inhibition of nuclear division.

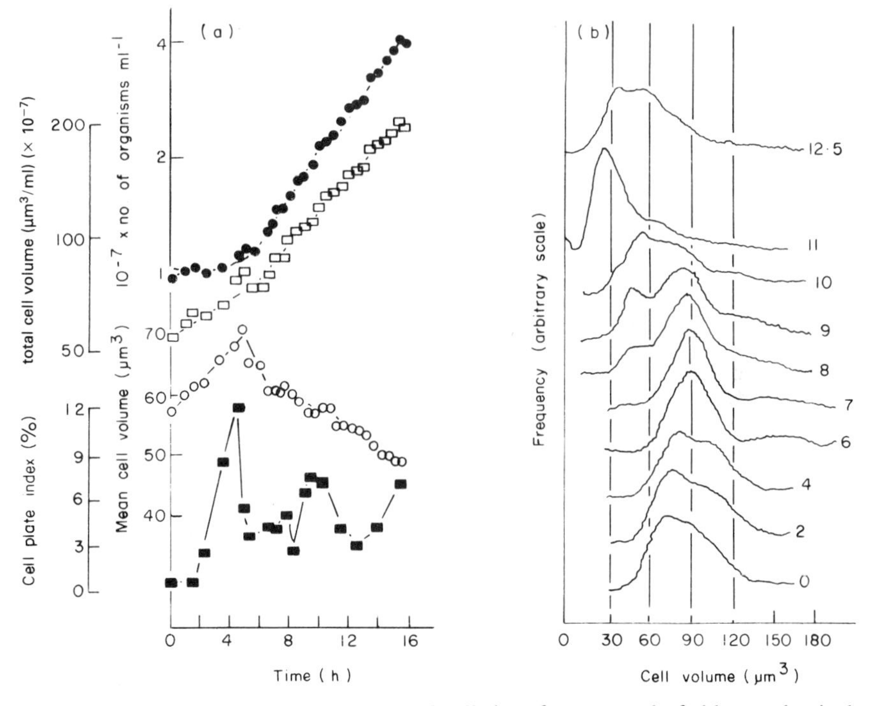

Fig. 8.7 Recovery of growth and normal cell size after removal of chloramphenicol from an inhibited culture of *Schizosaccharomyces pombe*. Chloramphenicol (2 mg ml^{-1}) was added to an exponentially growing culture containing glycerol (1%, w/v) as carbon source; 20 h later the cells were harvested by centrifugation, washed and resuspended in fresh medium that lacked the inhibitor. (a) shows the resulting changes in cell numbers; (●) mean cell volume (○) and total cell volume (□) i.e. [no. of cells ml^{-1}] × [mean cell volume (μm^3)]. The cell plate index (■) is the percentage of cells showing cell plates. In (b) the accompanying changes in cell volume distributions, determined with a Coulter Counter and Channelyzer, are shown. All distributions are normalized with respect to areas under the curves; the accompanying numbers are sampling times (h). (Reproduced with permission from Quinton and Poole, 1977.)

In addition to this control over mitosis, further analysis of the *wee* mutants (Nurse, 1975; Nurse and Thuriaux, 1977; Fantes and Nurse, 1978) and of the wild-type (Nurse and Thuriaux, 1977) has revealed an extra sizer control, this time over the initiation of DNA synthesis. In *wee* mutant strains, and in the wild-type grown under conditions in which the cells are small, DNA replication takes place in cells of the same size, suggesting that there is a minimal size below which the cell cannot initiate DNA replication. It is argued that this control is cryptic in wild-type cells, because the cell size at S-phase (which is directly determined by control over mitosis in the previous cycle), is always above the minimal size that a cell must attain to initiate DNA synthesis. This may explain why the *wee* mutants, in contrast to the wild-type, have a significant G_1 period (occupying about 0·3 of the cycle) during which they grow after cell division to the critical size for DNA synthesis to proceed. A further control that may be imposed on the initiation of S is that it is dependent upon completion of the previous mitosis (Nurse *et al.*, 1976), coupled with a short minimum time spent in G_1 (Fantes and Nurse, 1978; Nasmyth *et al.*, 1979). Both controls are operative in both large and small cells, but only one of them is important in determining the timing of S-phase in a particular cell. The "choice" is dictated by size (or, more correctly, mass) of a cell at division (Nasmyth *et al.*, 1979). In cells undergoing division at a small mass, the timing of S is determined by the necessity to achieve a critical cell mass. In cells undergoing division at a large mass, the timing of S is determined by the requirement for the completion of both the previous mitosis and a short G_1. Fig. 8.8 shows that the transition from one control mechanism to the other occurs when the protein content of the cell at division is about 15 pg.

Alberghina *et al.* (1979) have presented a quantitative description of such a "two-threshold" model of division control. Calculations of the average DNA and protein contents in cells during exponential growth and at nuclear division are consistent with the model.

Fantes and Nurse (1977) have suggested that the existence of two cell size controls may be adequate to explain the fact that many cell types (*Schiz. pombe*, as well as budding yeast, *Physarum* and mammalian cells) arrest in G_1 on entering the stationary phase. It is envisaged that one or more nutrients become growth-limiting and that the critical cell size required for nuclear division falls. Cells at birth become progressively smaller and eventually reach a size where the control over DNA synthesis becomes operative. The cells thus accumulate prior to DNA synthesis, i.e. in G_1.

Further complexity in cell cycle controls is suggested by the finding

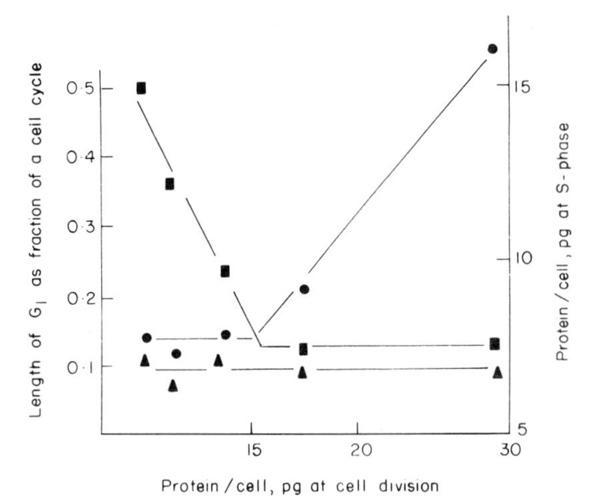

Fig. 8.8 The relationship between cell mass at division and the duration of G1 in *Schizosaccharomyces pombe*. ■, length of G1; ▲, length of S-phase; ●, protein content per cell at S-phase. (Reproduced with permission from Nasmyth *et al.*, 1979.)

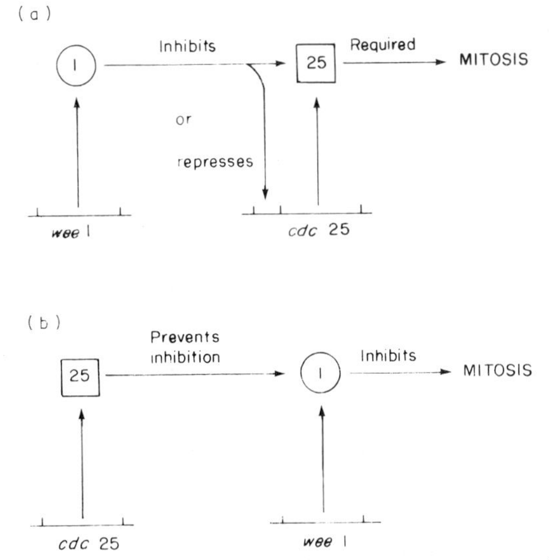

Fig. 8.9 Models to explain the interaction between *wee* genes and *cdc* 25. (a) Functional *cdc* 25 product is required for mitosis. Either the synthesis or activity of this product is inhibited by functional *wee*1 product. (b) Functional *wee*1 product inhibits mitosis. The inhibition is prevented by functional *cdc* 25 product. ① is the *wee*1 gene product, and [25] the *cdc* 25 product. (Reproduced with permission from Fantes, 1977.)

of a possible control in G_1 phase that regulates entry into S (Nasmyth, 1979b). Under poor nutrient conditions (e.g. nitrogen-limited growth in a chemostat) the cycle time is extended. The length of $(S + G_2)$ remains constant, despite great variation in growth rate; variation in G_1 alone "takes up the slack", as has been documented in many other eukaryotic cells (Prescott, 1976a,b; also see Section II.2B).

Recently, Fantes (1979) has suggested how two cell cycle genes (Chapter 7) may interact in the control of cell division in this organism. The mitotic defect caused by the *cdc* 25 mutation can be almost completely suppresed by *wee*-1. The *wee*-1^+ product is considered to be an inhibitor of mitosis (Nurse, 1977) whilst the *cdc* 25^+ gene is apparently required for mitosis. The possible epistatic interactions between these two gene products in the regulation of mitosis are shown in Fig. 8.9.

IV. The Control of Growth and Division in Prokaryotes

A. INTRODUCTION AND PERSPECTIVE

Systematic studies of the mechanism of growth and division of prokaryotes have a long history, but despite extensive experimental work and numerous models, the precise mechanism remains obscure. The aim of this section is to give an illustrative rather than comprehensive account of these endeavours, drawing largely on work with *E. coli*. Significant advances have been made with other bacteria, however, particularly with *Caulobacter* (Kurn and Shapiro, 1975; Shapiro, 1976; Osley and Newton, 1978, 1980) and with *Bacillus* and *Streptococcus* (for reviews, see Daneo-Moore and Shockman, 1977; Sargent, 1979).

At first sight, the division cycles of prokaroytes and of eukaroytes appear to be fundamentally different, the most marked distinction being the occurrence of continuous DNA replication in the former under certain growth conditions. Recently, however, Cooper (1979) has pointed out that the S period of eukaryotes and the C period of prokaryotes are analagous (times of DNA synthesis) as are the G_2 and D periods (when defined as the time between the termination of DNA synthesis and mitosis or cell division). Furthermore these periods tend to be invariant, at least in comparison with G_1 in eukaryotes and its equivalent in slowly-growing bacteria.

Most published models for control of bacterial growth and division are deterministic. Thus, Koch (1980) favours a growth-controlled model for *E. coli*, although for another bacterium, *Staphylococcus albus*, α and β plots are in accord with the transition probability model (Shields, 1978; Chapter 1).

The pioneering work from Maaløe's laboratory in the 1960s (Maaløe and Kjeldgaard, 1966) produced two important concepts regarding the regulation of macromolecular synthesis in bacterial cultures:

(1) The constancy of efficiency of ribosomes in protein synthesis, and the dependence of synthesis on the number of ribosomes present at any growth rate;

(2) The view that if the extension rate of a macromolecule is constant, then alterations in growth rate must affect the frequency of initiation of synthesis of that macromolecular species. The most clearly-documented case of the latter theorem is the multisite initiation of DNA replication in fast-growing bacteria (Cooper and Helmstetter, 1968). As described in Chapter 3, Part 2, arguments continue over the lengths of the C and D periods in different bacteria and, in particular, over how the model must be adapted to explain the relationship between DNA replication and the cell cycle in slowly-growing cultures. These uncertainties, however, should not detract from the usefulness of the model in formulating models for control of growth and division, all of which implicate some coupling of the chromosome replication cycle to division.

The Cooper–Helmstetter model assumes that the prokaryotic cell proceeds step-by-step through the cycle. However, as Koch (1977) points out, the facts are also consistent with a radically different mode of operation, in which cell division does not depend on the previous completion of chromosome replication but rather on the generation of signals which may independently trigger chromosomal replication and cell division (see Section IV.C3).

The systematic studies of Schaechter *et al.* (1958) on *Salmonella typhimurium* and Helmstetter *et al.* (1968) on *E. coli* showed that average cell mass increased as an exponential function of the growth rate. This observation rules out a very simple and long-standing suggestion regarding division control, namely that a cell divides when it reaches a certain size. However, with modification, this idea can provide an explanation of the growth rate dependence of cell size (see Section II).

This is a topic that has been reviewed many times. Among the most useful and recent of reviews are those by Daneo-Moore and Shockman (1977), Koch (1977), Davern (1979), Donachie (1979) and Helmstetter *et al.* (1979).

B. CONTROL OF INITIATION OF CHROMOSOME REPLICATION

1. *Initiation at Critical Cell Mass*

The elements of the Cooper–Helmstetter model, taken with the established relationship between growth rate and cell mass, were combined by Donachie (1968) in a model where chromosome initiations occur at the constant cell mass or multiples of that mass (Fig. 8.10). Early experimental results were entirely consistent with this view (Donachie

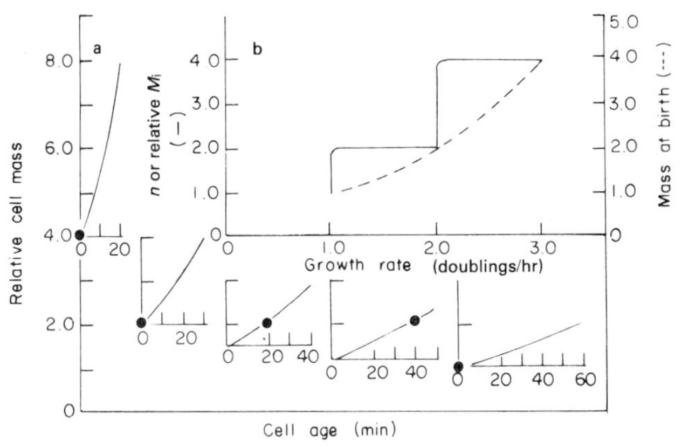

Fig. 8.10 Donachie's (1968) model for the relationship of cellular mass and time of initiation of chromosome replication. (a) Individual cells from cultures at each of five representative doubling times of 20, 30, 40, 50 and 60 min are shown increasing exponentially in mass during their respective cell cycles. The relative mass at birth (time 0) is taken from the plot in (b), where the relative mass at birth is assumed to increase with growth rate in a manner proportional to M, the average mass cell of the entire population. In (a) the solid circles indicate the cell age in minutes at chromosome initiation as derived from the Cooper–Helmstetter model. (b) Both the relative mass at chromosome initiation (M_i) and the number of chromosome origins per cell per cycle (n) coincide. Therefore the ratio of M_i to chromosome origins per cell is constant at all growth rates. (Reproduced with permission from Daneo-Moore and Shockman, 1977.)

et al., 1968). A similar model was also proposed and refined by Pritchard (1968) and Pritchard *et al.* (1969). Donachie's (1968) model is "positive" in the sense that it assumes that initiation takes place when the ratio of cell mass to the number of initiation sites reaches a critical value (Maaløe and Kjeldgaard, 1966; Bremer *et al.*, 1979), so that the time of initiation within the cell cycle varies with growth rate (Fig. 8.10). The alternative, negative models postulate the intermittent production of an inhibitor by the transient expression of a gene early in the DNA replication cycle. The dilution of this inhibitor to an ineffective level by further growth (Pritchard *et al.*, 1969) or, alternatively, the continuous synthesis of an anti-inhibitor (Rosenberg *et al.*, 1969) would result in initiation of replication. These alternative postulates have not been experimentally resolved (Fantes *et al.*, 1975). Both have been incorporated by Shuler *et al.* (1979), Nishimura and Bailey (1980) and Alberghina and Mariana (1980) into mathematical models for growth of a single cell.

In Donachie's model, an exponential increase in cell volume and thus mass is assumed during the division cycle (Abbo and Pardee, 1960; Cummings, 1965; Koppes *et al.*, 1978a). However, there is some convincing evidence for linear growth in volume during the cycle (e.g. Kubitschek, 1968a,b; Meyer *et al.*, 1979; Bugeja *et al.*, 1980). Even so, if cell density fluctuates during the cycle, as has been demonstrated for *E. coli* (Poole, 1977a,b) under certain growth conditions (Poole and Pickett, 1978), the increase in cell mass may still be exponential (Fig. 8.11).

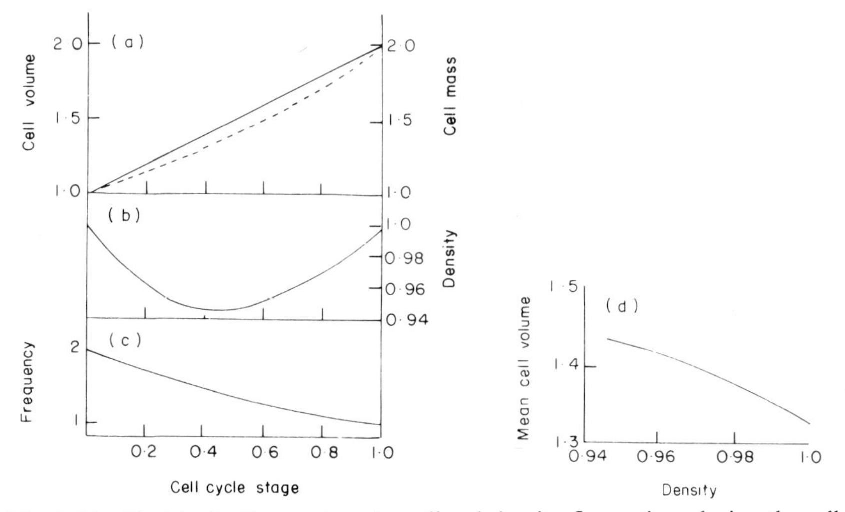

Fig. 8.11 Model of cell growth and predicted density fluctuations during the cell cycle of *Escherichia coli*. A linear increase in cell volume (——) and an exponential increase in cell mass (– – –) (a) leads to a fluctuation in density (b) during the cell cycle. From the relative number of cells at each stage of the cell cycle (c), the mean volume can be calculated as a function of density (d). Cell volume, cell mass and density are in arbitrary units. The model is derived from, and consistent with, experimental data on density fluctuations in the cycle. Cells from an exponential culture were centrifuged to equilibrium in a density gradient of Ludox (colloidal silica) and polyvinylpyrrolidone. (Reproduced with permission from Poole, 1977b.)

Daneo-Moore and Shockman (1977) have used known values of (C + D) for *Streptococcus faecalis* to calculate that mean cell mass increases with growth rate far less than would be anticipated from Donachie's (1968) model. A similar analysis of Sargent's (1975a,b,c) data for another Gram-positive organism, *Bacillus subtilis*, also failed to validate the application of this model. However, reasonably good fits were found using published data from various laboratories for *Aerobacter aerogenes*, *Salmonella typhimurium*, *E. coli* and the blue-green alga *Anacystis nidulans*. It seems probable, therefore, that the exponential increase in cell mass with growth rate observed by Schaechter *et al.* (1958) is restricted to some prokaryotes only, thus invalidating the general application of Donachie's (1968) model.

2. *Other Models*

As early as 1963, Jacob *et al.* proposed that the bacterial cytoplasmic membrane provided the enzymes and regulatory elements needed for initiation, replication and termination of DNA synthesis, as well as acting to segregate the newly-replicated chromosomes. Helmstetter (1974a,b) has proposed that the cell envelope controls the *timing* of initiation of replication by envelope elongation between the envelope-bound origins of sister chromosomes. The roles of surface growth in these phenomena have received particular attention in the review of Daneo-Moore and Shockman (1977). Since that review appeared, Pierucci (1978) has shown that new sites of envelope synthesis are inaugurated coincident with initiation of DNA replication. This model is similar to that of Shockman *et al.* (1974) for *Streptococcus faecium* and Case and Marr's (1976) model for *Salmonella typhimurium*. Supporting evidence for the involvement of membranes also comes from the identification of membrane-associated proteins such as protein D (Gudas *et al.*, 1976), which appears to be synthesized between the end of one round of replication and the beginning of the next (see Section IV.C2b).

The involvement of surface growth in chromosome replication and particularly in chromosome segregation requires further study. Despite extensive experimental work, the identity of the surface structure in the multilayered envelope that separates the two chromosome strands is unknown. The evidence for conservation of areas of cytoplasmic membrane is equivocal (Chapter 5). The cell wall, however, especially the peptidoglycan, is an attractive candidate and, in view of its close proximity to the membrane in Gram-positive organisms, could provide a vehicle for DNA segregation. Likewise, in Gram-negative organisms, the adhesion sites (Bayer, 1975) seen between inner and outer membranes and the peptidoglycan layer could provide anchor points for DNA to the outermost envelope components. These possibilities and the contributions of turnover and remodelling of the wall to chromosome replication are discussed in greater depth by Daneo-Moore and Shockman (1977).

C. CONTROL OF CELL DIVISION

1. *The Role of Chromosome Replication*

(a) *Chromosome replication as the timer for division.* The existence of two apparent constants, C and D, in the Cooper–Helmstetter model suggests a means by which the timing of division could be controlled, namely that division could occur (C + D) min after initiation of chromosome replication or D min after its termination.

There are several observations that lend support to this idea. First, it is at the time that chromosome replication is completed that the cell becomes committed to division. At about this time, approximately 20 min before division, cell length is invariant over a wide range of growth rates (Donachie *et al.*, 1976; Grover *et al.*, 1977) and, a short while later, nuclear separation occurs. Secondly, inhibition of DNA synthesis in *E. coli* blocks cell division leading to the formation of filamentous cells (Clark, 1968; Helmstetter and Pierucci, 1968). However, in *Bacillus subtilis*, division continues in spite of inhibition of DNA synthesis (Donachie *et al.*, 1971). The evidence (Donachie, 1979) that chromsome replication *per se* is *not* directly responsible for the timing of division is more persuasive:

(i) Certain *E. coli* mutants are capable of division in the absence of DNA replication, although only one of the sibling cells contains DNA (see also Tang and Helmstetter, 1980).

(ii) A mutation (*dna* A_{ts}) that allows completion, but not initiation, of rounds of chromosome replication permits cells to divide. Significantly, however, if rounds of replication which have been initiated are prevented from terminating (by judicious use of temperature-sensitive mutations), cell division *is* blocked and filamentous cells are produced (Hirota *et al.*, 1968a,b).

(iii) The D period can be reduced to as little as 7 min if termination of chromosome replication is delayed. Under these conditions cells behave as though they had initiated at least some of the processes leading to cell division *before* termination of chromosome replication had occurred.

(iv) Premature termination of chromosome replication by inhibition of protein synthesis does not lead to premature septation.

It seems unlikely, therefore, that chromosome replication is responsible for the *timing* of division, even though it is normally a prerequisite for division. However, in a recent model, based on the correlations in slowly-growing *E. coli* between cell size, age and the timing of initiation of chromosome replication, Koppes *et al.* (1980) suggest that chromosome replication is the pacemaker for division. It is proposed that cells initiate chromosome replication at a particular cell size, which is independent of size at birth. Two processes are then started, and proceed at a rate that increases with cell size: (1) preparation for the next initiation of chromosome replication, and (2) preparation for division, which will occur after a period equal to (C + D).

(b) *Blockage of division by initiation of replication.* The suggestion that initiation of chromosome replication is a block to subsequent cell division in *B. subtilis* was first made by Gross *et al.* (1968). The experiments of Hirota *et al.* (1968a,b) with *E. coli* also suggested that if initiation of replication does not occur, cell division can occur in the absence of DNA synthesis.

Donachie (1974) has suggested that this idea may also provide an explanation for the curious division-blocking effects of introducing into *E. coli* exogenous, u.v.-damaged DNA. It is proposed in the "blocked replicon" model that initiation, but not completion, of replication can occur in these plasmids and that the cell is unable to distinguish between initiation of replication of the plasmid and that of its own chromosome. The findings of Scott (1970), working with a temperature-sensitive mutant of plasmid P, are consistent with this hypothesis. Cells which carry the plasmid are able to grow but not divide at the restrictive temperature; presumably the mutation in P allows initiation but not completion of replication under the restrictive condition.

2. *The Requirement for Protein Synthesis*

The termination of chromosome replication is normally a necessary, but not sufficient, condition for cell division. The early experiments of Pierucci and Helmstetter (1969) strongly suggested that a constant period of protein synthesis is also required before division can take place.

(a) *"Protein X"*. The existence of a "division protein", protein X, in

E. coli was suggested by the experiments of Smith and Pardee (1970). These experiments and the remarkable analogy between the results and those obtained with heat-shocked *Tetrahymena* (see Chapter 1) are discussed by Mitchison (1971). In view of later results (Wu and Pardee, 1973), the analogy seems even closer, since heat-shocked *E. coli* undergo a shape change. Pardee (1974b) suggested that protein X, a membrane protein that increases in amount when division is inhibited, may normally form a "girdle" that must be built up during the cycle, constricting the cell at division.

It is now clear that protein X is involved in both DNA repair (Gudas, 1976; Gudas and Pardee, 1976) and cell septation (Satta and Pardee, 1978). The degree to which septation is inhibited following damage to DNA correlates well with the amount of protein X that is formed (Table 8.3), suggesting that protein X is produced as a consequence of DNA damage and is an inhibitor of cell divison (see, however, Huisman *et al.*, 1980a). Such a connnection would ensure that daughter cells are produced by division only if they are able to receive a complete copy of the genome.

As described in Chapter 7, protein X has now been identified (Gudas and Mount, 1977; McEntee, 1977) as the *rec A* protein, whose synthesis is one of several manifestations of the SOS response (Radman, 1974) to treatments that impede the normal progression of the replication fork. Expression of the SOS response depends on the presence of functional *rec* A^+ and *lex* A^+ genes. The *rec* A^+ gene product is generally considered to be the inducer of the entire SOS system. The inhibition of septation associated with the SOS response has been explained by George *et al.* (1975) as the result of induction of a Repair Associated Division Inhibitor (RADI) (see Chapter 7). Recent data in support of the RADI hypothesis comes from Huisman *et al.* (1980b).

(b) *"Protein D"*. Gudas *et al.* (1976) have identified on SDS gels "protein D" of mol. wt 80 000, that is synthesized and incorporated into the outer membrane of *E. coli* B/r for a brief fraction of the cycle. There appears to be a close correlation between the timing of protein D synthesis and the initiation of DNA synthesis. On the basis of the protein's binding to DNA *in vitro* and the effect of antibiotics on its synthesis, it was proposed that the protein links murein synthesis, protein synthesis and DNA initiation and also acts as an attachment site for DNA to the cell envelope.

A similar membrane protein (mol. wt 76 000), whose synthesis is apparently restricted to a brief period of the cycle, has been described by Churchward and Holland (1976a). A membrane protein of identical

Table 8.3. *Summary of the relation between induction of protein X and inhibition of cell division in* Escherichia coli *and other bacteria*

Determination	Protein X induced	Cell division inhibited	DNA degraded	Reference
A. Chemical and physical treatment of wild-type strains				
Nalidixic acid	Yes	Yes	Yes	Inouye and Pardee (1970)
Mitomycin	Yes	Yes	Yes	Inouye and Pardee (1970)
Bleomycin	Yes	Yes	Yes	Gudas and Pardee (1976)
5-Diazouracil	Yes	Yes	?	Inouye and Pardee (1970)
u.v. irradiation	Yes	Yes	Yes	Inouye and Pardee (1970)
Thymine starvation	Yes	Yes	Yes	Inouye and Pardee (1970)
B. Treatment of mutants				
Rec A⁻ with nalidixic acid	No	No	Yes	Inouye (1971)
Rec A⁻ thymine starved	No	No	Yes	Inouye (1971)
Lex A⁻ with nalidixic acid	No	No	Yes	Gudas (1976)
Lex A⁻ thymine starved	No	No	Yes	Gudas (1976); Howe and Mount (1975)

C. DNA temperature-sensitive mutants at nonpermissive temperatures				
E. coli MK 74T2 ts27	No	No	Yes	Inouye and Pardee (1970)
S. typhimurium 11G *dnaA*	No	No	No	Satta and Pardee, unpublished data
K. pneumoniae Mir M7 *dnaA*	No	No	No	Satta and Pardee, unpublished data
K. pneumoniae Mir M7 + nalidixic acid	No	No	No	Satta and Pardee, unpublished data
K. pneumoniae + bleomycin	Yes	Yes	Yes	Satta and Pardee all cited by Satta and Pardee (1978)
D. Division temperature-sensitive mutants at nonpermissive conditions				
E. coli DM936 *Rec A tsl*	Yes	Yes		Gudas (1976); Mount *et al.* (1973)
E. coli DM511 *tsl*	Yes	Yes	No	Gudas (1976); Mount *et al.* (1973)
E. coli WP 44S *tif*	Yes	Yes	No	Gudas (1976); Mount *et al.* (1973)
E. coli tim-1	Yes	No	No	Gudas (1976); Kirby *et al.* (1967)

(Modified and reproduced with permission from Satta and Pardee, 1978).

molecular weight can be induced by thymine starvation (Churchward and Holland, 1976b). It was subsequently shown (Boyd and Holland, 1977) that this protein is in fact the product of the *feu B* locus and is a binding protein essential for enterochelin-mediated iron transport. Much of the evidence suggesting a role for this protein in DNA replication was invalidated by the demonstration (Boyd and Holland, 1977) that synthesis of the protein was induced merely by filtering a bacterial culture and resuspending the cells in fresh medium.

(c) *Termination proteins*. The requirement for synthesis of "termination proteins" has been reported by Jones and Donachie (1973) and incorporated into a model (Donachie *et al.*, 1973) that views the cell cycle of *E. coli* as two clocks, one for DNA replication and one for division (Fig. 8.12). However, a recent search for any polypeptide that is synthesized uniquely at a particular point in the cycle was negative (Lutkenhaus *et al.*, 1979). Models similar to that of Jones and Donachie

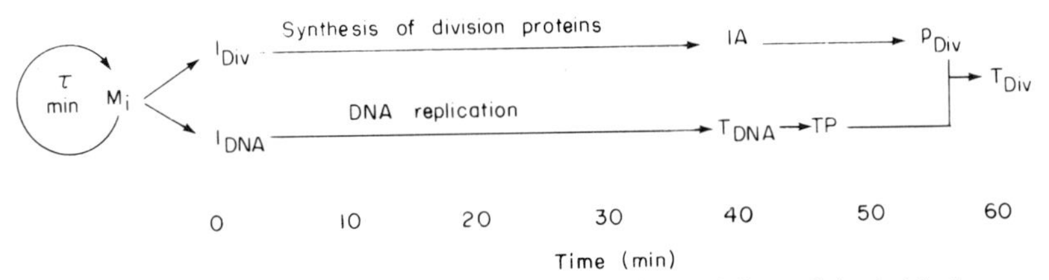

Fig. 8.12 Model of the cell cycle in *Escherichia coli*. Doubling of the initiation mass (M_i) takes place every mass doubling time (τ min). At each doubling, two processes are initiated approximately simultaneously. These are the initiation of DNA replication (I_{DNA}) and the initiation of the sequences of events leading to division (I_{Div}). Termination of chromosome replication (T_{DNA}) at 40 min induces the synthesis of termination protein (TP). The first 40 min of the division sequence involves protein synthesis which is then followed by the initiation of assembly (IA). (N.B. Assembly could actually start earlier.) After 15–20 min more the cell has reached a stage (P_{Div}) where interaction between some septum "primordium" and termination protein leads to cell division (T_{Div}). This view of the timing of the main events in the cycle is therefore of a periodic event that occurs at intervals equal to the mass doubling time of the cell and at multiples (2^n) of a constant cell mass (M_i). This event triggers two parallel but separate sequences of events which take *constant* periods of time to complete, largely independent of the rate of cell growth. One of these processes is chromosome replication and the synthesis of the termination protein, and requires 40–45 min to complete. The other is a sequence of protein synthesis, followed by another process which may be assembly of some septum precursor. This sequence requires nearly 60 min and, at the end of it, there is an interaction between the septum precursor(s) and the termination protein to give the final septum and cell division. The last event takes only a few minutes. (Reproduced with permission from Donachie *et al.*, 1973.)

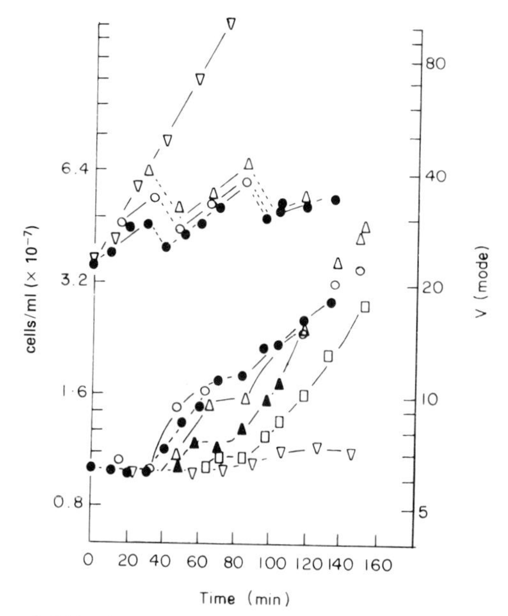

Fig. 8.13 Effect of shifts from 42 to 30°C on growth and division in a synchronous culture of a mutant temperature-sensitive for the *fts* A^+ gene product. A synchronous culture was prepared by selecting small cells after sucrose gradient centrifugation; the selected fraction was inoculated into nutrient broth plus thymine at 42°C. At intervals thereafter, portions were shifted down to 30°C. Lower curves show cell numbers per milliliter (▽, culture kept at 42°C throughout; ●, ○, △, ▲, □, subcultures shifted to 30°C at 0, 15, 30, 45 and 60 min respectively). Upper curves show modal cell volume in the 42°C control (▽) and in the subcultures shifted to 30°C after 0 (●), 15 (○) and 30 (△) min. (Reproduced with permission from Donachie *et al.*, 1979.)

(1973) have been proposed for *Streptococcus faecalis* (Shockman *et al.*, 1974) and *Bacillus subtilis* (Miyakawa *et al.*, 1980).

There is now good genetic evidence, however, for the functioning of particular proteins during the cycle. Strains carrying a mis-sense mutation at the *fts A* locus, and which thus synthesize a temperature-sensitive *fts A* protein stop division immediately on transfer to the restrictive temperature (Walker *et al.*, 1975; Fletcher *et al.*, 1978), suggesting that the *fts A* gene product is required in an active form throughout the entire septation process. However, Donachie *et al.* (1979) have shown, using cells of a strain temperature-sensitive for the synthesis of the *fts* A^+ gene product, that synthesis is essential only during a short period immediately before cell separation (Fig. 8.13). Because the *fts* A^+ gene product has recently been identified as a polypeptide of mol. wt 50 000 (Lutkenhaus and Donachie, 1979) direct measurements of its synthesis during the cell cycle are now possible. It is already clear (Donachie *et al.*, 1979) that accumulation of the protein to critical levels during prolonged growth is not required: a short burst of synthesis at a critical point in the cycle is sufficient for septation to occur. Tormo *et al.* (1980) have proposed that the *fts A* gene product exhibits the functions of a termination protein, whose action extends from the end of the replication cycle to the end of septation, and that it may also have a structural role.

Other proteins have been implicated as requirements for division. De Pedro *et al.* (1975) suggest that the *div A* product must be built up prior to division and is consumed at a division site when its function is executed. Spratt's (1975, 1977) study of mutants of *E. coli* with thermolabile penicillin-binding proteins implicates them as essential components of the division process.

3. *Independent Triggering of Chromosome Initiation and Cell Division*

In contrast to the Cooper–Helmstetter model, in which the timing of cell division is controlled by the constancy of the D period, the earlier model of Koch and Schaechter (1962) assumed that nuclear division resulted from the attainment by the cell of a critical size and that cell division is tied to the cell achieving a second, larger, critical size. Koch (1977) argues that it would be poor economy for the cell to have major cellular processes such as RNA and protein synthesis limited by minor processes (in terms of utilization of cell resources) such as DNA replication and cell division. He suggests that "ultimately the ability of the organism to produce protoplasm from the resources of the environment must limit growth and must time the initiation of chromosome replication, nuclear division and cell divison". The assumption that the autocatalytic properties of macromolecular biosynthesis are the deterministic part of the control of the cell division process is strengthened by several findings:

(i) the observed positive correlation of cycle length of sister cells and the negative correlation of the cycle lengths of mother cells with their daughters;
(ii) the finding that cells synchronized by a modification of the technique of Cutler and Evans (1966) remain synchronized for up to 21 generations (Kelley, 1974). Failure to lose synchrony implies that there are no local variations in specific rates of biosynthesis and that the cell division process apportions cytoplasm extremely evenly between every pair of daughters.

A similar hypothesis may be advanced to explain the persistent synchrony observed by Anagnostopoulos (1976) on synchronizing *E. coli* by periodic glucose starvation (Fig. 8.14), although Anagnostopoulos has preferred to view the results as the selection of cells with an unusually narrow distribution of (long) doubling times.

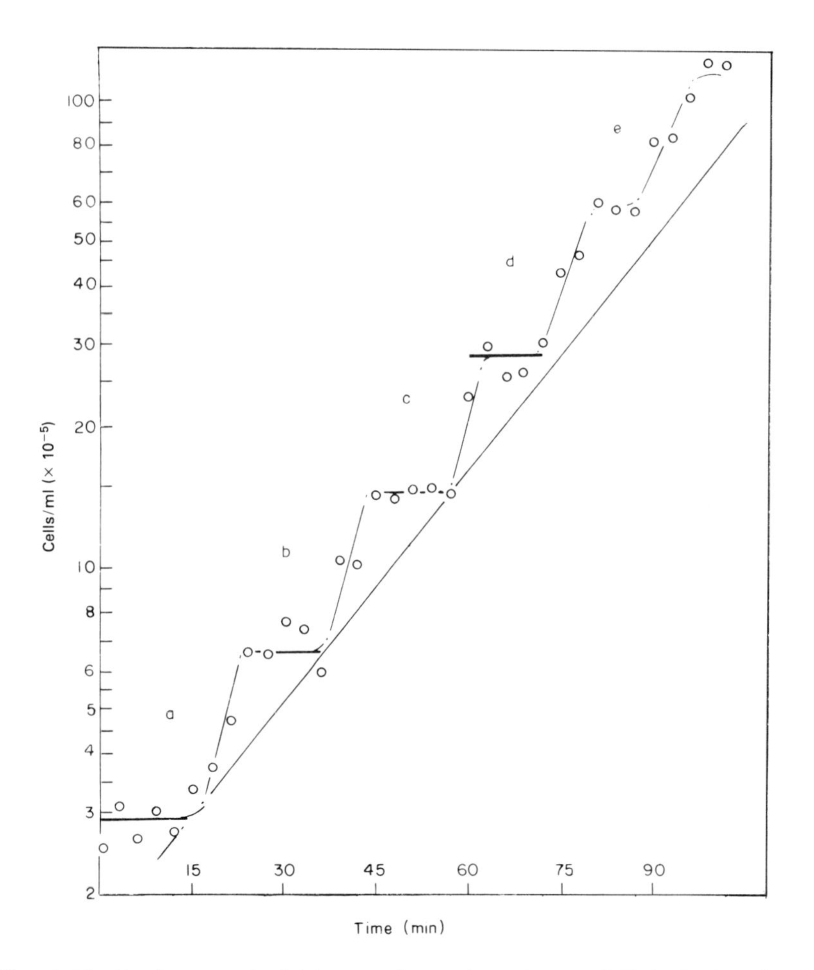

Fig. 8.14 Persistence of division synchrony in cultures of *Escherichia coli* B/r/1 grown under conditions of periodic glucose starvation and then heated to kill one of the resultant sub-populations. The last five (*a* to *e*) of 13 synchronous divisions that occurred in the experiment are shown. The synchrony indices, calculated by the method of Scherbaum (1963) are for (a) 0·70, (b) 0·67, (c) 0·57, (d) 0·45 and (e) 0·40. Cell counts are viable counts. (Reproduced with permission from Anagnostopoulos, 1976.)

Computer simulations by Koch (1977) of the autoradiographic data of Chai and Lark (1970) and of Forro (unpublished) on slowly growing cultures of *E. coli* exclude all previous models. Of particular significance was the finding that the coefficient of variation of size at initiation of DNA replication was large, compared to that for size at division.

More recent, conflicting estimates of these variations are given by Koppes *et al.* (1978a,b) and Koppes and Nanninga (1980). Koppes *et al.* (1978b) found that in three slowly-growing substrains of *E. coli* B/r, the standard deviation of the length at which chromosome initiation occurred was similar to that at which separation occurred. The results suggest the occurrence of a deterministic phase in the cell cycle (reminiscent of that proposed by Smith and Martin, 1973) in which, after initiation of chromosome replication, the cells proceed towards division. Subsequent investigation of the variability of length and age at successive cell cycle events in slowly-growing, synchronized *E. coli* (Koppes *et al.*, 1980) confirmed that the "B"-period (the interval between birth and initiation of DNA replication) was the most variable period of the cycle, and favoured a model in which chromosome initiation occurs at a particular cell size rather than by some random event (cf. a transition probability; Chapter 1) in the "B"-period.

Koch (1977) points to the experiments of Nagata and Meselson (1968) as evidence that initiation events are controlled well in a culture as a whole and as regular events in real time, rather than in time counted from the particular chromosome initiation or cell division event creating a particular cell cycle. More recently, Newman and Kubitschek (1978) have confirmed the remarkably small variation in the time interval from the synthesis of a region of the chromosome to the time of synthesis of the same region of the chromosome in the daughter cell. These experiments, in Koch's view, reinforce the idea that, basically, the cell cycle of prokaryotes could be very precise, even though the timing of divisions is quite imprecise.

Koch's model is consistent with the data of Meacock and Pritchard (1975) which show that an increase in the length of C is accompanied by a decrease in D. By itself this finding clearly contradicts the simple view that termination of a round of replication provides a signal for division to occur a fixed time later (Helmstetter and Pierucci, 1968; Clark, 1968). However, the kinetics of transition of exponential cultures moving between steady states have provided evidence that an event late in the replication cycle, and not at initiation, determines the timing of division.

D. CONTROL OF CELL ELONGATION

Studies of the variation in cell size and dimensions and of the synthesis of envelope compounds at different growth rates have led to the general view that the rate of cell elongation or envelope synthesis increases discretely during the bacterial cell cycle (Sargent, 1975a,b,c; Hoffman *et al.*, 1972; see, however, Koppes *et al.*, 1980). Many models have been proposed for the control of length extension; all postulate a constant rate of elongation (Previc, 1970; Pritchard, 1974; Donachie *et al.*, 1973) that changes concomitantly with a particular event in the cycle. It is the identity of this event that will be considered here.

1. *Termination of Chromosome Replication*

Pritchard and his colleagues have proposed that the termination of chromosome replication is a signal for a doubling in the rate of cell elongation. This idea stems from experiments with thymine-requiring auxotrophs which, when grown with limiting thymine concentrations, undergo C over a prolonged period. Since it is believed that cell division can only occur after completion of a round of replication (Clark, 1968; Helmstetter and Pierucci, 1968) and since initiation of rounds of replication take place at doublings of cell mass (see Section IV.B1), the consequence of lower replication velocity is an increase in size at division, which is accomodated by an increase in cell diameter rather than length (Meacock and Pritchard, 1975; Pritchard *et al.*, 1978; Zaritsky and Woldringh, 1978). The hypothesis (Meacock and Pritchard, 1975; Pritchard, 1974; Zaritsky and Pritchard, 1973) explains the observed changes in cell width and length in relation to growth and division of *E. coli*. This is that cell length at a time 20 min (Meacock *et al.*, 1978; Trueba and Woldringh, 1980) or not (Begg and Donachie, 1978) of growth-rate related changes in cell diameter.

2. *The Attainment of a Critical Cell Length*

An alternative model, that of Donachie *et al.* (1976) was formulated on the basis of the observed relationship between cell length, cell growth and division of *E. coli*. This is that cell length at a time 20 min before division is almost constant at all growth rates and furthermore that this length is twice the theoretical minimum length (λ) of a cell in any given growth condition (Fig. 8.15). The rate of cell elongation changes at length 2λ; under constant growth conditions, the increase in rate is a doubling. The calculated length in question was originally given by Donachie *et al.* (1976) as 2·8 μm, but Cullum and Vicente

(1978) have suggested that 3·5 to 4·5 μm is a better estimate. Of more importance than the actual size of a cell when the rate of elongation doubles is the question of how the rate increase occurs. Experiments to test the suggestions (Zaritsky and Pritchard, 1973; Zaritsky, 1975) that this is due to termination of a round of replication (see above)

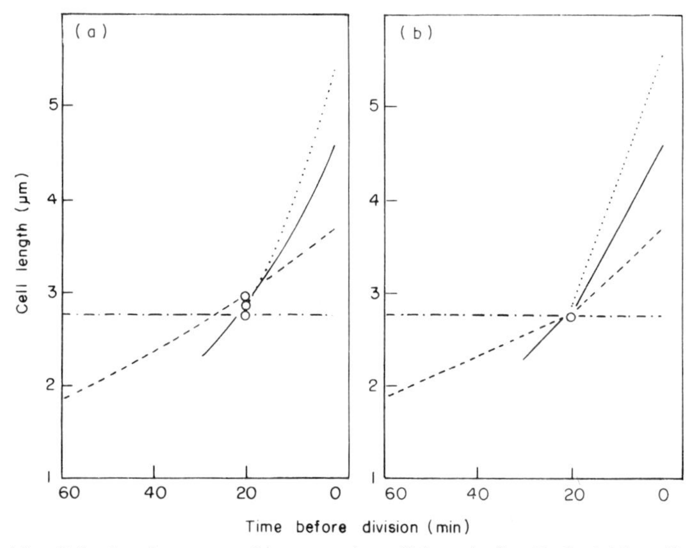

Fig. 8.15 Calculated course of increase in cell length for *Escherichia coli* over cell cycles of various durations, assuming that length doubles over the course of each cycle, that the relationship between mean population cell length and growth rate is given by the straight-line function $\bar{L} = (2 + 2R/3)$ μm (where $\bar{L}$ is mean cell length and R is the number of doublings in cell number h^{-1}) and that populations have the theoretical age distribution (Powell, 1956). The figure shows cell elongation according to two sets of assumptions: (a) that the rate of elongation increases continuously according to a logarithmic function, and (b) that the rate of elongation is constant but doubles 20 min before division. For clarity, only four growth rates are shown, namely those for generation times of 20 min (. . .), 30 min (——), 60 min (– – –) and infinity (— · —). The open circles represent cell length at 20 min before division. (Reproduced with permission from Donachie *et al.*, 1976.)

gave negative results. Instead, it was suggested (Donachie *et al.*, 1976) that the minimum "unit cell" of length λ has a fixed number of sites of elongation. Growth in length of such a unit cell takes place at these sites at a rate proportional to the number of sites, as well as to the growth rate in cell mass until a new length is completed (at 2λ). Septation is then initiated between the two completed unit cells and each of the two units begins to grow independently from new sets of growth sites. Complete septation takes about 20 min in all growth conditions. Independently, Grover *et al.* (1977) came to the conclusion that *E. coli* exhibits a linear rate of elongation that doubles about 17 min before division. This model is well supported by theoretical considerations (Rosenberger *et al.*, 1978a). The elements of these models, namely that cells divide in a fixed period of time after reaching a critical size have some similarity with models proposed for yeast cells (see Section II.B) and for *Bacillus subtilis* (Sargent, 1975a,b,c). We should note, however, that the model of Donachie *et al.* (1976) does not invoke cell length *per se* as the trigger for septation. Other controls such as the requirement for specific proteins may be intimately involved.

A similar "unit cell" hypothesis has been proposed for *Streptococcus faecalis* (Edelstein *et al.*, 1980) in which the size of the polar shells of the cell wall is independent of growth rate.

3. *The Attainment of a Critical, Minimal Density*

Rosenberger *et al.* (1978b) have proposed that cell density could control the rate of surface growth. A critical density, ρ_d, is considered to lead to the addition of envelope growth zones. The model is in excellent agreement with the dimensions of bacteria growing over a wide range of doubling times and with the measured variations in buoyant density that occur during the cell cycle of *E. coli* K12 (Poole, 1977b). The estimated value of d (the time interval between the doubling in rate of elongation and cell division) was 44 min, in close agreement with the value of 49 min obtained previously by Rosenberger *et al.* (1978a).

4. *Nutrient Uptake*

Kubitschek (1968a,b) measured the volume distributions of cell samples from synchronous cultures of *E. coli* using a Coulter counter and multichannel analyzer and showed that cell volume increased linearly through the cell cycle. Measurements of solute uptake rates through the cell cycles of *E. coli* (Kubitschek *et al.*, 1971) and also of yeasts (Kubitschek and Claymen, 1976; Kubitschek and Edvenson, 1977) are in accord with the idea that linear growth extension with a doubling in growth rate at some point in the cycle, is a common mode of growth. Kubitschek suggests that the doubling in growth rate arises from constant uptake of nutrients and the doubling in synthesis or function of nutrient uptake sites late in the cell cycle, although Ho and Shuler (1977) have shown that an equally tenable mechanism for *E. coli* may be via feedback control of nutrient uptake rates.

V. Histone Modifications and Cell Cycle Control

Post-synthetic, covalent modifications of histones by phosphorylation, acetylation and methylation have been observed and may be important in the functions of these nuclear proteins. Such changes have been studied as a function of the cell cycle (see Prescott, 1976; Ord *et al.*, 1975) and in some cases roles for such modifications in the control of cell cycle traverse have been suggested.

A. HISTONE F-1

Phosphorylation of histone F-1 (very lysine-rich histone) occurs in both S and G_2 in mammalian cells and *Physarum polycephalum*. The phosphorylation that occurs in S presumably reflects histone synthesis, which requires some phosphorylation for the maintenance of a constant overall phosphate content per molecule, and on this basis its significance for cell cycle control has been questioned (Bradbury *et al.*, 1974a). Most F-1 phosphorylation in both *Physarum* (Bradbury *et al.*, 1973) and mammalian cells (Gurley *et al.*, 1973) occurs in late G_1. F-1 histone remains phosphorylated in mitosis and is dephosphorylated in G_1 (Lake *et al.*, 1972; Marks *et al.*, 1973). The precise timing ($\pm 0{\cdot}005$ of a generation time) of phosphorylation in *Physarum* at chromosome condensation has been proposed as the initiation step for mitosis (Bradbury *et al.*, 1974b). The role of phosphorylation of F-1 histone in chromosome condensation was subsequently and dramatically confirmed by addition of F-1 histone phosphokinase prepared from Ehrlich ascites cells to *Physarum* with the consequent acceleration of entry into mitosis of G_2 cells (Bradbury *et al.*, 1974a; Inglis *et al.*, 1976).

B. OTHER HISTONES

Phosphorylation of histone H1 is one of the major chemical changes associated with growth. In *Physarum*, nuclear histone kinase activity rises sharply as a result of activation or transport of the enzyme (Mitchelson *et al.*, 1978), and precedes H1 phosphorylation in prophase

(Bradbury *et al.*, 1974b). These changes are regarded as an important part of the sequences of events that determine the progress of the nucleus through G2 (Bradbury *et al.*, 1974a,b; Inglis *et al.*, 1976; Corbett *et al.*, 1980).

Histones can also be modified by acetylation (D'Anna *et al.*, 1980a). Very recently, Chahal *et al.* (1980) have shown that tetra-acetylated H4 exhibits dramatic changes during the *Physarum* cell cycle with maxima in the mid S and mid G2 phases. Highly acetylated H4 (2–4 acetates per molecule) is inversely correlated with H1 phosphorylation and the initiation of chromsome condensation in prophase (Fig. 8.16). Taken together, these results indicate that chromsome condensation

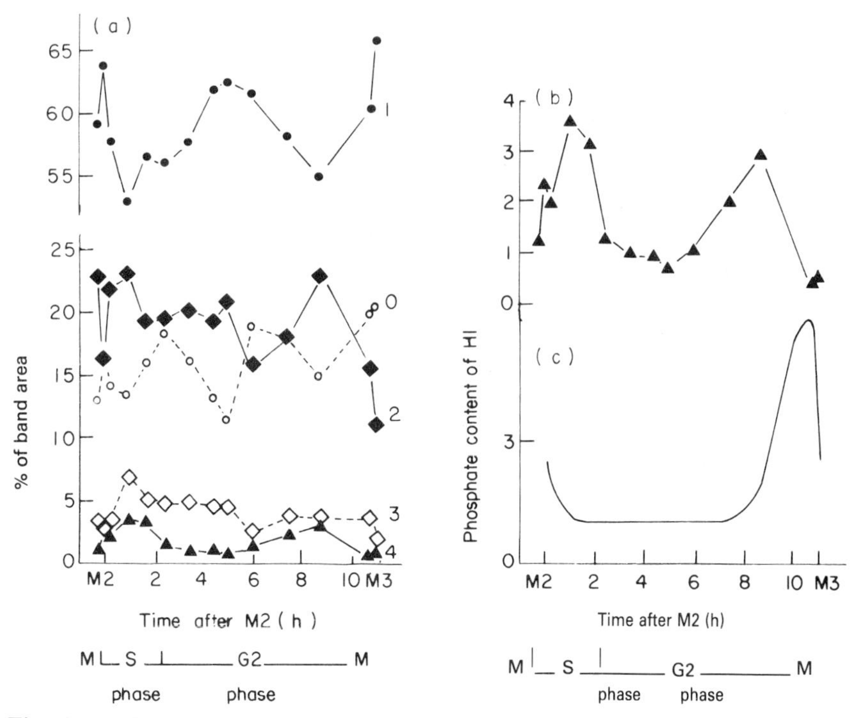

Fig. 8.16 Acetylated H4 through the cell cycle of *Physarum polycephalum*. (a) ○ - ○ Percentage of total H4 in the non-acetylated form; ●—● percentage of total H4 in the mono-acetylated form, Ac_1H4; ◆—◆ percentage of total H4 in the di-acetylated form, Ac_2H4; ◇—◇ percentage of total H4 in the tri-acetylated form, Ac_3H4; ▲—▲ percentage of total H4 in the tetra-acetylated form, Ac_4H4; (b) Percentage of Ac_4H4, shown on a larger scale. (c) Relative phosphate content of *Physarum* histone H1 through the cell cycle. (Reproduced with permission from Chahal *et al.*, 1980.)

may involve the coordinated modification of H1 by phosphorylation and of H4 by deacetylation.

Gurley *et al.* (1978) found that, in CHO cells, three of the five histone fractions were significantly phosphorylated; H1, H2a and H3. Histone H4 was phosphorylated to only a limited extent and, as Fig. 8.17 shows, no phosphorylation of H2b was detected. H1 phosphorylation was detectable during G_1 and S as well as mitosis ($H1_M$, "superphosphorylation") where many phosphorylation sites are

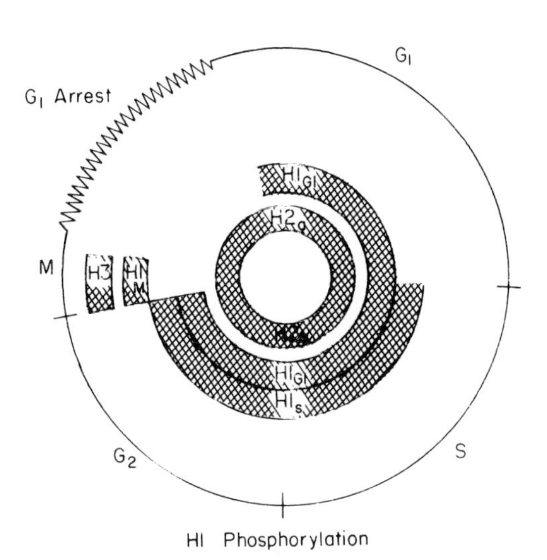

Fig. 8.17 Histone phosphorylation in the CHO Chinese hamster cell cycle. Cells progress through the phases of the cell cycle clockwise from G_1 to division at the end of mitosis (M). The shaded bands indicate the phases of the cycle in which the histones (H1, H2a and H3) are phosphorylated. (Reproduced with permission from Gurley *et al.*, 1978.)

involved on all H1 molecules. Only during mitosis was histone H3 seen to be phosphorylated, whilst histone H2a was rapidly phosphorylated throughout the cycle. Similar patterns of phosphorylation are seen in a wide variety of mammalian cell types.

The phosphorylation of H1 histones that occurs in mitosis ($H1_M$) and of H3 are strictly mitotic events in CHO cells and not related to the condensation of chromatin in G_2. This appears to be in contrast with the *Physarum* results in which it is claimed that H1 phosphorylation is complete before chromosome condensation. Gurley *et al.* (1978) have pointed out, however, that the experiments of Bradbury's group do not distinguish between interphase-type H1 phosphorylation and $H1_M$ phosphorylation, the latter (together with H3 phosphorylation) being favoured by Gurley *et al.* (1978) as possible triggers of mitosis. In addition, the omission of phosphatase inhibitors from the *Physarum* preparations may have resulted in loss of $H1_M$ and H3 phosphates.

A new class of H1-like protein has been described in Chinese hamster cells grown in the presence of sodium butyrate (d'Anna *et al.*, 1980a) and named BEP (butyrate-enhanced protein). Like histone H1 from these cells, it is phosphorylated in interphase and "superphosphorylated" in mitosis, the latter being coincident with chromosome condensation and occurrence of H1 phosphorylation (Gurley *et al.*, 1980b).

VI. Regulation of Cell Division by Ca^{2+} and Mg^{2+}

The possible involvement of cellular calcium levels, in conjunction with the levels of magnesium ions (e.g. McKeehan and Ham, 1978), cyclic nucleotides or calcium binding proteins (Means and Dedman, 1980), has received considerable attention.

Magnesium, because of its relatively high intraceullar concentration and its roles in macromolecular synthesis and transphosphorylation reactions may be considered to be a function-maintaining element. Calcium's special status as a regulator is suggested by the extraordinarily steep gradients of the ion that cells maintain across their boundary membranes.

Magnesium and calcium act antagonistically in many biological processes (Hughes, 1981). Rubin (1975a) has postulated that many effects observed in response to Ca^{2+} concentrations are the result of competitive binding of these two ions. For example, calcium inhibits polymerization of tubulin in a way that is regulated by the Mg^{2+} concentration (Rosenfeld *et al.*, 1976). Although Ca^{2+} has received more attention as a potential factor in growth control, Rubin (1976) has pointed out that when mammalian cells are deprived of Ca^{2+} with EGTA they shrink from their lateral attachments and assume an abnormal shriveled appearance, in no way resembling that of physiologically inhibited cells. The limited growth inhibitory effects of severe Ca^{2+} deprivation may be indirect results of its creating a deficit of free Mg^{2+} within the cell (Rubin and Koide, 1976).

A. CALCIUM

The concept that Ca^{2+} is an important growth regulator (for a review see Berridge, 1976) is not new; Mazia (1937) considered Ca^{2+} to be responsible for triggering division after fertilization. The notion is still attractive, as indicated by the following examples of Ca^{2+} involvement in growth and division:

(i) Ca^{2+} is known to regulate the interaction of actin and myosin (e.g. Nachmias and Asch, 1976) and microtubule formation (Weisenberg, 1972). Microtubules themselves have been implicated in cell cycle control (see Section VIII).

(ii) In *Physarum polycephalum*, cyclic uptake and release of Ca^{2+} occurs during mitosis and these fluctuations correlate with specific structural and kinetic events in the mitotic cycle (Holmes and Stewart, 1977).

(iii) Calcium ions are required for initiation of DNA synthesis (Rixon and Whitfield, 1976; Boynton *et al.*, 1976; Swierenga *et al.*, 1976).

(iv) Growth stimulation of various mammalian cells by growth factors (Frank 1973), Concanavalin A (Freedman *et al.*, 1975) or phytohaemaglutinin (Whitney and Sutherland, 1973) leads to early increases in $^{45}Ca^{2+}$ uptake. Conversely, lymphocytes may be stimulated to proliferate by increasing the availability of Ca^{2+} (Luckasen *et al.*, 1974; Maino *et al.*, 1974) by using the ionophore A23187 (see Section VIII.B).

(v) 3T3 cells have a calcium-sensitive restriction point at which cells arrest in G_1 (Paul and Ristow, 1979). Interestingly, though, SV-40 virus-transformed cells, which grow without restriction in monolayer cultures, appear to have no such restriction point and to have lost the capacity to sense intracellular Ca^{2+} levels. Other examples may be cited in which relaxation of growth control mechanisms is paralleled by decreasing Ca^{2+} requirements (Balk *et al.*, 1973; Boynton and Whitfield, 1976).

(vi) Levels of the intracellular Ca^{2+} receptor, calmodulin, increase and reach maximal levels in early S and are maintained until division (Chafouleas *et al.*, 1980).

(vii) Long and Williamson (1979) have implicated Ca^{2+} binding to the cell surface by glycosaminoglycans as a controlling element in cell proliferation.

(viii) There are many ways in which Ca^{2+} interacts with the intracellular levels of cyclic nucleotides, themselves implicated in cycle regulation (see Section VII). Adenyl cyclase activity may be inhibited by Ca^{2+} or, at very low Ca^{2+} levels, stimulated, the latter effect being modulated by calcium-dependent proteins (e.g. calmodulin; Means and Dedman, 1980). In addition, Ca^{2+} can stimulate guanyl cyclase. This, together with the influence of cyclic GMP and cyclic AMP on calcium homeostasis, generates a complex network of feedback interactions involving the two cyclic nucleotides and calcium (Berridge, 1975a). In view of this, Berridge (1975b) has suggested that an intracellular increase in Ca^{2+}, modulated by cyclic AMP at a specific stage of the cell cycle, may be a universal stimulus for cell growth. In Berridge's model (Fig. 8.18), it is proposed that the critical switch in G_1 between (i) progression through the cycle to mitosis or (ii) differentation, is Ca^{2+}-dependent. A high level of cyclic AMP during G_1 will assist in lowering the level of Ca^{2+}, and also initiate differentiation. The cyclic nucleotides are proposed to play a secondary role in that they are related to Ca^{2+} only through the feedback relationships noted

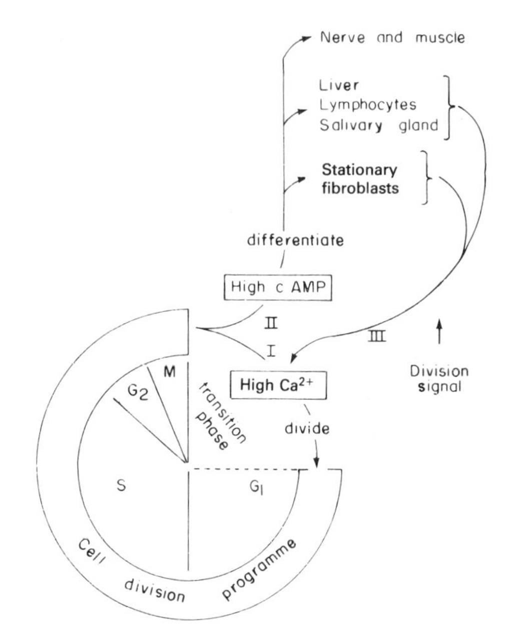

Fig. 8.18 The proposed role of calcium and cyclic AMP during the transition phase of the cell cycle early in G_1. A high level of calcium stimulates cells to divide (I) whereas cyclic AMP switches them out of the cell cycle and induces them to differentiate (II). Many differentiated cells can divide if provided with the appropriate division stimulus (III). If a cell commits itself to divide it embarks upon a set cell division programme which terminates with mitosis (M). (Reproduced with permission from Berridge, 1975b.)

earlier. Cyclic AMP plays an obscure role. The high levels of cyclic GMP which are associated with division stimuli in many cells would be a direct consequence of the Ca^{2+} signal.

Whitfield *et al.* (1979) propose that calcium, in cooperation with cyclic AMP, calmodulin and protein kinases, generates a signal for the initiation of DNA synthesis. Calcium deprivation is reported to stop reversibly proliferative development in that part of G_1 in which cyclic AMP transiently rises and the synthesis of the four deoxyribonucleotides begins.

Calcium has been reported not to have an essential role in cell growth and division in the yeast *Schiz. pombe* (Walker, 1978), although EGTA (a Ca^{2+} chelator) inhibits cell division (Walker and Duffus, 1979) and addition of Ca^{2+} to A23187-treated cells allows division of about 70% of the inhibited cells (Duffus and Paterson, 1974a).

In *Tetrahymena*, oscillations in the amounts of both Ca^{2+} and Mg^{2+} were found during "free-running" synchrony after heat-shock induction, but only during growth in a complex proteose peptone medium (Walker and Zeuthen, 1980). Peaks were observed to coincide with cell division. However, substantial fluctuations (upto a twofold variation) in Mg^{2+} content per cell also occurred in a control asynchronous culture and with approximately the same periodicity as that seen in the synchronized culture, although these were not considered significant. When cells that had been given six heat-shocks in proteose peptone were transferred to a defined inorganic medium, no such oscillations were observed, despite a high degree of division synchrony. Thus the claims (London *et al.*, 1979; Charp and Whitson, 1980) that a burst in intracellular Ca^{2+} is necessary to trigger cell division in *Tetrahymena* was not supported in this work.

B. MAGNESIUM

Magnesium ions have been shown to be essential for cell division in yeast. This requirement has been demonstrated by exposing cells to a variety of Mg^{2+}-chelating agents (Duffus and Paterson, 1974a,b; Penman and Duffus, 1975; Ahluwalia *et al.*, 1978; Walker and Duffus, 1979) and transferring cells to Mg^{2+}-deficient media (Walker and Duffus, 1980). Under such conditions, *Schiz. pombe* continues to elongate, but nuclear division and cell plate formation are prevented, suggesting that Mg^{2+} limitation arrests cells late in the G_2 phase. Addition of Mg^{2+} (but not Ca^{2+}, Sr^{2+}, Be^{2+}, Mn^{2+} or Zn^{2+}) to Mg^{2+} deficient cultures induces synchronous cell division. Analyses of total cell Mg^{2+} in cultures of both *Schiz. pombe* and *Kluyveromyces fragilis*,

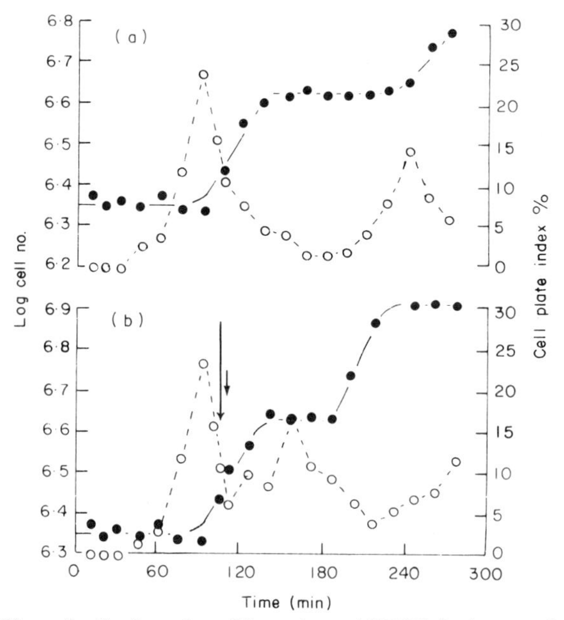

Fig. 8.19 Effect of a 5 min pulse of ionophore A23187 during synchronous growth of *Schizosaccharomyces pombe* in Ca^{2+}-free minimal medium. (a) Control culture synchronized by lactose gradient size-selection. (b) As in (a), but where indicated, A23187 was added (final concentration 4·0 μg/ml) to the culture and pulsed for 5 min. Long arrow, beginning of pulse. Short arrow represents the time when cells were reinoculated into fresh Ca-free medium after filtration and washing. ●, log cell no.; ○, cell plate index. (Reproduced with permission from Walker and Duffus, 1980.)

synchronized by various induction and selection procedures, revealed a steady fall in Mg^{2+} concentration, terminating in a rapid influx of magnesium just before cell division. This observation has led to the proposal that falling Mg^{2+} concentration may act as a transducer of cell size, eventually triggering tubulin polymerization and spindle formation, and a membrane change that permits rapid uptake of Mg^{2+} to a concentration that brings about spindle formation. Some support for the proposal comes from the finding that a short (5 min) pulse of the ionophore A23187 applied just before division accelerates cell division (Fig. 8.19).

Changes in the concentration of Mg^{2+} elicit at least three elements of the "co-ordinate response" in chick embryo fibroblasts (Rubin, 1976). This response (Rubin, 1975a,b) is induced in cell cultures by a variety of unrelated agents in the absence of serum and is thus distinguished from the "pleiotypic response" (Hershko *et al.*, 1971) by its insensitivity to serum. In addition, protein synthesis and in turn DNA synthesis are very sensitive to small changes in intracellular Mg^{2+} within a physiological range (Rubin *et al.*, 1979). Regulation of the availability of Mg^{2+} within the cell thus presents a plausible mechanism for growth control and metabolism.

VII. Cyclic Nucleotides and the Cell Cycle

Both cyclic AMP and cyclic GMP, functioning in association with hormone action and calcium levels, have been implicated as controlling elements of the eukaryotic cell cycle (for reviews, see Pastan *et al.*, 1975; Abou-Sabe, 1976; Whitfield *et al.*, 1979). In the view of other researchers, however (Bourne *et al.*, 1975; Coffino *et al.*, 1975; review by Baserga, 1976) cyclic AMP is not an essential regulator of growth. This conclusion is based in part on experiments with lymphoma cells; mutant cells which lack the cyclic AMP-dependent protein kinase are capable of normal growth and are also insensitive to dibutyryl cyclic AMP, which arrests the growth of normal cells. This view tends to be strenghened by the very diverse effects that cyclic AMP exerts, depending on the cell type under investigation. In addition, cyclic AMP fails to affect significantly the long term growth of cells in which it causes morphological changes supposedly characteristic of slowly-growing cultures (Kimball *et al.*, 1975). Nevertheless the involvement of cyclic nucleotides continues to attract attention. The account that follows is necessarily very selective from an enormous literature. The widespread suggestion that lowering of cyclic AMP levels is an initiator of proliferation of quiescent cells will not be considered (see review by Pastan *et al.*, 1975). It should be noted, however, that alternative viewpoints exist (Boynton *et al.*, 1978; Krishnaraj, 1978).

A. FLUCTUATION IN LEVELS OF CYCLIC NUCLEOTIDES DURING ANIMAL CELL CYCLES

1. *Cyclic AMP*

Prescott (1976) has described the changes in cAMP levels that occur during the cell cycles of cultured animal cells and the remarkable oscillations that occur during cycles of cleavage in sea urchin embryos. Since that review appeared, there has been ample confirmation that fluctuations in the cellular content of both cAMP and its specific protein kinases are a normal feature of cell cycle traverse (Byus *et al.*, 1977; Hibasam *et al.*, 1977; Costa *et al.*, 1976, 1978; Wang *et al.*, 1978; Kaiser *et al.*, 1979; Haddox *et al.*, 1980). There is general agreement that cAMP levels are at a minimum during mitosis. Zeilig *et al.* (1976) have confirmed this conclusion (see Fig. 8.20) and also demonstrated transient elevations in cAMP levels during late S-phase. A number of reports confirm a peak in cAMP levels during G_1 (for references see Zeilig *et al.*, 1976). The timing of these peaks in relation

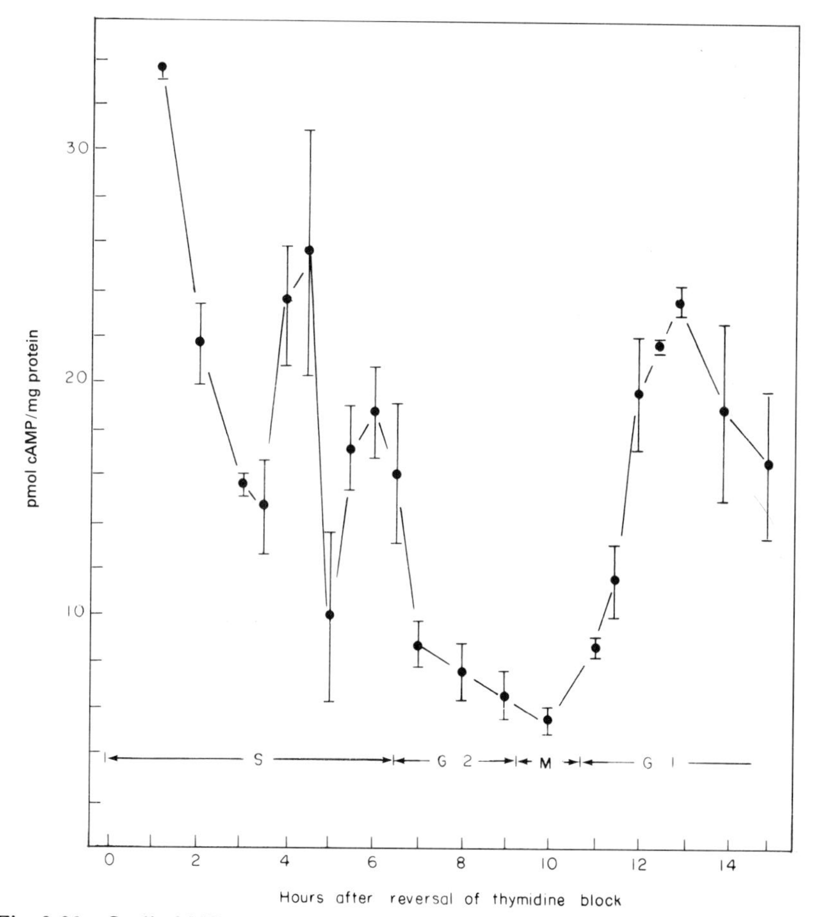

Fig. 8.20 Cyclic AMP content measured at short intervals in HeLa cells synchronized by the double thymidine block method. Cyclic AMP content was determined at 0·5 h intervals, as indicated, after reversal of the second thymidine block. Results represent the means and standard errors of three cell suspensions derived at zero time from a common culture. (Reproduced with permission from Zeilig *et al.*, 1976.)

to DNA synthesis is variable, however, suggesting that increased cAMP levels are not the immediate signal for initiation of S-phase, although they may be intimately involved in the events leading to DNA synthesis.

2. *Cyclic GMP*

Cyclic GMP has also been implicated in a regulatory role. A transient increase in cGMP was first reported in early G_1 of partially-synchronized mouse fibroblasts (Seifert and Rudland, 1974a,b). In a fast-growing rat hepatoma cell line, cGMP levels varied reciprocally with relation to cAMP levels (Zeilig and Goldberg, 1977). A close correlation appeared to exist between changes in the ratio of the two nucleotides and traverse of discrete cell cycle stages (Fig. 8.21). A reciprocal

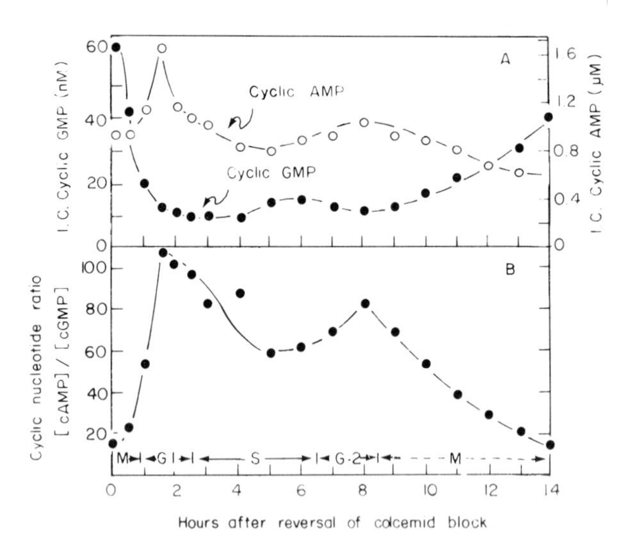

Fig. 8.21 Concentrations of intracellular (I.C.) cyclic GMP and cyclic AMP during the cell cycle of Novikoff hepatoma cells, synchronized with a colcemid block. (Reproduced with permission from Zeilig and Goldberg, 1977.)

relationship between cAMP and cGMP also occurs during cleavage of *Mactra* eggs (Geilenkirchen *et al.*, 1977). In both these instances, cAMP levels were minimal at mitosis.

B. INFLUENCE OF EXTRACELLULAR CYCLIC NUCLEOTIDES ON CYCLE TRAVERSE IN ANIMAL CELLS

There have been conflicting reports of the effects of experimentally elevated levels of cAMP on the cell cycle. The extensive literature has been critically reviewed in recent years (Chlapowski *et al.*, 1976; Friedman *et al.*, 1976; Prescott, 1976a; Rebhun, 1977). For example, prolongation of, or arrest in, S (van Wijk *et al.*, 1972; Dipasquale and McQuire, 1977), G_2 (Willingham *et al.*, 1972; Voorhees *et al.*, 1973; Zeilig *et al.*, 1976) and G_1 (Burger *et al.*, 1972; Pardee, 1974b; Coffino *et al.*, 1975; Coffino and Gray, 1978) have all been observed in response to dibutyryl cAMP or other agents that elevate cAMP levels. It is difficult to explain these disparate observations, but it is clear that there are numerous pitfalls in the interpretation of such experiments. Of particular concern is the possibility that catabolites of cAMP may exert inhibitory effects on cell growth. However, the experiments of Zeilig *et al.* (1976), showing that elevation of cAMP levels in S or G_2 caused arrest of cells in G_2, did incorporate measures to ensure that the observed effects were specifically due to cAMP, and suggest the existence of a cAMP-mediated checkpoint in G_2.

C. CYCLIC NUCLEOTIDES AND THE DIVISION CYCLES OF MICRO-ORGANISMS

There is an increasing body of literature describing fluctuations in the levels of cyclic nucleotides during the division cycles of micro-organisms.

A study of the regulation of intracellular events by cyclic nucleotides has been made in *Tetrahymena pyriformis*. Cyclic AMP, adenyl cyclase, cAMP phosphodiesterase and cAMP-dependent protein kinase activity have all been demonstrated in this organism (for references, see Dickinson *et al.*, 1976). Both cAMP and cGMP exhibit dramatic oscillations in cultures of selection synchronized cells, with maxima coincident with cell division (Dickinson *et al.*, 1976; Graves *et al.*, 1976). Blockage of cells in G_2 by a single hypoxic shock is accompanied by extraordinarily high levels of intracellular cAMP (Dickinson *et al.*, 1977). After release from hypoxia, the cells exhibit a modulation of cAMP similar to that seen in selection-synchronized cells (Fig. 8.22). It is suggested that a decline in cAMP allows progression into G_1, and that elevated levels constitute a signal for inhibition of division. In view of the belief that cAMP and cGMP are antagonistic (the former acting

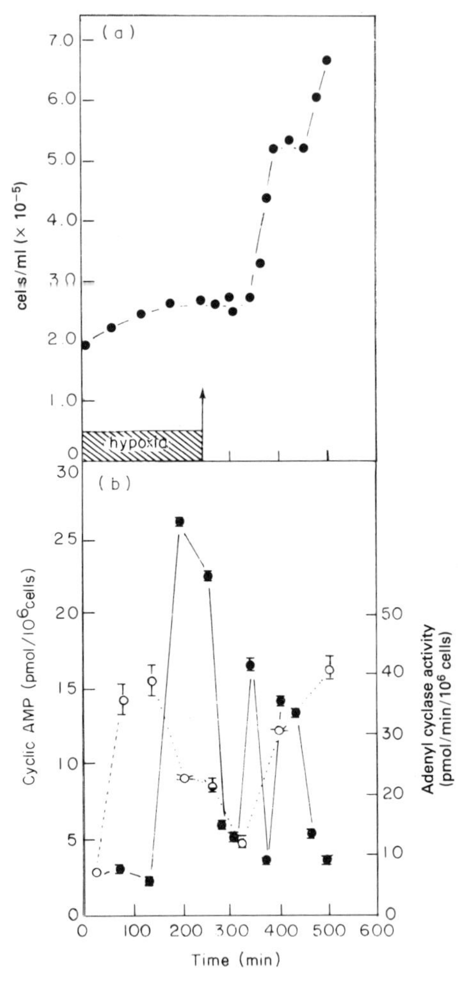

Fig. 8.22 Measurement of cyclic AMP and adenyl cyclase activity in hypoxic-shocked *Tetrahymena pyriformis*. (a) Cell density. The arrow indicates the point at which re-shaking started; (b) (■—■) Cyclic AMP (pmol/10^6 cells) and (○· · · ·○) adenyl cyclase activity pmol/min/10^6 cells measured in hypoxic-shocked cells. These results represent the mean of at least two experiments. (Reproduced with permission from Dickinson *et al.*, 1977.)

as a growth inhibitor and the latter a growth promotor (Seifert and Rudland, 1974a, b)) it is somewhat surprising that the levels of both nucleotides varied in the same fashion during induction synchrony (Gray, N. C. C. *et al.*, 1977).

Dickinson *et al.* (1976) also reported that the activity of adenylate cyclase in extracts (expressed per cell) oscillates during synchronous growth. Kassis and Zeuthen (1979) have confirmed these results with heat-synchronized *Tetrahymena*, but showed that enzyme activity remains constant when expressed with respect to cell protein. Kassis and Zeuthen also found cAMP to oscillate but at levels (20–70 pmol/10^6 cells) much higher than the unusually low levels reported by Dickinson *et al.* (1976). Furthermore, in the heat synchronized system, cell division was associated with minimal levels of cAMP, as it is in higher animal cells (see Section VIII.A1) and *S. cerevisiae* (Watson and Berry, 1977).

Levels of cAMP (Lovely and Threlfall, 1976) and adenylate cyclase (Lovely and Threlfall, 1978) show a rapid rise and fall just before mitosis during synchronous plasmodial growth of *Physarum polycephalum*. Similarly, cGMP (Lovely and Threlfall, 1976) and guanylate cyclase activity (Lovely and Threlfall, 1979) oscillate, each rising to maxima in S and late G_2. In contrast, Daniel *et al.* (1980) and Garrison and Barnes (1980) have reported only small fluctuations in the cyclic nucleotide levels, which are unrelated to any unique position in the cycle and are presumably not required for normal progress through the cycle.

Although cAMP occurs in various unicellular algae, it has been implicated in cell cycle control only in diatoms. In *Cylindrotheca fusiformis*, abrupt changes in cAMP and cGMP closely precede or accompany the uptake of silicic acid, wall formation and cell septation but no role can as yet be assigned to these changes (Borowitzka and Volcani, 1977).

Transient increases in the intracellular levels of cGMP have been observed, with a periodicity of one generation time, in certain strains of *E. coli* and *Bacillus licheniformis* (Cook *et al.*, 1980) and taken to suggest a role for the nucleotide in regulation of the cell cycle. In *Caulobacter crescentus*, evidence for a "start"-like event has been obtained, arrest at which can be overcome by exogenous dibutryl cAMP (Kurn *et al.*, 1977). In *Nocardia restricta*, cAMP metabolism appears to be regulated in such a way that cAMP levels are maximal at the onset of DNA replication (Lefebvre *et al.*, 1980).

VIII. Centrioles, Microtubules and Cilia

As early as 1898, it was suggested (Henneguy, 1898; von Lenhossek, 1898) that the formation of intracellular cilia from centrioles forced cells to become quiescent by removing the centriole from the mitotic cycle. The notion was revived by Rash *et al.* (1969) and several more recent investigations are in accord with the idea (Archer and Wheatley, 1971; Dingemans, 1969; Mori *et al.*, 1979). Other evidence (Sluder, 1979) indicates a role for microtubules in the timing of mitosis and the duration of the cycle as well as in the execution of mitotic events.

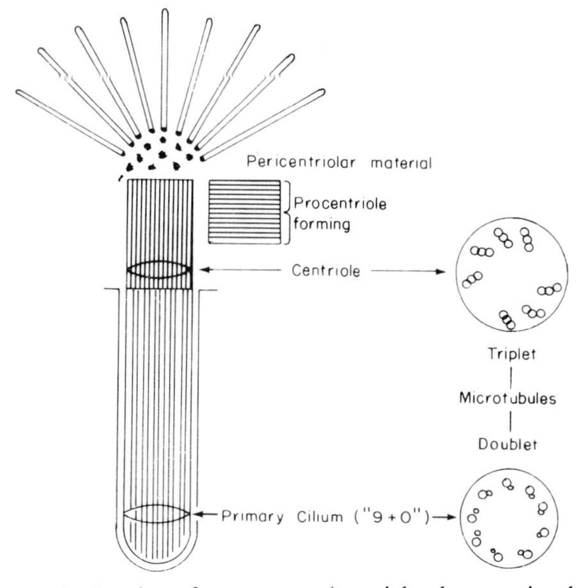

Fig. 8.23 Schematic drawing of centrosome (centriole plus associated structures). Each shaded cylinder represents a centriole, 0·25 μm by 0·5 μm. A cilium is formed at the distal end, and a procentriole (daughter centriole) and the pericentriolar material are found at the proximal end. The pericentriolar material organizes the polymerization of cytoplasmic and mitotic microtubules. The cilium formed by the interphase centriole is "9 + 0", consisting of nine peripheral microtubular doublets without a central doublet. (Reproduced with permission from Tucker *et al.*, 1979.)

In apparent contradiction to these proposals, however, Fonte *et al.* (1971) found that cells of rapidly growing tissues, and not just quiescent cells, exhibited ciliated centrioles (see Fig. 8.23). The recent results of Tucker *et al.* (1979) reconcile these observations. Using indirect immunofluorescence with an antitubulin antibody, Tucker *et al.* (1979) confirmed that centriole ciliation occurs when cells are arrested (e.g. by conditions of low serum or high cell density) in G_1. When stimulated to divide, the unduplicated centriole deciliated within 1–2 h, reciliated by 6–8 h and later underwent a further deciliation coincident with the initiation of DNA synthesis. Centrioles duplicate at about this time

(Robbins *et al.*, 1968; Rattner and Phillips, 1973; Snyder and Liskay, 1978). The deciliated and duplicated centrioles subsequently separated in preparation for mitosis. Thus, centrioles appear to be ciliated in G_1, near the time that quiescent cells become committed to DNA synthesis. Similarly, genetic studies of yeast (Chapter 7) have shown that the replication of the spindle pole body (analogous to the centriolar complex) occurs just prior to initiation of DNA synthesis and is regarded as a key event in "start" of the cycle.

Teng *et al.* (1977) have reported that the anti-microtubule agents, colcemide, vinblastin and colchicine potentiate the action of known mitogens in increasing the number of cells moving from G_1 to S. The underlying mechanism is obscure at present, but it has been suggested that microtubules anchor membrane components, including receptors for growth stimulants. Destruction of microtubule structures would facilitate mobility of membrane components, signal transfer and initiation of DNA synthesis.

IX. Roles for Membranes in the Control of Growth and Division

> "The cell biologist's passionate concern with the cell membrane can be summarized in a statement that sounds sarcastic: Whenever our ordinary explanations break down, we like to say that the final explanation will be found in the cell membrane". (Mazia, 1974).

Over the past few years, many models have been proposed that implicate the cell surface as a primary site for growth control, since it alone is in intimate contact with the diverse extracellular factors that influence cell proliferation (Nicolson, 1976; McGuire and Barber, 1976). One of the consequences of implicating cAMP in control of the cell cycle (Section VII) is that the membrane may be involved (Andersson, 1979), since adenylate cyclase and cAMP phosphodiesterase are both located in the surface membrane of eukaryotic cells. Furthermore, there are many examples correlating the control of proliferation with transient alterations in membrane composition and properties (Pardee, 1975). The supply of lipids for membrane synthesis is required for sustained proliferation (Cornell *et al.*, 1977; Cornell and Horwitz, 1980).

Proposals for the role of the cell membrane in regulating growth and division include the following:

(i) The control of division and malignant transformation by the control of uptake of essential nutrients (Bhargava, 1975, 1977; Bhargava *et al.*, 1979).

(ii) The excretion from the cell of mitosis-inhibiting proteins (Onda, 1979, 1980).

(iii) The stimulation of mitosis by mechanical stressing of the membrane (Curtis and Seehar, 1978).

(iv) In ciliates, a requirement for specific patterns of surface structures for oral development and cell division (De Terra, 1974).

(v) The storage or release by membranes of a growth initiator (Blanquet, 1978).

(vi) The role of structural transitions of the membrane in regulating division (Chernavskii *et al.*, 1977; Palamarchuk *et al.*, 1978).

(vii) The role of membranes in the contractile ring theory of division (Greenspan, 1977; Pujara and Lardner, 1979; Akkas, 1980).

Summary

No detailed, unifying hypothesis for the control of cellular growth and division has emerged. In eukaryotic cells, the primary regulatory events appear to reside in G_1. Thus, current studies focus on the kinetics of occupancy of this phase (in deference to the transition probability model) and on the identity of events that constitute the "start" event in yeast and the analogous restriction point in higher eukaryotes.

Simple "sizer" models of control, which propose that cells divide when they reach a critical size, do not suffice to explain cell cycle regulation. Constituent events of cell cycles *do*, however, show a size dependence; these include the initiation of DNA synthesis in bacteria, and fission yeast, the traverse of "start" in budding yeast and mitosis in fission yeast. Elements of the sizer concept, modified by a growth-rate dependence of the critical size, have been combined with evidence that cell division follows the initiation of chromosome replication in a deterministic fashion and after a relatively fixed period of time. This "sizer-timer" model forms an attractive hypothesis for cell cycle regulation in both bacteria and some lower eukaryotes.

A complete description of the control of division will necessitate explanations of the roles of other factors that have been implicated as key regulators or prerequisites for division. These include the synthesis of specific proteins, histone modifications, the intracellular levels of calcium, magnesium and cyclic nucleotides and the structure and function of microtubules and membranes.

9. Implications of Cell Division Cycle Studies

"To think of a man without duration is as ridiculous as thinking of him without an inside"

Sir Arthur Eddington

I. Cancer Chemotherapy and the Cell Cycle

Antitumour agents produce two types of responses in treated cells: (a) lethality, and (b) delay in normal cell cycle transit. These effects cannot be entirely studied in cell-free systems, as they evidently involve lesions in critical biosynthetic pathways, subcellular organelles or membranes, as well as in bypass and repair mechanisms. Concentrations producing effects at the cellular level are often far in excess of those reported to inhibit specific biochemical reactions as studied with purified enzymes. For these reasons, the development of chemotherapeutic protocols of practical value depends on cell cycle investigations using cells growing in culture or transplantable tumours in animals.

The most important cells in a tumour are the malignant stem cells (McCullough *et al.*, 1964; Bruce and Valeriote, 1968; Park *et al.*, 1971; Hamburger and Salmon, 1977) which are capable of unlimited proliferation and migration, thus permitting both growth of the primary tumour and the initiation of distant metastases (Hill, 1978a). It is unlikely that studies of model systems accurately reflect the kinetic behaviour of this malignant stem cell population; this is one major factor which limits the clinically-predictive usefulness of such studies. Most, but not all, drugs are more effective at killing proliferating cells than quiescent cells; bleomycin and BCNU are exceptional in this respect.

The use of a quantitative spleen-colony forming assay in the mouse (Bruce *et al.*, 1966) has enabled assessment of selective toxicity of chemotherapeutic agents for malignant stem cells and the kinetic classification of anticancer agents to be made. Thus Class I agents are equally toxic for both nonproliferating (normal) and cycling (lymphoma) cells; Class II are more toxic to lymphoma cells. Class II agents are also termed "Phase-Specific", as cells of the proliferating population are killed during a specific phase of the cell division cycle (Madoc-Jones and Mauro, 1974), and short exposures (< 24 h) do not affect nonproliferating cells. In Class III, the slopes of the survival curves for normal and lymphoma colony-forming cells show as much as a sixfold difference; these agents are termed "Cycle-Specific" since although they kill both proliferating and quiescent cells, dividing cells are much more sensitive and are killed at all stages of their cell division cycle. Flow cytofluorimetry has greatly facilitated elucidation of cell cycle stage specificities (Tobey and Crissman, 1972). Extensions of these early studies (for review see Hill, 1978a) have provided basic information for the development of safer regimes of cancer chemotherapy with minimal toxicity to normal bone marrow.

The original proposals of phase-specificity have been modified more recently to accomodate data from synchronous cultures of cell-lines indicating that some agents act at more than one phase of the cell division cycle. Concentration-dependent phase specificity is also encountered; low doses often favour delay in cell cycle transit, whereas high doses may be lethal (Fig. 9.1). Table 9.1 presents a summary of data on antitumour agents: these fall into four classes; antimetabolites, antibiotics, alkylating agents and antimitotic drugs.

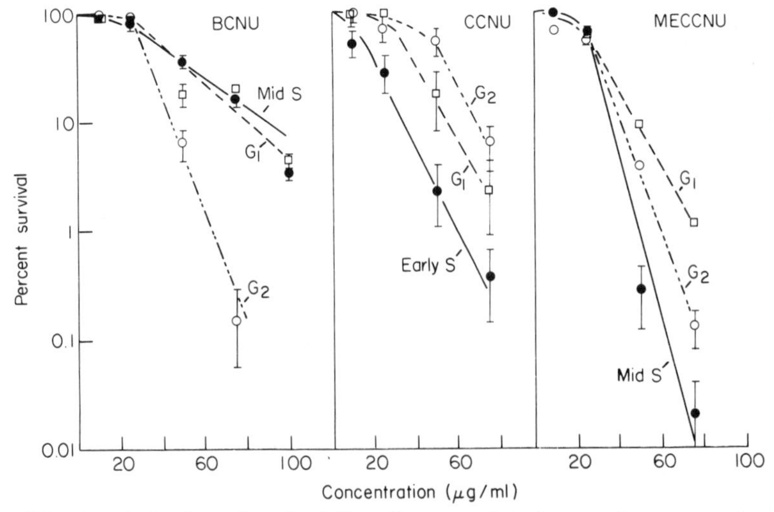

Fig. 9.1 Survival of synchronized T_1 cells exposed to increased concentrations of nitrosourea derivatives at selected points of the cell cycle. Abbreviations are given in Table 9.1. (Reproduced with permission from Drewinko and Barlogie, 1976.)

Table 9.1. *Cell Cycle Effects of Antitumour Agents*

	Kinetic Class	Phase of Delay or Arrest	Phase of Lethal Effects	Reference
Acronycine		G_2		Reddy *et al.* (1977)
Actinomycin D	III	G_1 $G_1/S/G_2$	G_1/S M	Kishimoto and Liberman (1964), Epanfanova *et al.* (1969), Bacchetti and Whitmore (1969), Bhuyan *et al.* (1972), Klein (1972), Bhuyan *et al.* (1976), Ross (1976), Watson and Chambers (1979)
AD-32 (*N*-trifluoroadriamycin)		G_2	S M	Krishman *et al.* (1977)
Adriamycin	III	SG_2	S S/G_2	Tobey (1972), Kim and Kim (1972), Barranco *et al.* (1973), Drewinko and Gottlieb (1973), Hittelman and Rao (1974), Sigdestad *et al.* (1979), Gohde *et al.* (1979a,b), Grdina *et al.* (1979), Bhuyan *et al.* (1980)
ANT(9,10-Anthracenedione,-1,4-bis{{2-{(2-hydroxyethyl)amino}-ethyl}amino}-diacetate				Evenson *et al.* (1979)
L-Asparaginase		G_1	S	Saunders (1972), Bosmann (1972), Harris *et al.* (1978)

Table 9.1. *(continued)*

	Kinetic Class	Phase of Delay or Arrest	Phase of Lethal Effects	Reference
5-Azacytidine	II	G_1 $G_1/S/G_2$	S	Nowell (1964), Li *et al.* (1970), Tobey (1972), Presant *et al.* (1975)
Azaleucine			S	Bosmann (1972)
Azaserine	II			Hill (1978a)
Azathioprine	II			Hill (1978a)
6-Azauridine		G_2		Burki (1971)
BCNU [1,3-bis-(2-chloroethyl)-1-nitrosourea]	III	SG_2	G_1/S G_2	Madoc-Jones and Mauro (1968), Tobey and Crissman (1975), Drewinko and Barlogie (1976), Nomura *et al.* (1978), Drewinko *et al.* (1979)
Bleomycin		S/G_2	G_2 M	Barranco and Humphrey (1971), Hittelman and Rao (1974), Ross (1976). Reddy *et al.* (1977), Barranco and Bolton (1977), Barranco *et al.* (1977a,b)
Camptothecin		$S/G_2/M$	S	Tobey (1972), Bhuyan *et al.* (1972), Li *et al.* (1972), Morais (1977)
CCNU [1-(2-chloroethyl)-3-cyclohexyl-1-nitrosourea]	III	G_2	G_1 G_1/S S M	Bhuyan *et al.* (1972), Tobey and Crissman (1975), Drewinko and Barlogie (1976)

Chartreusin		G_2				Bhuyan *et al*. (1978)
Chlorambucil	III		G_1		M	Bhuyan *et al*. (1972)
Chloramphenicol				G_2		Lieberman *et al*. (1973)
Chlorozotocin		SG_2				Tobey *et al*. (1975)
Cis-acid [4-(3-2-chloroethyl)-3-nitrosoureido)-*cis*-cyclohexane carboxylic acid]		G_2		S G_2		Rao and Rao (1976), Drewinko *et al*. (1977)
Cis-Platinum [*cis*-diaminedichloroplatinum]	III		G_1	G_2		Rosenberg *et al*. (1969b), Drewinko (1975), Drewinko and Barlogie (1976), Sigdestad *et al*. (1979)
Colcemid		M (metaphase)				Gelfant (1963), Ross (1976)
Cycloheximide		G_2	G_1/S			Bacchetti and Sinclair (1970), Mauro and Madoc-Jones (1970), Sentein (1979)
Cyclophosphamide (cytoxan)	III	G_2				Klein (1972), Gomez and Stutzman (1978)
Cytosine arabinoside	II	G_1/S/G_2		S		Bremerskov *et al*. (1970), Karon and Shirakawa (1970), Bhuyan *et al*. (1972), Ross (1976), Bhuyan *et al*. (1976), Reddy *et al*. (1977), Jones *et al*. (1977), Aglietta and Colly (1979), Paterson *et al*. (1979), Malec *et al*. (1979), Hromas *et al*. (1980)
DAG [1,2:5,6dianhydrogalactitol]		G_1 S G_2	G_1	G_2	M	Palyi (1975), Barranco and Fluournoy (1977), Goh *et al*. (1977)

Table 9.1. (*continued*)

	Kinetic Class	Phase of Delay or Arrest	Phase of Lethal Effects	Reference
Daunomycin	III	G_2	G_1 S	Brehaut (1969), Kim *et al.* (1968), Silvestrini *et al.* (1970)
Dianhydrodulcitol	III			Hill (1978a)
Dianhydromannitol	III			Hill (1978a)
Dibromodulcitol	III			Hill (1978a)
Dibromomannitol	III			Hill (1978a)
Dihydroxyanthraquinone				Traganos *et al.* (1980a)
Dimethylmyleran	III		M	Tobey and Crissman (1972)
DTIC [5-(3,3-dimethyl-1-triazeno)imidazole-4-carboxamide]	III	G_1S SG_2/M	G_1 G_1/S S	Gerulath *et al.* (1974)
Ellipticine			G_1 M	Bhuyan *et al.* (1972), Traganos *et al.* (1980b)
Emetine	II			Hill (1978a)
Ethyl methylsulphonate			G_1 M	Tobey and Crissman (1972)
5-FU [5-fluorouracil]	III	G_1/S	G_1 S G_2 M	Baserga (1968), Wheeler *et al.* (1972), Bhuyan *et al.* (1972), Ross (1976), Bhuyan *et al.* (1977a), Jellingh *et al.* (1979)
5-FUDR [5-fluoro deoxyuridine]		G_1/S S	G_1 S G_2	Rueckert and Mueller (1960), Epifanova (1969), Bhuyan *et al.* (1972)
Griseofulvin		M (metaphase)	S	Dustin (1963), Mauro and Madoc-Jones (1970)

Guanazole	II	G_1 G_1/S				Wheeler *et al*. (1972)
Homoharringtonine						Baaske and Heinstein (1977)
Hydrocortisone		G_1 G_1/S				Klein (1972), Eaves and Bruce (1974)
Hydroxyurea	II	G_1/S	G_1	S		Sinclair (1965), McCullough and Till (1971), Ross (1976), Herken *et al*. (1978), Tomita
ICRF 159 [1,2 *bis* (3,5-dioxopiperazine-lyl propane)]	II	G_2/M (prophase)			G_2M	Sharpe *et al*. (1970), Creighton (1975), van Putten and Lelieveld (1976)
Isophosphamide	III					Rentschler *et al*. (1978)
Leucanthone				S/G_2 G_2/M		Kimler and Schneiderman (1977), Kimler *et al*. (1978)
Maytansine					M	Wolpert-Defilippes *et al*. (1975), Rao *et al*. (1979)
Melphalan	III	G_2				Barlogie and Drewinko (1975)
6-Mercaptopurine	II	G_1		S		Vandevoorde and Hansen (1970), Bergsagel (1971)
Methotrexate	II	G_1/S	G_1	G_1/S S		Rueckert and Mueller (1960), Tobey and Crissman (1972), Harris *et al*. (1978), Weinstein *et al*. (1979), Torres (1979)
Methyl CCNU	III	G_2		S		Tobey and Crissman (1972), Drewinko and Barlogie (1976)
Methyl GAG [methylglyoxal-*bis*(guanylhydrazone)]		G_1				Heby *et al*. (1977), Pathak *et al*. (1977), Sunkara *et al*. (1977), Pathak *et al*. (1979)
Methylxanthine						Rowley *et al*. (1980)
Mithramycin		G_2				Tobey (1972)

Table 9.1. *(continued)*

	Kinetic Class	Phase of Delay or Arrest	Phase of Lethal Effects			Reference
Mitomycin C	III	S/G_2	G_1	G_2	M	Nowell (1964), Djordjevic and Kim (1968), Wheeler *et al.* (1970), Krishan (1975)
MMS [methylmethane sulphonate]		G_1	G_1			Nias *et al.* (1970)
Neocarzinostatin		$S\ G_2$	G_1	G_2	M	Bhuyan *et al.* (1972), Ebina *et al.* (1975), Rao and Rao (1976), Ishida *et al.* (1979), Iseki *et al.* (1980), Berry and Collins (1980)
Nitrogen mustard	III	G_2		G_1/S	M	Caspersson *et al.* (1963), Sakamoto and Elkind (1969), Mauro and Madoc-Jones (1970)
Nocodazole						Kunkel (1980)
Nogalamycin						Bhuyan *et al.* (1980)
Nor-HN2 [nor nitrogen mustard]	III					Hill (1978a)
NY 3170[1-propargyl-5-chloropyrimidin-2-one]		M (metaphase)				Wibe *et al.* (1979)
Peptichemio [multipeptide complex of *m*(di-(2chloroethyl)amino) phenylalanine]		$S\ G_2$	G_1	G_2		Barlogie *et al.* (1977)

Phenylalanine mustard				G_1	M	Bhuyan *et al*. (1972)
Phleomycin		G_2M	G_1	S/G_2G_2		Djordjevic and Kim (1967)
Predisone		G_1/S				Ernst and Killman (1970)
Procarbazine	II					Hill (1978a)
Puromycin		G_2M	G_1S			Kishimoto and Liberman (1964), Mauro and Madoc-Jones (1970)
Pyrazofurin	II	G_1/S				Hill (1978b), Olah *et al*. (1980)
Retinoic acid		G_1				Dion and Gifford (1980)
Rhodium (II) butyrate		S		S		Rao *et al*. (1980)
Streptozotocin		S G_2	G_1	S G_2	M	Bhuyan *et al*. (1972), Tobey *et al*. (1975)
Sulphur mustard			G_1/S		M	Mauro and Elkind (1968), Mauro and Madoc-Jones (1970)
6-Thioguanine	II	G_1 G_1/S	G_1	S		Wheeler *et al*. (1972)
Thiotepa	III		G_1		M	Bhuyan *et al*. (1972)
Trenimon (2,3,5-triethyleneimino-benzoquinone)						Grunicke *et al*. (1978)
Triethylenemelamine	III	G_2				Wheeler *et al*. (1970)
Triethylenetriphosphoramide		G_2				Wheeler *et al*. (1970), Kato *et al*. (1978)
Uracil mustard			G_1/S		M	Mauro and Madoc-Jones (1970)
Vinblastine	II	M (metaphase)	G_1	S	M	Palmer *et al*. (1960), Madoc-Jones and Mauro (1968), Ross (1976), Chirife and Studzins (1978)

Table 9.1. *(continued)*

	Kinetic Class	Phase of Delay or Arrest	Phase of Lethal Effects	Reference
Vincristine	II	M (metaphase)	S	Cardinali and Mehrota (1963), Madoc-Jones and Mauro (1968), Duffill *et al.* (1977), Alberts *et al.* (1977), Hill and Whelan (1980)
Vindesine	II			Hill and Whelan (1980)
VM26 [4′-dimethyl-epipodophyllotoxin-9-(4,6-*o*-ethenylidine-β-D-glucan pyranoside]	II	G_2/M (metaphase)	G_2	Stahelin (1970), Lieberman *et al.* (1973), Misra and Roberts (1975), Rao and Rao (1976), Gotzos *et al.* (1979)
VP-16-213 [4′-dimethyl-epipodophyllotoxin-9-(4,6-*o*-ethenylidene-β-D-glucanopyranoside]	II	G_2	S G_2	Grieder *et al.* (1974), Drewinko and Barlogie (1976), Barlogie and Drewinko (1978)
Yoshi-864 [3,3′iminodi-1-propanoldimethane sulphonate(ester)]	III	G_2	G_1	Drewinko (1975), Drewinko and Barlogie (1976)

Various types of drug treatment, irradiation or nutritional insufficiency may disturb the random temporal distribution of cell divisions both in tissue culture and *in vivo* for a limited period of time (van Putten *et al.*, 1976; Belt *et al.*, 1980; Camplejohn, 1980; Moran and Straus,

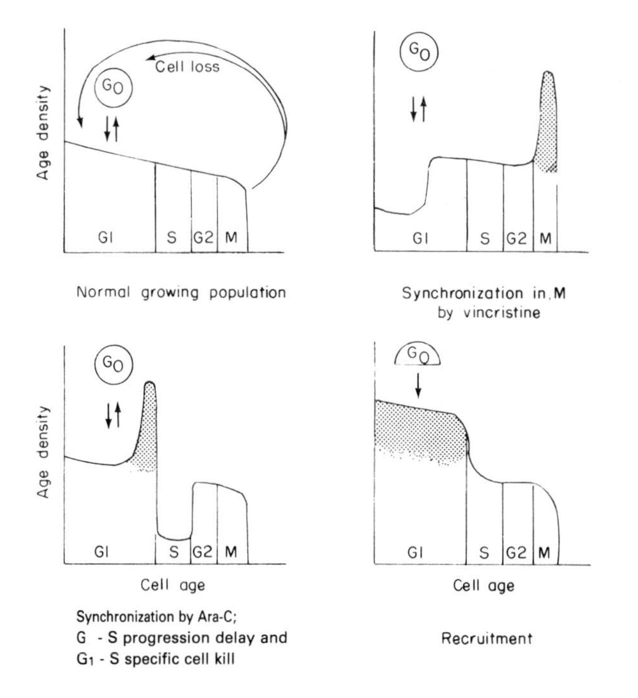

Fig. 9.2 Diagrammatic representation of cell population kinetics of a random dividing cell system. Synchronization (by vincristine and Ara-C) and recruitment are also represented. The areas of the age density diagrams are intended to roughly indicate the relative sizes of the populations in the different cell cycle phases. (Reproduced with permission from van Putten *et al.*, 1976.)

1980; Volm *et al.*, 1980; Wayss *et al.*, 1980; Zywietz and Jung, 1980). Synchronization *in vivo* is achieved by at least two mechanisms (Fig. 9.2). Selective killing of cells in a sensitive phase of the cell division cycle leads to a predominance of cells in the insensitive phase. Alternatively, delay or blockage of cell cycle traverse may occur. A further indirect factor contributing to synchronization is the simultaneous recruitment of "resting" cells into the mitotic cycle as a consequence of cell killing (Vogler *et al.*, 1976). Synchronization *in vivo* in early S has been achieved with hydroxyurea and cytosine arabinoside, whilst vincristine yields accumulation at mitosis, and bleomycin gives progressive delay in G_2. Metabolism or excretion of the drugs leads to the release of the block and the more or less synchronous passage of cells through the mitotic cycle. Many other drugs and also radiation give partial synchronization (Nicolini, 1976). The consequent time-dependent variation in sensitivity of the cells to antitumour agent(s) (used singly, or more usually in various combinations) or to regimens of fractionated radiotherapy may then be exploited (van Putten *et al.*, 1976).

The alternative rationales for the use of combinations of chemotherapeutic agents are as follows (Bhuyan *et al.*, 1976):

1. Phase and cycle-specific drugs; e.g. cytosine arabinoside (ara-C) and a non-cycle specific drug (e.g. BCNU).
2. Drugs with different phase-specific cytotoxicities, e.g. ellipticine or neocarzinostatin (maximum toxicity in M and G_1) + ara-C, hydroxyurea or vincristine (all of which are S-phase specific) (Bhuyan *et al.*, 1972).
3. Drugs acting at the same phase but causing cytotoxicity by differing mechanisms; e.g. ara-C (DNA polymerase inhibitor) used with hydroxyurea or 5-hydroxypicolinaldehyde (ribonucleotide reductase inhibitors). Additive effects are not always produced.
4. Sequential combinations in which the first drug accumulates cells in a particular phase (e.g. vincristine or colcemid, causing mitotic arrest). The block is released and the second drug is given when cells reach the sensitive phase of their division cycle, e.g. ara-C is given in S-phase (Vadlmudi and Goldin, 1971).
5. A combination of drugs in which the first blocks progression at one phase of the cycle (e.g. ara-C blocks at G_1/S interface). The second drug, e.g. actinomycin D is chosen on the basis of maximum cytotoxicity to cells at this phase. This rationale has been experimentally vindicated for L 1210 and DON cells (Bhuyan *et al.*, 1976). Another example of this kind of combination is the use of daunorubicin or doxorubicin after ara-C in the treatment of leukaemia (Edelstein *et al.*, 1973; Whittaker *et al.*, 1977).

Apart from suggesting effective drug combinations, clinically-relevant predictions from such experiments (Hill, 1978a) include the importance of treating small tumours rather than large ones, the necessity for intensive chemotherapy until the tumour is eradicated, and the reduced toxicity of intermittent periods of treatment which each last no longer than 36 h (< 2 cell cycle times) (Price and Goldie, 1971). A number of key parameters of tumour growth cannot be reliably estimated at present (e.g. the proportion of stem cells in human solid tumours), and the primary test for promising new agents (e.g. the P388 leukaemia screen) may not detect those most effective against solid tumours containing a large proportion of non-cycling cells. An alternative which may prove more useful is to test new cytotoxic drugs against xenografts (human tumours growing in immune-deficient rodents) (Hill, 1978a). Methods of development of new anti-cancer drugs are reviewed in a monograph (Sanders and Carter, 1977). Several promising approaches to the problem of making toxic anti-cancer agents more target-specific are currently being tested (Hoult, 1980). A novel approach to combination chemotherapy is provided by the experiments of Pardee and Dubrow (1977) in which suboptimal nutritional conditions or specific drugs are used to reversibly shift normal cells into the quiescent state. Some malignant cells do not react like normal cells and continue to grow: simultaneous use of a regulative protecting drug and a growth-toxic anticancer drug is indicated. The integration of combination chemotherapy with surgery and/or radiotherapy and/or immunotherapy provides the basis for cancer treatment in general; evidently these advances rely on increasing appreciation of cell cycle kinetics.

The number of possible combinations of drugs, dosage, and time schedule is so numerous that several computer-aided techniques have been developed in order to evolve protocols that optimize drug dosage regimens for experimental *in vivo* synchronization of cells in animals and for cancer chemotherapy. An extensive literature describes the various approaches to mathematical modelling of cell cycle kinetics (Valleron, 1975). These include Monte Carlo simulation (Nicolini, 1976) and control systems analyses (Padulo and Arbib, 1974). Some recent approaches to these problems are found in the following papers: Woo *et al*. (1975); Durand (1976); Gray (1976); Krug (1976); Alberghina (1977); Boyarsky (1977); Green and Bauer (1977); Kim *et al*. (1977); Lee (1977); Maertelaer and Galand (1977); Moebs (1977); Ravkin and Pryanish (1977); Rotenberg (1977); Chuang and Soong (1978); Dorian *et al*. (1978); Lee (1978); Skehan and Friedman (1979); White and Thames (1979); Zajicek *et al*. (1979); Hagander (1980).

II. Radiation, Heat and the Cell Cycle

A. IONIZING RADIATION

Variations in sensitivity to radiation and heat at different stages of the cell cycle have been observed in bacteria, lower eukaryotes, and in the cells of higher organisms. A reasonable solution to the problem of cancer therapy is suggested by these observations, namely: synchronize the population of malignant cells and then irradiate at that stage of the cycle at which they are most radiation-sensitive. Two difficulties hinder this approach: (a) not all the malignant cells are "cycling" as some are withdrawn into the G_0 state, and (b) the degree of *in vivo* synchronization achieved in most cases is modest. Even so, a more quantitative understanding of differential response of tissues and tumours with different cell population kinetics has already helped to achieve optimization of fractionation during radiotherapy.

The timing of exposure of a cell to radiation with respect to its cell cycle traverse will produce various responses in terms of survival, chromosome aberrations, DNA synthesis and repair, and division delay (Quastler, 1963; Sinclair, 1968; Nias and Lajtha, 1968). Survival curves for mammalian cells *in vitro* (Fig. 9.3; Terasima and Tolmach, 1961), indicate that in general, whenever there is a long G_1 period, there appears to be a resistant state in early G_1 followed by a decline in survival towards S—except in a human kidney cell line (Vos *et al.*, 1966). Resistance begins to rise as soon as cells enter S to a maximum in late-S. Survival then falls sharply in G_2 to the value found in mitosis (Frindel and Tubiana, 1971). Some strains of L cells have a radiosensitive S-phase (Whitmore *et al.*, 1965). However age response functions depend strongly on the dose level applied, and thus discussion of cell-cycle dependent response always needs specification of the irradiation dose (Fig. 9.4; Fidorra and Linden, 1977; Mitchell *et al.*, 1979). The main features of radiosensitivity of mammalian cells *in vitro* are paralleled *in vivo*. Thus, Kallman (1963) showed that survival of mice after whole body irradiation was dependent on the interval between a first conditioning dose and subsequent doses. The effects of split dose regimes have been widely studied since the pioneering experiments of Elkind *et al.* (1961), who demonstrated that cyclic variation of radiosensitivity *in vivo* after a conditioning dose may be considered to depend on three independent phenomena: (a) selective killing of radiosensitive cells, (b) repair of sublethal damage and, (c)

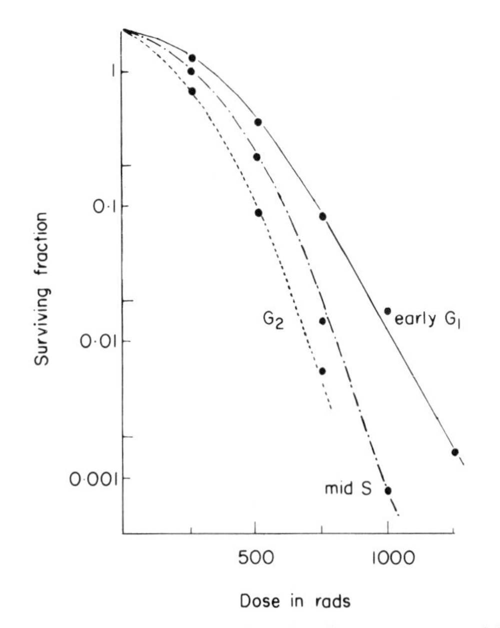

Fig. 9.3 Survival of synchronized L-P59 cells after treatment with X rays at different phases of the cell cycle. (Reproduced with permission from Thompson and Humphrey, 1969.)

change in the radiosensitivity of the surviving cells as they synchronously progress through the cell cycle after a temporary division delay.

The repair of sublethal and potentially lethal damage is also cell cycle-dependent. Thus, in Chinese hamster cells, recovery occurs mainly in S cells (Sinclair and Morton, 1963) with some recovery from irradiation in G_1, but none from G_2- or M-irradiated cells (Sinclair, 1968); the repair processes are themselves radiation sensitive (Ueno *et al.*, 1979). The variation of radiosensitivity over the cell cycle is often similar for hypoxic and well-oxygenated cells (Kluuv and Sinclair, 1968; Hall *et al.*, 1968; Legrys and Hall, 1969), although cell cycle-dependent variations in the oxygen enhancement effect has occasionally been demonstrated (Pettersen *et al.*, 1977). Cysteamine has a differential X-ray-protective effect on Chinese hamster cells during the cell cycle, protecting the sensitive mitotic cells the most, and relatively insensitive S cells the least (Sinclair, 1968). 2-Nitroimidazole is used as a radiosensitizer for hypoxic tumour cells (Petersen, 1978). Inhibitors of DNA synthesis added after irradiation enhance cell killing (Weiss and Tolmach, 1967). Thioguanine-resistant

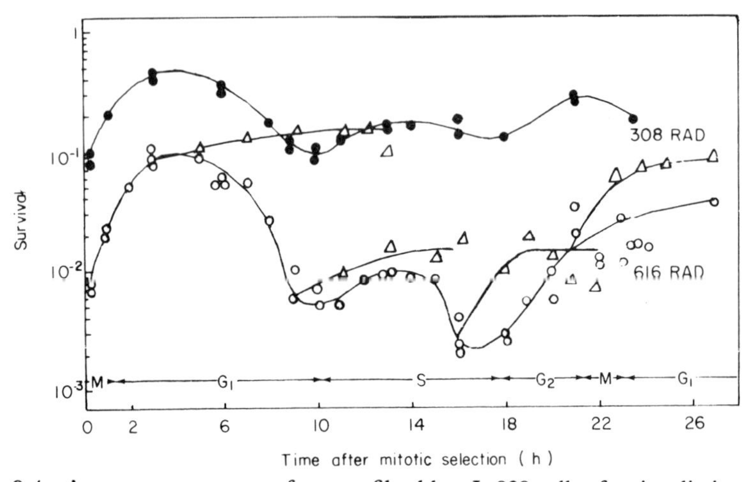

Fig. 9.4 Age response curves of mouse fibroblast L-929 cells after irradiation with 200 kVp X rays (308 rad, and 616 rad, respectively), at different time in the cell cycle. Cell survival is plotted versus time after mitotic detachment. Zero time is taken as middle of mitosis. Each symbol represents the mean of about 10 independent colony tests. Recovery curves after fractionated irradiation with 2 doses of 308 rad (heavy curves). Between the two exposures the cells were maintained at 22°C. (Reproduced with permission from Fidorra and Linden, 1977.)

clones are induced in synchronous CHO cells after exposure to ionizing radiation (Burki, 1980).

The extent of chromosomal damage incurred on irradiation also shows a marked cell cycle dependency (Nordenson, 1978; Vanbuul *et al.*, 1978; Natarajan and Meyers, 1979; Leonard and Decat, 1979). Euler and Hevesy (1942) demonstrated a decreased incorporation of radioactivity into DNA of the Jensen Sarcoma of the rat after irradiation, and similar findings have been documented with many animals. In regenerating liver, ionizing radiations disturb DNA synthesis by inhibiting enzyme formation (Beltz and Applegate, 1959; Bollum *et al.*, 1960; Okada, 1960). Radiation in G_2 leaves DNA synthesis unchanged for the duration of M and G_1; decreased precursor uptake into DNA for several hours after irradiation has been described by Yamada and Puck (1961) and by Whitmore *et al.* (1961). On the other hand, the effect of irradiation on DNA in G_1 is more rapid (Mak and Till, 1963; Lesher, 1967). Biphasic response curves for irradiation of mammalian cells in S have been observed both *in vitro* (Lajtha *et al.*, 1958) and *in vivo* (Ord and Stocken, 1958; Looney *et al.*, 1963).

Increased precursor uptake into DNA following irradiation has frequently been demonstrated; this may be due to S-phase arrest, increased initiation of DNA synthesis, or selective killing of cells in stages other than in S phase (Frindel and Tubiana, 1971; Keng and Wheeler, 1980a,b).

The delay of mitosis produced by ionizing radiation, first demonstrated in 1927 by Canti and Spear, shows a cell cycle-dependency which is roughly the inverse of that for survival (Puck and Yamada, 1962; Froese, 1966; Yu and Sinclair, 1967; Terasima and Tolmach, 1963b) and is dose-dependent (Yamada and Puck, 1961; Elkind *et al.*, 1963; Schneider and Whitmore, 1963; Caldwell *et al.*, 1965; Aleschenko *et al.*, 1967). The relative efficiencies of neutrons, gamma rays and alpha-rays at producing division delay in bean root tips are approximately in the proportion of 4 : 2 : 1; this ratio reflects the relative extents of ionization produced by the different radiations (Grey *et al.*, 1940). Recent studies on cell cycle variations in sensitivity to irradiation include experiments with ultrasound (Fu *et al.*, 1980), α-particles (Raju *et al.*, 1980b; Nusse, 1980), β-rays from incorporated [^{3}H]-thymidine (Pollack *et al.*, 1979), fast electrons (Akimov and Moskalik, 1977), negative pions (Schlag *et al.*, 1978; Heyder and Pohlit, 1979), fast neutrons (Gragg *et al.*, 1978), high energy argon ions (Hall *et al.*, 1977; Raju *et al.*, 1980a) and heavy-ion beams (Blakely *et al.*, 1980).

Detailed information on the molecular mechanisms involved in cellular damage by irradiation and its repair come from studies with radioresistant and radiosensitive mutants of bacteria and lower eukaryotes (Haynes *et al.*, 1978), as these systems are more amenable to genetic studies than are the mammalian cells. Recent cell cycle research in this area includes studies with root meristem (Evans, 1978), *Caulobacter crescentus* (Iba *et al.*, 1977a,b) and yeast (Holtz and Pohlit, 1977; Magni *et al.*, 1977).

B. ULTRAVIOLET IRRADIATION

Exposure of synchronous HeLa S3 and Chinese hamster cell cultures to u.v. light at selected times during their cell division cycle indicates that they are resistant in G_1, sensitive through S and most sensitive in the middle of DNA synthesis; sensitivity then decreases in G_2 (Sinclair and Morton, 1965; Djordjevic and Tolmach, 1967; Han and Sinclair, 1969; Burg *et al.*, 1977; Riddle and Hsie, 1978; Downes *et al.*, 1979). Studies with 5′-fluoro-2′deoxyuridine and hydroxyurea suggest that cellular sensitivity is not directly correlated with disruption to DNA synthesis itself but rather to some concomitant process. The major difference between cell responses to X-rays and to u.v. irradiation is that X-ray survival has two minima, one in M and a second in later G_1, whereas response to u.v. shows only one peak of sensitivity. The G_2 population become more resistant to u.v. but more sensitive to X-rays as it ages. Timing of maximum survival to both types of radiation can be shifted by using inhibitors of DNA and protein synthesis (Han and Sinclair, 1969). Cell cycle blocks and the delay by u.v. irradiation have been shown in mouse L cells (Domon and Rauth, 1968); both the duration of delay and proportion of blocked cells are proportional to exposure, and related to cell cycle stage at the time of exposure, with the S-phase cell the most sensitive in this respect (Djordjevic and Tolmach, 1967; Domon and Rauth, 1968).

Haematoporphyrin has a photodynamic activation effect on near u.v. irradiation of human (NHIK 3025) cells *in vitro* (Christensen and Moan, 1979; Christensen *et al.*, 1979). Maximal sensitivity to activation was observed in mid S-phase in synchronous cultures prepared by mitotic selection.

Damage to DNA by u.v. irradiation is of three main kinds: damage or removal of bases, strand breakage, and covalent cross-linking of the DNA. Photochemical alteration of ribosomal RNA synthesis (Koch *et al.*, 1976) and membrane damage are also important factors (Alper, 1970).

Three major repair mechanisms have been proposed (Haynes *et al.*, 1978); excision, mutagenic error-prone and recombinational processes have been clearly differentiated by the use of specific bacterial and yeast (*rad*) mutants defective in these (Cole *et al.*, 1976; Averbeck *et al.*, 1978). Cell-cycle dependence of DNA repair has been followed (Giulotto *et al.*, 1978; Burg *et al.*, 1979; Collins *et al.*, 1980a,b). In synchronous cultures, G_2 resistance to X-rays is paralleled by resistance to the combined effect of near u.v. light (365 nm) and 8-methoxypsoralen, a compound that photo-binds to pyrimidine bases (Henriques *et al.*, 1977); excision repair probably acts mainly in G_1, whereas efficient recombinational repair operates in G_2. Marked division delays in u.v. irradiated cells have been described for *Schiz. pombe* (Gill, 1965; Hannan *et al.*, 1976) and *Physarum polycephalum* (McCorquodale and Guttes, 1977; Tyson *et al.*, 1979). In *Schiz. pombe*, Swann (1962) described a tenfold difference in u.v. sensitivity at different stages of the cell cycle; essentially similar results have been obtained by Gill (1965) and Bullock (1979). Using 2-phenylethanol to shift the timing of the S-phase, Williams (1979) has resolved two u.v.-sensitive phases (Fig. 9.5).

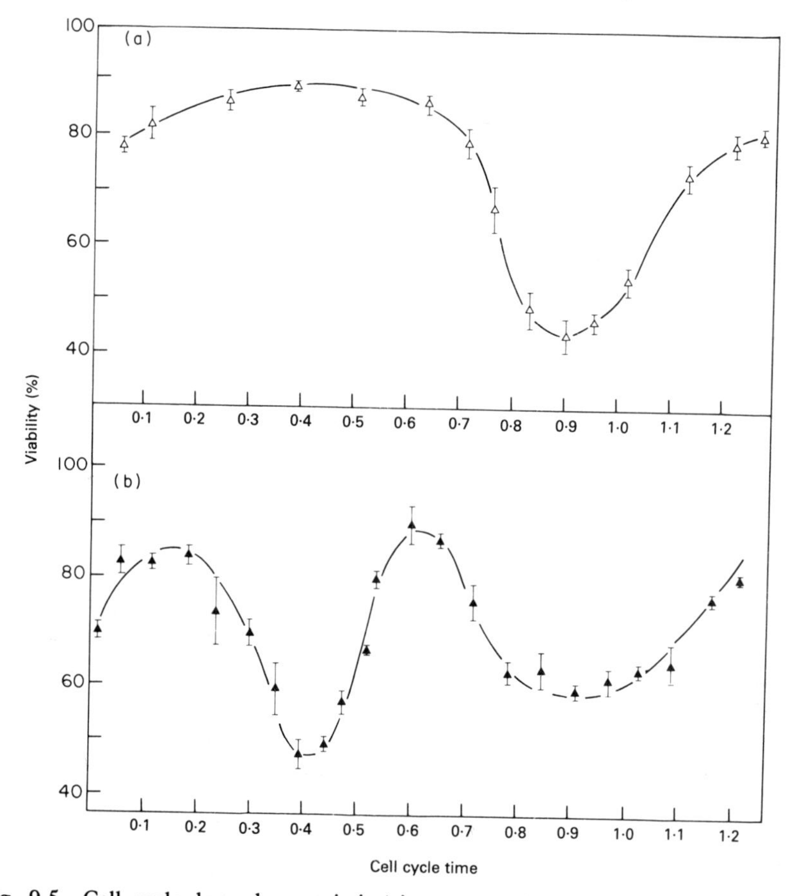

Fig. 9.5 Cell cycle-dependent variation in u.v. sensitivity of *Schizosaccharomyces pombe* grown in the absence (a) and presence (b) of 0·2% 2-phenylethanol. (a) Each data point represents the mean of six independent determinations of survival after u.v. irradiation (6 cm from a 6 W low pressure H_g lamp for 3 min), error bars are SE of mean, 1·0 was the mid-point of first synchronous division. (b) As in (a) but seven determinations were made. See Fig. 9.6 for cell cycle markers. (Reproduced with permission from Williams, 1979.)

Ultraviolet irradiation of bromosubstituted chromosomes (using 5-bromo-2′-deoxyuridine) of naturally synchronous cells in Allium root meristems has been employed in an attempt to produce phenocopies of cell cycle mutants (De la Torres and Gonzales, 1979). Radiation-induced mutations in synchronous CHO cells have been described by Burki *et al.* (1980).

C. HEAT

The usefulness of sub-lethal heat shocks in studies of the mechanism and control of cell division has been discussed in Chapter 1; selective effects on cancer cells have evoked the widespread current interest in hyperthermia as an antitumour agent (Cavaliere *et al.*, 1967; Pettigrew *et al.*, 1974). Enhancement of the therapeutic ration for cytotoxic drugs is also an attractive possible consequence of the observed synergistic effects of hyperthermia and drug therapy (Robinson *et al.*, 1974).

The sensitivity to hyperthermia of both normal and malignant cells is cell cycle dependent (Ross-Riveros and Alpen, 1978) with cells in late S and mitosis most susceptible (Dewey *et al.*, 1971; Westra and Dewey, 1971; Palzer and Heidelberger, 1973; Kim *et al.*, 1974; Dewey and Highfield, 1976; Bhuyan *et al.*, 1977a). Hyperthermia blocks cell cycle traverse at the S/G_2 boundary or in G_2 (Sisken *et al.*, 1965; Kal *et al.*, 1975; Dewey *et al.*, 1977; Kase and Hahn, 1976; Schlag and Lück-Hule, 1976; Sapareto *et al.*, 1977), or in mitosis (Juul and Kemp, 1933; Selawry *et al.*, 1957; Martin and Schloerb, 1964; Sisken *et al.*, 1965; Dewey *et al.*, 1971; Westra and Dewey, 1971; Highfield and Dewey, 1975). These effects are reversible under favourable growth conditions (Selawry *et al.*, 1957; Martin and Schloerb, 1964; Sapozink *et al.*, 1973; Kal *et al.*, 1975). Heat treatments which are mild in terms of killing (>20% survival) induce long division delays in Chinese hamster ovary cells (Gerweck and Dewey, 1976). A log-linear relationship between cell lethality and chromosomal aberrations has been shown in cells treated with heat during S-phase (Dewey *et al.*, 1971). Changes in centrosome structure induced by hyperthermia have also been reported (Barrau *et al.*, 1978). Heat treatment prior to X-ray irradiation increases radiation-induced cell lethality possibly due to chromosomal damage by heat (Dewey *et al.*, 1978; Tomasovic *et al.*, 1979; Bichel *et al.*, 1979). Potentiation of the destructive effect of heat (42°C) on synchronized cancer cells in culture has also been achieved using a specific antiserum (Jasiewicz and Dickson, 1976). Growth of the fission yeast *Schiz. pombe* with 2 phenylethanol shifts forward S-phase from its usual timing just before cell division, but leaves the most heat-sensitive stage unaltered (Fig. 9.6, Bullock and Coakley, 1976, 1978, 1979). An important implication is that DNA damage is not the dominant cause of cell killing by heat.

Dependence of freeze-thaw damage on cell cycle stage in cultured mammalian cells and in a yeast has also been shown (Terasima and Yasukawa, 1977; McGann and Kruuv, 1977; Cottrell, 1980). The

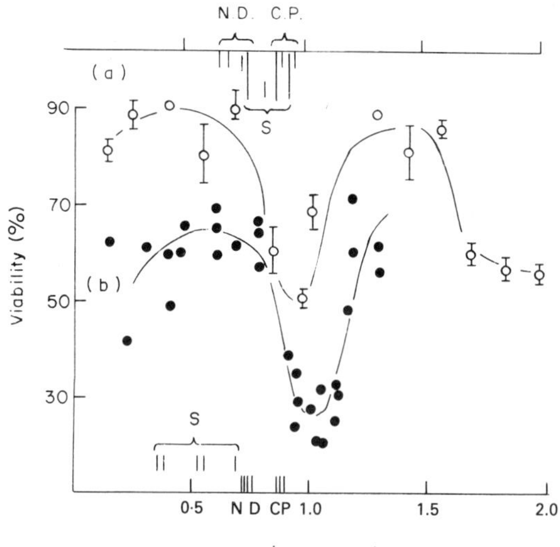

Fig. 9.6 Cell cycle-dependent variation in heat sensitivity of *Schizosaccharomyces pombe* grown in the absence (a) and presence (b) of 0·2% 2-phenylethanol. Survival values are after 15 min, (49°C) heat shocks at different stages of the cell cycle. Shown also are the position of each S phase, nuclear division and cell plate peak determination. The survival values in the absence of 2-PE show error bars of 1 S.E.M. The cell cycle time of 1·0 was taken as the middle of the first synchronous division. (Reproduced with permission from Bullock and Coakley, 1979.)

lethal and mutagenic actions of hydrostatic pressure on yeast also show stage-specific changes during the cell cycle (Rosin and Zimmerman, 1977).

III. Chemical Mutagenesis—Cell Cycle Dependence

The induction of both nuclear and non-Mendelian mutants in lower eukaryotes by chemical mutagenesis shows cell cycle dependence. Thus N-methyl-N′-nitro-N-nitrosoguanidine (MNNG) induction of nuclear mutants in *Chlamydomonas reinhardii* (Lee and Jones, 1973) and *Saccharomyces cerevisiae* (Dawes and Carter, 1974; Kee and Haber, 1975) is maximal during nuclear S-phase. In some strains of yeast, mitochondrial mutants are induced at a cell cycle stage different from the nuclear S-phase, at a phase where mitochondrial DNA replication is postulated to occur (Dawes and Carter, 1974), although less equivocal results were obtained for a different yeast strain (Hamill, 1976).

The induction of non-Mendelian, streptomycin-resistant mutants in *C. reinhardii* shows a maximum during the chloroplast DNA S-period (Lee and Jones, 1973). It is proposed that these effects result from elevated mutagenesis of DNA within or in close proximity to replication complexes (Cerdá-Olmedo *et al.*, 1968). The lethal effects of MNNG are also most pronounced during the nuclear S-phase of *C. reinhardii* (Lee and Jones, 1976); demonstration of differential action of the mutagen at different stages of the cell cycle is not possible at high mutagen concentrations (Plusquellec and LeGal, 1972; Schimmer and Loppes, 1975). Similar studies have been carried out with yeast (Barale *et al.*, 1980a,b) and with synchronous CHO cells (Goth-Goldstein and Burki, 1980).

IV. Chemical Carcinogenesis—Cell Cycle Dependence

A good positive correlation exists between the ability of a compound to inhibit DNA synthesis and its carcinogenicity as assessed by the Ames test (Painter and Howard, 1978). However, it has been shown that measuring the inhibition of DNA synthesis at some arbitrary point after application of a compound is not as informative as measuring the consequent perturbation of cell division cycle kinetics (Bartholomew *et al.*, 1979). In this study, flow cytofluorimetry and measurements of [^{3}H]-thymidine incorporation showed that the rate of traverse of the S-phase in mouse liver cell strains is slowed by a carcinogenic diol-epoxide derivative of benzo(a)pyrene. The parent compound, or its 7,8-diol, had very little effect on cell cycle progression in a cell line which could not rapidly produce the diol-epoxide. Differential sensitivities of synchronized HeLa S3 cells to radiation carcinogen-induced and chemical carcinogen-induced mutagenesis have been demonstrated (Watanabe and Horikawa, 1980).

V. Differentiation and Development

The choice between continued cell division cycling or entry into a pathway of differentiation is taken early in the eukaryotic cell cycle, at some time in the G_1 phase. This is so for several different systems. In one of the best understood systems, *S. cerevisiae* (Hartwell, 1974), alternative developmental programs available to the newly-divided cell include the reversible loss of proliferative ability as a consequence of depletion of nutrients, sexual conjugation after synchronization of the cell cycle with that of an individual of opposite mating type, or sporulation and meiosis (Hirschberg and Simchen, 1977). In *Tetrahymena*, the division/conjugation decision is taken in G_1 (Wolfe, 1976) and the same is true of *Euplotes crassus* (Dini and Luporini, 1979). Before the decision point, organisms are undifferentiated in the sense that they are capable of proceeding to different developmental pathways. But cell differentiation is not always the dichotomous situation so far presented: cell division and cell differentiation coexist during embryogenesis and in the tissues of adult metazoans, where partially differentiated cells continue to divide (e.g. erythroblasts, myoblasts etc.) and differentiated stem cell lines maintain the cellular populations of continuously renewed tissues (Buckley and Konigsberg, 1977). Evidently timers are essential for the genesis of developmental sequence (Soll, 1979).

These problems, although strictly speaking outside the scope of this monograph, are in a very real sense analogous to those we have considered within the cell division cycle. Thus the processes of differentiation, pattern formation and morphogenesis which together constitute development in biological systems are spatial-temporal ones. Inherited information has to be translated into positional information (Wolpert, 1969), or in other words, into a map which can be read by cells (Goodwin and Cohen, 1969) and a clock for temporal coordination. Thus it is proposed that a map arises from wave-like propagation of activity from localized pacemakers. Temporal organization of an individual cell is converted by functional intercellular interactions into a spatial ordering. The idea that oscillatory processes in an organism underly the biological clock is an old one, but the idea that they may also be involved in the spatial organization is rather new (Hejnowicz, 1975) arising from the work of Turing (1952), Othmer and Scriven (1974), and Jorné (1974).

New understanding of the dynamic behaviour of living systems, particularly of the mechanisms by which stable and astable dynamic states are developed, maintained, and controlled, gives fresh impetus to studies of developmental biology (Haken, 1977, 1978, 1979). Dissipative structures often occur in open thermodynamic systems operating far from equilibrium (Glansdorff and Prigogine, 1971; Nicolis and Prigogine, 1977). In biological systems these can arise as a consequence of the coupling of allosteric enzymes with transport processes via appropriate diffusion gradients (Goldbeter, 1973). These dissipative structures are seen in simple systems (a horizontal fluid layer heated from below gives an organized convective motion which takes the form of a hexagonally arrayed pattern or Bénard instability). The experimental demonstration of dissipative structure formation in chemical systems is very well documented, and the most familiar example is the Belousov-Zhabotinski reaction in which the formation of propagating wavefronts is initiated from distinct centres in a thin layer of unstirred reactants (Zaikin and Zhabotinski, 1970).

The occurrence of dynamic spatial structures (i.e. heterogeneous spatio-temporal distribution of NADH in glycolysing cell-free extracts of yeast (Boiteux and Hess, 1980) provides an example of a system intermediate in complexity between the chemical systems and the morphogenetic patterning of embryos described by Turing (1952). Whether the structures generated in extracts have any counterparts *in vivo* however seeems rather unlikely; diffusion coefficients might have to be drastically reduced to produce patterning in domains of subcellular dimensions and the processes of cyclosis and cytosolic flow provide a well-mixed milieu (but see Hunding, 1981, 1982).

Morphogenetic fields in protozoa and developing embryos give symmetry and patterns which can be modelled using Laplace's field equation; the molecular basis for the cellular processes involved in spatial organization of cells and organisms remains unknown (Goodwin, 1980). If diffusion-controlled chemical gradients are not responsible we must look to electrical phenomena. Ion currents generated by the movement of ions across membranes in such a way as to produce an imbalance of electrical charge are responsible for the performance of diverse work functions (Harold, 1977). The electric field in the developing eggs of brown algae has a polarity which may provide localized growth patterning (Nuccitelli and Jaffe, 1977). These exciting recent observations suggest new approaches to the problems of epigenetic information flow in the spatio-temporal unfolding of differentiating systems to produce structure and form.

Summary

Far reaching implications of cell division cycle studies include studies of cancer chemotherapy, effects of ionizing and u.v. irradiation and hyperthermia on cycle traverse. Differential sensitivities to physical and chemical agents at different times in the cell cycle have important consequences in chemical mutagenesis and carcinogenesis—and uses in medical and environmental sciences. An understanding of the basic mechanisms involved in the cell division cycle must be the key to the unravelling of the controls involved in different developmental programmes utilized in processes of cellular differentiation.

References

Aaronson, R. P. and Blobel, G. (1975) *Proc. natn. Acad. Sci. USA* **72**, 1007–1011.
Abbo, F. E. and Pardee, A. B. (1960) *Biochim. biophys. Acta* **39**, 478–483.
Abraham, K. A., Pryme, I. F., Åbro, A. and Dowben, R. M. (1973) *Exp. Cell Res.* **82**, 95–102.
Abou-Sabe, M. (1976) (Ed.) "Cyclic Nucleotides and the Regulation of Cell Growth". Dowden, Hutchison and Ross, Inc., Stroudsberg, Penn., USA.
Adams, J. (1977) *Expl. Cell Res.* **106**, 267–275.
Adams, R. L. P. and Lindsay, J. G. (1967) *J. biol. Chem.* **242**, 1314–1318.
Adair, G. M., Thompson, L. H. and Lindl, P. A. (1978) *Somat. Cell Genet.* **4**, 27–44.
Adar, F. (1978) *In* "Frontiers of Biological Energetics" (P. Dutton, J. S. Leigh and A. Scarpa, Eds) Vol. I, pp. 592–599. Academic Press, New York and London.
Adler, H. I. and Hardigree A. A. (1964) *J. Bact.* **87**, 720–726.
Adler, H. I. and Hardigree, A. A. (1972) In "Biology and Radiobiology of Anucleate Systems" (S. Bonotto, R. Kirchman, R. Goutier, and J. R. Maisin, Eds.) **1**, pp. 51–56, Academic Press, New York and London.
Adler, H. I., Fischer, W. D., Cohen, A. and Hardigree, A. A. (1967) *Proc. natn. Acad. Sci. USA* **57**, 321–326.
Adler, H. I., Fisher, W. D. and Hardigree, A. A. (1969) *Trans N.Y. Acad. Sci.* **31**, 1059–1070.
Adler, K. (1976) *Plant Science Letters* **6**, 261–266.
Adolf, G. R. and Swetly, P. (1978) *Biochim. biophys. Acta* **518**, 334–344.
Agabian, N., Evinger, M. and Parker, G. (1979) *J. Cell. Biol.* **81**, 123–136.
Aggeler, J., Kapp, L. N. and Werb, Z. (1978) *J. Cell Biol.* **79**, A10.
Aglietta, M. and Colly, L. (1979) *Cancer Res.* **39**, 2727–2732.
Ahluwalia, B., Duffus, J. H., Paterson, L. J. and Walker, G. M. (1978) *J. gen. Microbiol.* **106**, 261–264.
Ahmed, N. and Rowbury, R. J. (1971) *J. gen. Microbiol.* **67**, 107–115.
Akimov, A. A. and Moskalik, K. G. (1977) *Vop Onkol.* **23**, 63–67.
Akkas, N. (1980) *J. biomech.* **13**, 459–461.
Alberghina, F. A. M., Sturani, E. and Gohlke, J. R. (1975) *J. biol. Chem.* **250**, 4381–4388.
Alberghina, L. (1977) *J. theor. Biol.* **69**, 633–643.
Alberghina, L. and Mariani, L. (1980) *J. math. Biol.* **9**, 389–398.
Alberghina, L., Mariani, L. and Zippel, R. (1979) *Differentiation* **15**, 135–138.
Alberts, B. and Sternglanz, R. (1977) *Nature, Lond.* **269**, 655–661.
Alberts, D. S., Durie, B. G. M. and Salmon, S. E. (1977) *Cancer Treat. Rep.* **61**, 381–388.
Albrecht-Beuhler, G. (1976) *J. Cell Biol.* **69**, 275–286.
Albrecht-Beuhler, G. and Goldman, R. D. (1976) *Exp. Cell Res.* **97**, 329–339.
Aldridge, J. and Pye, E. K. (1976) *Nature, Lond.* **259**, 670–671.
Aleschenko, A. V., Linchina, L. P. and Frankfurt, O. S. (1967) *Cytologia* **9**, 217–222.
Allen, J. S., Filip, C. C., Gustafson, R. A., Allen, R. G. and Walker, J. R. (1974) *J. Bact.* **117**, 978–986.
Allin, E. P. and Guetard, D. (1979) *Cell Molec. Biol.* **24**, 231–235.
Allred, L. E. and Porter, K. R. (1977) *In* "Growth Kinetics and Biochemical

Regulation of Normal and Malignant Cells" (B. Drewinko and R. M. Humphrey, Eds.) 295–314. The Williams and Wilkins Co., Baltimore, USA.
Alper, T. (1970) *Proc. 2nd Symp. on Microdosimetry, Stresa, Italy* (H. G. Ebert, Ed.) p. 5. Euratom, Brussels.
Altman, P. L. and Katz, D. D. (1976) *Cell Biology I*. Fed. Amer. Societies for Exptl. Biology, Bethesda Md.
Anagnostopoulos, G. D. (1971) *J. gen. Microbiol.* **65**, 23–33.
Anagnostopoulos, G. D. (1975) *Arch. Microbiol.* **104**, 95–96.
Anagnostopoulos, G. D. (1976) *Arch. Microbiol.* **107**, 199–203.
Anagnostopoulos, G. D. and Salvesen, G. (1977) *Microbios. Lett.* **4**, 27–33.
Andersen, H. A. and Engberg, J. (1975) *Expl. Cell Res.* **92**, 159–163.
Andersen, H. A. and Nielsen, S. J. (1979) *J. Cell Sci.* **35**, 17–24.
Andersen, K. B. and von Meyenburg, K. (1980) *J. Bact.* **144**, 114–123.
Andersson, G. (1979) *J. theoret. Biol.* **77**, 1–18.
Andronov, A. A., Witt, A. A. and Chaikin, S. E. (1966) "Theory of Oscillators". Pergamon Press, Oxford.
Antipa, G. A. (1980) *Acta Protozool.* **19**, 1–14.
Archer, F. L. and Wheatley, D. N. (1971) *J. Anat.* **109**, 277–292.
Armstrong, J. J., Surzycki, S. J., Moll, B. and Levine, R. P. (1971) *Biochemistry* **10**, 692–701.
Ashihara, T., Chang, S. D. and Baserga, R. (1978) *J. Cell Physiol* **96**, 15–22.
Asato, Y. (1979) *J. Bact.* **140**, 65–72.
Atkinson, A. W., Gunning, B. E. S. and John, P. C. L. (1972) *Planta* **107**, 1–32.
Atkinson, A. W., John, P. C. L. and Gunning, B. E. S. (1974) *Protoplasma* **81**, 77–109.
Atkinson, D. E. and Walton, G. M. (1967) *J. biol. Chem.* **242**, 3239–3241.
Aufderheide, K. (1976) *Genet. Res.*, Camb. **27**, 171–177.
Autissier, F. and Kepes, A. (1971) *Biochem. biophys. Acta* **249**, 611–615.
Autissier, F. and Kepes, A. (1972) *Biochimie* **54**, 93–101.
Autissier, F., Jaff, A. and Kepes, A. (1971) *Mol. gen. Genet.* **112**, 275–288.
Averbeck, D. E., Moustacchi, E. and Bisaghi, E. (1978) *Biochim. biophys. Acta* **518**, 464–481.
Axelrad, A. A. and McCulloch, E. A. (1958) *Stain Technol.* **33**, 67–71.
Ayad, S. R., Fox, M. and Winstanley, D. (1969) *Biochem. biophys. Res. Comm.* **37**, 551–558.
Baaske, D. M. and Heinstein, P. (1977) *Antim Ag. Ch.* **12**, 298–300.
Bacchetti, S. and Sinclair, W. K. (1970) *Radiat. Res.* **44**, 788–800.
Bacchetti, S. and Whitmore, G. F. (1969) *Biophys. J.* **9**, 1427–1455.
Bachmann, B. J. and Low, K. B. (1980) *Microbiol. Rev.* **44**, 1–56.
Back, F. (1976) *Int. Rev. Cytol.* **45**, 25–64.
Bacon, J. S. D., Farmer, V. C., Jones, D. and Taylor, I. F. (1969) *Biochem. J.* **114**, 557–567.
Badger, A. M. and Cooperband, S. R. (1976) *In* "Methods in Cell Biology" (D. M. Prescott, Ed.) Vol. 14, pp. 319–325. Academic Press, New York and London.
Bagi, G., Csatorday, K. and Farkas, G. L. (1979) *Arch. Microbiol.* **123**, 109–111.
Bahr, G. F. and Zeitler, E. (1962) *J. Cell Biol.* **15**, 489–501.
Bailey, J. E., Fazel-Madjlessi, J., McQuilty, D. N., Lee, L. Y., Allred, J. C. and Oro, J. A. (1977) *Science* **198**, 1175–1176.
Bakalkin, G. Y., Kalnov, S. L., Galkin, A. V., Zubatov, A. S. and Luzikov, V. N. (1978) *Biochem. J.* **170**, 569–576.

Baker, B. S., Carpenter, A. T. C., Esposito, H. S., Esposito, R. E. and Sandler, L. (1976) *Ann. Rev. Genet.* **10**, 53–134.
Baldwin, W. W. and Wegener, W. S. (1976) *Canad. J. Microbiol.* **22**, 390–394.
Balin, A. K., Goodman, D. B. P., Rasmussen, H. and Cristofalo, V. J. (1978) *J. Cell Biol.* **78**, 390–400.
Balk, S. D., Whitfield, J. F., Yendale, T. and Brown, A. C. (1973) *Proc. natn. Acad. Sci. USA* **70**, 675–679.
Ball, W. J. and Atkinson, D. E. (1975) *J. Bact.* **121**, 975–982.
Banks, G. R. (1973) *Nature Lond. New Biol.* **245**, 196–199.
Banks G. R. (1974) *Eur. J. Biochem.* **47**, 499–507.
Barale, R., Rusciano, D. and Loprieno, N. (1980a) *Mutat. Res.* **74**, 176.
Barale, R., Rusciano, D. and Loprieno, N. (1980b) *Mutat. Res.* **74**, 251.
Barford, J. P. and Hall, R. J. (1976) *Expl. Cell Res.* **102**, 276–284.
Barlogie, B. and Drewinko, B. (1975) *J. Cell Biol.* **67**, 20a.
Barlogie, B. and Drewinko, B. (1978) *Eur. J. Canc.* **14**, 741–745.
Barlogie, B., Drewinko, B., Gohde, W. and Bodey, G. P. (1977) *Cancer Res.* **37**, 2583–2588.
Barner, H. D. and Cohen, S. S. (1956) *J. Bacteriol.* **72**, 115–123.
Barnes A., Nurse, P. and Fraser, R. S. S. (1979) *J. Cell Sci.* **35**, 41–51.
Barnett, A. (1969) *Science* **164**, 1417–1419.
Barranco, S. C. and Bolton, W. E. (1977) *Cancer Res.* **37**, 2589–2591.
Barranco, S. C. and Fluournoy, D. R. (1977) *J. natn. Cancer Inst*, **58**, 657–663.
Barranco, S. C. and Humphrey, R. M. (1971) *Cancer Res.* **31**, 1218–1223.
Barranco, S. C., Gerner, E. W., Burk, K. H. and Humphrey, R. M. (1973) *Cancer Res.* **33**, 11–16.
Barranco, S. C., Haenelt, B. R. and Bolton, W. E. (1977a) *Cell Tiss.* K. **10**, 335–340.
Barranco, S. C., Haenelt, B. R. and Bolton, W. E. (1977b) *J. natn. Canc. Inst.* **59**, 1685–1689.
Barrau, M. D., Blackburn, G. R. and Dewey, W. C. (1978) *Cancer Res.* **38**, 2290–2294.
Bartholomew, J. C., Pearlman, A. L., Landolph, J. R., and Straub, K. (1979) *Cancer Res.* **39**, 2538–2543.
Baserga, R. (1962) *J. Cell Biol.* **12**, 633–637.
Baserga, R. (1968) *Cell Tissue Kinet.* **1**, 167–191.
Baserga, R. (1976) "Multiplication and Division in Mammalian Cells". Biochemistry of Disease. A series of Monographs, Vol. 6. Marcel Dekker, New York.
Baserga, R. (1978) *J. Cell Physiol.* **95**, 367–376.
Baserga, R. and Nicolini, C. (1976) *Biochim. biophys. Acta* **458**, 109–134.
Bashford, C. L., Chance, B., Lloyd, D. and Poole, R. K. (1980) *Biophys. J.* **29**, 1–12.
Basilico, C. and Meiss, H. K. (1974) *In* "Methods in Cell Biology" (D. M. Prescott, Ed.) Vol. 8, pp. 1–22. Academic Press, New York and London.
Baudhuin, P., Leroy-Houyet, M. A., Quintart, J. and Berthet, P. (1979) *J. Microscopie* **115**, 1–17.
Baugh, L. C. and Thompson, G. A. Jr. (1975) *Expl. Cell Res.* **94**, 111–121.
Baumgartel, D. M. and Howell, S. H. (1977) *Biochemistry* **16**, 3182–3189.
Bauza, M. T., De Loach, J. R., Aguanno, J. J. and Larrabee, A. R. (1976) *Arch. Biochem. Biophys.* **174**, 344–349.
Bayer, M. E. (1975) *In* "Membrane Biogenesis". (A. Tzagoloff, Ed.) pp. 393–427. Plenum Press, New York.

Bayne-Jones, S. and Adolph, E. F. (1933) *J. cell comp. Physiol.* **2**, 329–348.
Beall, P. T., Hazlewood, C. F. and Rao, P. N. (1976) *Science* **192**, 904–907.
Beck, B. D. and Park, J. T. (1976) *J. Bact.* **126**, 1250–1260.
Beck, B. D. and Park, J. T. (1977) *J. Bact.* **130**, 1292–1302.
Beck, D. P. and Levine, R. P. (1974) *J. Cell Biol.* **63**, 759–772.
Beck, H. P. (1978) *Cell Tiss. Kinet* **11**, 139–148.
Beck, J. C. and Levine, R. P. (1973) *J. Cell Biol.* **59**, 20a Abstr.
Beck, J. C. and Levine, R. P. (1977) *Biochim. Biophys. Acta* **489**, 360–369.
Bedard, D. P., Singer, R. A. and Johnston, G. C. (1980) *J. Bact.* **141**, 100–105.
Bedell, G. W., Werth, and A. Soll, D. R. (1980) *Expl. Cell Res.* **127**, 103–115.
Beeson, J. and Sueoka, N. (1979) *J. Bacteriol.* **139**, 911–916.
Begg, K. J. (1978) *J. Bacteriol.* **135**, 307–310.
Begg, K. J. and Donachie, W. D. (1973) *Nature Lond. New Biol.* **245**, 38–39.
Begg, K. J. and Donachie, W. D. (1977) *J. Bact.* **129**, 1524–1536.
Begg, K. J. and Donachie, W. D. (1978) *J. Bacteriol.* **133**, 452–458.
Begg, K. J., Hatfull, G. F. & Donachie, W. D. (1980) *J. Bacteriol.* **144**, 435–437.
Beiderman, B., Whitney, J. O. & Thaler, M. M. (1979) *Gastroenterol.* **76**, 1275.
Bello, L. J. (1968) *Biochim, biophys. Acta* **157**, 8–15.
Bello, L. J. (1969) *Biochim. biophys. Acta* **179**, 204–213.
Belt, R. J., Haas, C. D., Kennedy, J. and Taylor, S. (1980) *Cancer* **46**, 455–462.
Beltz, R. E. and Applegate, R. L. (1959) *Biochem. biophys. Res. Comm.* **1**, 298–301.
Bendetti, P. A. and Lenci, F. (1977) *Photochem. Photobiol.* **26**, 315–320.
Bendetti, P. A., Bianchini, G. and Chiti, G. (1976) *Appl. Optics* **15**, 2554–2558.
Berger, N. A., Petzold, S. J. and Berger, S. J. (1979) *Biochem. biophys. Acta* **564**, 90–104.
Bergeron, J. J. M., Warmsley, A. M. H. and Pasternak, C. A. (1970) *Biochem. J.* **119**, 489–492.
Bergsagel, D. E. (1971) *Can. Med. Assoc. J.* **104**, 31–36.
Berk, A. J. and Clayton, D. A. (1974) *J. Mol. Biol.* **86**, 801–824.
Berlin, R. D. and Oliver, J. M. (1980) *J. Cell Biol.* **85**, 660–671.
Berlin, R. D., Oliver, J. M. and Walter, R. J. (1978) *Cell* **15**, 327–341.
Bernstein, E. (1960) *Science* **131**, 1528–1529.
Bernstein, E. (1964) *J. Protozool.* **11**, 56–74.
Berridge, M. J. (1975a) *Cyclic Nucleotide Res.* **6**, 1–98.
Berridge, M. J. (1975b) *Cyclic Nucleotide Res.* **1**, 305–320.
Berridge, M. J. (1976) *Symp. Soc. Exp. Biol.* **30**, 219–231.
Berry, D. E. and Collins, J. M. (1980) *Cancer Res.* **40**, 2405–2410.
Bersier, D. and Braun, R. (1974) *Biochim. biophys. Acta* **340**, 463–471.
Bertaux, O., Frayssinet, C. and Valencia, R. (1976) *C. r. acad. Sci.* **282**, 1293–1296.
Betz, R. and Duntze, W. (1979) *Eur. J. Biochem.* **95**, 469–475.
Bhargava, P. M. (1975) *In* "Regulation of Growth and Differentiated Function in Eukaryote Cells". (G. P. Talwar, Ed.) pp. 79–96. Raven Press, New York.
Bhargava, P. M. (1977) *J. Theoret. Biol.* **68**, 101–137.
Bhargava, P. M., Dwarakanath, V. N. and Prasad, K. S. N. (1979) *Cell and Molec. Biol.* **25**, 85–94.
Bhuyan, B. K., Scheidt, L. G. and Fraser, T. J. (1972) *Cancer Res.* **32**, 398–407.
Bhuyan, B. K., Fraser, T. J. and Day, K. J. (1976) *Cancer Tr. Rep.* **60**, 1813–1818.
Bhuyan, B. K., Day, K. J., Edgerton, C. E. and Ogunbase, O. (1977a) *Cancer Res.* **37**, 3780–3784.
Bhuyan, B. K., Blowers, C. L., Neil, G. L., Bond, V. H. and Day, K. J. (1977b) *Cancer Res.* **37**, 3204–3208.

Bhuyan B. K., Robinson, M. I., Shugars, K. D., Bond, V. H. and Dion, R. L. (1978) *Cancer Res.* **38**, 2734–2739.
Bhuyan, B. K., Blowers, C. L. and Shugars, K. D. (1980) *Cancer Res.* **40**, 3437–3442.
Bichel, P., Overgaard, J. and Nielsen, O. S. (1979) *Eur. J. Canc.* **15**, 1191–1196.
Biely, P. (1978) *Archiv. Microbiol.* **119**, 213–214.
Biely, P., Kovarik, J. and Bauer, S. (1973) *Arch. Microbiol.* **94**, 365–371.
Birch, B. and Turnock, G. (1977) *FEBS Lett.* **84**, 317–319.
Bird, R. E., Louarn, J., Martuscelli, J. and Caro, L. (1972) *J. mol. Biol.* **70**, 549.
Bishop, N. I. and Senger, H. (1971) *Methods Enzymol.* **23**, 53–66.
Bisson, L. and Thorner, J. (1977) *J. Bact.* **132**, 44–50.
Blakely, E., Ngo, F., Chang, P., Lommel, L., Kraftwey, W., Kraft, G., Roots, R. and Tobias, C. (1980) *Radiat. Res.* **83**, 368.
Blau, S. and Mordoh, J. (1972) *Proc. natn. Acad. Sci. USA* **69**, 2895–2898.
Blanquet, P. R. (1978) *J. theoret. Biol.* **70**, 345–399.
Blanquet, P. R., Decaestecker, A. M. and Collyn-d'Hoaghe, M. (1977) *Cytobiol.* **16**, 27–51.
Blenkinsopp, W. K. (1968) *Exp. Cell Res.* **50**, 265–276.
Blum, H., Poole, R. K. and Ohnishi, T. (1980) *Biochem. J.* **190**, 385–393
Blumenthal, A. B. and Clark, E. J. (1977) *Exp. Cell. Res.* **105**, 15–26.
Blumenthal, L. K. and Zahler, S. A. (1962) *Science* **135**, 724.
Blumenthal, A. B., Kriegstein, H. J. and Hogness, O. S. (1973) *Cold Spring Harbor Symp. Quant. Biol.* **38**, 205–223.
Bogenhagen, D., Gillum, A. M., Martens, P. A. and Clayton, D. A. (1978) *Cold Spring Harbor Symp. Quant. Biol.* **43**, 253–262.
Boiteux, A. and Hess, B. (1980) *Ber. Bunsenges Phys. Chem.* **84**, 392–398.
Boiteux, A., Goldbeter, A. and Hess, B. (1975) *Proc. natn. Acad. Sci. USA* **72**, 3829–3833.
Bollum, F. J., Anderegg, J. W., McElyn, A. B. and Potter, V. R. (1960) *Cancer Res.* **20**, 138–143.
Bols, N. C. and Zimmerman, A. M. (1977) *Expl. Cell Res.* **108**, 259–268.
Bonatti, S., Simili, M. and Abbondandolo, A. (1972) *J. Bact.* **109**, 484–491.
Bootsma, D., Budke, L. and Vos, O. (1964) *Expl. Cell Res.* **33**, 301–309.
Borowitzka, L. J. and Volcani, B. E. (1977) *Arch. Microbiol.* **112**, 147–152.
Borun, T. W., Scharff, M. D. and Robbins, E. (1967) *Proc. natn. Acad. Sci. USA* **58**, 1977–1983.
Bosmann, H. B. (1972) *Biochem. Pharmacol.* **21**, 1977–1988.
Bosmann, H. B. and Winston, R. A. (1970) *J. Cell Biol.* **45**, 23–33.
Bostock, C. J. (1970) *Expl. Cell Res.* **60**, 16–26.
Bostock, C. J., Donachie, W. D., Masters, M. Y. and Mitchison, J. M. (1966) *Nature Lond.* **210**, 808–810.
Bouché, J.-P., Rowen, L. and Kornberg, A. (1978) *J. biol. Chem.*, **253**, 765–769.
Bourguignon, L. Y. W. and Palade, G. E. (1976) *J. Cell Biol.* **69**, 327–344.
Bourne, H. R., Coffino, P. and Tomkins, G. M. (1975) *J. Cell Physiol.* **85**, 611–619.
Boyarsky, A. (1977) *Int. J. Syst.* **8**, 447–456.
Boyd, A. R. and Holland, I. B. (1977) *FEBS Lett.* **76**, 20–24.
Boyd, A. R. and Holland, I. B. (1979) *Cell* **18**, 287–296.
Boynton, A. L. and Whitfield, J. F. (1976) *Proc. natn. Acad. Sci. USA* **73**, 1651–1654.
Boynton, A. L., Whitfield, J. F. and Isaacs, R. J. (1976) *In Vitro* **12**, 120–123.

Boynton, A. L., Whitfield, J. F., Isaacs, R. J. and Tremblay, R. G. (1978) *Life Sciences* **22**, 703–710.
Bradbury, E. M., Inglis, R. J., Matthews, H. R. and Sarner, N. (1973) *Eur. J. Biochem.* **33**, 131–139.
Bradbury, E. M., Inglis, R. J. and Matthews, H. R. (1974a) *Nature Lond.* **247**, 257–261.
Bradbury, E. M., Inglis, R. J., Matthews, H. R. and Langan, T. A. (1974b) *Nature Lond.* **249**, 553–556.
Brady, J. (1978) "Biological Clocks". Edward Arnold, London.
Brandt, P. (1975) *Planta* **124**, 105–107.
Brandt, W. F. and Von Holt, C. V. M. (1976) *FEBS Lett.* **65**, 386–390.
Braun, R. and Evans, T. E. (1969) *Biochim. biophys. Acta* **182**, 511–522.
Braun, R., Mittermayer, C. and Rusch, H. P. (1965) *Proc. natn. Acad. Sci. USA* **53**, 924–931.
Braun, R., Mittermayer, C. and Rusch, H. P. (1966) *Biochim. biophys. Acta* **114**, 27–35.
Bravo, R. and Celis, J. E. (1980) *J. Cell Biol.* **84**, 795–802.
Bray, G. and Brent, T. P. (1972) *Biochim. biophys. Acta*, **269**, 184–191.
Breakefield, X. O. and Landman, O. E. (1973) *J. Bact.* **113**, 985–998.
Brehaut, L. A. (1969) *Cell Tiss. Kinet.* **2**, 311–318.
Bremer, H., Churchward, G. and Young, R. (1979) *J. theoret. Biol.* **81**, 533–545.
Bremerskov, V., Kadeu, P. and Mittermayer, C. (1970) *Eur. J. Cancer* **6**, 379–392.
Briles, E. B. and Tomasz, A. (1970) *J. Cell Biol.* **47**, 786–790.
Britten, R. J. and Davidson, E. H. (1969) *Science* **165**, 349–357.
Brodsky, W. Ya. (1975) *J. theoret. Biol.* **55**, 167–200.
Brodsky, W. Ya., Veksler, A. M., Litinskaya, L. L., Nechaeva, N. V., Novikova, T. E. and Fateyeva, V. I. (1979) *Cytologia* **21**, 976–978.
Brooks, R. F. (1979) *Cell Biol. Int. Reps.* **3**, 707–716.
Brooks, R. F., Bennett, D. C. and Smith, J. A. (1980) *Cell* **19**, 493–504.
Brubaker, L. C. and Evans, W. H. (1973) *J. Lab. Clin. Med.* **73**, 1036–1038.
Bruce V. G. (1965) *In* "Circadian Clocks" (J. Aschoff, Ed.) pp. 125–138. North Holland, Amsterdam.
Bruce V. G. (1970) *J. Protozool.* **17**, 328–333.
Bruce, W. R. and Valeriote, F. A. (1968) *In* "The Proliferation and Spread of Neoplastic Cells", *21st Ann. Symp. Cancer Res.* at M. D. Anderson Hosp. and Tumor Inst., Houston, Texas, pp. 409–420. Williams and Wilkins, Baltimore.
Bruce, W. R., Meeker, B. E. and Valeriote, F. A. (1966) *J. natn. Cancer Inst.* **37**, 233–245.
Brugal, G. and Chassery, J. M. (1977) *Histochemie* **52**, 241–258.
Bruns, P. J. and Sanford, Y. M. (1978) *Proc. natn. Acad. Sci. USA* **75**, 3355–3358.
Brutlag, D., Schekman, R. and Kornberg, A. (1971) *Proc. natn. Acad. Sci. USA.* **68**, 2826–2829.
Bücking-Throm, E., Duntze, W., Hartwell, L. H. and Manney, T. R. (1973) *Exp. Cell Res.* **76**, 99–110.
Buckley, D. E. and Anagnostopoulos, G. D. (1975) *Arch. Microbiol.* **105**, 169–172.
Buckley, D. E. and Anagnostopoulos, G. D. (1976) *Arch. Microbiol.* **109**, 143–146.
Buckley, P. A. and Konigsberg, I. R. (1977) *Proc. natn. Acad. Sci. USA* **74**, 2031–2035.
Bugeja, V. C., Saunders, P. T. and Bazin, M. (1980) *Soc. Gen. Microbiol. Quart.* **7**, 88.

Buhse, H. E. and Rasmussen, L. (1974) *C. r. Trav. Lab. Carlsberg.* **40**, 59–67.
Buhse, H. E., Stamler, S. J. and Corliss, J. O. (1973) *Trans. Amer. micros. Soc.* **92**, 95–105.
Bullock, J. G. (1979) Ph.D. Thesis, University of Wales.
Bullock, J. G. and Coakley, W. T. (1976) *Expl. Cell Res.* **103**, 447–449.
Bullock, J. G. and Coakley, W. T. (1978) *J. therm. Biol.* **3**, 159–162.
Bullock, J. G. and Coakley, W. T. (1979) *Expl. Cell Res.* **121**, 441–445.
Burg, K., Collins, A. R. S. and Johnson, R. T. (1977) *J. Cell Sci.* **28**, 29–48.
Burg, K., Collins, A. R. S. and Johnson, R. T. (1979) *Act. Bio. Med.* **38**, 1271–1275.
Burger, M. M. (1969) *Proc. natn. Acad. Sci. USA* **62**, 994–1001.
Burger, M. M., Bombik, B. M., Breckenridge, B. M. and Sheppard, J. R. (1972) *Nature Lond. New Biol.* **239**, 161–163.
Burki, H. J. (1980) *Radiat. Res.* **81**, 76–84.
Burki, H. J., Lam, C. K. and Wood, R. D. (1980) *Mutat. Res.* **69**, 347–356.
Burki, H. R. (1971) *Cancer Res.* **31**, 1188–1191.
Burland, T. G. and Dee, J. (1979) *Genet. Res.* **34**, 33–40.
Burland, T. G. and Dee, J. (1980) *Mol. gen. Genet.* **179**, 43–48.
Burnett-Hall, D. G. and Waugh, W. A. O'N. (1967) *Biometrics* **23**, 693–716.
Burns, F. J. and Tannock, I. F. (1970) *Cell Tiss. Kinet.* **3**, 321–334.
Burns, V. W. (1961) *Expl. Cell Res.* **23**, 582–594.
Burns, V. W. (1964) *In* "Synchrony in Cell Division and Growth" (E. Zuethen, Ed.) pp. 433–439. Interscience, London.
Burstin, S. J. and Basilico, C. (1975) *Proc. natn. Acad. Sci. USA* **72**, 2540–2544.
Burstin, S. J., Meiss, H. K. and Basilico, C. (1974) *J. Cell Physiol.* **84**, 397–408.
Busby, W. F. and Lewin, J. C. (1967) *J. Physiol.* **3**, 127–131.
Busch, G. E. and Rentzepis, P. M. (1977) *Trends Biochem. Sci.* **2**, N253–N255.
Byers, B, and Goetsch, L. (1973) *Cold Spring Harbor Symp. Quant. Biol.* **38**, 123–131.
Byers, B. and Goetsch, L. (1975) *J. Bact.* **124**, 511–523.
Byers, B. and Goetsch, L. (1976) *J. Cell Biol.* **70**, 35a.
Byrnes, J. J., Downey, K. M., Black, V. L. and So, A. G. (1976). *Biochemistry* **15**, 2817–2823.
Byus, C. V., Klimpel, G. K., Lucas, D. O. and Russell, D. H. (1977) *Nature Lond.* **268**, 63–64.
Cabib, E. and Bowers, B. (1971) *J. biol. Chem.* **246**, 152–159.
Cabib, E. and Bowers, B. (1975) *J. Bact.* **124**, 1586–1593.
Cabib, E. and Farkăs, V. (1971) *Proc. natn. Acad. Sci. USA.* **68**, 2052–2056.
Cadenas, E. and Garland P. B. (1979) *Biochem. J.* **184**, 45–50.
Cairns, J. (1963) *J. mol. Biol.* **6**, 208–213.
Cairns, J. (1966) *J. mol. Biol.* **15**, 372–373.
Caldwell, W. L., Lamberton, L. L. and Bewley, D. K. (1965) *Nature Lond.* **208**, 168–170.
Callan, H. G. (1973) *Cold Spring Harbor Symp. Quant. Biol.* **38**, 195–203.
Callan, H. G. (1976) *Biol. Zbl.* **95**, 531–545.
Calleja, G. B., Zucker, M., Johnson, B. F. and Yoo, B. Y. (1980) *J. theoret. Biol.* **84**, 523–545.
Calvayrac, R. and Lefort-Tran, M. (1976) *Protoplasma* **89**, 353–358.
Calvayrac, R., Butow, R. A. and Lefort-Tran, M. (1972) *Exp. Cell Res.* **71**, 422–432.

Calvayrac, R., Bertaux, O., Lefort-Tran, M. and Valencia, R. (1974) *Protoplasma* **80**, 355–370.
Cameron, I. L. (1966) *Nature Lond.* **209**, 630–631.
Cameron, I. L. and Bols, N. C. (1975) *J. Cell Biol.* **67**, 518–522.
Camplejohn, R. S. (1980) *Cell Tiss. Kinet* **13**, 327–335.
Canti, R. G. and Spear, F. G. (1927) *Proc. Roy. Soc.* B, **102**, 92–101.
Cantor, M. H. and Burton, M. D. (1975) *J. Protozool.* **22**, 135–139.
Cantor, M. H. and Klotz, J. (1971) *Experientia* **27**, 801–803.
Cantor, M. H., Patterson, N. D. and Koprowski, C. (1978) *J. Protozool.* **25**, A11.
Capaldo, F. N. and Barbour, S. D. (1975) *J. mol. Biol.* **91**, 53–66.
Caplan, A., Ord, M. G. and Stocken, L. A. (1978) *Biochem. J.* **174**, 475–483.
Cardinali, G. and Mehrota, T. N. (1963) *Proc. Am. Ass. Cancer Res.* **4**, 10–11.
Carroll, J. W., Thomas, J., Dunaway, C. and O'Kelley, J. C. (1970) *Photochem. Photobiol.* **12**, 91–98.
Carter, B. L. A. and Dawes, I. W. (1975) *Expl. Cell Res.* **92**, 253–258.
Carter, B. L. A. and Halvorson, H. O. (1973) *Expl. Cell Res.* **76**, 152–158.
Carter, B. L. A. and Jagadish, M. N. (1978a) *Expl. Cell Res.* **112**, 15–24.
Carter, B. L. A. and Jagadish, A. N. (1978b) *Expl. Cell Res.* **112**, 373–383.
Carter, B. L. A. and Sudbery, P. E. (1980) *Genetics.* **96**, 561–566.
Carter, B. L. A., Lorincz, A. and Johnston, G. C. (1978) *J. gen. Microbiol.* **106**, 221–225.
Cartledge, T. G. and Lloyd, D. (1972) *Biochem. J.* **127**, 693–703.
Case, M. J. and Marr, A. G. (1976) *Abstr. Ann. Meet. Am. Soc. Microbiol.* **197**, p. 127.
Caspersson, T., Faber, S., Foky, G. E. and Killander, D. (1963) *Expl. Cell Res.* **32**, 529–552.
Castellazzi, M. (1976) *J. Bact.* **127**, 1150–1156.
Castellazzi, M., George, J. and Buttin, G. (1972a) *Mol. gen. Genet.* **119**, 139–152.
Castellazzi, M., George, J. and Buttin, G. (1972b) *Mol. gen. Genet.* **119**, 153–174.
Castor, L. N. (1980) *Nature Lond.* **287**, 857–859.
Cavaliere, R., Ciocatto, E. C., Giovanella, B. C., Heidelberger, C., Johnson, R. O., Margottini, M., Mondovi, B., Moricca, G. and Rossi-Fanelli, A. R. (1967) *Cancer* **20**, 1351–1381.
Cavalier-Smith, T. (1974) *J. Cell Science* **16**, 529–556.
Cercek, L., Cercek, B. and Ockey, C. H. (1978). *Biophys. J.* **23**, 395–405.
Cerdá-Olmedo, E., Hanawalt, P. C. and Guerola, N. (1968) *J. mol. Biol.* **33**, 705–719.
Chaffin, W. L. and Sogin, S. J. (1976) *J. Bact.* **126**, 771–776.
Chafouleas, J. G., Bolton, W. E., Boyo, A. E., Deoman, J. R. and Means, A. R. (1980) *J. Cell Biol.* **87**, X825a.
Chagla, A. H. and Griffiths, A. J. (1978) *J. gen Microbiol.* **108**, 39–43.
Chahal, S. J., Matthews, H. R. and Bradbury, E. M. (1980) *Nature Lond.* **287**, 76–79.
Chai, N. C. and Lark, K. G. (1970) *J. Bact.* **104**, 401–409.
Chakravarty, N. (1976) *In* "Measurement of Oxygen" (H. Degn, I. Balslev and R. Brook, Eds.) pp. 11–25. Elsevier, Amsterdam.
Chambon, P. (1975) *Ann. Rev. Biochem.* **44**, 613–638.
Champoux, J. J. (1978) *Ann. Rev. Biochem.* **47**, 449–479.
Champoux, J. J. and Dulbecco, R. (1972) *Proc. natn. Acad. Sci. USA* **69**, 143–146.
Chan, R. K. (1977) *J. Bact.* **130**, 766–774.

Chan, K.-Y. and Cheng, C. Y. (1977) *FEMS Microbiol. Letts.* **2**, 177–180.
Chance, B. (1940) *J. Franklin Inst.* **229**, 613–640.
Chance, B. (1955) *In* "The Harvey Lectures, Series XLIX", pp. 145–175. Academic Press, New York and London.
Chance, B. and Erecińska, M. (1971) *Arch biochem. Biophys.* **143**, 675–687.
Chance, B. and Hollunger, G. (1963) *J. biol. Chem.* **238**, 418–431.
Chance, B. and Williams, G. R. (1956) *Adv. Enzymol.* **17**, 65–134.
Chance, B., Williams, G. R. and Hollunger, G. (1963). *J. biol. Chem.* **278**, 439–444.
Chance, B., Estabrook, R. W. and Ghosh, A. (1964a) *Proc. natn. Acad. Sci. USA.* **51**, 1244–1251.
Chance, B., Gibson, Q. H., Eisenhardt, R. E. and Lonberg-Holm, K. K. (1964b) "Rapid Mixing and Sampling Techniques in Biochemistry". Academic Press, New York and London.
Chance B., Pye, K. and Higgins, J. (1967) *IEEE Spectrum* **4**, 79–87.
Chance, B., Pye, E. K., Ghosh, A. K. and Hess, B. (1973a) "Biological and Biochemical Oscillators". Academic Press, New York and London.
Chance, B., Williamson, G., Lee, I. Y., Mela, L., De Vault, D., Ghosh, A. and Pye, E. K. (1973b) *In* "Biological and Biochemical Oscillators" (B. Change, K. Pye, A. K. Ghosh and B. Hess, Eds.) pp. 285–300. Academic Press, New York and London.
Chance, B., Barlow, C., Nakase, Y., Takeda, M., Mayevsky, A., Fischett, R., Graham, N. and Sorge, J. (1978) *Am. J. Physiol.* **235**, H809–H920.
Chandler, M., Bird, R. E. and Caro, L. (1975) *J. Mol. Biol.* **94**, 127–132.
Chang, F. N., Navickas, I. J., Au, C. and Budzilowicz, C. (1978) *Biochim. biophys. Acta* **518**, 89–94.
Chang, H. L. and Baserga, R. (1977) *J. Cell Physiol.* **92**, 333–344.
Chapman-Andresen, L. and Nilsson, J. R. (1968) *C. r. Trav. Lab. Carlsberg* **36**, 405–432.
Chapman, A. G. and Atkinson, D. E. (1977) *Adv. Micro. Physiol.* **15**, 253–306.
Chapman, A. G., Fall, L. and Atkinson, D. E. (1971) *J. Bact.* **108**, 1072–1086.
Chapman, J. D., Webb, R. G. and Borsa, J. (1971) *J. Cell Biol.* **49**, 229–233.
Charp, P. A. and Whitson, G. L. (1980) *J. Cell Biol.* **87**, CC 6a.
Charret, R. (1969) *Expl. Cell Res.* **54**, 353–361.
Charret, R. and André, J. (1968) *J. Cell Biol.* **39**, 369–381.
Chatterjee, J., Seshadri, M. and Ray, P. R. (1979) *Indian J. Expl. Biol.* **17**, 847–848.
Chen, D. J. C. and Wang, R. J. (1977) *J. Cell Biol.* **75**, 19a.
Chernavskii, D. S., Palamarutuk, E. K., Polezhaev, A. A., Solyanik, P. N. and Burlakova, E. B. (1977) *Biosystems* **9**, 187–193.
Chiang, K. S. and Sueoka, N. (1967) *Proc. natn. Acad. Sci. USA.* **57**, 1506–1513.
Chiappino, M. L. and Volcani, B. E. (1977) *Protoplasma* **93**, 205–221.
Chin, B. and Berstein, I. A. (1968) *J. Bact.* **96**, 330–337.
Ching, L. M., Gavin, J. B., Marbrook, J. and Skinner, M. (1976) *Aust. J. Expl. Biol.* **54**, 137–147.
Chirife, A. M. and Studzins, G. P. (1978) *Proc. Soc. exp. Micro.* **157**, 206–210.
Chlapowski, F. J., Kelly, L. A. and Butcher, R. W. (1976) *Adv. Cylic Nucleotide Res.* **6**, 246–338.
Christensen, M., Schweppe, J. S. and Jungmann, R. A. (1979) *Expl. Cell Res.* **124**, 15–24.
Christensen, T. and Moan, J. (1979) *Cancer Res.* **39**, 3735–3737.
Christensen, T., Moan, J., Wibe, E. and Oftebro, R. (1979) *Br. J. Canc.* **39**, 64–68.
Chu, E. H. Y. (1978) *J. Cell Physiol* **95**, 365–366.

Chua, N.-H., Blobel, G., Siekevitz, P. and Palade, G. E. (1973) *Proc. natn. Acad. Sci. USA* **70**, 1554–1558.
Chua, N.-H., Blobel, G., Siekevitz, P. and Palade, G. E. (1976) *J. Cell Biol.* **71**, 497–514.
Chuang, S. N. and Soong, T. T. (1978) *B. math. Biol.* **40**, 499–512.
Chung, K. L., Hawirko, R. Z. and Isaac, P. K. (1965) *Can. J. Microbiol.* **11**, 953–957.
Churchward, G. and Bremer, H. (1977) *J. Bact.* **130**, 1206–1213.
Churchward, G. G. and Holland, I. B. (1976a) *J. mol. Biol.* **105**, 245–261.
Churchward, G. G. and Holland, I. B. (1976b) *FEBS Lett.* **62**, 347–350.
Ciejek, E. and Thorner, J. (1979) *Cell* **18**, 623–635.
Ciésla, Z., Bagdasalian, M., Szcurkiewicz, W., Przygońsea, M. and Klopotowski, T. (1972) *Mol. gen. Genet.* **116**, 107–125.
Clain, E. and Brulfert, A. (1980). *Planta* **150**, 26–31.
Clark, D. J. (1968) *Cold Spring Harbor Symp. Quant. Biol.* **33**, 823–838.
Clark, D. J. and Maaløe, O. (1967) *J. mol. Biol.* **23**, 99–112.
Clarke, D. (1978) Ph.D. Thesis, University of Wales.
Clarke, L. and Carbon, J. (1980) *Proc. natn. Acad. Sci. USA* **77**, 2173–2177.
Clark-Walker, G. D. and Miklos, G. L. G. (1974) *Eur. J. Biochem.* **41**, 359–365.
Clason, A. E. and Burdon, R. H. (1969) *Nature Lond.* **223**, 1063–1064.
Cleaver, J. E. (1967) "Thymidine Metabolism and Cell Kinetics". North Holland, Amsterdam.
Cleffmann, G. and Frankel, J. (1978) *Expl. Cell Res.* **117**, 191–194.
Cleffmann, G., Reuter, W. O. and Seyfert, H. M. (1979a) *H-S Z. Physiol.* **360**, 243.
Cleffmann, G., Reuter, W. O. and Seyfert, H. M. (1979b) *J. Cell Sci.* **37**, 117–124.
Clowes, F. A. L. (1965) *J. exp. Bot.* **16**, 581–586.
Clutterbuck, A. J. (1973) *Genet. Res.* **21**, 291–296.
Coffino, P. and Gray, J. W. (1978) *Cancer Res.* **38**, 4285–4288.
Coffino, P., Gray, J. W. and Tomkins, G. N. (1975) *Proc. natn. Acad. Sci. USA* **72**, 878–882.
Cohn, M. and Horibata, K. (1959) *J. Bact.* **78**, 613–623.
Cole, R. M. (1965) *Bact. Rev.* **29**, 326–344.
Cole, R. M. and Hahn, J. J. (1962) *Science* **135**, 722–724.
Cole, R. S., Levitan, D. and Sinden, R. R. (1976) *J. mol. Biol.* **103**, 39–59.
Collard, J. G., de Wildt, A., Oomen-Meulemans, E. P. M., Smeekens, J., Emmelot, P. and Inbar, M. (1977) *FEBS Lett.* **77**, 173–178.
Collins, A. R., Downes, C. S. and Johnson, R. T. (1980a) *Radiat. Res.* **83**, 423–424.
Collins, A. R., Downes, C. S. and Johnson, R. T. (1980b) *J. Cell Physiol.* **103**, 179–191.
Collins, J. M., Berry, D. E. and Cobbs, C. S. (1977) *Biochem.* **16**, 5438–5444.
Collyn-d'Hoaghe, M., Valleron, A. J. and Malaise, E. P. (1977) *Expl. Cell Res.* **106**, 405–407.
Comings, D. E. and Kafekuda, T. (1968) *J. mol. Biol.* **33**, 255–229.
Cone, R. A. (1972) *Nature Lond. New Biol.* **236**, 39–43.
Cook, J. B. and Cook, B. (1962) *Expl. Cell Res.* **28**, 524–530.
Cook, J. R. (1966) *J. Cell Biol.* **29**, 369–373.
Cook, J. R., Haggard, S. S. and Harris, P. (1976) *J. Protozool.* **23**, 368–373.
Cook, J. R. and James, T. W. (1964) *In* "Synchrony in Cell Division and Growth" (E. Zuethen, Ed.) pp. 485–495. Wiley (Interscience), New York.
Cook, W. R., Kalb, V. F., Peace, A. A. and Bernlohr, R. W. (1980) *J. Bacteriol.* **141**, 1450–1453.

Coombs, J., Darley, W. M., Holm-Hansen, O. and Volcani, B. E. (1967a) *Pl. Physiol.* **42**, 1601–1606.
Coombs, J., Halicki, P. J., Holm-Hansen, O. B. and Volcani, B. E. (1967b) *Expl. Cell Res.* **47**, 315–328.
Cooper, S. (1979) *Nature Lond.* **280**, 17–19.
Cooper, S. and Helmstetter, C. E. (1968) *J. mol. Biol.* **31**, 519–540.
Cooper, T. G., Britton, C., Brand, L. and Sumrada, R. (1979) *J. Bact.* **137**, 1447–1448.
Corbett, J. (1964) *Expl. Cell Res.* **33**, 155–160.
Corbett, S., Bradbury, E. M. Matthews, H. R. (1980) *Expl. Cell Res.* **128**, 127–132.
Cornell, R. B. and Horwitz, A. F. (1980) *J. Cell. Biol.* **86**, 810–819.
Cornell, R., Grove, G. L., Rothblat, G. H. and Horwitz, A. F. (1977) *Expl. Cell. Res.* **109**, 299–307.
Cortat, M., Matile, P. and Wiemken, A. (1972) *Arch. Mikrobiol.* **82**, 183–205.
Cosgrove, W. B. (1971) *In* "Developmental Aspects of the Cell Cycle" (I. L. Cameron, Ed.) pp. 1–21. Academic Press Inc., New York and London.
Cosgrove, B. and Skeen, M. J. (1970) *J. Protozool.* **17**, 172–177.
Costa, M., Gerner, E. W. and Russell, D. H. (1976) *J. biol. Chem.* **251**, 3313–3319.
Costa, M., Fuller, D. J. M., Russell, D. H. and Gerner, E. W. (1977) *Biochim. biophys. Acta* **479**, 416–426.
Costa, M., Gerner, E. W. and Russell, D. X. (1978) *Biochim. biophys. Acta* **538**, 1–10.
Cottrell, S. F. (1977) *J. Cell Biol.* **75**, A15.
Cottrell, S. F. (1979) *Expl. Cell Res.* **118**, 398–401.
Cottrell, S. F. (1980) *J. Cell Biol.* **87**, A3.
Cottrell, S. F. and Avers, C. J. (1970) *Biochem. biophys. Res. Comm.* **38**, 973–980.
Cottrell, S. F. and Avers, C. J. (1971) *In* "Autonomy and Biogenesis of Mitochondria and Chloroplasts" (N. K. Boardman, A. W. Linnane and R. M. Smillie, Eds.) pp. 481–491. North Holland, Amsterdam.
Cottrell, S. F., Rabinowitz, M. and Getz, G. S. (1973) *Biochemistry* **12**, 4374–4378.
Cottrell, S. F., Rabinowitz, M. and Getz, G. S. (1975) *J. Biol. Chem.* **250**, 4087–4094.
Cottrell, S. F., Getz, G. S. and Rabinowitz, M. (1978) *J. Cell Biol.* **79**, 320a.
Cowan, A. E. and Young, P. G. (1978) *Expl. Cell Res.* **112**, 79–87.
Cox, C. G. and Gilbert, J. B. (1970) *Biochem. biophys. Res. Comm.* **38**, 750–757.
Cozzarelli, N. R. (1977) *Ann. Rev. Biochem.* **46**, 641–668.
Crabtree, G. R., Munck, A. and Smith, K. A. (1980) *J. Immunol.* **125**, 13–17.
Craine, B. L. and Rupert, C. S. (1979) *J. Bact.* **137**, 740–745.
Crane, M.St. J. and Thomas D. B. (1976) *Nature Lond.* **261**, 205–208.
Crane, M.St. J., Clarke, J. B. and Thomas, D. B. (1977) *Expl. Cell Res.* **107**, 89–94.
Creanor, J. (1978a) *J. Cell Sci.* **33**, 385–397.
Creanor, J. (1978b) *J. Cell Sci.* **33**, 399–411.
Creanor, J. and Mitchison, J. M. (1979) *J. gen. Microbiol.* **112**, 385–388.
Creighton, A. M. (1975) *In* "Proceedings of the XIth Int. Cancer Conf., Florence" (P. Bucalossi, Ed.) **3**, 423. Excepta Medica, Amsterdam.
Crippa, M. (1966) *Expl. Cell Res.* **42**, 371–375.
Crissman, H. A. and Tobey, R. A. (1974) *Science* **184**, 1297–1298.
Crissman, H. A., Mullaney, P. F. and Steinkamp, J. A. (1975) *In* "Methods in Cell Biology" (D. A. Prescott, Ed.) Vol. 9, pp. 179–246. Academic Press, New York and London.

Crum, L. A., Coakley, W. T. and Deeley, J. O. T. (1979) *Biochim. biophys. Acta* **554**, 90–101.

Cryer, D. R., Goldthwaite, C. D., Zinker, S., Lam, K. B., Storm, E., Hirschberg, R., Blamire, J., Finkelstein, D. B. and Marmur, J. (1973) *Cold Spring Harbor Sym. Quant. Biol.* **38**, 17–29.

Csatorday, K. and Horvath, G. (1977) *Arch. Microbiol.* **111**, 245–246.

Cullen, M. H., Rees, G. M., Nancekievill, D. G. and Amess, J. A. L. (1979) *Br. J. Haem.* **42**, 527–534.

Cullum, J. and Vicente, M. (1978) *J. Bact.* **134**, 330–337.

Culotti, J. R. and Hartwell, L. M. (1971) *Expl. Cell Res.* **67**, 389–401.

Cummings, D. J. (1965) *Biochim. biophys. Acta* **85**, 341–350.

Cummings, D. J. (1977) *J. mol. Biol.* **117**, 273–277.

Cummins, J. E. and Day, A. W. (1975) *Can. J. genet. Cytol.* **18**, 556.

Cunningham, D. O. and Pardee, A. B. (1969) *Proc. natn. Acad. Sci. USA* **64**, 1049–1056.

Curtis, A. S. G. and Seehar, G. M. (1978) *Nature Lond.* **274**, 52–53.

Cutler, R. G. and Evans, J. E. (1966) *J. Bact.* **91**, 469–476.

Cutler, R. G. and Evans, J. E. (1967a) *J. mol. Biol.* **26**, 81–90.

Cutler, R. G. and Evans, J. E. (1967b) *J. mol. Biol.* **26**, 91–105.

Daneo-Moore, L. and Shockman, L. O. (1977) *In* "The Synthesis, Assembly and Turnover of Cell Surface Components" (G. Poste and G. L. Nicholson, Eds.) pp. 597–715. Elsevier/North Holland Biomedical Press.

Daniel, J. W., Oleinick, N. L. and Brewer, E. N. (1980) *J. Cell Biol.* **87**, cc7a.

Daniels, M. J. (1969) *Biochem. J.* **115**, 697–701.

D'Anna, J. A., Tobey, R. A. and Gurley, L. R. (1980a) *Biochemistry* **19**, 2656–2671.

D'Anna, J. A., Gurley, L. R., Becker, R. R., Barham, S. S., Tobey, R. A. and Walters, R. A. (1980b) *Biochemistry* **19**, 4331–4341.

Darby, V. and Holland, I. B. (1979) *Mol. gen. Genet.* **176**, 121–128.

Darley, W. M. and Volcani, B. E. (1971) *Meth. Enzymol.* **23A**, 85–96.

Darley, W. M., Sullivan, C. W. and Volcani, B. E. (1976) *Planta* **130**, 159–167.

Darzynkiewicz, Z., Evenson, D. P., Staiano-Coico, L., Sharpless, T. K. and Melamed, M. L. (1979) *J. Cell Phys.* **100**, 425–438.

Davern, C. I. (1979) *In* "Structure and Replication of Genetic Material. Cell Biology: A Comprehensive Treatise". (D. M. Prescott and L. Goldstein, Eds.) Vol. 2, pp. 131–169 Academic Press, New York and London.

Davis, B. and Merrett, M. J. (1974) *Pl. Physiol* **53**, 575–580.

Davison, M. T. and Garland, P. B. (1977) *J. gen. Microbiol.* **98**, 147–153.

Dawes, I. W. Carter, B. L. A. (1974) *Nature Lond.* **250**, 709–712.

Dawson, P. S. S. (1965) *Can. J. Microbiol.* **11**, 893–903.

Dawson, P. S. S. (1966) *Nature Lond.* **210**, 375–380.

Dawson, P. S. S. (1968) *In* "Continuous Cultivation of Microorganisms" (Proc. 4th Symp.) pp. 71–85. Academia, Prague.

Dawson, P. S. S. (1977) *FEMS Microbiol. Lett.* **1**, 21–23.

Dawson, P. S. S. and Steinhauer, L. P. (1980) *Biotech. Bioeng.* **22**, 137–156.

Dawson, P. S. S. and Westlake, D. W. S. (1975) *Can. J. Microbiol.* **21**, 1013–1019.

Day, A. W. and Cummins, J. E. (1975) *Genet. Res. Camb.* **25**, 263–266.

Dean, R. T. (1975) *Nature Lond.* **257**, 414–416.

Dean, R. J. (1978) "Cellular Degradative Processes", p. 80 Chapman and Hall, London.

De Balros, A. V., De Castro, J. F. and De Castro, F. J. (1973) *J. Cell Physiol.* **81**, 149–152.

Defer, N., Tichonicky, L., Kruh, J., Zoutewelle, G. and Vanwijk, R. (1979) *Biochemie* **61**, 855–859.
Degn, H. (1972) *J. chem. Education* **49**, 302–307.
Degn, H., Olsen, L. F. and Perram, J. W. (1979) *Ann. N.Y. Acad. Sci.* **316**, 623–637.
De Laat, S. W., Van der Saag, P. T. and Shinitzky, M. (1977) *Proc. natn. Acad. Sci. USA* **74**, 4458–4461.
De Laat, S. W., Van der Saag, P. T., Elson, E. L. and Schlessinger, J. (1980) *Proc. natn. Acad. Sci. USA* **77**, 1526–1528.
de la Torre, C., Sacristangarate, A. and Navarrete, M. H. (1979) *Cytobios.* **24**, 25–31.
de la Torre, C. and Gonzalez, A. (1979) *Photochem. Photobiol.* **29**, 977–981.
Delcastillo, L., Outstrin, M. L. and Wright, M. (1978) *Mol. gen. Genet.* **164**, 145–154.
Dell'Orco, R. T., Crissman, H. A., Steinkamp, J. A. and Kraemer, P. M. (1975) *Expl. Cell Res.* **92**, 271–274.
Delrey, F., Santos, T., Garcia-Acha, I. and Nombela, C. (1979) *J. Bact.* **139**, 924–931.
Demerec, M., Adelberg, E. A., Clark, A. J. and Hartman, P. E. (1966) *Genetics* **54**, 61–76.
Dennis, P. P. (1971) *Nature Lond. New Biol.* **232**, 43–48.
De Pamphilis, M. L. and Wassarman, P. M. (1980) *Ann. Rev. Biochem.* **49**, 627–666.
de Pedro, M. A. and Cánovas, J. L. (1977) *J. gen. Microbiol.* **99**, 283–290.
de Pedro, M. A., Llamas, J. E. and Cánovas, J. L. (1975). *J. gen. Microbiol.* **91**, 307–314.
De Terra, N. (1974) *Science* **184**, 530–537.
Dewey, W. C. and Fuhr, M. A. (1976) *Expl. Cell. Res.* **99**, 23–30.
Dewey, W. C. and Highfield, D. P. (1976) *Radiat. Res.* **65**, 511–528.
Dewey, W. C., Westra, H. H. and Nagasawa, H. (1971) *Int. J. Rad. Biol.* **20**, 505–520.
Dewey, W. C., Hopwood, L. E., Sapareto, S. A. and Gerweck, L. E. (1977) *Radiology* **123**, 463–474.
Dewey, W. C., Sapareto, S. A. and Betten, D. A. (1978) *Radiat. Res.* **76**, 48–59.
Dharmalingham, K. and Jayaraman, J. (1973) *Arch. biochem. Biophy.* **157**, 197–202.
Dickinson, J. R., Graves, M. G. and Swoboda, B. E. P. (1976) *FEBS Lett.* **65**, 152–154.
Dickinson, J. R., Graves, M. G. and Swoboda, B. E. P. (1977) *Eur. J. Biochem.* **78**, 83–87.
Dietzel, I., Kolb, V. and Boos, W. (1978) *Arch. Microbiol.* **118**, 207–218.
Dingemans, K. P. (1969) *J. Cell Biol.* **43**, 361–367.
Dini, F. and Luporini, P. (1979) *Develop. Biol.* **69**, 506–516.
Dion, L. D. and Gifford, G. E. (1980) *Proc. Soc. Exp. Biol. Med.* **163**, 510–514.
Dipasquale, A. and McQuire, J. (1977) *J. Cell Phys.* **93**, 395–405.
Dittrich, W. and Gohde, W. (1969) *Z. Naturforsch.* **24b**, 360.
Djordjevic, B. and Kim, J. H. (1967) *Cancer Res.* **27**, 2255–2261.
Djordjevic, B. and Kim, J. H. (1968) *J. Cell Biol.* **38**, 477–483.
Djordjevic, B. and Tolmach, L. J. (1967) *Radiat. Res.* **32**, 327–346.
Dodge, J. D. and Vickerman, K. (1980) *In* "The Eukaryotic Microbial Cell" (G. W. Gooday, D. Lloyd and A. P. J. Trinci, Eds.) *Symp. Soc. Gen. Microbiol.* **30**, pp. 77–102. Cambridge, Cambridge University Press.

Doida, Y. and Okada, S. (1967a) *Expl. Cell Res.* **48**, 540–548.
Doida, Y. and Okada, S. (1967b) *Nature Lond.* **216**, 272–273.
Domingo, J., Serratosa, J., Vidal, C. and Rius, E. (1978) *Nature Lond.* **273**, 50–52.
Domon, M. and Rauth, A. M. (1968) *Radiat Res.* **35**, 350–368.
Donachie, W. D. (1968) *Nature Lond.* **219**, 1077–1079.
Donachie, W. D. (1974) *In* "Mechanism and Regulation of DNA Replication" (A. R. Kolber and M. Katinyama, Eds.) pp. 431–445. Plenum Press, New York.
Donachie, W. D. (1979) *In* "The Developmental Biology of Prokaryotes" (J. H. Parish, Ed.). Studies in Microbiol. Vol. 1 (N. G. Carr, J. L. Ingraham and S. C. Rittenberg, series Eds.) pp. 11–35. Blackwell Scientific Publications.
Donachie, W. D. and Masters, M. (1966) *Genet. Res.* **8**, 119–124.
Donachie, W. D. and Masters, M. (1969) *In* "The Cell Cycle. Geneenzyme Interactions" (Padilla, G. M., Whitson, G. L. and Cameron, I. L., Eds.) pp. 37–76. Academic Press, London and New York.
Donachie, W. D., Hobbs, D. G. and Masters, M. (1968) *Nature Lond.* **219**, 1079–1080.
Donachie, W. D., Martin, D. T. M. and Begg, K. J. (1971) *Nature Lond. New Biol.* **231**, 274–276.
Donachie, W. D., Jones, N. C. and Teather, R. (1973) *Symp. Soc. Gen. Microbiol.* **23**, 9–44.
Donachie, W. D., Begg, K. J. and Vicente, M. (1976) *Nature Lond.* **264**, 328–333.
Donachie, W. D., Begg, K. J., Lutkenhaus, J. F., Salmond, G. P. C., Martinez-Salas, E. and Vicente, M. (1979) *J. Bact.* **140**, 388–394.
Donch, J., Green, M. H. L. and Greenberg, J. (1968) *J. Bact.* **96**, 1704–1710.
Dorian, R., Bernheim, J. L., and Mendelsohn, J. (1978) *Cell Tiss. Kinet* **11**, 33–44.
Douzou, P. (1977) "Cryobiochemistry, An Introduction". Academic Press, London and New York.
Downes, C. S., Collins, A. R. S. and Johnson, R. T. (1979) *Biophys. J.* **25**, 129–150.
Dressler, D. (1975) *Ann. Rev. Microbiol.* **29**, 525–559.
Drewinko, B. (1975) *In* "Cancer Chemotherapy. Fundamental Concepts and Recent Advances". 19th Clin. Conf. on Cancer, 1974, Univ. Texas. pp. 63–77. Year Book Med. Publishers Inc., Chicago.
Drewinko, B. and Barlogie, B. (1976) *Cancer Treat. Rep.* **60**, 1707–1717.
Drewinko, B. and Gottlieb, J. A. (1973) *Cancer Res.* **28**, 2437–2442.
Drewinko, B., Green, C. and Loo, T. L. (1977) *Cancer Treat. Rep.* **61**, 1513–1518.
Drewinko, B., Barlogie, B. and Freireich, E. J. (1979) *Cancer Res.* **39**, 2630–2636.
Dubbs, D. R. and Kit, S. (1976) *Somatic Cell Genet*, **2**, 11–15.
Dubois, C. and Rampini, C. (1978) *Biochimie* **60**, 1307–1313.
Duffill, M. B., Appleton, D. R., Dyson, P., Shuster, S. and Wright, N. A. (1977) *Brit. J. Derm.* **96**, 493–502.
Duffus, J. H. and Paterson, L. J. (1974a) *Nature Lond.* **251**, 626–627.
Duffus, J. H. and Paterson, L. J. (1974b) *Z. allg. Mikrobiol.* **14**, 727–729.
Duker, N. J. and Grant, C. L. (1980) *Expl. Cell Res.* **125**, 493–497.
Dulbecco, R. and Elkington, J. (1975) *Proc. natn. Acad. Sci. USA.* **72**, 1584–1588.
Dunn, J. H., Jervis, H. H., Wilkins, J. H., Meredith, M. J., Smith, K. T., Flora, J. B. and Schmidt, R. R. (1977) *Biochim. biophys. Acta* **485**, 301–313.
Duntze, W., Mackay, V. and Manney T. R. (1970) *Science* **168**, 1472–1473.
Durand, R. E. (1976) *Cell Tiss. Kinet*, **9**, 403–412.
Dustin, P. (1963) *Int. Rev. Cytol.* **14**, 1–39.
Dwek, R. D., Kobrin, L. H., Grossman, N. and Ron, E. Z. (1980) *J. Bact.* **144**, 17–21.

Eagon, R. G. (1962) *J. Bact.* **83**, 736–737.
Eaves, A. C. and Bruce, W. R. (1974) *Cancer Chemother. Rep.* **58**, 813–820.
Ebina, T., Ohtsuki, K., Seto, M. and Ishida, N. (1975) *Eur. J. Cancer* **11**, 155–158.
Echetebu, C. O. and Plesner, P. (1977) *J. gen Microbiol.* **103**, 389–392.
Ecker, R. E. and Kokaisl, G. (1969) *J. Bact.* **98**, 1219–1226.
Edelstein, M., Vietti, T. and Valerioff, F. (1973) *Cancer Res.* **34**, 293–297.
Edelstein, E. M., Rosenzweig, M. S., Daneo-Moore and Higgins, M. L. (1980) *J. Bacteriol.* **143**, 499–505.
Edenberg, H. J. and Huberman, J. A. (1975) *Ann. Rev. Genet.* **9**, 245–284.
Edlund, T., Gustafsson, P. and Wolf-Watz, H. (1976) *J. mol. Biol.* **108**, 295–303.
Edmunds, L. N. Jr. (1964) *Science* **145**, 266–268.
Edmunds, L. N. Jr (1965) *J. Cell comp. Physiol.* **66**, 147–158.
Edmunds, L. N. Jr. (1966) *J. Cell Physiol.* **67**, 35–44.
Edmunds, L. N. Jr. (1971) *In* "Biochronometry" (M. Menaker, Ed.) 594–611. Nat. Acad. Sci., Washington D.C.
Edmunds, L. N. Jr. (1974) *In* "Mechanisms of Regulation of Plant Growth" (R. L. Bieleski, A. R. Ferguson and M. M. Cresswell, Eds.) 287–297. Roy. Soc. N.Z., Wellington, New Zealand.
Edmunds, L. N. Jr. (1975) *In* "Les Cycles Cellulaires at leur Blocage chez Plusiers Protistes", 53–67. Colloques Int. C.N.R.S. no. 240, Paris.
Edmunds, L. N. Jr. (1978) *In* "Ageing and Biological Rhythms" (H. V. Samis Jr. and S. Capobianco, Eds.) Plenum Press, New York.
Edmunds, L. N. Jr. and Cirillo, V. P. (1974) *Int. J. Chronobiol.* **2**, 233–246.
Edmunds, L. N. Jr. and Funch, R. (1969a) *Planta (Berl.)* **87**, 134–163.
Edmunds, L. N. Jr. and Funch, R. (1969b) *Science* **165**, 500–503.
Edmunds, L. N. Jr., Chuang, L., Jarrett, R. M. and Terry, O. W. (1971) *J. Interdisc. Cycle Res.* **2**, 121–132.
Edmunds, L. N. Jr., Jay, M. E., Kohlmann, A., Liu, S. C., Merriam, V. H. and Sternberg, H. (1976) *Arch. Microbiol.* **108**, 1–8.
Edmunds, L. N. Jr., Apter, R. I., Rosenthal, P. J., Shen, W.-K. and Woodward, J. R. (1979) *Photochem. Photobiol.* **30**, 595–601.
Edwards, C. (1980a) *J. gen. Microbiol.* **119**, 277–279.
Edwards, C. (1980b) *Soc. Gen. Microbiol. Quart.* **8**, 30.
Edwards, C. (1981) "The Microbial Cell Cycle". Nelson, Surrey.
Edwards, C. and Jones, C. W. (1977) *J. gen. Microbiol.* **99**, 383–388.
Edwards, C. and McCann, R. J. (1981) *J. gen. Microbiol.* **125**, 47–53.
Edwards, C., Statham, M. and Lloyd, D. (1975) *J. gen. Microbiol.* **88**, 141–152.
Edwards, C., Spode, J. A. and Jones, C. W. (1978) *Biochem. J.* **172**, 253–260.
Edwards, D. R. W., Taylor, J. B., Wakeling, W. F., Watts, F. Z. and Johnston, I. R. (1978) *Cold Spring Harb. Symp. Quant. Biol.* **43**, 577–586.
Edwards, S. W. and Lloyd, D. (1977a) *Biochem. J.* **162**, 39–46.
Edwards, S. W. and Lloyd, D. (1977b) *J. gen. Microbiol.* **102**, 135–144.
Edwards, S. W. and Lloyd, D. (1978) *J. gen. Microbiol.* **108**, 197–204.
Edwards, S. W. and Lloyd, D. (1980) *FEBS Lett.* **109**, 21–26.
Edwards, S. W., Evans, J. B. and Lloyd, D. (1982) **125**, 459–462.
Ehmann, U. K. and Lett, J. T. (1972) *Expl. Cell Res.* **48**, 540–548.
Ehmann, U. K., Williams, J. R., Nagle, W. A., Brown, J. A., Belli, J. A. and Lett, J. T. (1975) *Nature Lond.* **258**, 633–636.
Ehret, C. F. and Dobra, K. W. (1977) *Proc. Int. Soc. Study Chronobiol. Int. Symp. 12th*, Washington D.C. 563–570.
Ehret, C. F. and Trucco, E. (1967) *J. theor. Biol.* **15**, 240–262.

Ehret, C. F. and Wille, J. J. (1970) *In* "Photobiology of Microorganisms" (P. Halldal, Ed.) 369–416. Wiley (Interscience) New York.
Eisinoff, M. L. and Rich, M. A. (1959) *Cancer Res.* **19**, 521–524.
Eisenberg, S., Griffith, J. and Kornberg, A. (1977) *Proc. natn. Acad. Sci. USA* **74**, 3198–3202.
Elgin, S. C. R. and Weintraub, H. (1975) *Ann. Rev. Biochem.* **44**, 725–774.
El-Khayat, G. H. (1980) Ph.D. Thesis, University of Wales.
Elkind, M., Sutton, H. and Moses, W. B. (1961) *J. Cell comp. Physiol.* **58** (Suppl.) 113–134.
Elkind, M., Han, A. and Volz, K. W. (1963) *J. natn. Cancer Inst.* **30**, 705–721.
Elliott, S. G. and McLaughlin, C. S. (1978) *Proc. natn. Acad. USA* **75**, 4384–4388.
Elliott, S. G. and McLaughlin, C. S. (1979) *J. Bact.* **137**, 1185–1190.
Elliott, S. G., Warner, J. R. and McLaughlin, C. S. (1979) *J. Bact.* **137**, 1048–1050.
Elorza, M. V., Lostau, C. M., Villaneuva, J. R. and Sentandreu, R. (1976) *Biochim. biophys. Acta* **454**, 263–272.
Elorza, M. V., Lostau, C. M., Villaneuva, J. R., Sentandreu, R. and Sanchez, E. (1977) *Biochim. biophys. Acta* **475**, 638–651.
Emmerson, P. T. and West, S. C. (1977) *Mol. gen. Genet.* **155**, 77–85.
Engelberg, J. (1964) *In* "Synchrony in Cell Division and Growth" (E. Zeuthen, Ed.) 497–508. Wiley (Interscience) New York.
Engelberg, J. (1968) *J. theoret. Biol.* **20**, 249–259.
Enger, M. D. and Tobey, R. A. (1969) *J. Cell Biol.* **42**, 308–315.
England, J. M., Pica-Mattoccia, L. and Attardi, G. (1974) *In* "Cell Cycle Controls" (Padilla, G. M., Cameron, I. L. and Zimmerman, A. Eds). 101–116. Academic Press, London and New York.
Epanfanova, O. I., Smolenskaya, I. N., Kurdyumova, A. G. and Sevestyanova, M. V. (1969) *Expl. Cell Res.* **58**, 401–410.
Erecińska, M. Stubbs, M., Miyata, Y., Ditre, C. M. and Wilson, D. F. (1977). *Biochim. biophys. Acta* **462**, 20–35.
Eremenko, T. and Volpe, P. (1975) *Eur. J. Biochem.* **52**, 203–210.
Eriksson, T. (1966) *Physiologia Pl.* **19**, 900–910.
Erlandson, R. A. and De Harven, E. (1971) *J. Cell Sci.* **8**, 353–397.
Ernst, P. and Killman, S. (1970) *Blood* **36**, 689–696.
Esposito, R. E. (1968) *Genetics, Princeton,* **59**, 191–210.
Esposito, M. S. and Esposito, R. E. (1969) *Genetics* **61**, 79–89.
Ettl, H. (1976) *Protoplasma* **88**, 75–84.
Euler, H. and Hevesy, G. (1942) *Skr. Biol. Medd.* **17**, 8–15.
Evans, J. B. (1975) *J. gen. Microbiol.* **91**, 188–190.
Evans, L. S. (1978) *Am. J. Bot.* **65**, 1084–1090.
Evenson, D. P. and Prescott, D. M. (1970) *Expl. Cell Res.* **61**, 71–78.
Evenson, D. P., Darzynkiewicz, Z., Staiano-Coico, L., Traganos, F. and Melamed, M. R. (1979) *Cancer Res.* **39**, 2574–2581.
Evinger, M. and Agabian, N. (1979) *Proc. natn. Acad. Sci. USA.* **76**, 175–178.
Falchuk, K. H. and Krishan, A. (1977) *Cancer Res.* **37**, 2050–2056.
Falchuk, K. H., Krishan, A. and Vallee, B. L. (1975) *Biochemistry* **14**, 3439–3444.
Fakan, S., Turner, G. N., Pagano, J. S. and Hancock, R. (1972) *Proc. natn. Acad. Sci. USA* **69**, 2300–2305.
Fan, H. and Penman, S. (1970a) *J. mol. Biol.* **50**, 655–670.
Fan, H. and Penman, S. (1970b) *Science* **168**, 135–138.
Fan, H. and Penman, S. (1971) *J. mol. Biol.* **59**, 27–42.

Fantes, P. A. (1977) *J. Cell Sci.* **24**, 51–67.
Fantes, P. A. (1979) *Nature Lond.* **279**, 428–430.
Fantes, P. A. and Nurse, P. (1977) *Expl. Cell Res.* **107**, 377–380.
Fantes, P. A. and Nurse, P. (1978) *Expl. Cell Res.* **115**, 317–329.
Farber, R. A. and Liskay, R. M. (1974) *Cytogenet. Cell Genet.* **13**, 384–396.
Fareed, G. C. and Davoli, D. (1977) *Ann. Rev. Biochem.* **46**, 471–522.
Feinendegen, L. E. and Bond, V. P. (1963) *Expl. Cell Res.* **30**, 393–404.
Feinendegen, L. E., Bond, V. P., Shreeve, W. W. and Painter, R. B. (1960) *Expl. Cell Res.* **19**, 443–459.
Fell, J. W. (1966) *Ant. van Leewenhoek* **32**, 99–102.
Ferretti, J. J. and Gray, E. D. (1968) *J. Bact.* **95**, 1400–1406.
Fiddes, J. C., Barrell, B. G. and Godson, G. N. (1978) *Proc. natn. Acad. Sci. USA* **75**, 1081–1085.
Fidorra, J. and Linden, W. A. (1977) *Radiat. Env.* **14**, 285–294.
Field, C. and Schekman, R. (1980) *J. Cell Biol.* **86**, 123–128.
Filfilan, S. A. and Sigee, D. C. (1977) *J. Cell Sci.* **27**, 81–90.
Finean, J. B., Coleman, R. and Michell, R. H. (1978) "Membranes and their Cellular Functions". Blackwell, Oxford.
Fink, K. (1975) *Biochim. biophys. Acta* **414**, 85–89.
Fink, K. (1980) *Expl. Cell Res.* **127**, 438–441.
Fink, K. and Turnock, G. (1977) *Eur. J. Biochem.* **80**, 93–96.
Finkelstein, D. B. and Strausberg, S. (1979) *J. biol. Chem.* **254**, 796–803.
Fischer, I. and Shalitin, C. (1977) *Biochim. biophys. Acta* **475**, 64–73.
Fleet, G. H. and Phaff, H. J. (1974) *J. biol. Chem.* **249**, 1717–1728.
Flegel, T. W. (1978) *Can. J. Microbiol.* **24**, 827–833.
Fletcher, G., Irwin, C. A., Henson, J. M., Fillingim, C., Malone, A. A. and Walker, J. R. (1978) *J. Bact.* **133**, 91–100.
Floros, J., Ashihara, T. and Baserga, R. (1978) *Cell Biol. Int. Reps.* **2**, 259–269.
Fonte, V. G., Searles, R. L. and Hilfer, R. S. (1971) *J. Cell Biol.* **49**, 226–229.
Forde, B. G. and John, P. C. L. (1973) *Expl. Cell Res.* **79**, 127–135.
Forde, B. G. and John, P. C. L. (1974) *Nature Lond.* **252**, 410–412.
Forde, B. G. and Sachsenmaier, W. (1979) *J. gen. Microbiol.* **115**, 135–143.
Forde, B. G., Gunning, B. E. S. and John, P. C. L. (1976) *J. Cell Sci.* **21**, 329–340.
Fox, C. Sheppard, J. R. and Burger, N. M. (1971) *Proc. natn. Acad. Sci. USA* **68**, 244–247.
Fox, T. and Pardee, A. B. (1971) *J. biol. Chem.* **246**, 6159–6165.
Fraley, R. T., Leuking, D. R. and Kaplan, S. (1978) *J. biol. Chem.* **253**, 458–464.
Fraley, R. T., Leuking, D. R. and Kaplan, S. (1979) *J. biol. Chem.* **254**, 1980–1986.
Francis, G. W., Strand, L. P., Lien, T. and Knutsen, G. (1975) *Arch. Microbiol.* **104**, 249–254.
Franck, U. F. (1978) *Angew. Chem. Int. Ed. Engl.* **17**, 1–15.
Francke, B. and Ray, D. S. (1971) *J. mol. Biol.* **61**, 565–586.
Franco, L., Johns, E. W. and Navlet, J. M. (1974) *Eur. J. Biochem.* **45**, 83–89.
Frank, W. (1973) *Z. Naturforsch.* **28**c, 322–328.
Frankel, J. (1962) *C. r. Trav. Lab. Carlsberg,* **33**, 1–52.
Frankel, J. (1964) *Expl. Cell Res.* **35**, 349–360.
Frankel J. (1967a) *J. exp. Zool.* **164**, 435–460.

Frankel, J. (1967b) *J. Cell Biol.* **34**, 841–858.
Frankel, J., Doerder, F. P., Jenkins, L. M., Nelsen, E. M. and De Bault, L. E. (1975) *In* "Molecular Biology of Nucleocytoplasmic Relationships" (S. Puiseaux-Dao, Ed.) 285–289. Elsevier, Amsterdam.
Frankel, J., Jenkins, L. M., Doerder, F. P. and Nelsen, E. A. (1976a) *Genetics*, **83**, 489–506.
Frankel, J., Jenkins, L. A. and De Bault, L. E. (1976b) *J. Cell Biol.* **71**, 242–260.
Frankel, J., Nelson, E. M. and Jenkins, L. M. (1977) *Develop. Biol.* **58**, 255–275.
Frankel, J., Frankel, A. K. and Mahler, J. (1980a) *J. Cell. Sci.* **43**, 59–74.
Frankel, J., Mohler, J. and Frankel, A. K. (1980b) *J. Cell Sci.* **43**, 75–91.
Fraser, R. S. S. and Carter, B. L. A. (1976) *J. mol. Biol.* **104**, 223–242.
Fraser, R. S. and Nurse, P. (1978) *Nature Lond.* **271**, 726–730.
Fraser, R. S. S. and Moreno, F. (1976) *J. Cell Sci.* **21**, 497–521.
Fraser, R. S. S. and Nurse, P. (1979) *J. Cell Sci.* **35**, 25–40.
Frazer, A. C. and Curtiss III, R. (1975) *Curr. Top. Microbiol. Immunol.* **69**, 1–84.
Frazier, E. A. J. (1973) *Develop. Biol.* **34**, 77–92.
Frederic, J. (1958) *Arch. Biol. (Liège)* **69**, 167–349.
Freedman, A. H., Raff, M. C. and Gomperts, B. (1975) *Nature Lond.* **255**, 378–382.
Freeman, D. A. and Crissman, H. A. (1975) *Stain Technol.* **50**, 279–284.
Fried, J., Perez, A. G. and Clarkson, B. D. (1976) *J. Cell Biol.* **71**, 172–181.
Fried, J., Perez, A. G. and Clarkson, B. (1980) *Expl. Cell Res.* **126**, 63–74.
Friedman, C. A., Kohn, K. W. and Erickson, L. C. (1974) *Biochemistry* **14**, 4018–4023.
Friedman, D. L., Johnson, R. A. and Zeilig, C. E. (1976) *In* "Advances Cyclic Nucleotide Research" (P. Greengard and G. A. Robinson, Eds.) Vol. 7, 69–111. Raven, New York.
Frindel, E. and Tubiana, M. (1971) *In* "The Cell Cycle and Cancer" (R. Baserga, Ed.) 391–447. Marcel Dekker, New York.
Fritz, P. J., White, E. L., Pruitt, K. M. and Vesell, E. S. (1973) *Biochemistry* **12**, 4034–4039.
Froelich, J. P., Sullivan, J. V. and Berger, R. L. (1976) *Analyt. Biochem.* **73**, 331–341.
Froese, G. (1966) *Int. J. Radiat Biol.* **10**, 357–367.
Fu, Y. K., Miller, M. W., Lange, C. S., Griffith, T. D. and Kaufman, G. E. (1980) *Ultrasoun. M.* **6**, 39–46.
Fujisawa, T. and Eisenstark, A. (1973) *J. Bact.* **115**, 168–176.
Fujiwara, Y. (1967) *J. Cell Physiol.* **70**, 291–300.
Fukuda, A., Miyakawa, K., Iba, H. and Okada, Y. (1976) *Virology (USA)* **71**, 583–592.
Fulton, C. (1977) *Ann. Rev. Microbiol.* **31**, 597–629.
Furcht, L. T. and Scott, R. E. (1974) *Expl. Cell Res.* **88**, 311–318.
Gaffal, K. P. and Kreutzer, D. (1977) *Protoplasma* **91**, 167–177.
Gaffney, E. V. and Nardone, R. M. (1968) *Expl. Cell Res.* **53**, 410–416.
Gaffney, E. V. (1975) *In* "Methods in Cell Biology" (D. M. Prescott, Ed.) Vol. 9, 71–84. Academic Press, New York and London.
Galdiero, F. (1973a) *Experientia* **29**, 496–497.
Galdiero, F. (1973b) *Arch. Microbiol* **94**, 125–132.
Galkin, A. V., Tsol, T. V. and Luzikov, V. N. (1979a) *FEBS Lett.* **103**, 111–113.
Galkin, A. V., Tsol, T. V. and Luzikov, V. N. (1979b) *FEBS Lett.* **105**, 373–375.
Galleron, C. and Durrand, A. M. (1979) *Protoplasma* **100**, 155–165.

Gallili, G. and Lampen, J. O. (1977) *Biochim. biophys. Acta* **475**, 113–122.
Galling, G. (1970) *Cytobiol* **2**, 359–375.
Gallwitz, D. and Mueller, G. C. (1969) *J. biol. Chem.* **244**, 5947–5952.
Gambari, R., Terada, M., Bank, A., Rifkind, R. A. and Marks, P. A. (1978) *Proc. natn. Acad. Sci. USA* **75**, 3801–3804.
Game, J. C. (1976) *Mol. gen. Genet.* **146**, 313–315.
Garrido, J. (1975) *Expl. Cell Res.* **94**, 159–175.
Garrison, P. N. and Barnes, L. D. (1980) *Biochim. biophys. Acta*, **633**, 114–121.
Gayoa, R. C., Yamamoto, L. T. and Markovitz, A. (1976) *J. Bacteriol.* **127**, 1208–1216.
Gazitt, Y. (1979) *J. Cell. Physiol.* **99**, 407–416.
Gear, A. R. L. (1976) *Analyt. Biochem.* **72**, 332–345.
Gear, A. R. L. and Bednarek, J. M. (1972) *Cell Biol.* **54**, 325–345.
Gebicki, J. M. and Hunter, F. E. (1964) *J. biol. Chem.* **239**, 631–639.
Gefter, M. L. (1975) *Ann. Rev. Biochem.* **44**, 45–78.
Gefter, M. L., Hirota, Y., Kornberg, T., Wechsler, J. A. and Barnoux, C. (1971) *Proc. natn Acad. Sci. USA* **68**, 3150–3153.
Geider, K. and Kornberg, A. (1974) *J. biol. Chem.* **249**, 3999–4005.
Geider, K., Beck, E. and Schaller, H. (1978) *Proc. natn. Acad. Sci. USA* **75**, 645–649.
Geiger, L. E. and Morris, D. R. (1980) *J. Bacteriol.* **141**, 1192–1198.
Geilenkirchen, W. L. A., Jansen, J., Coosen, R. and Van Wijk, R. (1977) *Cell Biol. Intern. Rep.* **1**, 419–426.
Gelfant, S. (1963) *Int. Rev. Cytol.* **14**, 1–39.
Gellert, M., Mizuuchi, K., O'Dea, M., Itoh, T. and Tomizawa, J.-I. (1977) *Proc. natn. Acad. Sci. USA* **74**, 4772–4776.
Gellert, M., O'Dea, M. H., Itoh, T. and Tomizawa, J.-I. (1976) *Proc. natn. Acad. Sci. USA* **73**, 4474–4478.
George, J., Castellazzi, M. and Buttin, G. (1975) *Mol. gen. Genet.* **140**, 309–332.
Gerecke, D., Jegsen, A. and Gross, R. (1976) *Experientia* **32**, 1088–1090.
Gerisch, G., Fromm, H., Huesgen, A. and Wick, U. (1975) *Nature Lond.* **255**, 547–549.
Gerner, E. W., Glick, M. C. and Warren, L. (1970) *J. Cell Physiol.* **75**, 275–280.
Gerner, E. W., Newsome, V. L. and Holmes, D. K. (1977) *J. Cell Biol.* **75**, A14.
Gerschenson, L. E., Strasser, F. F. and Rounds, D. E. (1965) *Life Sci.* **4**, 927–935.
Gerulath, A. H., Barranco, S. C. and Humphrey, R. M. (1974) *Cancer Res.* **34**, 1921–1925.
Gerweck, L. E. and Dewey, W. C. (1976) *Proc. 1st Inst. Symp. Cancer Ther. Hyperthermia Radiat* p. 16.
Giacomoni, D. and Finkel, D. (1972) *J. mol Biol.* **70**, 725–728.
Gibbs, S. P., Lüttke, A. and Cattolico, R. A. (1976) *J. Cell Biol.* **70**, 215a.
Gibson, J. F., Hall, D. O., Thurnley, J. F. and Whatley, F. (1966) *Proc. natn. Acad. Sci. USA* **56**, 987–990.
Gibson, Q. H. and Milnes, L. (1964) *Biochem. J.* **91**, 161–166.
Giddings, T. H. and Staehelin, L. A. (1978) *Cytobiol.* **16**, 235–249.
Gilbert, D. A. (1978a) *Biosystems* **10**, 227–233.
Gilbert, D. A. (1978b) *Biosystems* **10**, 235–240.
Gilbert, D. A. (1978c) *Biosystems* **10**, 241–245.
Gilbert, W. and Dressler, D. H. (1968) *Cold Spring Harbor Symp. Quant. Biol.* **33**, 473–484.

Gill, B. F. (1965) Ph.D. Thesis, University of Edinburgh.
Gillies, R. J. and Shulman, R. G. (1980) *J. Cell Biol.* **87**, A6.
Gillott, M. A. and Triemer, R. E. (1978) *J. Cell Sci.* **31**, 25–35.
Gingold, E. C., Grant, W. O., Wheals, A. G. and Wren, M. (1976) *Mol. gen. Genet.* **149**, 115–119.
Giulotto, E., Mottura, A., Decarli, L. and Nuzzo, F. (1978) *Expl. Cell Res.* **113**, 415–420.
Glansdorff, P. and Prigogine, I. (1971) "Thermodynamic Theory of Structure, Stability and Fluctuations". Wiley, New York.
Glas, U. and Bahr, G. F. (1966) *J. Cell Biol.* **29**, 507–523.
Glassberg, J., Franck, M. and Stewart, C. R. (1977) *Virology* **78**, 433–441.
Godman, G. C., Miranda, A. F., Deitch, A. D. and Tanenbaum, S. W. (1975) *J. Cell Biol.* **64**, 644–667.
Godson, G. N. (1977) *J. mol. Biol.* **117**, 353–367.
Godson, G. N. (1978) *Col Spring Harbor Symp. Quant. Biol.* **43**, 367–377.
Goh, T. S., Lampkin, B. C., Helmsworth, M., Gruppo, R. A. and Higgins, G. (1977) *Proc. Am. Assn. Canc.* **18**, 224.
Gohde, W., Meistrich, M., Meyn, R., Schumann, J., Johnston, D. and Barlogie, B. (1979) *J. Histochem. Cytochem.* **27**, 470–473.
Gojdics, M. (1953) "The Genus Euglena", 1–268. Madison, Wisconsin: University of Wisconsin Press.
Goldberg, A. L. (1971) *Nature Lond. New Biol.* **234**, 51–52.
Goldberg, A. L. and Dice, J. F. (1974) *Ann. Rev. Biochem.* **43**, 835–869.
Goldberg, A. L. and St. John, A. C. (1976) *Ann. Rev. Biochem.* **45**, 747–803.
Goldberg, A. L., Olden, K. and Prouty, W. F. (1975) *In* "Intracellular Protein Turnover" (R. T. Schimke and N. Katunuma, Eds.) 17–55. Academic Press, London and New York.
Goldberg, R. B. and Chargaff, E. (1971) *Proc. natn. Acad. Sci. USA* **68**, 1702–1706.
Goldbeter, A. (1973) *Proc. natn. Acad. Sci. USA* **70**, 3255–3259.
Goldbeter, A. and Nicolis, G. (1976) *Prog. theor. Biol.* **4**, 65–160.
Gotzos, V., Cappelli, B. and Despond, J. M. (1979) *Histochem. J.* **11**, 691–707.
Gould, A. R. (1977) *Planta* **137**, 29–36.
Gould, A. R. (1979) *J. Cell Sci.* **39**, 235–245.
Gould, S. J. and Lewontin, R. C. (1979) *Proc. Roy. Soc. Lond. B.* **205**, 581–598.
Goulian M. (1971) *Ann. Rev. Biochem.* **40**, 855–898.
Goyns, M. H. (1980) *Experientia* **36**, 936–937.
Golombek, J. and Wintersberger, E. (1974) *Expl. Cell Res.* **86**, 199–202.
Goldthwait, D. and Jacob, F. (1964) *C. r. Acad. Sci. (Paris)* **259**, 661–664.
Gomez, G. A. and Stutzman, L. (1978) *Eur. J. Canc.* **14**, 1051–1055.
Gonzolez, P. and Nardone, R. M. (1968) *Expl. Cell Res.* **50**, 599–615.
Gooch, V. D. and Packer, L. (1974) *Biochim. biophys. Acta* **346**, 245–260.
Goodwin, B. C. (1963) "Temporal Organization in Cells". Academic Press, London and New York.
Goodwin, B. C. (1966) *Nature, Lond.* **209**, 479–481.
Goodwin, B. C. (1969a) *Eur. J. Biochem.* **10**, 511–514.
Goodwin, B. C. (1969b) *Eur. J. Biochem.* **10**, 515–522.
Goodwin, B. C. (1969c) *Symp. Soc. gen. Microbiol.* (Meadow, P. M. and Pirt, S. J., Eds.) **19**, 223–236.
Goodwin, B. C. (1976) "Analytical Physiology of Cells and Developing Organisms". Academic Press, London and New York.

Goodwin, B. C. (1980) *In* "The Eukaryotic Cell". (G. W. Gooday, D. Lloyd and A. P. J. Trinci, Eds.) *Soc. Gen. Microbiol. Symp.* No. 30, 377–403. Cambridge University Press, Cambridge.
Goodwin, B. C. and Cohen, M. H. (1969) *J. theoret. Biol.* **25**, 49–107.
Gordon, C. N. (1977) *J. Cell Sci.* **24**, 81–93.
Gordon, C. N. and Elliott, S. G. (1977) *J. Bact.* **129**, 97–100.
Gorman, J., Tauro, P., LaBerge, M. and Halvorson, H. (1964) *Biochem. biophys. Res. Commun.* **15**, 43–49.
Gorovsky, M. A. and Keevert, J. B. (1975) *Proc. natn. Acad. Sci. USA* **72**, 2672–2676.
Gorovsky, M. A., Glover, C., Johmann, C. A., Keevert, J. B., Mathis, D. J. and Samuelson, M. (1977) *Cold Spring Harbor Symp. Quant. Biol.* **42**, 493–503.
Goth-Goldstein R., and Burki, M. J. (1980) *Mutat. Res.* **69**, 127–137.
Gragg, R. L., Humphrey, R. M., Thames, H. D. and Meyn, R. E. (1978) *Radiat Res.* **76**, 283–291.
Graham, C. F. (1961) *In* "The Regulation of Mammalian Reproduction". National Institutes of Health, USA.
Graham, C. F. and Morgan, R. W. (1966) *Devel. Biol.* **14**, 439–460.
Graham, C. F., Arms, K. Y. and Gurdon, J. B. (1966) *Devel. Biol.* **14**, 349–381.
Graham, J. M., Sumner, M. C. B., Curtis, O. H. and Pasternak, C. A. (1973) *Nature Lond.* **246**, 291–295.
Grant, D., Swinton, D. C. and Chiang, K. S. (1978) *Planta* **141**, 259–267.
Gratzner, H. G., Pollack, A., Ingram, D. J. and Leif, R. C. (1976) *J. Histochem. Cytochem.* **24**, 34–39.
Grasman, J. and Jansen, M. J. W. (1979) *J. math. Biol.* **7**, 171–197.
Graves, J. A. M. (1972) *Expl. Cell Res.* **72**, 393–403.
Graves, M. G., Dickinson, J. R. and Swoboda, B. L. P. (1976) *FEBS Lett.* **69**, 165–166.
Gray, J. W. (1976) *Cell Tiss. Kinet.* **9**, 499–516.
Gray, J. W., Carver, J. H., George, Y. S. and Mendelson, M. L. (1977) *Cell Tiss. Kinet.* **10**, 97–109.
Gray, L. H., Mottram, J. C., Read, J. and Spear, F. G. (1940) *Brit. J. Radiat* **13**, 371–392.
Gray, N. C. C., Dickinson, J. R. and Swoboda, B. E. P. (1977) *FEBS Lett.* **81**, 311–314.
Grdina, D. J., Sigdestad, C. P., and Peters, L. J. (1979) *Brit. J. Cancer* **39**, 152–158.
Grdina, D. J., Sigdestad, C. P. and Jovonovi, J. A. (1979) *Int. J. Radiat* **5**, 1305–1308.
Green, E. W. and Schaechter, M. (1972) *Proc. natn. Acad. Sci. USA* **69**, 2312–2316.
Green, P. B. and Bauer, K. (1977) *J. theor. Biol.* **68**, 299–315.
Green, N. H. L., Greenberg, J. and Donch, J. (1969) *Genet. Res.* **14**, 155–162.
Greenspan, H. P. (1977) *J. theor. Biol.* **65**, 79–99.
Greksák, M., Haricová, M. and Weissová, K. (1971) *Abstr. FEBS Meet. 7th,* **Abstr.** No. 637.
Greksák, M., Nejedlý, K. and Zborowski, J. (1977) *Folia microbiol.* **22**, 30–34.
Grieder, A., Maurer, R. and Stahelin, H. (1974) *Cancer Res.* **34**, 1788–1793.
Griffin, D. H., Timberlake, W. E. and Cheney, J. C. (1974) *J. gen. Microbiol.* **80**, 381–388.
Gröbner, P. (1979) *H-S. Z. Physl.* **360**, 1152–1153.
Gröbner, P. and Sachsenmaier, W. (1976) *FEBS Lett.* **71**, 181–184.
Groppi, V. E. and Coffino, P. (1980) *Fed. Proc.* **39**, p. 349.

Gross, J. D., Karamata, D. and Hampstead, P. G. (1968) *Cold Spring Harbor Symp. Quant. Biol.* **33**, 307–312.
Gross, P. R. and Fry, B. J. (1966) *Science* **153**, 749–751.
Grosschedl, R. and Hobom, G. (1979) *Nature Lond.* **277**, 621–627.
Grossman, L., Braun, A., Feldberg, R. and Mahler, I. (1975) *Ann. Rev. Biochem.* **44**, 19–43.
Grover, N. B., Woldringh, C. L., Zaritsky, A. and Rosenberger, R. T. (1977) *J. theoret. Biol.* **67**, 181–193.
Grummt, F. (1978) *Proc. natn. Acad. Sci. USA* **75**, 371–375.
Grunicke, H., Gantner, G., Holzwebeer, F., Ihlenfeldt, M. and Puschendorf, B. (1978) *Adv. Enzyme Reg.* **17**, 291–305.
Gudas, L. J. (1976) *J. mol. Biol.* **104**, 567–587.
Gudas, L. J. and Mount, D. W. (1977) *Proc. natn Acad. Sci. USA.* **74**, 5280–5284.
Gudas, L. J. and Pardee, A. B. (1974). *J. Bact.* **117**, 1216–1223.
Gudas, L. J. and Pardee, A. B. (1976) *J. mol. Biol.* **101**, 459–478.
Gudas, L. J., James, R. and Pardee, A. B. (1976) *J. biol. Chem.* **251**, 3470–3479.
Gudjonsson, H. and Johnsen, S. (1978) *Expl. Cell Res.* **112**, 289–295.
Gull, K. and Trinci, A. P. J. (1974) *Trans. Br. mycol. Soc.* **63**, 457–460.
Gurel, O. and Rössler, O. E. (Eds.) (1979) "Bifurcation Theory and Applications in Scientific Disciplines". *Ann. N. Y. Acad. Sci.* **316**.
Gurley, L. R., Walters, R. A. and Tobey, R. A. (1973) *Biochem. biophys. Res. Comm.* **50**, 744–750.
Gurley, L. R., D'Anna, J. A., Barham, S. S., Deaven, L. L. and Tobey, R. A. (1978) *Eur. J. Biochem.* **84**, 1–15.
Guttes, E. and Guttes, S. (1964) *In* "Methods in Cell Physiology" (D. M. Prescott, Ed.) Vol. 1. 43–54. Academic Press, New York and London.
Guttes, E. and Guttes, S. (1969) *J. Cell Biol.* **43**, 229–236.
Guttes, E. and Telatnyk, M. M. (1971) *Experientia* **27**, 772–774.
Guttes, E. W., Hanawalt, P. C. and Guttes, S. (1967) *Biochim. biophys. Acta* **142**, 181–194.
Gyurasits, E. B. and Wake, R. G. (1973) *J. mol. Biol.* **73**, 55–63.
Haars, L. and Hampel, A. (1975) *J. Cell Biol.* **67**, 151A.
Hachman, H. J. and Lezius, A. G. (1976) *Eur. J Biochem.* **61**, 325–330.
Haddox, M. K., Magun, B. E. and Russell, D. H. (1980) *Proc. natn. Acad. Sci. USA.* **77**, 3445–3449.
Haest, C. W. M., Verklerg, A. J., de Gier, J., Scheek, R., Ververgaert, P. H. J. and Van Deenen, L. L. M. (1974) *Biochim. biophys. Acta.* **356**, 17–26.
Hagander, P. (1980) *Math. Biosci.* **48**, 241–265.
Hagar, W. G. and Punnett, T. R. (1973) *Science* **182**, 1028–1029.
Haken, H. (1977) "Synergetics, A Workshop". Springer, Berlin.
Haken, H. (1978) "Synergetics; An Introduction". Springer, Berlin.
Haken, H. (1979) "Synergetics; Far from Equilibrium". Springer, Berlin.
Hakenbeck, P. and Messer, W. (1974) *Ann. Microbiol.* **125** B, 163–166.
Hakenbeck, R. and Messer, W. (1977a) *J. Bact.* **129**, 1234–1238.
Hakenbeck, R. and Messer, W. (1977b) *J. Bact.* **129**, 1239–1244.
Hale, A. H., Winkelhake, J. L. and Weber, H. J. (1975) *J. Cell Biol.* **64**, 398–407.
Hall, E. J., Brown, J. M. and Canavagh, J. (1968) *Radiat Res.* **35**, 622–634.
Hall, E. J., Bird, R. P., Rossi, H. H., Coffey, R., Varga, J. and Lam, Y. M. (1977) *Radiat. Res.* **70**, 469–479.
Hall, L. and Turnock, G. (1976) *Eur. J. Biochem.* **62**, 471–477.

Halvorson, H. O. (1977) *In* "Cell Differentiation in Microorganisms, Plants and Animals" (L. Nover and K. Mothes, Eds.) 361–376. Gustav Fischer Verlag, Jena.

Halvorson, H. O., Bock, R. M., Tauro, P., Epstein, R. and Laberge, M. (1966) *In* "Cell Synchrony" (I. L. Cameron and G. M. Padilla, Eds.) pp. 102–116. Academic Press, London and New York.

Halvorson, H. O., Carter, B. L. A. and Tauro, P. (1971a) *In* "Methods in Enzymology" (K. Moldave and G. Grassman, Eds.) Vol. 21D, pp. 462–471. Academic Press, New York and London.

Halvorson, H. O., Carter, B. L. A. and Tauro, P. (1971a) *Adv. microbial Physiol.* **6**, 47–106.

Hamburger, A. and Salmon, S. (1977) *Science* **197**, 461–463.

Hamburger, K. (1962) *C. r. Trav. Lab. Carlsberg,* **32**, 359–370.

Hamburger, K. (1975) *C. r. Trav. Lab. Carlsberg,* **40**, 175–185.

Hamburger, K. and Zeuthen, E. (1957) *Expl. Cell Res.* **13**, 443–453.

Hamburger, K., Kramhøft, B., Nissen, S. B. and Zeuthen, E. (1977) *J. Cell Sci.* **24**, 69–79.

Hamill, M. (1976) Ph.D. Thesis, University of Wales.

Hamlin, J. L. (1978) *Expl. Cell Res.* **112**, 255–232.

Han, A. and Sinclair, W. K. (1969) *Biophys. J.* **9**, 1171–1192.

Hand, R. (1975a) *J. Cell Biol.* **64**, 89–97.

Hand, R. (1975b) *J. Cell Biol.* **67**, 761–773.

Hand, R. and Tamm, I. (1974) *J. mol. Biol.* **82**, 175–183.

Hannan, M. A., Miller, D. R. and Nasim, A. (1976) *Radiat. Res.* **68**, 469.

Harford, N. (1975) *J. Bact.* **121**, 853–847.

Harold, F. M. (1977) *Ann. Rev. Microbiol.* **31**, 181–205.

Harris, H., Watkins, J. F., Ford, C. E. and Schoefl, G. I. (1960) *J. Cell Sci.* **1**, 1–30.

Harris, R. E., Weetman, R. M., Provisor, D. S. and Baehner, R. L. (1978) *P. Am. Ass. Ca.* **19**, 328.

Harrison, D. E. F. (1970) *J. Cell Biol.* **45**, 514–521.

Hartmann, W., Tan, I., Hutterman, A. and Kühlwein, H. (1977) *Arch. Microbiol.* **114**, 13–18.

Hartridge, H. and Roughton, F. J. W. (1923) *Proc. R. Soc. Biol.* **104**, 376–394.

Hartwell, L. H. (1970) *J. Bact.* **104**, 1280–1285.

Hartwell, L. H. (1971a) *J. mol. Biol.* **59**, 183–194.

Hartwell, L. H. (1971b) *Expl. Cell Res.* **69**, 265–276.

Hartwell, L. H. (1972) *J. mol. Biol.* **67**, 339.

Hartwell, L. H. (1973a) *J. Bact.* **115**, 966–974.

Hartwell, L. H. (1973b) *Expl. Cell Res.* **76**, 111–117.

Hartwell, L. H. (1974) *Bact. Rev.* **38**, 164–198.

Hartwell, L. H. (1976) *J. mol. Biol.* **104**, 803–817.

Hartwell, L. H. (1978) *J. Cell Biol.* **77**, 627–637.

Hartwell, L. H. (1980) *J. Cell Biol.* **85**, 811–822.

Hartwell, L. H. and Unger, M. W. (1977) *J. Cell Biol.* **75**, 422–435.

Hartwell, L. H., Culotti, J. and Reid, B. (1970) *Proc. natn. Acad. Sci. USA* **66**, 352–359.

Hartwell, L. H., Mortimer, R. K., Culotti, J. and Culotti, U. (1973) *Genetics* **74**, 267–286.

Hartwell, L. H., Culotti, J., Pringle, J. R. and Reid, B. J. (1974) *Science* **183**, 46–51.

Harvey, R. J. (1968) *In* "Methods in Cell Physiology" (D. M. Prescott, Ed.) Vol. 3, 1–24 Academic Press, New York and London.
Hassell, J. A. and Engelhardt, D. L. (1976) *Biochemistry* **15**, 1375–1381.
Hastings, J. W. (1967) *In* "The Molecular Basis of Circadian Rhythms" 53–54. Dahlem Konferenzen, Berlin.
Hastings, J. W. and Schweiger, H.-G. (1976) "The Molecular Basis of Circadian Rhythms". Dahlem Konferenzen, Berlin.
Hastings, J. W. and Sweeney, B. M. (1960) *J. gen. Physiol.* **43**, 697–706.
Hastings, J. W. and Sweeney, B. M. (1964) *In* "Synchrony in Cell Division and Growth" (E. Zeuthen, Ed.) 307–321. Wiley (Interscience), New York.
Hatzfeld, J. and Buttin, G. (1975) *Cell* **5**, 123–129.
Haugli, F. B. and Dove, W. F. (1972) *Mol. gen. Genet.* **118**, 109–124.
Hawley, E. S. and Wagner, R. P. (1967) *J. Cell Biol.* **35**, 489–499.
Hayashibe, M. and Katohda, S. (1973) *J. gen. appl. Microbiol.* **19**, 23–39.
Hayashibe, M., Sando, N. and Osumi, M. (1970) *J. gen. appl. Microbiol.* **16**, 171–179.
Hayashibe, M., Abe, N. and Matsui, M. (1977) *Archiv. Microbiol.* **114**, 91–92.
Haynes, R. H., Prakash, L., Resnick, M. A., Cox, B. S., Moustacchi, E. and Boyd, J. B. (1978) *In* "DNA Repair Mechanisms" (P. C. Hanawalt, E. C. Friedberg, and C. F. Fox, Eds.) 405–411. Academic Press, New York and London.
Heath, I. B. (1980) *Mycologia* **72**, 229–250.
Heath, R. L., Coulson, C. L. and Chimiklis, P. (1973) *Anal. Biochem.* **53**, 555–563.
Heby, O., Gray, J. W., Lindl, P. A., Marton, L. J. and Wilson, C. B. (1976) *Bioch. biophys. Res. Commun.* **71**, 99–105.
Heby, O., Marton, L. J., Wilson, C. B. and Gray, J. W. (1977) *Eur. J. Cancer* **13**, 1009–1017.
Heisenberg, W. (1927) *Z. Physiol.* **41**, 239.
Hejnowicz, Z. (1975) *J. theoret. Biol.* **54**, 345–362.
Hellerqvist, C. G. (1979) *J. Cell Biol.* **82**, 682–687.
Helmstetter, C. E. (1967) *J. mol. Biol.* **24**, 417–427.
Helmstetter, C. E. (1969) *In* "Methods in Microbiology" (J. R. Norris and D. W. Ribbons, Eds.) Vol. 1, 327–363. Academic Press, London and New York.
Helmstetter, C. E. (1974a) *J. mol. Biol.* **84**, 1–19.
Helmstetter, C. E. (1974b) *J. mol. Biol.* **84**, 21–36.
Helmstetter, C. E. and Cooper, S. (1968) *J. mol. Biol.* **31**, 507–518.
Helmstetter, C. E. and Cummings, D. J. (1963) *Proc. natn. Acad. Sci. USA* **50**, 767–774.
Helmstetter, C. E. and Cummings, D. J. (1964) *Biochim. biophys. Acta* **82**, 608–610.
Helmstetter, C. E. and Pierucci, O. (1968) *J. Bact.* **95**, 1627–1633.
Helmstetter, C. E. and Uretz, R. B. (1963) *Biophys. J.* **3**, 35–47.
Helmstetter, C. E., Cooper, S., Pierucci, O. and Revelas, E. (1968) *Cold Spring Habor Symp. Quant. Biol.* **33**, 809–822.
Helmstetter, C. E., Pierucci, O., Weinberger, M., Holmes, H. and Tang, M.-S. (1979) *In* "Bacteria: A Treatise on Structure and Function, Mechanisms of Adaptation" (J. R. Sokotch and K. N. Ornston, Eds.) Vol. 7 517–519 Academic Press, New York and London.
Hendrick, S. L., Wu, J. S. R. and Johnson, L. F. (1980) *Proc. natn. Acad. Sci. USA* **77**, 5140–5144.
Henneguy, L. F. (1898) *Arch. Anat. microsc. Morphol. exp.* **1**, 481–496.
Henriques, J. A. P., Chanet, R., Averbeck, D. and Moustacchi, E. (1977) *Mol. gen. Genet.* **158**, 63–72.

Henry, T. J. and Knippers, R. (1974) *Proc. natn. Acad. Sci. USA* **71**, 1549–1553.
Henson, J. M., Chu, A., Irwin, C. A. and Walker, J. R. (1979) *Genetics* **92**, 1041–1059.
Herdman, M., Faulkner, B. M. and Carr, N. G. (1970) *Arch. Microbiol.* **73**, 238–249.
Hereford, L. M. and Hartwell, L. H. (1971) *Nature Lond. New Biol.* **234**, 171–172.
Hereford, L. M. and Hartwell, L. H. (1973) *Nature Lond. New Biol.* **244**, 129–131.
Hereford, L. M. and Hartwell, L. H. (1974) *J. mol. Biol.* **84**, 445–461.
Herken, R., Merker, H. J. and Krcwke, R. (1978) *Teratology* **18**, 103–117.
Hermolin, J. and Zimmerman, A. M. (1976) *J. Protozool.* **23**, 594–600.
Hershko, A. P., Mamont, P., Shields, R. and Tomkins, G. (1971) *Nature Lond. New Biol.* **245**, 175–177.
Hess, B. (1968) *Nova Acta Leopoldina* **33**, 195–230.
Hess, B., Brand, K. and Pye, E. K. (1966) *Biochem. biophys. Res. Comm.* **23**, 102–107.
Hesse, M. (1972) *Z. Pfl. Physiol.* **67**, 58–77.
Hesse, M., Kulandaivelu, G. and Boger, P. (1977) *Arch. Microbiol.* **112**, 141–145.
Heyder, H. O. and Pohlit, W. (1979) *Radiat. Environm. Biophys.* **16**, 251–260.
Heywood, P. (1977) *J. Cell Sci.* **26**, 1–8.
Hibasami, H., Tanaka, M., Nagai, J. and Ikeda, T. (1977) *Aust. J. exp. Biol. med. Sci.* **55**, 379–383.
Hicks, J. B. and Herskowitz, I. (1976) *Nature Lond.* **260**, 246–248.
Higgins, J. (1967) *Ind. Engin. Chem.* **59**, 19–62.
Higgins, M. L. and Shockman, G. D. (1970) *J. Bacteriol.* **101**, 643–648.
Higgins, M. L. and Shockman, G. D. (1971) *CRC Crit. Rev. Microbiol.* **1**, 29–72.
Higgins, M. L. and Shockman, G. D. (1976) *J. Bact.* **127**, 1346–1358.
Higgins, M. L., Pooley, H. M. and Shockman, G. D. (1971) *J. Bact.* **105**, 1175–1183.
Highfield, D. P. and Dewey, W. C. (1972) *Expl. Cell Res.* **75**, 314–320.
Highfield, D. P. and Dewey, W. C. (1975) *In* "Methods in Cell Biology" (D. M. Prescott, Ed.) Vol. 9, 85–101. Academic Press, New York and London.
Hildebrandt, A. and Duspiva, F. (1969) *Z. Naturforsch.* **24B**, 747–750.
Hildebrand, C. E. and Tobey, R. A. (1975) *Biochim. biophys. Res. Commun.* **63**, 134–139.
Hill, B. T. (1978a) *Biochim. biophys. Acta* **516**, 389–417.
Hill, B. T. (1978b) *In* "On the Vinca Alkaloids in the chemotherapy of Malignant Disease" (R. Lucas, Ed.) *Proc. 5th Eli Lilly Symp.* Eli Lilly, Basingstoke.
Hill, B. T. and Whelan, R. D. H. (1980) *Cancer Tr. Rev.* **7**, 5–15.
Hinks, R. P., Daneo-Moore, L. and Shockman, G. D. (1978) *J. Bact.* **134**, 1074–1080.
Hirai, A., Nishimura, T. and Iwanmura, T. (1979) *Pl. Cell Physiol.* **20**, 93–102.
Hirota, Y. and Ricard, M. (1972) *In* "Biology and Radiobiology of Anucleate Systems" (S. Bonollo, Ed.) 29–50. Academic Press, New York and London.
Hirota, Y., Ryter, A. and Jacob, F. (1968a) *Cold Spring Harbor Symp. Quant. Biol.* **33**, 677–693.
Hirota, Y., Jacob, F., Ryter, A., Buttin, G. and Nakai, T. (1968b) *J. mol. Biol.* **35**, 175–192.
Hirsch, H. R. (1980) *Mech. Age D.* **12**, 15–23.
Hirschberg, J. and Simchen, G. (1977) *Expl. Cell Res.* **105**, 245–252.
Hirschberg, J., Goitein, R. and Marcus, M. (1980) *Eur. J. Cell. Biol.* **22**, 596.
Hittelman, W. N. and Rao, P. N. (1974) *Cancer Res.* **34**, 3433–3439.

Ho, S. V. and Shuler, M. L. (1977) *J. theor. Biol.* **68**, 415–435.
Hoch, S. O. and McVey, E. (1977) *J. biol. Chem.* **252**, 1881–1887.
Hodge, L. D., Robbins, E. and Scharff, M. D. (1969) *J. Cell Biol.* **40**, 497–507.
Hodge, L. D., Mancini, P., Davis, F. M. and Heywood, P. (1977) *J. Cell Biol.* **72**, 194–208.
Hoffman, R. A. and Britt, W. B. (1979) *J. Histochem. Cytochem.* **27**, 234–240.
Hoffman, B., Messer, W. and Schwartz, U. (1972) *J. Supramol. Str.* **1**, 29–37.
Holland, I. B. and Darby, V. (1976) *J. gen. Microbiol.* **92**, 156–166.
Holley, R. (1972) *Proc. natn. Acad. Sci. USA* **69**, 2840–2841.
Holley, R. W. (1975) *Nature Lond.* **258**, 487–490.
Holmes, A. M. and Johnston, I. R. (1975) *FEBS Lett.* **60**, 233–243.
Holmes, R. P. and Stewart, P. R. (1977) *Nature Lond* **269**, 592–594.
Holt, C. E. and Gurney, E. G. (1969) *J. Cell Biol.* **40**, 484–496.
Holtz, G. W. and Pohlit, W. (1977) *Int. J. Radiat Biol.* **31**, 121–129.
Hopkins, H. A., Sitz, T. O. and Schmidt, R. R. (1970) *J. Cell Physiol* **76**, 231–233.
Horáková, K., Fusková, A. and Ceglédyová, N. (1976) *Neoplasma* **23**, 499–505.
Horan, P. K. and Wheeless, L. L. Jr. (1977) *Science* **198**, 149–157.
Horgen, P. A. and Silver, J. C. (1978) *Ann. Rev. Microbiol.* **32**, 249–284.
Hori, T. (1977) *Jap J. Genet.* **52**, 53–64.
Horiuchi, K. Ravetch, J. V. and Zinder, N. D. (1978) *Cold Spring Harbor Symp. Quant. Biol.* **43**, 389–399.
Hosoda, J. and Matthews, E. (1968) *Proc. natn. Acad. Sci. USA* **61**, 997–1004.
Hossack, J. A., Rose, A. H. and Dawson, P. S. S. (1979) *J. gen. Microbiol.* **113**, 199–202.
Hoult, R. (1980) *New Scientist* **88**, 707–708.
Hourcade, D., Dressler, D. and Wolfson, J. (1973) *Proc. Natl. Acad. Sci. USA* **70**, 2926–2930.
Housman, D. and Huberman, J. A. (1975) *J. mol. Biol.* **94**, 173–181.
Howard, A. and Pelc, S. R. (1953) *Heredity* **6**, 261–273.
Howard, F. L. (1932) *Ann. Bot.* **46**, 461–477.
Howard-Flanders, P. and Boyce, R. P. (1966) *Radiat. Res. Suppl.* **6**, 156–184.
Howard-Flanders, P., Simson, E. and Theriot, L. (1964) *Genetics* **49**, 237–246.
Howe, W. E. and Mount, D. W. (1975) *J. Bact.* **124**, 1113–1121.
Howe, W. E. and Mount, D. W. (1978) *J. Bact.* **133**, 1278–1281.
Howell, J. A., Tsuchiya, H. M. and Frederickson, A. G. (1967) *Nature Lond* **214**, 582–584.
Howell, S. H. (1974) *In* "Cell Cycle Controls" (G. M. Padilla, I. L. Cameron and A. Zimmerman, Eds.) 235–249. Academic Press, New York, San Francisco, London.
Howell, S. H. and Naliboff, J. A. (1973) *J. Cell Biol.* **57** 760–722.
Howell, S. H. and Walker, L. L. (1977) *Devel. Biol.* **56**, 11–23.
Howell, S. H., Blaschko, W. J. and Drew, C. M. (1975) *J. Cell Biol.* **67**, 126–135.
Howell, S. H., Posakony, J. W. and Hill, K. R. (1977) *J. Cell Biol.* **72**, 223–241.
Hromas, R. A., Markel, D. E. and Scholes, V. E. (1980) *Res. Comm. CP* **30**, 365–368.
Hua, S. A. and Markovitz, A. (1972) *J. Bact* **110**, 1089–1099.
Huberman, J. A. and Riggs, A. D. (1968) *J. mol. Biol.* **32**, 327–341.
Huberman, J. A., Tsai, A. and Deich, R. A. (1973) *Nature Lond.* **241**, 32–36.
Hughes, M. N. (1981) "The Inorganic Chemistry of Biological Processes". 2nd edition John Wiley, Chichester.
Huisman, O, D'Ari, R. and George, J. (1980a) *J. Bact.* **142**, 819–828.

Huisman, O., D'Ari, R. and George, J. (1980b) *Mol. gen. Genet.* **177**, 629–636.
Hunding, A. (1981) *J. theoret. Biol.* **89**, 353–385.
Hunding, A. (1982) *In* "Synergetics: Phénomènes Non-Linéaires de la Dynamique Chimique". Springer, Berlin.
Hunt, R. C., Gold, E. and Brown, J. C. (1975) *Biochim. biophys. Acta* **413**, 453–458.
Hutter, K. J. and Eipel, H. E. (1978a) *FEMS Microbiol. Lett.* **3**, 35–38.
Hutter, K. J. and Eipel, H. E. (1978b) *Ant. van Leeuwenhoek.* **44**, 269–282.
Hutter, K. J. and Eipel, H. E. (1979) *J. gen. Micro.* **113**, 369–375.
Huzyk, L. and Clark, D. J. (1971) *J. Bacteriol.* **108**, 74–81.
Hynes, N. E. and Phillips, S. L. (1976) *J. Bact.* **128**, 502–505.
Hynes, R. and Bye, J. M. (1974) *Cell* **3**, 113–120.
Iba, H., Fukuda, A. and Okada, Y. (1975) *Jap. J. Microbiol.* **19**, 441–446.
Iba, H., Fukuda, A. and Okada, Y. (1977a) *J. Bact.* **129**, 1192–1197.
Iba, H., Fukuda, A. and Okada, Y. (1977b) *J. Bact.* **131**, 369–371.
Iba, H., Fukuda, A. and Okada, Y. (1978) *J. Bact.* **135**, 647–655.
Ikuma, H. and Bonner, W. D. Jr. (1967) *Pl. Physiol.* **42**, 1535–1544.
Imanaka, H., Gillis, J. R. and Slepecky, R. A. (1967) *J. Bact.* **93**, 1624–1630.
Ingles, C. J. (1978) *Proc. natn. Acad. Sci. USA* **75**, 405–409.
Inglis, R. D., Langan, T. A., Matthews, H. R., Hardie, D. G. and Bradbury, E. A. (1976) *Expl. Cell Res.* **97**, 418–425.
Ingram, L. O. and Fisher, W. D. (1973a) *J. Bact.* **113**, 1006–1014.
Ingram, L. O. and Fisher, W. D. (1973b) *J. Bact.* **113**, 999–1005.
Ingram, L. O. and Thurston, E. L. (1970) *Protoplasma* **71**, 55–75.
Ingram, L. O. and van Baalen, C. (1970) *J. Bact.* **102**, 784–789.
Ingram, L. O., Thurston, E. L. and van Baalen, C. (1972) *Arch. Mikrobiol.* **81**, 1–12.
Ingram, L. O., Olson, G. L. and Blackwell, M. M. (1975) *J. Bact.* **123**, 743–746.
Inman, R. B. and Schnös, M. (1971) *J. mol. Biol.* **56**, 319–325.
Inouye, M. (1969) *J. Bact.* **99**, 842–850.
Inouye, M. (1971) *J. Bact.* **106**, 539–542.
Inouye, M. and Pardee, A. B. (1970) *J. Bact.* **101**, 770–776.
Iseki, S., Ebina, T. and Ishida, N. (1980) *Cancer Res.* **40**, 3786–3791.
Isenberg, I. (1979) *Ann. Rev. Biochem.* **48**, 159–191.
Ishida, R., Nishimoto, T. and Takash, T. (1979) *Cell Struct. Func.* **4**, 235–249.
Ishiguro, S., Yamaguchi, H., Oka, Y. and Miyamoto, H. (1978) *Cell Struct. Func.* **3**, 331–336.
Israel, D. W., Gronostajski, R. M., Yeung, A. T. and Schmidt, R. R. (1977) *J. Bact.* **130**, 793–804.
Ito, K., Sato, T. and Yura, T. (1977) *Cell* **11**, 551–559.
Iwanij, V., Chua, N.-H. and Siekevitz, P. (1975) *J. Cell Biol.* **64**, 572–585.
Jackson, V., Granner, D. K. and Chalkey, R. (1975) *Proc. natn. Acad. Sci. USA* **72**, 4440–4444.
Jackson, V., Granner, D. and Chalkey, R. (1976) *Proc. natn. Acad. Sci. USA* **73**, 2266–2269.
Jacob, F. and Monod, J. (1961) *J. mol. Biol.* **3**, 318–356.
Jacob, F., Brenner, S. and Cuzin, F. (1963) *Cold Spring Harbor Symp. Quant. Biol.* **28**, 329–348.
Jagadish, M. N. and Carter, B. L. A. (1977) *Nature Lond.* **269**, 145–147.
Jagadish, M. N. and Carter, B. L. A. (1978) *J. Cell Sci.* **31**, 71–78.
Jagadish, M. N., Lorincz, A. and Carter, B. L. A. (1977) *FEMS Microbiol. Lett.* **2**, 235–237.

Jain, V. K. (1972) *Biophysik.* **8**, 133–140.
Jakoi, E. R., Elrod, L. H. and Padilla, G. M. (1976) *J. Microscopie Biol. Cell* **25**, 211–216.
Jalouzot, R., Briane, D., Ohlenbusch, H. H., Wilhelm, M. L. and Wilhelm, F. X. (1980) *Eur. J. Biochem.* **104**, 423–431.
James, E. R. and Gudas, L. J. (1976) *J. Bact.* **125**, 374–375.
James, T. W., Hemond, P., Czer, G. and Bohman, R. (1975) *Expl. Cell Res.* **94**, 267–276.
Jarrett, R. M. and Edmunds, L. N. Jr. (1970) *Science* **167**, 1730–1733.
Jarvick, J. and Botstein, D. (1973) *Proc. natn. Acad. Sci. USA* **70**, 2046–2050.
Jasiewicz, M. L. and Dickson, J. A. (1976) *J. Thermal. Biol.* **1**, 221–225.
Jay, E., Roychoudhury, R. and Wu, R. (1976) *Biochem. biophys. Res. Commun.* **69**, 678–686.
Jazwinski, S. A. and Edelman, G. M. (1976) *Proc. natn. Acad. Sci. USA* **73**, 3933–3936.
Jazwinski, S. M. and Edelman, G. M. (1979) *Proc. natn. Acad. Sci., USA* **76**, 1223–1227.
Jeggo, P. A., Unrau, P., Banks, G. R. and Holliday, R. (1973) *Nature Lond. New Biol.* **242**, 14–16.
Jellingh, W., Camplejo R., Schultze, B. and Maurer, W. (1979) *Urol. Res.* **7**, 39.
Jenkins, S. T. and Bennett, P. M. (1976) *J. Bact.* **125**, 1214–1216.
Jimenez de Asua, L., Levi-Montalcini, R., Shields, R. and Jacobelli, S. (1980) Eds. Control Mechanisms in Animal Cells. Specific Growth Factors. Raven Press, New York.
John, P. C. L., Cole, E. M. A., Keenan, P. and Rollins, M. J. (1980) *Soc. Gen. Microbiol. Quart.* **8**, 26–27.
Johnsen, S. and Stokke, T. (1977) *Expl. Cell Res.* **109**, 53–61.
Johnsen, S., Stokke, T. and Prydz, H. (1975) *Expl. Cell Res.* **93**, 245–251.
Johnson, B. F. (1965) *Expl. Cell Res.* **39**, 613–624.
Johnson, B. F. (1968) *J. Bact* **95**, 1169–1172.
Johnson, B. F. and Gibson, E. J. (1966a) *Expl. Cell Res.* **41**, 580–591.
Johnson, B. F. and Gibson, E. J. (1966b) *Expl. Cell Res.* **41**, 297–306.
Johnson, B. F. and Greenberg, J. (1975) *J. Bact.* **122**, 570–574.
Johnson, B. F., Yoo, B. Y. and Calleja, G. B. (1973) *J. Bact.* **115**, 358–366.
Johnson, B. F., Yoo, B. Y. and Calleja, G. B. (1974) *In* "Cell Cycle Controls" (I. L. Cameron, A. Zimmerman, Eds) 153–166. Academic Press, New York and London.
Johnson, B. F., Calleja, G. B., Boisclair, I. and Yoo, B. Y. (1979) *Expl. Cell Res.* **123**, 253–259.
Johnston, G. C. (1977) *J. Bact.* **132**, 730–739.
Johnston, G. C. and Singer, R. A. (1978) *Cell* **14**, 951–958.
Johnston, G. C. and Singer, R. A. (1980) *Mol. gen. Genet.* **178**, 357–360.
Johnston, G. C., Pringle, J. R. and Hartwell, L. H. (1977) *Expl. Cell Res.* **105**, 79–98.
Johnston, G. C., Ehrhardt, C. W., Lorincz, A. and Carter, B. L. A. (1979) *J. Bact.* **137**, 1–5.
Johnston, G. C., Singer, R. A., Sharrow, S. O. and Slater, M. L. (1980) *J. gen. Microbiol.* **118**, 479–484.
Johnston, L. H. and Game, J. C. (1978) *Mol. gen. Genet.* **161**, 205–214.
Johnston, L. H. and Nasmyth, K. A. (1978) *Nature Lond.* **274**, 891–893.
Jonak, G. J. and Baserga, R. (1979) *Cell* **18**, 117–123.

Jones, N. C. and Donachie, W. D. (1973) *Nature Lond. New Biol.* **243**, 100–103.
Jones, P. A., Baker, M. S., Bertram, J. S. and Benedict, W. F. (1977) *Cancer Res.* **37**, 2214–2217.
Jorcano, J. L. and Ruiz-Carrillo, A. (1979) *Biochemistry* **18**, 768–774.
Jordan, E. G., Severs, N. J. and Williamson, D. H. (1976) *In* "Progress in Differentiation Research" (N. Müller-Beral *et al.*, Eds.). North Holland Publishing Co., Amsterdam.
Jordan, E. G., Severs, N. J. and Williamson, D. H. (1977) *Expl. Cell Res.* **104**, 446–449.
Jorné, J. (1974) *J. theoret. Biol.* **43**, 375–380.
Jovin, T. M. (1976) *Ann. Rev. Biochem.* **45**, 889–920.
Juul, T. and Kemp, T. (1933) *Strahlentherapie* **48**, 457–499.
Kachel, V. (1976) *J. Histochem. Cytochem.* **24**, 211–230.
Kader, J. and Lloyd, D. (1979) *J. gen. Micro.* **114**, 455–461.
Kaiser, N., Bourne, H. R., Insel, P. A. and Coffino, P. (1979) *J. Cell Physiol.* **101**, 369–374.
Kal, H. B., Hatfield, M. and Hahn, G. M. (1975) *Radiology* **117**, 215–217.
Kallman, R. F. (1963) *Nature Lond.* **197**, 577–560.
Kalnov, S. L., Novikova, L. A., Zubatov, A. S. and Luzikov, V. N. (1979) *FEBS Lett.* **101**, 355–358.
Kamentsky, L. A., Melamed, M. R. and Derman, H. (1965) *Science* **150**, 630.
Kane, A., Basilico, C. and Baserga, R. (1976) *Expl. Cell Res.* **99**, 165–173.
Kaplan, S., Cain, B. and Deal, C. (1980) *Soc. gen. Microbiol. Quart.* **8**, 24–25.
Kapp, L. N. and Klevecz, R. R. (1976) *Expl. Cell Res.* **101**, 154–158.
Kapp, L. N. and Painter, R. B. (1977) *Expl. Cell Res.* **107**, 429–431.
Kapp, L. N. and Painter, R. B. (1979) *Biochim. biophys. Acta* **562**, 222–230.
Karakashian, M. W. and Hastings, J. W. (1962) *Proc. natn. Acad. Sci. USA* **48**, 2130–2137.
Karon, M. and Shirakawa, S. (1970) *J. natn. Cancer Inst.* **45**, 861–867.
Kasamatsu, H. and Vinograd, J. (1973) *Nature Lond. New Biol.* **241**, 103–105.
Kase, K. R. and Hahn, G. M. (1976) *Eur. J. Cancer* **12**, 481–491.
Kassir, Y. and Simchen, G. (1978) *Genetics* **90**, 49–68.
Kassis, S. and Zeuthen, E. (1979) *Expl. Cell Res.* **124**, 73–78.
Kates, J. R. and Jones, R. F. (1964) *J. Cell comp. Physiol.* **63**, 157–164.
Kato, H. and Yosida, T. H. (1970) *Expl. Cell Res.* **60**, 459–464.
Kato, T., Ishikawa, K., Nemoto, R. and Irwin, R. J. (1978) *Tohoku J. exp. Me.* **125**, 163–167.
Kauffman, S. A. (1974) *Bull. math. Biol.* **36**, 171–181.
Kauffman, S. A. and Wille, J. J. (1975) *J. theor. Biol.* **55**, 47–93.
Kauffman, S. A. and Wille, J. J. (1977) *In* "The Molecular Basis of Circadian Rhythms" (J. W. Hastings and H. G. Schweiger, Eds.).
Kay, D. G., Singer, R. A. and Johnson, G. C. (1980) *J. Bact.* **141**, 1041–1046.
Kazymin, S. D. and Sherban, S. D. (1980) *Biochemistry* (USSR) **45**, 20–26.
Kee, S. G. and Haber, J. E. (1975) *Proc. natn. Acad. Sci. USA* **72**, 1179–1183.
Keiding, J. and Andersen, H. A. (1978) *J. Cell Sci.* **31**, 13–23.
Keller, W. (1972) *Proc. natn. Acad. Sci. USA* **69**, 1560–1564.
Keller, W. (1975) *Proc. natn. Acad. Sci. USA* **72**, 2550–2554.
Kelley, M. S. (1974) Thesis: The University of Texas Health Science Center at Dallas, 1–110.
Kendall, F., Swenson, R., Borun, T., Rowinski, J. and Nicolini, C. (1977) *Science* **196**, 1106–1109.

Keng, P. C. and Wheeler, K. T. (1980a) *Radiat. Res.* **83**, 633–643.
Keng, P. C. and Wheeler, K. T. (1980b) *P. Am. Ass. Cancer* **21**, pp. 42.
Keng, P. C., Li, C. K. N. and Wheeler, K. T. (1980) *Cell Biophys.* **2**, 191–206.
Kepes, A. and Autissier, F. (1972) *Biochim. biophys. Acta.* **265**, 443–469.
Kepes, F. and Kepes, A. (1980) *Ann. Microbiol.* **131**, 3–16.
Kessler, D. (1967) *Expl. Cell Res.* **45**, 676–680.
Khachatourians, G. G. and Clark, D. J. (1970) *Bact. Proc.* **71**.
Khachatourians, G. G., Clark, D. J., Adler, H. T. and Hardigree, A. A. (1973) *J. Bact.* **116**, 226–229.
Killander, D. and Zetterberg, A. (1965) *Expl. Cell Res.* **38**, 272–284.
Kim, J. H. and Perez, A. G. (1965) *Nature Lond.* **207**, 974–975.
Kimball, R. F. and Perdue, S. W. (1962) *Expl. Cell Res.* **27**, 405–415.
Kimball, R. F., Perdue, S. W. and Hsie, A. L. (1975) *Expl. Cell Res.* **95**, 416–424.
Kim, J. H., and Stambuck, B. K. (1966) *Expl. Cell Res.* **44**, 631–634.
Kim, J. H., Gelbard, A. A. and Djordevic, B. (1968) *Cancer Res.* **28**, 2437–2442.
Kim, J. H., Kim, S. H. and Hahn, E. (1974) *Am. J. Roetgen Radium Ther. Nucl. Med.* **121**, p. 860.
Kim, K. Woo, K. B. and Perry, S. (1977) *Ann. Biomed. Engineer.* **5**, 12–33.
Kim, M., Wheeler, B. and Perry, S. (1978) *Cell Tiss. Kinet.* **11**, 497–512.
Kim, S. H. and Kim, J. H. (1972) *Cancer Res.* **32**, 323–325.
Kimler, B. F. and Schneiderman, M. H. (1977) *P. Am. Ass. Cancer Res.* **18** p. 11.
Kimler, B. F., Schneiderman, M. H. and Leeper, D. B. (1978) *Cancer Res.* **38**, 809–814.
King, G. A., Archambeau, J. O. and Klevecz, R. R. (1980) *Radiat. Res* **84**, 290–295.
King, D. W. and Barnhisel, M. L. (1967) *J. Cell Biol.* **33**, 265–272.
Kirby, E. P., Jacob, F. and Goldthwait, D. A. (1967) *Proc. natn. Acad. Sci. USA* **58**, 1903–1910.
Kishimoto, S. and Liberman, I. (1964) *Expl. Cell Res.* **36**, 92–101.
Kit, S. (1980) *J. Cell Physiol.* **95**, 410–416.
Kit, S. and Jorgensen, G. N. (1976) *J. Cell Physiol* **88**, 57–64.
Kjaergaard, L. and Joergensen, B. B. (1979) *Biotech. Bioeng.* **21**, 147–151.
Kjosbakken, J. and Colvin, J. R. (1975) *Can. J. Microbiol.* **21**, 111–120.
Klein, H. O. (1972) *Rev. Eur. Etudes clin. Biol.* **17**, 835–838.
Kletzien, R. F., Miller, M. R. and Pardee, A. B. (1977) *Nature Lond.* **270**, 57–59.
Klevecz, R. R. (1969a) *J. Cell Biol.* **43**, 207–219.
Klevecz, R. R. (1969b) *Science* **159**, 634–636.
Klevecz, R. R. (1972) *Analyt. Biochem.* **49**, 407–415.
Klevecz, R. R. (1975) *In* "Methods in Cell Biology" (D. M. Prescott, Ed.) Vol. 10, 157–172. Academic Press, New York and London.
Klevecz, R. R. (1976) *Proc. natn. Acad. Sci. USA* **73**, 4012–4016.
Klevecz, R. R. (1977) *In* "Growth, Nutrition and Metabolism of Cells in Culture" (G. H. Rothblat and V. J. Cristofalo, Eds.) Vol. III, pp. 149–196. Academic Press, London and New York.
Klevecz, R. R. and Kapp, L. N. (1973) *J. Cell Biol.* **58**, 564–573.
Klevecz, R. R. and Ruddle, F. H. (1968) *Science* **159**, 634–636.
Klevecz, R. R. and Stubblefield, E. (1967) *J. expl. Zool.* **165**, 259–268.
Klevecz, R. R., Kros, J. and Gross, S. D. (1978) *Expl. Cell Res.* **116**, 285–290.
Klevecz, R. R., Kros, J. and King, G. A. (1980) *Cytog. C. Genen.* **26**, 236–243.
Klevecz, R. R., King, G. A. and Shymko, R. M. (1981) *J. Supramolec. Struct.* (In press).
Klug, A., Rhodes, O., Smith, J., Finch, J. T. and Thomas, J. D. (1980) *Nature Lond.* **287**, 509–516.

Kluuv, J., and Sinclair, W. K. (1968) *Radiat. Res.* **36**, 45–54.
Knecht, E., Hernández-Yago, J., Martinez-Ramón, A. and Grisoliá, S. (1980) *Expl. Cell Res.* **125**, 191–199.
Knowles, C. J. (1977) *Symp. Soc. gen. Microbiol.* **27**, 241–283.
Knutsen, G. (1965) *Biochim. biophys. Acta* **103**, 495–502.
Knutsen, G., Lien, T., Schreiner, Ø. and Vaage, R. (1973) *Expl. Cell Res.* **81**, 26–30.
Knutton, S. (1976) *Expl. Cell Res.* **102**, 109–116.
Knutton, S., Sumner, M. C. B. and Pasternak, C. A. (1975) *J. Cell Biol.* **66**, 568–576.
Ko, C. and Goldstein, L. (1978) *J. Protozool.* **25**, 261–264.
Koch, A. L. (1975) *J. Bact.* **124**, 435–444.
Koch, A. L. (1977) *Adv. microbiol. Physiol.* **16**, 49–98.
Koch, A. L. (1980) *Nature Lond.* **286**, 80–82.
Koch, A. L. and Blumberg, G. (1976) *Biophys. J.* **16**, 389–405.
Koch, A. L. and Boniface, J. (1971) *Biochem. biophys. Acta* **225**, 239–247.
Koch, A. L. and Schaechter, M. (1962) *J. gen. Microbiol.* **29**, 435–454.
Koch, H., Waller, H. and Kiefer, J. (1976) *Biochem. biophys. Acta* **454**, 436–446.
Kohen, E. and Kohen, C. (1977) *Expl. Cell Res.* **107**, 261–268.
Kohen, E., Kohen, C. and Thorell, B. (1976) *Expl. Cell Res.* **101**, 47–54.
Koike, K. and Wolstenholme, D. R. (1974) *J. Cell Biol.* **61**, 14–25.
Kolb-Bachofen, V. and Vogell, W. (1975) *Expl. Cell Res.* **94**, 95–105.
Kolenbra, P. E. and Hohman, R. J. (1977) *J. Bact.* **130**, 1345–1356.
Kolter, R. and Helinski, D. R. (1979) *Ann. Rev. Genet.* **13**, 355–391.
Komarek, J., Ruzicka, J. and Simmer, J. (1968) *Biol. Pl. Praha* **10**, 177–189.
Konicek, J. (1977) *Folia microbiol. Praha* **22**, 451.
Konrad, C. G. (1963) *J. Cell Biol.* **19**, 267–277.
Kopecká, M. (1977) *Folia. Microbiol. Praha.* **22**, 440–441.
Kopecká, M., Horak, J. and Marsikova, H. (1977) *Folio. Microbiol. Praha* **22**, 426.
Koppes, L. J. H. and Nanninga, N. (1980) *J. Bact.* **143**, 89–99.
Koppes, L. J. H., Overbeeke, N. and Nanninga, N. (1978a) *J. Bact.* **133**, 1053–1061.
Koppes, L. J. H., Woldringh, C. L. and Nanninga, N. (1978b) *J. Bact.* **134**, 423–433.
Koppes, L. J. H., Meyer, M., Oonk, H. B., De Jong, M. A. and Nanninga, N. (1980) *J. Bact.* **143**, 1241–1252.
Kornberg, A. (1969) *Science* **163**, 1410–1418.
Kornberg, A. (1977) *Biochem. Soc. Trans.* **5**, 359–374.
Kornberg, A. (1978) *Cold Spring Harbor Symp. Quant. Biol.* **43**, 1–9.
Kornberg, A., Scott, J. F. and Bertch, L. L. (1978) *J. biol. Chem.* **253**, 3298–3304.
Kosugi, Y., Ikeee, J., Sekine, M., Musha, T., Shitara, N., Kohno, T. and Takakura, K. (1978) *IEEE Biomed.* **25**, 429–434.
Kováč, L. Galeotti, T. and Hess, B. (1968) *Biochim. biophys. Acta* **153**, 715–717.
Kramhøft, B. and Zeuthen, E. (1971) *C. r. Trav. Lab. Carlsberg* **38**, 351–368.
Kramhøft, B. and Zeuthen, E. (1975) *In* "Methods in Cell Biology" (D. M. Prescott, Ed.) Vol. 12, 373–380. Academic Press, New York and London.
Kramhøft, B., Nissen, S. B. and Zeuthen, E. (1976) *Carlsberg Res. Commun.* **41**, 15–25.
Kramhøft, B., Hamburger, K., Nissen, S. B. and Zeuthen, E. (1978) *Carlsberg Res. Commun.* **43**, 227–239.
Krishan, A. (1975) *J. Cell Biol.* **66**, 188–193.

Krishan, A., Frei, E. and Paika, K. (1977) *Proc. Am. Ass. Cancer Res.* **18**, 188.
Krishnaraj, R. (1978) *Current Sci.* **47**, 361–365.
Krokan, H., Wist, E. and Prydz, H. (1977) *Biochim. biophys. Acta.* **475**, 553–561.
Krug, H. (1976) *Acta Histochem.* **56**, 140–155.
Krynicka, I., Krupska, U., Malec, J. and Sitarska, E. (1980) *Neoplasma* **27**, 193–196.
Kubitschek, H. E. (1968a) *Biophys. J.* **8**, 792–804.
Kubitschek, H. E. (1968b) *Biophys. J.* **8**, 1401–1412.
Kubitschek, H. E. (1969) *In* "Methods in Microbiology" (J. R. Norris and D. W. Ribbons, Eds.) Vol. 1, pp. 593–610. Academic Press, London and New York.
Kubitschek, H. E. (1971) *Cell Tiss. Kinet.* **4**, 113–118.
Kubitschek, H. E. and Claymen, R. V. (1976) *J. Bact.* **127**, 109–113.
Kubitschek, H. E. and Edvenson, R. W. (1977) *Biophys. J.* **20**, 15–22.
Kubitschek, H. E., Bendigkeit, H. E. and Loken, M. R. (1967) *Proc. natn. Acad. Sci. USA* **57**, 1611–1617.
Kubitschek, H. E., Freeman, M. L. and Silver, S. (1971) *Biophys. J.* **11**, 787–797.
Kudo, T., Nagai, K. and Tamura, G. (1977) *Agric. biol. Chem. (Tokyo)* **41**, 97–107.
Kuempel, D. L. and Duerr, S. A. (1978) *Cold Spring Harbor Symp. Quant. Biol.* **43**, 563–567.
Kuempel, P. L. and Veomett, G. E. (1970) *Biochem. biophys. Res. Commun.* **41**, 973–980.
Kuempel, P., Masters, M. and Pardee, A. B. (1965) *Biochem. biophys. Res. Commun.* **18**, 858–867.
Küenzi, M. T. and Fiechter, A. (1969) *Arch. Mikrobiol.* **64**, 396–407.
Kuhl, A. and Lorenzen, H. (1964) *In* "Methods in Cell Physiology" (D. M. Prescott, Ed.) Vol. 1, 159–187. Academic Press, New York and London.
Kuhns, W. J. (1978) *J. Cell Biol.* **79**, Abstr. CC29.
Kung, F. C. and Glaser, D. A. (1977) *Appl. Envir.* **34**, 328–329.
Kung, F. C., Raymond, J. and Glaser, D. A. (1976) *J. Bact.* **126**, 1089–1095.
Kunisawa, P. and Cohen-Bazire, G. (1970) *Arch. Microbiol.* **71**, 49–59.
Kunkel, W. (1980) *Z. Allg. Mikr.* **20**, 315–324.
Kurn, N. and Shapiro, L. (1975) *In* "Current Topics in Cellular Regulation" (B. L. Horecker and E. R. Stadtman, Eds.) Vol. 9, 41–64. Academic Press, New York.
Kurn, N., Shapiro, L. and Agabian, N. (1977) *J. Bact.* **131**, 951–959.
Kuroiwa, T., Kawano, S. and Hizume, M. (1977) *J. Cell Biol.* **72**, 687–694.
Kuroiwa, T., Hizume, M. and Kawano, S. (1978) *Cytologia* **43**, 119–136.
Kurz, W. G. W., Larue, T. A. and Chatson, K. B. (1975) *Can. J. Microbiol.* **21**, 984–988.
Kuzmich, M. J. and Zimmerman, A. M. (1972) *Expl. Cell Res.* **72**, 441–452.
Ladygin, V. G., Semenova, G. A. and Tageeva, S. V. (1974) *Tsitologiya* **16**, 1203–1209.
Lafarge-Frayssinet, C., Bertaux, O., Valencia, R. and Frayssinet, C. (1978) *Biochim. biophys. Acta* **539**, 435–444.
Laffler, T. G., Wilkins, A., Selvig, S., Warren, N., Kleinschmidt, A. and Dove, W. F. (1979) *J. Bact.* **138**, 499–504.
Lai, C. S., Hopwood, L. E. and Swartz, H. M. (1980a) *Fed. Proc.* **39**, p. 2098.
Lai, C. S., Hopwood, L. E. and Swartz, H. M. (1980b) *Biochim. biophys. Acta* **602**, 117–126.
Lajtha, L. G. (1963) *J. Cell comp. Physiol.* **62**, 143–145.
Lajtha, L. G., Oliver, R., Berry, R. and Noyes, W. D. (1958) *Nature Lond.* **182**, 1788–1790.

Lake, R. S. (1973) *Nature Lond. New Biol.* **242**, 145–146.
Lake, R. S., Goidl, J. A. and Salzman, N. P. (1972) *Expl. Cell Res.* **73**, 113–121.
Landy-Otsukă, F. and Scheffler, I. E. (1978) *Proc. natn. Acad. Sci. USA* **75**, 5001–5005.
Lark, K. G. (1958) *Can. J. Microbiol.* **4**, 179–189.
Lark, K. G. (1972) *J. mol. Biol.* **64**, 47–60.
Lark, K. G. and Lark C. (1960) *Biochim. biophys. Acta* **43**, 520–530.
Lark, K. G. and Renger, H. (1969) *J. mol. Biol.* **42**, 221–235.
Laurent, S. J. (1973) *J. Bact.* **116**, 141–145.
Laval-Martin, D. L., Shuch, D. J. and Edmunds, L. N. Jr. (1979) *Pl. Physiol.* **63**, 495–502.
Lavin, M. F., Kikuchi, T., Counsilman, C., Jenkins, A., Winzor, D. J. and Kidson, C. (1976) *Biochemistry* **15**, 2409–2414.
Ledoigt, G. and Calvayrac, R. (1979) *J. Protozool.* **26**, 632–643.
Lee, L. S. (1977) *Math. Biosci.* **34**, 111–130.
Lee, L. S. (1978) *Math. Biosci.* **42**, 199–217.
Lee, R. W. and Jones, R. F. (1973) *Mol. gen. Genet.* **121**, 99–108.
Lee, R. W. and Jones, R. F. (1976) *Mol. gen. Genet.* **147**, 283–289.
Leedale, G. F. (1967) "Euglenoid Flagellates" 1–242. Prentice Hall, Englewood Cliffs, New Jersey.
Lefebvre. G., Raval, G. and Gay, R. (1980) *Biochim. biophys. Acta* **632**, 26–34.
Leff, J. and Lam, K. B. (1976) *J. Bact.* **127**, 354–361.
Legrys, G. A. and Hall, E. J. (1969) *Radiat. Res.* **37**, 161–172.
Lehman, I. R. (1974) *Science* **186**, 790–797.
Lehtonen, E. (1980) *J. Embryol. expl. Morph.* **58**, 231–249.
Leighton, P. M. and Donachie, W. D. (1970) *J. Bact.* **102**, 810–814.
Leonard, A. and Decat, G. (1979) *Can. J. Genet.* **21**, 473–478.
Lesher, S. (1967) *Radiat. Res.* **32**, 510–519.
Lepoint, A. (1977) *Virchows. Arch. Biol.* **25**, 53–60.
Levine, A. J. (1978) *J. Cell Physiol.* **95**, 387–392.
Lewin, B. (1974) *In* "Gene Expression" Vol. 1, 272–377. Wiley, London.
Lewin, R. A. (1952) *J. gen. Microbiol.* **6**, 233–248.
Lewin, J. C., Reimann, B. E., Busby, W. F. and Volcani, B. E. (1966) *In* "Cell Synchrony Studies in Biosynthesis Regulation" (I. L. Cameron and G. M. Padilla, Eds.) 169–188. Academic Press, New York and London.
Li, L. H., Olin, E. J., Fraser, T. J. and Bhuyan, B. K. (1970) *Cancer Res.* **29**, 2770–2775.
Li, L. H., Fraser, T. J., Olin, E. J. and Bhuyan, B. K. (1972) *Cancer Res.* **32**, 2643–2650.
Liao, H. and Thorner, J. (1980) *Proc. natn. Acad. Sci. USA* **77**, 1898–1902.
Liberman, D. F., Roti Roti, J. L. and Lange, C. S. (1973) *Expl. Cell Res.* **77**, 351–355.
Lien, T. and Knutsen, G. (1976) *Arch. Microbiol.* **108**, 189–194.
Lien, T. and Knutsen, G. (1979) *J. Phycol.* **15**, 191–200.
Lilley, D. M. J. and Pardon, J. F. (1979) *Ann. Rev. Genet.* **13**, 197–233.
Lin, E. C. C., Hirota, Y. and Jacob, F. (1971) *J. Bact.* **108**, 375–385.
Lindahl, P. E. (1948) *Nature Lond.* **161**, 648–649.
Lindahl, P. E. (1956) *Biochim. biophys. Acta* **21**, 411–415.
Lindahl, P. E. and Sörenby, L. (1966) *Expl. Cell Res.* **43**, 424–434.
Lindsey, J. K., Vance, B. D., Keeter, J. S. and Scholes, V. E. (1971) *J. Phycol.* **7**, 65–71.

Ling, V. (1977) *J. Cell Physiol.* **91**, 209–224.
Lipke, P. N., Taylor, A. and Ballou, C. E. (1976) *J. Bact.* **127**, 610–618.
Lipmann, F. (1941) *Adv. Enzymology* **1**, 99–162.
Liskay, R. M. (1974) *J. Cell Physiol.* **84**, 49–56.
Liskay, R. M. (1978) *Expl. Cell Res.* **114**, 69–77.
Liskay, R. M. and Meiss, M. K. (1977) *Somatic Cell Genet.* **3**, 343–347.
Liskay, R. M. and Prescott, D. M. (1978) *Proc. natn. Acad. Sci. USA* **75**, 2873–2877.
Liskay, R. M., Kornfield, B., Fullerton, P. and Evans, R. (1980) *J. Cell Physiol.* **104**, 461–467.
Littlefield, J. W. (1962) *Expl. Cell Res.* **26**, 318–326.
Liu, C.-C., Burke, R. L., Hibner, U., Barry, J. and Alberts, B. (1978) *Cold Spring Harbor Symp. Quant. Biol.* **43**, 469–487.
Livingston, D. M. and Richardson, C. C. (1975) *J. biol. Chem.* **250**, 470–478.
Livingston, D. M., Hinkle, D. C. and Richardson, C. C. (1975) *J. biol. Chem.* **250**, 461–469.
Livingston, D. M. and Kupfer, D. M. (1977) *J. mol. Biol.* **116**, 249–260.
Lloyd, D. (1974) "The Mitochondria of Micro-organisms". 553 pp. Academic Press, London and New York.
Lloyd, D. and Ball, J. (1979) *J. gen. Microbiol.* **114**, 463–466.
Lloyd, D. and Edwards, S. W. (1977) *Biochem. J.* **162**, 581–590.
Lloyd, D. and Edwards, S. W. (1979) *In* "Kinetics of Physiochemical Oscillations" (U. K. Frank and E. Wicke, Eds.) Vol. 2, 392–403. F. J. Mainz, Aachen.
Lloyd, D. and Edwards, S. W. (1980a) *Ber. Bunsenges Phys. Chem.* **84**, p. 417.
Lloyd, D. and Edwards, S. W. (1980b) *Soc. Gen. Microbiol. Quarterly*, **8**, 25.
Lloyd, D. and Edwards, S. W. (1982). *In* "Synergetics: Phénomènes Non-Linéaires de la Dynamique". Springer, Berlin.
Lloyd, D. and Poole, R. K. (1979) *In* "Techniques in Metabolic Research" B202, 1–27. Elsevier, North Holland Press, Amsterdam.
Lloyd, D. and Turner, G. (1980) *Symp. Soc. gen. Microbiol.* **30**, 143–179.
Lloyd, D., Turner, G., Poole, R. K., Nicholl, W. G. and Roach, G. I. (1971) *Subcellular Biochem.* **1**.
Lloyd, D., John, L., Edwards, C. and Chagla, A. H. (1975) *J. gen Microbiol* **88**, 153–158.
Lloyd, D., John, L., Hamill, M., Phillips, C., Kader, J. and Edwards, S. W. (1977) *J. gen. Microbiol.* **99**, 223–227.
Lloyd, D., Morgan, N. A., John, L. and Venables, S. E. (1978a) *J. gen. Microbiol.* **105**, 1–10.
Lloyd, D., Phillips, C. A. and Statham, M. (1978b) *J. gen. Microbiol* **106**, 19–26.
Lloyd, D., Edwards, C., Edwards, S. W., El'Khayat, G., Jenkins, S. J., John, L., Phillips, C. A. and Statham, M. (1978c). *Trends Bioch. Sci.* N138–139.
Lloyd, D., Edwards, S. W. and Williams, J. L. in press. *FEMS Microbiol. Lett.*
Loeb, L. A. (1974) *In* "The Enzymes" (P. D. Boyer, Ed.) Vol. 10, 173–209. Academic Press, London and New York.
Loffler, M., Postius, S. and Schneider, F. (1980) *H.-S. Z. Physiol.* **361**, p. 1313.
Lomniczi, B., Bosch, F. X., Hay, A. J. and Skehel, J. J. (1977) *J. gen. Virol* **35**, 187–190.
London, J. F., Charp, P. A. and Whitson, G. L. (1979) *J. Cell Biol.* **83**, 9a.
Long, W. F. and Williamson, F. B. (1979) *IRCS Med. Sci. Biochem.* **7**, 429–434.
Loomis, W. F., Wahrmann, J. S. and Luzzati, D. (1973) *Proc. natn. Acad. Sci. USA* **70**, 425–429.

Looney, W. B., Pardue, M. L. and Baughart, F. W. (1963) *Nature Lond.* **198**, 804–805.
Lord, P. G. and Wheals, A. E. (1980) *J. Bact.* **142**, 808–818.
Lorenzen, H. (1964) *In* "Synchrony in Cell Division and Growth" (E. Zeuthen, Ed.) 571–578. Interscience, New York.
Lorenzen, H. (1970) *In* "Photobiology of Microorganisms" (P. Halldal, Ed.) 187–212. Wiley (Interscience), New York.
Lorenzen, H. and Hesse, M. (1974) *In* "Algal Physiology and Biochemistry" (W. D. P. Stewart, Ed.) 894–908. Blackwell, Oxford.
Lorenzen, H. and Venkataraman, G. S. (1972). "Methods in Cell Physiology" Vol. 5, 373–383. Academic Press, New York and London.
Lorincz, A. and Carter, B. L. A. (1979) *J. gen. Microbiol.* **113**, 287–295.
Lovely, J. R. and Threlfall, R. J. (1976) *Biochem. biophys. Res. Commun.* **71**, 789–795.
Lovely, J. R. and Threlfall, R. J. (1978) *Biochem. biophys. Res. Commun.* **85**, 579–584.
Lovely, J. R. and Threlfall, R. J. (1979) *Biochem. biophys. Res. Commun.* **86**, 365–370.
Lövlie, A. (1963) *C. r. Trav. Lab. Carlsberg,* **33**, 377–413.
Lowdon, M. and Vitols, V. (1973) *Arch. Biochem. Biophys.* **158**, 177–184.
Lubochinsky, G. and Burger, M. M. (1969) *Abstr. 6th Meet. Fed. Eur. Biochem. Soc., Madrid* **345**, No. 1121.
Lucid, S. W. and Griffin, M. J. (1977) *Biochem. biophys. Res. Commun.* **74**, 113–118.
Luckasen, J. R., White, J. G. and Kersey, T. H. (1974) *Proc. natn. Acad. Sci. USA* **71**, 5088–5090.
Lueking, D. R., Fraley, R. T. and Kaplan, S. (1978) *J. biol. Chem.* **253**, 451–457.
Ludeke, K. (1946) *J. appl. Physiol.* **17**, 600–607.
Lutkenhaus, J. F. and Donachie, W. D. (1979) *J. Bact.* **137**, 1088–1094.
Lutkenhaus, J. F., Moore, B. A., Masters, M. and Donachie, W. D. (1979) *J. Bact.* **138**, 352–360.
Lutkenhaus, J. F., Wolf-Watz, H. and Donachie, W. D. (1980) *J. Bact.* **142**, 615–630.
Maaløe, O. and Kjeldgaard, N. O. (1966) "Control of Macromolecular Synthesis". W. A. Benjamin, New York.
Macdonald, H. R. and Miller, R. G. (1970) *Biophys. J.* **10**, 834–842.
MacKnight, A. D. C. and Leaf, A. (1977) *Physiol. Rev.* **57**, 510–573.
Maclean, N. (1965) *Nature Lond.* **207**, 322–323.
Madoc-Jones, H. and Mauro, F. (1968) *J. Cell Physiol.* **72**, 185–196.
Madoc-Jones, H. and Mauro, F. (1974) *In* "Handbuch der Experimentellen Pharmakologie" XXXVII/I (A. Sartorelli and D. Johns, Eds.) 205–219. Springer-Verlag, New York.
Maertelaer, V. D. and Galand, P. (1977) *Cell Tiss. Kinet.* **10**, 35–42.
Magnaval, R., Bertaux, O. and Valencia, R. (1979) *Expl. Cell Res.* **121**, 251–265.
Mahogaokar, S., Orengo, A. and Rao, P. N. (1980) *Expl. Cell Res.* **125**, 87–94.
Mahler, H. R., Assimos, K. and Lin, C. C. (1975) *J. Bact.* **123**, 637–641.
Maino, V. C., Green, N. M. and Crumpton, A. J. (1974) *Nature Lond.* **251**, 324–327.
Magni, G. E., Panzeri, L. and Sora, S. (1977) *Mutat. Res.* **42**, 223–233.
Mak, S. and Till, J. E. (1963) *Radiat. Res.* **20**, 600–618.
Makhlin, E. E., Kudryavtseva, M. V. and Kudryavtsev, B. N. (1979) *Expl. Cell Res.* **118**, 143–150.
Malec, J., Kornacka, L. and Sawecka, J. (1979) *Arch. Immun.* **27**, 539–545.

Maness, P. F. and Edelman, G. M. (1978) *Proc. natn. Acad. Sci. USA* **75**, 1304–1308.
Manney, T. R. and Meade, J. H. (1977) *In* "Microbial Interactions, Series B" (J. L. Reissig, Ed.) 281–321. J. Wiley and Sons, New York.
Mannino, R. J. and Burger, M. M. (1975) *Nature Lond.* **256**, 19–22.
Mano, Y. (1970) *Devptl. Biol.* **22**, 433–460.
Mano, Y. (1971a) *J. Biochem. (Tokyo)* **69**, 11–25.
Mano, Y. (1971b) *Arch. biochem. Biophys.* **146**, 237–248.
Mano, Y. (1975) *Biosystems* **7**, 51–65.
Manor, H. and Haselkorn, R. (1967) *Nature Lond.* **214**, 983–986.
Mansour, J. D., Henry, J. and Shapiro, L. (1980) *J. Bact.* **141**, 262–269.
Marcus, M. A. and Lewis, A. (1977) *Science* **195**, 1328.
Marcus, P. I. and Robbins, E. (1963) *Proc. natn. Acad. Sci. USA* **50**, 1156–1164.
Marino, W., Ammer, S. and Shapiro, L. (1976) *J. mol. Biol.* **107**, 115–130.
Marks, D. B., Paik, W. K. and Borun, J. W. (1973) *J. biol. Chem.* **248**, 5660–5667.
Marsh, R. C. and Worcel, A. (1977) *Proc. natn. Acad. Sci. USA* **74**, 2720–2724.
Martegani, E., Levi, M., Trezzi, F. and Alberghina, L. (1980) *J. Bact.* **142**, 268–275.
Martin, D. M. and Godson, G. N. (1977) *J. mol. Biol.* **117**, 321–335.
Martin, R. J. and Schloerb, P. R. (1964) *Cancer Res.* **24**, 1997–2000.
Martin, D., Tomkins, G. M. and Granner, D. (1969) *Proc. natn. Acad. Sci. USA* **62**, 248–255.
Martindale, D. W. and Pearlman, R. E. (1979) *Genetics* **92**, 1079–1092.
Martinéz-Salas, E. and Vicente, M. (1980) *J. Bact.* **144**, 532–541.
Marunouchi, T. and Nakano, M. M. (1980) *Cell Struct. Funct.* **5**, 53–67.
Maruyama, Y. (1964) *In* "Synchrony in Cell Division and Growth" (E. Zeuthen, Ed.) 421–432, 593–598. Wiley (Interscience), New York.
Maruyama, Y. and Yanagita, T. (1956) *J. Bact.* **71**, 542–546.
Marvin, D. A. (1968) *Nature Lond.* **219**, 485–486.
Masamune, Y. (1975) *J. gen. Appl. Microbiol.* **21**, 135–148.
Masters, M. and Broda, P. (1971) *Nature Lond. New Biol.* **232**, 137–140.
Masters, M. and Donachie, W. D. (1966) *Nature Lond.* **209**, 476–479.
Masters, M. and Pardee, A. B. (1965) *Proc. natn. Acad. Sci. USA* **54**, 64–70.
Mastro, A. M. (1979) *J. Cell Physiol.* **99**, 349–357.
Matney, T. S. and Suit, J. C. (1966) *J. Bact.* **92**, 960–966.
Matsumoto, S. and Funakoshi, H. (1978) *Cell Struct. Funct.* **3**, 173–179.
Matthys-rochon, R. (1979) *Biol. Cell* **35**, 313–320.
Mattick, J. S. and Hall, R. H. (1977) *J. Bact.* **130**, 973–982.
Matur, A. and Berry, D. R. (1978) *J. gen. Microbiol.* **109**, 205–213.
Maul, G. G., Maul, H. M., Scogna, J. E., Lieberman, M. W., Stein, G. S., Hsu, B. Y. and Borun, T. W. (1972) *J. Cell Biol.* **55**, 433–447.
Mauro, F. and Elkind, M. M. (1968) *Cancer Res.* **28**, 1150–1155.
Mauro, F. and Madoc-Jones, H. (1970) *Cancer Res.* **30**, 1397–1408.
May, J. W. (1962) *Expl. Cell Res.* **27**, 170–172.
May, R. (1973) "Stability and Complexity in Model Ecosystems". University Press, Princeton.
Mazia, D. (1937) *J. Cell comp. Physiol.* **10**, 291–304.
Mazia, D. (1974) *Sci. Am.* **230**, No. 1, 54–64.
Mazia, D. (1977) *In* "Mitosis—Facts and Questions" (M. Little, C. Patzelt, H. Ponstengl, O. Schroeter and H. P. Zimmerman, Eds.) Springer-Verlag, Berlin, Heidelberg.

McCorquodale, M. M. and Guttes, E. (1977) *Expl. Cell Res.* **104**, 279–285.
McCullough, E. A. and Till, J. E. (1971) *Am. J. Path* **65**, 601–619.
McCullough, E. A., Siminovitch, L. and Till, J. E. (1964) *Science* **144**, 844–846.
McCullough, W. and John, P. C. L. (1972) *Biochim. Biophys. Acta* **269**, 287–290.
McCully, E. K. and Robinow, C. F. (1971) *J. Cell Sci.* **9**, 475–507.
McEntee, K. (1977) *Proc. natn. Acad. Sci. USA* **74**, 5275–5279.
McEwen, C. R., Stallard, R. W. and Jukos, E. T. (1968) *Anal. Biochem.* **23**, 369–377.
McFarlane, E. S. (1980) *Arch. Microbiol.* **124**, 243–247.
McGann, L. E. and Kruuv, J. (1977) *Cryobiology* **14**, 503–505.
McGhee, J. O. and Felsenfeld, G. (1980) *Ann. Rev. Biochem.* **49**, 1115–1156.
McGuire, R. F. and Barber, R. J. (1976) *J. Supramol. Struct.* **4**, 259–269.
McHenry, C. and Kornberg, A. (1977) *J. biol. Chem.* **252**, 6478–6484.
McKeehan, W. L. and Ham, R. G. (1978) *Nature Lond.* **275**, 756–758.
McKenna, W. G. and Masters, M. (1972) *Nature Lond.* **240**, 536–539.
McMacken, R., Ueda, K. and Kornberg, A. (1977) *Proc. natn. Acad. Sci. USA* **74**, 4190–4194.
McMurrough, I. and Rose, A. M. (1969) *Biochem. J.* **105**, 189–203.
McQuillan, K., Roberts, R. P. and Britten, R. J. (1959) *Proc. natn. Acad. Sci. USA* **45**, 1437–1447.
Meacock, P. A. and Pritchard, R. H. (1975) *J. Bact.* **122**, 931–942.
Meacock, P. A., Pritchard, R. H. and Roberts, E. M. (1978) *J. Bact.* **133**, 320–328.
Means, A. R. and Dedman, J. R. (1980) *Nature London.* **285**, 73–77.
Mechali, M. and De Recondo, A. M. (1975) *Eur. J. Biochem.* **58**, 461–466.
Meijer, M., Beck, E., Hansen, F. G., Bergmans, H. E. N., Messer, W., von Meyenburg, K. and Schaller, H. (1979) *Proc. natn. Acad. Sci. USA* **76**, 580–584.
Meinert, J. C., Ehret, C. F. and Antipa, G. A. (1975) *Microbiol. Ecol.* **2**, 201–214.
Meiss, H. K. and Basilico, C. (1972) *Nature Lond. New Biol.* **239**, 66–68.
Meistrich, M. L., Meyn, R. E. and Barlogie, B. (1977) *Expl. Cell Res.* **105**, 169–177.
Mel, H. C. (1964) *J. theor. Biol.* **6**, 159–181.
Melero, J. A. (1979) *J. Cell Physiol.* **98**, 17–30.
Melero, J. A. and Fincham, V. (1978) *J. Cell Physiol.* **95**, 295–306.
Melero, J. A. and Smith, A. E. (1978) *Nature Lond.* **272**, 725–727.
Meloni, M., Perra, M. and Costa, M. (1980) *Expl. Cell Res.* **126**, 465–469.
Mendelson, N. H. (1972) *J. Bact.* **111**, 298–300.
Mendelson, N. H. and Cole, R. M. (1972) *J. Bact.* **112**, 994–1003.
Menge, U. and Kiermayer, O. (1977) *Protoplasma* **91**, 115–123.
Mergenhagen, D. and Schweiger, H. G. (1974) *Pl. Sci. Lett.* **3**, p. 387.
Messer, W. (1972) *J. Bact.* **112**, 7–12.
Meyer, M., De Jong, M. A., Demets, R. and Nanninga, N. (1979) *J. Bact.* **138**, 17–23.
Meyer, R. R., Shlomai, J., Kobori, J., Bates, D. L., Rowen, L., McMaken, R., Ueda, K. and Kornberg, A. (1978) *Cold Spring Harbor Symp. Quant. Biol.* **43**, 289–293.
Meyn, R. E., Meistrich, M. L. and White, R. A. (1980) *J. nat. Cancer Inst* **64**, 1215–1219.
Michaels, A., Schobert, B. and Herrin, D. (1980) *J. Cell Biol.* **87**, A188.
Michaels, G. A., Whitlock, S. O., Horowitz, P. and Levin, P. (1977) *Biochem. biophys. Res. Comm.* **75**, 480–487.
Mihara, S. and Hase, E. (1971) *Pl. Cell Physiol.* **12**, 225–236.
Milcarek, C. Zahn, K. (1978) *J. Cell Biol.* **79**, 833–838.

Miller, R. G. and Phillips, R. A. (1969) *J. Cell Physiol.* **73**, p. 191.

Miller, G. G., Schaer, J. C., Gautschi, J. R. and Schindler, R. (1979) *Mol. Cell Biochem.* **27**, 7–15.

Miller, M. R., Castellot, J. J. and Pardee, A. B. (1978) *Biochemistry* **17**, 1073–1080.

Millward, D. J., Garlick, P. J., Stewart, R. J. C., Nnanyelugo, D. O. T. and Waterlow, J. C. (1975) *Biochem. J.* **150**, 235–243.

Minassian I. and Bell, L. G. E. (1976) *J. Cell Sci.* **22**, 521–530.

Mindich, L. and Dales, S. (1972) *J. Cell Biol.* **55**, 32–41.

Minet, M., Nurse, P., Thuriaux, P. and Mitchison, J. M. (1979) *J. Bact.* **137**, 440–446.

Ming, P.-M. L., Chang, H. L. and Baserga, R. (1976) *Proc. natn. Acad. Sci. USA* **73**, 2052–2055.

Minor, P. D. and Smith, J. A. (1974) *Nature Lond.* **248**, 241–243.

Minorsky, N. (1962) "Nonlinear Oscillations". Van Nostrand, Princeton.

Mirelman, D., Yashouv-Gan, Y., Nuchamovitz, Y., Rozenhak, S. and Ron, E. Z. (1978) *J. Bact.* **134**, 458–461.

Misra, N. C. and Roberts, D. (1975) *Cancer Res.* **35**, 99–105.

Mita, S., Yasuda, H., Marunouchi, T., Ishiko, S. and Yamada, M. (1980) *Expl. Cell Res.* **126**, 407–416.

Mitchell, J. L. A. (1971) *Planta (Berl.)* **100**, 244–257.

Mitchell, B. F. and Tupper, J. T. (1977) *Expl. Cell Res.* **106**, 351–355.

Mitchell, J. B., Bedford, J. S. and Bailey, S. M. (1979) *Radiat. Res.* **79**, 520–536.

Mitchelson, K., Chambers, T., Bradbury, E. M. and Matthews, H. R. (1978) *FEBS Lett.* **92**, 339–342.

Mitchison, J. M. (1957) *Expl. Cell Res.* **13**, 244–262.

Mitchison, J. M. (1970) *In* "Methods in Cell Physiology" (D. M. Prescott, Ed.) Vol. 4, 131–165. Academic Press, New York and London.

Mitchison, J. M. (1971) "The Biology of the Cell Cycle" Cambridge University Press, Cambridge, UK.

Mitchison, J. M. (1974) *In* "Cell Cycle Controls" (G. M. Padilla, I. L. Cameron and A. Zimmerman, Eds.) 125–142. Academic Press, New York.

Mitchison, J. M. (1977a) *In* "Growth Kinetics and Biochemical Regulation of Normal and Malignant Cells" (B. Drewinko and R. M. Humphrey, Eds.) 23–33, Williams and Wilkins, Co., Baltimore, USA.

Mitchison, J. M. (1977b) *In* "Mitosis—Facts and Questions" (M. Little, C. Petzelt, H. Ponstingh, D. Schroeter and H. P. Zimmermann, Eds.) 1–13. Springer-Verlag, Berlin, Heidelberg.

Mitchison, J. M. (1977c) *In* "Cell Differentiation in Microorganisms, Plants and Animals" (L. Nover and K. Mothes, Eds.) 377–401. Gustav Fischer Verlag, Jena.

Mitchison, J. M. and Carter, B. L. A. (1975) *In* "Methods in Cell Biology" (D. M. Prescott, Ed.) Vol. 9, 201–219. Academic Press, New York and London.

Mitchison, J. M. and Creanor, J. (1969) *J. Cell Sci.* **5**, 373–391.

Mitchison, J. M. and Creanor, J. (1971) *Expl. Cell Res.* **67**, 368–374.

Mitchison, J. M. and Lark, K. G. (1962) *Expl. Cell Res.* **28**, 452–455.

Mitchison, J. M. and Vincent, W. S. (1965) *Nature Lond.* **205**, 987–989.

Mitchison, J. M. and Walker, P. M. B. (1959) *Expl. Cell Res.* **16**, 49–58.

Mitchison, J. M. and Wilbur, K. M. (1962) *Expl. Cell Res.* **26**, 144–157.

Mitchison, J. M., Cummins, J. E., Cross, P. R. and Creanor, J. (1969) *Expl. Cell Res.* **57**, 411–422.

Mittermayer, C., Braun, R. and Rusch, H. P. (1964) *Biochim. biophys. Acta* **91**, 399–405.

Mittermayer, C., Braun, R., Chayka, T. G. and Rush, H. P. (1966) *Nature Lond.* **210**, 1133–1137.

Miyakawa, Y., Komano, T. and Marayama, Y. (1980) *J. Bact.* **141**, 502–507.

Miyamoto, H., Rasmussen, L. and Zeuthen, E. (1973) *J. Cell Sci.* **13**, 889–900.

Miyamoto, T., Watanabe, M., Takabe, Y. and Terasima, T. (1976) *Cell Struct. Funct.* **1**, 177–185.

Miyata, M. and Miyata, H. (1978) *J. Bact.* **136**, 558–564.

Miyata, H., Miyata, M. and Ito, M. (1978a) *Cell Struct. Funct.* **3**, 39–46.

Miyata, H., Miyata, M. and Ito, M. (1978b) *Cell Struct. Funct.* **3**, 153–159.

Miyata, H., Miyata, M. and Ito, M. (1979) *Cell Struct. Funct.* **4**, 81–89.

Miyata, M., Imura, K. and Miyata, H. (1980) *J. gen. appl. Microbiol.* **26**, 109–118.

Moebs, W. D. C. (1977) *Math. Biosci.* **34**, 237–250.

Moens, P. B. and Rapport, E. (1971) *J. Cell Biol.* **50**, 344–361.

Molineux, I. J. and Gefter, M. L. (1975a) *J. Mol. Biol.* **98**, 811–825.

Molineux, I. J. and Gefter, M. L. (1975b) *Proc. natn. Acad. Sci. USA* **71**, 3858–3862.

Molineux, I. J., Friedman, S. and Gefter, M. L. (1974) *J. biol. Chem.* **249**, 6090–6098.

Moll, R. and Wintersberger, E. (1976) *Proc. natn. Acad. Sci. USA* **73**, 1863–1867.

Molloy, G. R. and Schmidt, R. R. (1970) *Biochem. biophys. Res. Comm.* **40**, 1125–1133.

Monod, J. and Jacob, F. (1961) *Cold Spring Harbor Symp. Quant. Biol.* **26**, 389–401.

Monod, J., Wyman, J. and Changeux, J.-P. (1965) *J. mol. Biol.* **12**, 88–118.

Moor, H. (1967) *Arch. Microbiol.* **57**, 135–146.

Mor, J. R. and Fiechter, A. (1968) *Biotech. Bioeng.* **10**, 159–176.

Mora, M., Darzynkiewicz, Z. and Baserga, R. (1980) *Expl. Cell Res.* **125**, 241–249.

Morais, R. (1977) *Can. J. Biochem.* **55**, 1180–1185.

Mori, Y., Akedo, H., Tanigaki, Y., Tanaka, K. and Okada, M. (1979) *Expl. Cell Res.* **120**, 435–439.

Moran, R. E. and Straus, M. J. (1980) *Cancer Tr. Rep* **64**, 81–86.

Morimoto, H. and James, T. W. (1969) *Expl. Cell Res.* **58**, 195–200.

Morris, N. R. (1976a) *Genet. Res.* **26**, 237–254.

Morris, N. R. (1976b) *Expl. Cell Res.* **98**, 204–210.

Morris, N. R. (1980) *In Symp. Soc. gen. Microbiol.* **30**, (G. W. Gooday, D. Lloyd, and A. P. J. Trinci, Eds.) 41–75.

Morris, N. R., Cramer, J. W. and Reno, D. (1967) *Expl. Cell Res.* **48**, 216–218.

Morris, N. R., Lai, M. H. and Oakley, C. E. (1979) *Cell* **16**, 437–442.

Morrison, D. C. and Morowitz, H. J. (1970) *J. mol. Biol.* **49**, 441–459.

Morton, G. T. and Berger, J. D. (1975) *J. Protozool.* **22**, p. 20A.

Moser, G. C. and Meiss, H. R. (1975) *J. Cell Biol.* **67**, p. 297A.

Moser, G. C. and Meiss, H. K. (1977) *Somatic Cell Genet.* **3**, 449–456.

Moser, G. C., Müller, H. and Robbins, E. (1975) *Expl. Cell Res.* **91**, 73–78.

Moses, H. L., Proper, J. A., Volkenant, M. E. and Swartzendruber, D. E. (1980) *J. Cell Physiol.* **102**, 367–378.

Mount, D. W., Low, K. B. and Edmiston, S. J. (1972) *J. Bact.* **112**, 886–893.

Mount, D. W., Walker, A. C. and Kosel, C. (1973) *J. Bact.* **116**, 950–956.

Moya, F. and Glaser, L. (1980) *J. biol. Chem.* **255**, 3258–3260.

Moyne, G., Bertaux, O. and Puvion, E. (1975) *J. Ultrastruct. Res.* **52**, 362–376.

Mueller, G. C. and Kajiwara, K. (1966) *19th Symp. Fundamental Cancer Res.* 452–474. Univ. Texas. Williams and Wilkins, Baltimore, Md.

Müller, J. and Dawson, P. S. S. (1968) *Can. J. Microbiol.* **14**, 1115–1126.
Müller, R. N., Haverbeke, Y. V., Blave, A., Aguilera, A., Michel, N., Miller-Faures, A. and Miller, A. O. A. (1980) *FEBS Lett.* **114**, 231–233.
Muller, W. E. G., Schroder, H. C., Arendes, J., Steffen, R., Zahn, R. K. and Dose, K. (1977) *Eur. J. Biochem.* **76**, 531–540.
Mycielski, R., Kociszewska-Kauc, B. and Bailkowska-Hobrzanska, H. (1977) *Acta Microbiol. Polonica* **26**, 129–135.
Myers, J. and Graham, J.-R. (1975) *Pl. Physiol.* **55**, 686–688.
Nachmias, V. T. and Asch, A. (1976) *Biochemstry* **15**, 4273–4278.
Nachtwey, D. S. and Dickinson, W. J. (1967) *Expl. Cell Res.* **47**, 581–595.
Nagata, T. (1963) *Proc. natn. Acad. Sci. USA* **49**, 551–559.
Nagai, K. and Tamura, G. (1973) *J. Bact.* **112**, 959–966.
Nagai, K., Kaneko, H. E. and Tamura, G. (1971) *Biochem. biophys. Res. Comm.* **42**, 669–675.
Nagai, K., Some, H. and Tamura, G. (1976) *Agric. biol. Chem.* **40**, 2237–2243.
Nagata, T. and Meselson, M. (1968) *Cold Spring Harbor Symp. Quant. Biol.* **33**, 553–557.
Naha, P. M. (1973) *Nature Lond. New Biol.* **241**, 13–14.
Naha, P. M. (1979a) *J. Cell Sci.* **35**, 53–58.
Naha, P. M. (1979b) *J. Cell Sci.* **40**, 33–42.
Naha, P. M., Meyer, A. and Hewitt, K. (1975) *Nature Lond.* **258**, 49–53.
Naha, P. M. and Sorrentino, R. (1980) *Cell Biol. Int. Rep.* **4**, 365–378.
Nakamo, M. M., Sekiguchi, T. and Yamada, M. (1978) *Somat. Cell Genet.* **4**, 169–178.
Nanjundiah, V., Hara, K. and Konijn, T. N. (1976) *Nature Lond.* **260**, 705.
Nasmyth, K. A. (1977) *Cell* **12**, 1109–1120.
Nasmyth, K. A. (1979a) *J. Cell Sci.* **36**, 155–168.
Nasmyth, K. A. (1979b) *J. mol. Biol.* **130**, 273–284.
Nasmyth, K. A. and Reed, S. I. (1980) *Proc. natn. Acad. Sci. USA* **77**, 2119–2123.
Nasmyth, K., Nurse, P. and Fraser, R. S. S. (1979) *J. Cell Sci.* **39**, 215–233.
Natarajan, A. T. and Meyers, M. (1979) *Human Genet.* **52**, 127–132.
Navalgund, L. G., Rossana, C., Muench, A. J. and Johnson, L. F. (1980) *J. biol. Chem.* **255**, 7386–7390.
Neal, W. K., Funkhouser, E. A. and Price, C. A. (1968) *J. Protozool.* **15**, 761–763.
Nechaeva, N. V., Yarygin, K. N., Fateeva, V. I., Novikova, T. E. and Brodsky, V. Ya. (1980) *Bull. exp. Biol. Med.* **90**, 211–213.
Nedelman, J. and Rubinow, S. I. (1980) *Cell Biophysiol.* **2**, 207–231.
Nelsen, E. M. (1970) *J. exp. Zool.* **175**, 69–84.
Nelsen, D. A., Beltz, W. R. and Rill, R. (1977) *Proc. natn. Acad. Sci. USA* **74**, 1343–1347.
Nelson, R. G. and Fangman, W. L. (1979) *Proc. natn. Acad. Sci. USA* **76**, 6515–6519.
Newlon, C. S. and Fangman, W. L. (1975) *Cell* **5**, 423–428.
Newlon, C. S., Luoesche, R. D. and Walter, S. K. (1979) *Mol. gen. Genet.* **169**, 189–194.
Newman, C. N. and Kubitschek, H. E. (1978) *J. mol. Biol.* **121**, 461–471.
Newman, J. and Hanawalt, P. C. (1968) *J. mol. Biol.* **35**, 639–642.
Newton, B. A. (1957) *J. gen. Microbiol.* **17**, 718–730.
Nias, A. H. W. and Lajtha, L. G. (1968) *In* "Actions Chemiques et Biologiques des Radiations". Vol. 12, p. 97. M. Haissinsky, Paris.
Nias, A. H. W., Fox, M. and Fox, B. W. (1970) *Cell Tiss. Kinet.* **3**, 207–215.

Nicolini, C. (1975) *J. natn. Cancer Inst.* 55, 821–826.
Nicolini, C. (1976) *Biochem. biophys. Acta.* **458**, 243–282.
Nicolini, C., Ajiro, K., Borun, T. and Baserga, R. (1975) *J. biol. Chem.* **250**, 3381–3385.
Nicolini, C., Kendall, F. and Giaretti, W. (1977) *Biophysiol. J.* **19**, 163–176.
Nicolis, G. and Portnow, J. (1973) *Chem. Rev.* **73**, 365–384.
Nicolis, G. and Prigogine, I. (1977) "Self-organization in Non-Equilibrium Systems". Wiley, New York.
Nicolson, G. L. (1976) *Biochim. biophys. Acta* **457**, 57–108.
Nilshammar, M. and Walles, B. (1974) *Protoplasma* **79**, 317–332.
Nishi, A. and Kogoma, T. (1965) *J. Bact.* **90**, 884–890.
Nishi, A., Okamura, S. and Yanagita, T. (1967) *J. gen. appl. Microbiol. Tokyo* **13**, 103–119.
Nishimoto, T. and Basilico, C. (1978) *Somatic. Cell Genet.* **4**, 323–340.
Nishimoto, T., Takahashi, T. and Basilico, C. (1980) *Somatic Cell Genet.* **6**, 465–476.
Nishimura, Y. and Bailey, J. E. (1980) *Nach. Biosci.* **51**, 305–328.
Nishioka, Y. and Eisenstark, A. (1970) *J. Bact.* **102**, 320–333.
Njus, D., Sulzman, F. M. and Hastings, J. W. (1974) *Nature Lond.* **248**, 116–120.
Noguchi, T. (1978) *Protoplasma* **95**, 73–88.
Noll, H. (1967) *Nature Lond.* **215**, 360–363.
Nomura, K., Hoshino, T., Knebel, K., Deen, D. F. and Barker, M. (1978) *Canc. Tr. Rep.* **62**, 747–754.
Noonan, K. D. and Burger, M. M. (1973) *J. biol. Chem.* **248**, 4286–4292.
Noonan, K. D., Levine, A. J. and Burger, M. M. (1973) *J. Cell Biol.* **58**, 491–497.
Nordenson, I. (1978) *Hereditas,* **89**, 163–167.
Normark, S. (1971) *J. Bact.* **108**, 51–58.
Normark, S., Boman, H. G. and Bloom, G. D. (1971) *Acta Pathol. Microbiol. Scand. B.* **79**, 651–664.
Nosoh, Y. and Takamiya, A. (1962) *Pl. Cell Physiol.* **3**, 53–66.
Novák, F. J., Schwammenhöferová, K., Čihaliková, J. and Ondřej, M. (1979) *Biolog. Plantarum (Praha)* **21**, 51–56.
Novick, A. and Weiner, M. (1957) *Proc. natn. Acad. Sci. USA* **43**, 553–566.
Novick, A. and Weiner, M. (1959) *In* "Symposium on Molecular Biology" (R. E. Zirkle, Ed.) 78–90. Univ. of Chicago Press, Chicago.
Nowakowski, M., Atkinson, P. H. and Summers. D. F. (1972) *Biochim. biophys. Acta.* **266**, 154–160.
Nowell, P. C. (1964) *Expl. Cell Res.* **33**, 445–449.
Nurse, P. (1975) *Nature Lond.* **256**, 547–551.
Nurse, P. (1977) *Biochem. Soc. Trans.* **5**, 1191–1193.
Nurse, P. (1980) *Nature Lond.* **286**, 9–10.
Nurse, P. and Thuriaux, P. (1977) *Expl. Cell Res.* **107**, 365–375.
Nurse, P. and Wiemken, A. (1974) *J. Bact.* **117**, 1108–1116.
Nurse, P., Thuriaux, P. and Nasmyth, K. (1976) *Mol. gen. Genet.* **146**, 167–178.
Nurse, P., Fantes, P. A. and Wheals, A. E. (1977) *Nature Lond.* **267**, p. 647.
Nuccitelli, R. and Jaffe, L. F. (1977) *Ann. Rev. Biophysiol. Bioeng.* **6**, 445–476.
Nusse, M. (1980) *Radiat Env.* **17**, 296.
Nüsslein, V., Otto, B., Bonhoeffer, F. and Schaller, H. (1971) *Nature Lond. New Biol.* **234**, 285–286.
Nygaard, O. F., Güttes, G. and Rusch, H. P. (1960) *Biochim. biophys. Acta* **38**, 298–306.

Oakley, B. R. and Bisalputra, T. (1977) *Can. J. Bot.* 55, 2789–2800.
Ober, K. (1974) *Arch. Microbiol.* **99**, 369–378.
Ober, K. (1975) *Arch. Microbiol.* **102**, 129–137.
Obrénovitch, A., Sené, C., Nègre, M. T. and Monsigny, M. (1978) *FEBS Lett.* **88**, 187–191.
O'Brien, R. L., Sanyal, A. B. and Stanton, R. H. (1973) *Expl. Cell Res.* **80**, 340–344.
Oertel, W. and Goulian, M. (1979) *J. Bact.* **140**, 333–341.
Ogawa, T. and Okazaki, T. (1979) *Nucleic Acids Res.* **7**, 1621–1633.
Ogawa, T. and Okazaki, T. (1980) *Ann. Rev. Biochem.* **49**, 421–457.
Ogur, M., Minckler, S. and McClary, D. O. (1953) *J. Bact.* **66**, 642–645.
Ohara, H., and Terasima, T. (1976) *Cell Struct. Funct.* **1**, 187–195.
Ohki, M. (1972) *J. mol. Biol.* **68**, 249–264.
Ohki, M. and Mitsui, H. (1974) *Nature Lond.* **252**, 64–66.
Ohki, M. and Sato, S. (1975) *Nature Lond.* **253**, 654–656.
Ohlsson-Wilhelm, B. M., Freed, J. J. and Perry, R. P. (1976) *J. Cell Physiol.* **89**, 77–88.
Ohlsson-Wilhelm, B. M., Leary, J. and Pacilio, M. (1979) *J. Cell Biol.* **83**, 4A.
Okada, S. (1960) *Nature Lond.* **185**, 193–194.
Okamura, S., Ishikawa, H., Suziki, N. and Yamada, E. (1977) *Cell Struct. Funct.* **2**, 229–240.
Okazaki, R., Okazaki, T., Sakebe, K., Sugimoti, K., Kainuma, R., Sugino, A. and Iwatsuki, N. (1968) *Cold Spring Harbor Symp. Quant. Biol.* **33**, 129–143.
Okazaki, R., Arisawa, M. and Sugino, A. (1971) *Proc. natn. Acad. Sci. USA* **68**, 2954–2957.
Okinaka, R. T. and Barnhart, B. J. (1978) *J. Cell Biol.* **79**, 383A.
Okita, T. W. and Volcani, B. E. (1980) *Expl. Cell Res.* **125**, 471–481.
Olah, E., Lui, M. S., Tzeng, D. Y. and Weber, G. (1980) *Cancer Res.* **40**, 2869–2875.
Oman, S., Grubič, Z. and Brzin, H. (1977) *Analyt. Biochem.* **83**, 211–216.
Onda, H. (1979) *J. theoret. Biol.* **77**, 367–377.
Onda, H. (1980) *J. theoret. Biol.* **85**, 771–787.
Ooka, T. (1976) *In* "Methods in Cell Biology" (D. M. Prescott, Ed.) Vol. 14, 287–295. Academic Press, New York and London.
Ooka, T. and Daillie, J. (1974) *Expl. Cell Res.* **84**, 219–222.
Orcival-Lafont, A. M., Pineau, B., Ledoigt, G. and Calvayrac, R. (1972) *Can. J. Bot.* **50**, 1503–1508.
Ord, M. J. (1968) *J. Cell Sci.* **3**, 483–491.
Ord, M. G. and Stocken, L. A. (1958) *Nature Lond.* **182**, 1787–1788.
Ord, M. G., Stocken, L. A. and Thrower, S. (1975) *Sub-Cell. Biochem.* **4**, 147–156.
Orr, E. and Rosenberger, R. F. (1976a) *J. Bact.* **126**, 895–902.
Orr, E. and Rosenberger, R. F. (1976b) *J. Bact.* **126**, 903–906.
Orr, E., Fairweather, N. F., Holland, I. B. and Pritchard, R. H. (1979) *Mol. gen. Genet.* **177**, 103–112.
Osafune, T. (1973) *J. Elect. Micros.* **22**, 51–61.
Osafune, T., Mihara, S., Hase, E. and Ohkuro, I. (1972a) *Pl. Cell Physiol.* **13**, 211–227.
Osafune, T., Mihara, S., Hase, E. and Ohkuro, I. (1972b) *Pl. Cell Physiol.* **13**, 981–989.
Osafune, T., Mihara, S., Hase, E. and Ohkuro, I. (1975a) *Pl. Cell Physiol.* **16**, 313–326.

Osafune, T., Mihara, S., Hase, E. and Ohkuro, I. (1975b) *J. Elect. Micros.* **24**, 33–39.
Osafune, T., Mihara, S., Hase, E. and Ohkuro, I. (1975c) *J. Elect. Micros.* **24**, 283–286.
Osafune, T., Mihara, S., Hase, E. and Ohkuro, I. (1975d) *J. Elect. Micros.* **24**, 247–252.
Osafune, T., Mihara, S., Hase, E. and Ohkuro, I. (1976) *J. Elect. Micros.* **25**, 261–269.
Osley, M. A. and Newton, A. (1977) *Proc. natn. Acad. Sci. USA* **74**, 124–128.
Osley, M. A. and Newton, A. (1978) *J. Bact.* **135**, 10–17.
Osley, M. A. and Newton, A. (1980) *J. Molec. Biol.* **138**, 109–128.
Osley, M. A., Sheffy, M. and Newton, A. (1977) *Cell* **12**, 393–400.
Osumi, M. and Sando, N. (1969) *J. Elect. Micros.* **18**, 47–56.
Osumi, M., Masuzawa, E. and Sando, N. (1968) *Jap. Wom. Univ. J.* **15**, 33–40.
Osumi, M., Ichinokawa, K. and Hirosawa, T. (1971) *Jap. Wom. Univ. J.* **18**, 65–74.
Othmer, H. G. (1975) *Math. Biosci.* **24**, 205–238.
Othmer, H. G. and Scriven, L. E. (1974) *J. theoret. Biol.* **43**, 83–112.
Otsuka, F. and Scheffle, I. E. (1978) *J. Supramol. Struct.* **S2**, 334.
Otto, B. (1977) *FEBS Lett.* **79**, 175–178.
Otto, B. and Knippers, R. (1976) *Eur. J. Biochem.* **71**, 617–622.
Otto, B., Baynes, M. and Knippers, R. (1977) *Eur. J. Biochem.* **73**, 17–24.
Oulevey, N., Deshusses, J. and Turian, G. (1970) *Protoplasma* **70**, 217–224.
Paau, A. S., Cowles, J. R. and Ord. J. (1977) *Cen. J. Micro.* **23**, 1165–1169.
Padilla, G. M. and Bragg, R. J. (1968) *J. Cell Biol.* **39**, 101a.
Padilla, G. M. and Cameron, I. L. (1964) *J. Cell Comp. Physiol.* **64**, 303–308.
Padilla, G. M. and Cook, J. R. (1964) *In* "Synchrony in Cell Division and Growth" (E. Zuethen, Ed.) 521–536. Wiley (Interscience), New York.
Padilla, G. M. and James, T. W. (1964) *In* "Methods in Cell Physiology" (D. M. Prescott, Ed.) Vol. 1, 141–157. Academic Press, New York.
Padilla, G. M., Cameron, I. L. and Elrod, L. H. (1966) *In* "Cell Synchrony" (I. L. Cameron and G. M. Padilla, Eds.) p. 269. Academic Press, New York and London.
Padmanabhan, R., Padmanabhan, R. and Green, M. (1976) *Biochem. biophys. Res. Commun.* **69**, 860–867.
Padulo, L. and Arbib, M. (1974) "System Theory". Saunders, Philadelphia.
Pages, J. M., Piovant, M., Lazdunski, A. and Lazdunski, C. (1975) *Biochimie* **57**, 303–313.
Pagoulatos, G. M. and Darnell, J. E. (1970) *J. Cell Biol.* **44**, 476–483.
Painter, R. B. and Drew, R. M. (1959) *Lab. Invest.* **8**, 278–285.
Painter, R. B. and Howard, R. A. (1978) *Mutat. Res.* **54**, 113–115.
Painter, R. B. and Schaefer, A. W. (1969) *J. mol. Biol.* **45**, 467–479.
Palamarchuk, Y. K., Polezhayev, A. A., Solyanik, G. I., Chernavskii, D. S. and Burlakova, Y. B. (1978) *Biofizika* **23**, 845–851.
Palmer, C. G., Livergood, D. and Warren, A. K. (1960) *Expl. Cell Res.* **20**, 198–201.
Palmer, J. D., Brown, F. A. Jr. and Edmunds, L. N. Jr. (1976) "An Introduction to Biological Rhythms". Academic Press, New York and London.
Palyi, I. (1975) *Cancer Chemother. Rep.* **59**, 493–499.
Palzer, R. J. and Heidelberger, C. (1973) *Cancer Res.* **33**, 422–427.
Pardee, A. B. (1962) *In* "The Bacteria" (I. C. Gunsalus and R. Y. Stainier, Eds.) Vol. III, 577–630. Academic Press, New York and London.

Pardee, A. B. (1974a) *Proc. natn. Acad. Sci. USA* **71**, 1286–1290.
Pardee, A. B. (1974b) *Eur. J. Biochem.* **43**, 209–213.
Pardee, A. B. (1975) *Biochim. biophys. Acta* **417**, 153–172.
Pardee, A. B. and Dubrow, R. (1977) *Cancer* **39**, 2747–2754.
Pardee, A. B., Dubrow, R., Hamlin, J. L. and Kletzien, R. F. (1978) *Ann. Rev. Biochem.* **47**, 715–750.
Park, C. H., Bergsagel, D. and McCullough, E. A. (1971) *J. natn. Cancer Inst.* **46**, 411–416.
Parsons, J. A. (1965) *J. Cell Biol.* **25**, 641–646.
Parsons, J. A. and Rustad, R. C. (1968) *J. Cell Biol.* **37**, 683–693.
Pastan, I. A., Johnson, G. S. and Anderson, W. B. (1975) *Ann. Rev. Biochem.* **44**, 491–522.
Pasternak, C. A. (1976a) *TIBS* **1**, 148–151.
Pasternak, C. A. (1976b) *J. theoret. Biol.* **58**, 365–382.
Pasternak, C. A., Warmsley, A. M. H. and Thomas, D. B. (1971) *J. Cell Biol.* **50**, 562–564.
Pasternak, C. A., Sumner, McB. and Collin, R. C. L. S. (1974) *In* "Cell Cycle Controls" (G. M. Padilla, I. L. Cameron and A. Zimmerman, Eds.) 117–124. Academic Press, New York and London.
Paterson, A. R., Jakobs, E. S., Lauzon, G. J. and Weinstein, W. M. (1979) *Cancer Res.* **39**, 2216–2219.
Pathak, S. N., Dave, C. and Krishan, A. (1977) *Pharmacology* **19**, 207.
Pathak, S. N., Dave, C. and Krishan, A. (1979) *J. Med. Res.* **70**, 777–792.
Paul, D. and Ristow, H. J. (1979) *J. Cell Physiol.* **98**, 31–40.
Paul, J. S. and Volcani, B. E. (1976) *Arch. Microbiol.* **110**, 247–252.
Paulin, J. J. (1975) *J. Cell Biol.* **66**, 404–413.
Pavlidis, T. (1969) *J. theoret. Biol.* **22**, 418–436.
Pavlidis, T. (1973) "Biological Oscillators: their Mathematical Analysis". Academic Press, New York and London.
Pavlidis, T. and Kauzmann, W. (1969) *Arch. biochem. Biophys.* **132**, 338–348.
Pedersen, F. S., Lund, E. and Kjelgaard, N. O. (1973) *Nature New Biol.* **234**, 13–15.
Pederson, T. (1972) *Proc. natn. Acad. Sci. USA* **69**, 224–228.
Pedrali-Noy, G., Spadari, S., Miller-Faurés, A., Miller, A. O. A., Kruppa, J. and Koch, G. (1980) *Nucleic Acids Res.* **8**, 377–387.
Pellegrini, M. (1980) *J. Cell Sci.* **43**, 137–166.
Pellegrini, M. and Pellegrini, L. (1976) *C. r. ser. D.* **282**, 357–360.
Pendland, J. C. and Aldrich, H. C. (1978) *J. Cell Biol.* **79**, p. 12a.
Penman, C. S. and Duffus, J. H. (1975) *J. gen. Microbiol.* **90**, 76–80.
Perasso, R. and Beisson, J. (1978) *Biol. Cellulaire* **32**, 275–290.
Perlman, D. and Rownd, R. H. (1976) *Nature Lond.* **259**, 281–284.
Perry, R. P. (1976) *Ann. Rev. Biochem.* **45**, 605–629.
Petersen, D. F. and Anderson, E. C. (1964) *Nature Lond.* **203**, 642–643.
Petersen, D. F., Anderson, E. C. and Tobey, R. A. (1968) *In* "Methods of Cell Physiology" (D. M. Prescott, Ed.) Vol. 3, 347–370. Academic Press, New York.
Peterson, E. C. and Berger, J. D. (1976) *Can. J. Zool.* **54**, 2089–2097.
Peterson, J. B. and Ris, H. (1976) *J. Cell Sci.* **22**, 219–242.
Petes, T. D. (1980) *Ann. Rev. Biochem.* **49**, 845–876.
Petes, T. D. and Newlon, C. S. (1974) *Nature Lond.* **251**, 637–639.
Petes, T. D., and Williamson, D. H. (1975a) *Cell* **4**, 249–253.

Petes, T. D. and Williamson, D. H. (1975b) *Expl. Cell Res.* **95**, 103–110.
Petes, T. D., Newlon, C. S., Byers, B. and Fangman, W. L. (1974) *Cold Spring Harbor Symp. Quant. Biol.* **38**, 9–16.
Petes, T. D., Byers, B. and Fangman, W. L. (1973) *Proc. natn. Acad. Sci. USA* **70**, 3072–3076.
Pettersen, E. O. (1978) *Radiat. Res.* **73**, 180–191.
Pettersen, E. O., Christensen, T., Bakke, O. and Oftenbro, R. (1977) *Int. J. Radiat. Biol.* **31**, 171–184.
Pettigrew, R. T., Galt, J. M., Ludgate, C. M. and Smith, A. N. (1974) *Br. Med. J.* **4**, 679–682.
Petty, K. M. and Dutton, P. L. (1976) *Biochim. biophys. Acta* **172**, 335–345.
Petzelt, C., Auel, D. and Sachsenmaier, W. (1980) *Cell Biol. I.* **4**, 579–583.
Pfau, J., Werthmüller, K. and Senger, H. (1971) *Arch. Mikrobiol.* **75**, 338–345.
Pfeiffer, S. E. (1968) *J. Cell Physiol.* **71**, 95–104.
Pfeiffer, S. E. and Tolmach, L. J. (1967) *Nature Lond.* **213**, 139–142.
Pfeiffer, S. E. and Tolmach, L. J. (1968) *J. Cell Physiol.* **71**, 77–94.
Phaff, H. J. (1971) *In* "The Yeasts" (A. H. Rose and J. S. Harrison, Eds.) Vol. 2, p. 135. Academic Press, London and New York.
Phillips, C. A. and Lloyd, D. (1978) *J. gen. Microbiol.* **105**, 95–103.
Phillips, I. R., Shephard, E. A., Stein, J. L., Kleinsmith, L. J. and Stein, G. S. (1979) *Biochim. biophys. Acta* **179**, 326–346.
Pickett, A. M. and Lester, J. C. (1979) *Lab. Proc.* **28**, 253–255.
Pickett-Heaps, J. O., Tipppit, D. H. and Andreozzi, J. A. (1979) *Biol. Cell* **35**, 199–203.
Pierron, G. and Sauer, H. W. (1980) *J. Cell Sci.* **41**, 105–113.
Pierucci, O. (1978) *J. Bact.* **135**, 559–574.
Pierucci, O. (1979) *J. Bact.* **138**, 453–460.
Pierucci, O. and Helmstetter, C. E. (1969) *Proc. Fedn. Am. Socs. exp. Biol.* **28**, 1755–1760.
Pine, M. J. (1972) *Ann. Rev. Microbiol.* **26**, 103–126.
Piñon, R. (1979) *Chromosoma* **70**, 337–352.
Piper, A. A., McCaffer, C. A., Milthorpe, B. K., Fox, R. N. and Tattersall, M. H. (1980) *Clin. Exp. Ph.* **7**, 65–66.
Piras, R. and Piras, M. M. (1975) *Proc. natn. Acad. Sci. USA* **72**, 1161–1165.
Pirson, A. and Lorenzen, H. (1958) *Z. Bot.* **46**, 53–66.
Pittendrigh, C. S. and Bruce, V. G. (1957) "Rhythmic and Synthetic Processes of Growth" p. 75. University Press, Princeton.
Plagemann, P. G. W. and Roth, M. F. (1969) *Biochemistry* **8**, 4782–4789.
Plesner, P. (1961) *Cold Spring Harbor Symp. Quant. Biol.* **26**, 159–162.
Plesner, P. (1964) *C. r. Trav. Lab. Carlsberg* **34**, 1–76.
Plusquellec, Y. and Le Gal, A. (1972) *C. r. Soc. Biol. Paris* **166**, 538–541.
Poccia, D. L., Levine, D. and Wang, J. C. (1978) *Develop. Biol.* **64**, 273–283.
Pochron, B. F. and Baserga, R. (1979) *J. Biol. Chem.* **254**, 6352–6356.
Polanshek, M. M. (1977) *J. Cell Sci.* **23**, 1–23.
Pollack, A., Bagwell, C. B. and Irvin, G. L. (1979) *Science* **203**, 1025–1027.
Pollack, M. S. and Price, C. A. (1971) *Anal. Biochem.* **42**, 38–47.
Poole, B. and Wibo, M. (1973) *J. biol. Chem.* **248**, 6221–6226.
Poole, R. K. (1977a) *FEMS Microbiol.* **1**, 305–307.
Poole, R. K. (1977b) *J. gen. Microbiol.* **98**, 177–186.
Poole, R. K. (1977c) *J. gen. Microbiol.* **103**, 19–27.

Poole, R. K. (1977d) *J. gen. Microbiol.* **99**, 369–377.
Poole, R. K. (1980) *In* "Diversity of Bacterial Respiratory Systems". **1** (C. J. Knowles, Ed.) 87–114. CRS Press Inc. Boca Raton, Florida.
Poole, R. K. and Lloyd, D. (1973) *Biochem. J.* **136**, 195–207.
Poole, R. K. and Lloyd, D. (1974) *Biochem. J.* **144**, 141–148.
Poole, R. K. and Pickett, A. M. (1978) *J. gen. Microbiol.* **107**, 399–402.
Poole, R. K. and Salmon, I. (1978) *J. gen. Microbiol.* **106**, 153–164.
Poole, R. K., Lloyd, D. and Kemp, R. B. (1973) *J. gen. Microbiol.* **77**, 209–220.
Poole, R. K., Lloyd, D. and Chance, B. (1974) *Biochem. J.* **138**, 201–210.
Poole, R. K., Scott, R. I. and Britnell, C. H. (1978) *Soc. Gen. Microbiol. Quart.* **6**, 22–23.
Poole, R. K., Blum, H., Scott, R. I., Collinge, A. and Ohnishi, T. (1980a) *J. gen. Microbiol.* **119**, 145–154.
Poole, R. K., Scott, R. I., Salmon, I. Gibson, J. C. and Misri, R. (1980b) *Soc. gen. Microbiol. Quart.* **8**, 21–22.
Poole, R. K., Scott, R. I. and Blum, H. (1981) *J. gen. Microbiol.* **124**, 181–185.
Pooley, H. M., Schlaeppi, J.-M. and Karamata, D. (1978) *Nature Lond.* **274**, 264–266.
Porter, K., Prescott, O. and Frye, J. (1973) *J. Cell Biol.* **57**, 815–836.
Posakony, J. W., England, J. M. and Attardi, G. (1977) *J. Cell Biol.* **74**, 468–491.
Powell, E. O. (1956) *J. gen. Micro.* **15**, 492–511.
Prakash, L., Hinkle, D. and Prakash, S. (1979) *Mol. gen. Genet.* **172**, 249–258.
Presant, C. A., Vietti, T. and Valeriote, F. A. (1975) *Cancer Res.* **35**, 1926–1930.
Prescott, D. M. (1955) *Expl. Cell Res.* **9**, 328–337.
Prescott, D. M. (1956) *Expl. Cell Res.* **11**, 86–98.
Prescott, D. M. (1960) *Expl. Cell Res.* **19**, 228–238.
Prescott, D. M. (1964) *In* "Synchrony in Cell Divsiion and Growth" (E. Zeuthen, Ed.) 71–97. Wiley (Interscience), New York.
Prescott, D. M. (1966) *J. Cell Biol.* **31**, 1–9.
Prescott, D. M. (1970) *Adv. Cell Biol.* **1**, 57–117.
Prescott, D. M. (1976a) "Reproduction of Eucaryotic Cells". Academic Press, New York and London.
Prescott, D. M. (1976b) *Adv. Genetics* **18**, 99–177.
Prescott, D. M. and Bender, M. A. (1962) *Expl. Cell Res.* **26**, 260–268.
Prescott, D. M. and Goldstein, L. (1967) *Science* **155**, 469–470.
Prescott, D. M. and Goldstein, L. (1968) *J. Cell Biol.* **39**, 404–414.
Prescott, D. M. and Kimball, R. F. (1961) *Proc. natn. Acad. Sci. USA* **47**, 686–693.
Prescott, D. M. and Kuempel, P. L. (1972) *Proc. natn. Acad. Sci. USA* **69**, 2842–2845.
Previc, E. P. (1970) *J. theoret. Biol.* **27**, 471–497.
Price, L. A. and Goldie, L. H. (1971) *Br. Med. J.* **4**, 336–339.
Prigogine, I. (1980) "From Being to Becoming; Time and Complexity in Physical Sciences". Freeman, San Francisco.
Prigogine, I. and Nicolis, G. (1971) *Quart. Rev. Biophys.* **4**, 107–148.
Prince, R. C., Baccarini-Melandri, A., Hauska, G. A., Melandri, B. A. and Crofts, A. R. (1975) *Biochim. biophys. Acta* **387**, 212–227.
Pringle, J. R. (1975) *In* "Methods in Cell Biology". (D. M. Prescott, Ed.) Vol. 12, 149-184. Academic Press, New York and London.
Pringle, J. R. (1978) *J. Cell Physiol.* **95**, 393–405.

Pritchard, R. H. (1968) *Heredity* **23**, 472–473.
Pritchard, R. H. (1974) *Phil. Trans. R. Soc. Biol.* **267**, 303–336.
Pritchard, R. H., Barth, P. T. and Collins, J. (1969) *Symp. Soc. gen. Microbiol.* **19**, 263–297.
Pritchard, R. H., Meacock, P. A. and Orr, E. (1978) *J. Bact.* **135**, 575–580.
Probst, H. and Maisenbacher, J. (1973) *Expl. Cell Res.* **78**, 335–344.
Probst, H. and Maisenbacher, J. (1975) *In* "Methods in Cell Biology". (D. M. Prescott, Ed.) Vol. 10, 173–184. Academic Press, New York and London.
Prouty, W. F. and Goldberg, A. L. (1972) *J. biol. Chem.* **247**, 3341–3352.
Puck, T. T. (1964) *Science* **144**, 565–566.
Puck, T. T. and Yamada, M. A. (1962) *Radiat. Res.* **16**, 589.
Pujara, P. & Lardner, T. J. (1979) *J. Biomech.* **12**, 293–299.
Pye, E. K. (1969) *Can. J. Bot.* **47**, 271–285.
Pye, E. K. and Chance, B. (1966) *Proc. natn. Acad. Sci. USA* **555**, 888–894.
Quastler, H. (1963) *In* "Actions Chimiques et Biologiques des Radiations". p. 149. M. Haissinsky, Paris.
Quintart, J. and Baudhuin, P. (1976) *Arch. Int. physiol. Biochim.* **84**, 409–410.
Quintart, J., Bartholeyns, J. and Baudhuin, P. (1979) *Biochem. J.* **184**, 133–141.
Quintart, J., Leroy-Houyet, M.-A. and Baudhuin, P. (1980) *J. Ultrastruct. Res.* **72**, 76–89.
Quinton, A. and Poole, R. K. (1977) *J. gen. Microbiol.* **103**, 271–275.
Rabito, C. A. and Tchao, R. (1980) *Am. J. Physiol* **238**, C43–C48.
Rackham, S. J. (1977) MSc. Thesis, University of Waikato, New Zealand.
Radman, N. (1974) *In* "Molecular and Environmental Aspects of Mutagenesis" 128–142. C. C. Thomas Pub. Co., Springfield, Ill.
Raju, M. R., Bain, E., Carpenter, S. G., Jett, J., Walters, R. A., Howard, J. and Powersri, P. (1980a) *Radiat. Res.* **84**, 152–157.
Raju, M. R., Johnson, T. S., Tokita, N., Carpenter, S. and Jett, J. H. (1980b) *Radiat. Res.* **84**, 16–24.
Rao, A. P. and Rao, P. N. (1976) *J. natn. Cancer Inst.* **57**, 1139–1143.
Rao, P. N. (1968) *Science* **160**, 774–776.
Rao, P. N. and Johnson, R. T. (1970) *Nature Lond.* **225**, 159–164.
Rao, P. N. and Johnson, R. T. (1974) *Adv. Cell mol. Biol.* **3**, 135–183.
Rao, M. V. N. and Prescott, D. M. (1967) *J. Cell Biol.* **33**, 281–285.
Rao, P. N. and Sunkara, P. S. (1978) *In* "Cell Cycle Regulation" (J. R. Jeter Jr., I. L. Cameron, G. M. Padilla and A. Zimmerman, Eds.) 133–147. Academic Press, London and New York.
Rao, P. N. and Sunkara, P. S. (1980) *Expl. Cell Res.* **125**, 507–511.
Rao, P. N., Freireic, E. J., Smith, M. L. and Loo, T. L. (1979) *Cancer Res.* **39**, 3152–3155.
Rao, P. N., Smith, M. L., Pathak, S., Howard, R. A. and Bear, J. L. (1980) *J. natn. Cancer Inst.* **64**, 905–911.
Rapaport, E. and Zamecnik, P. C. (1976) *Proc. natn. Acad. Sci. USA* **73**, 3984–3988.
Rapp, P. E. (1979) *J. exp. Biol.* **81**, 281–306.
Rash, J. E., Shay, J. W. and Biesell, J. J. (1969) *J. Ultrastruct. Res.* **29**, 470–484.
Rasse-Messenguy, F. and Fink, G. R. (1973) *Genetics* **75**, 459–464.
Rasmussen, L. (1963) *C. r. Trav. Lab. Carlsberg* **33**, 53–71.
Rasmussen, L., Cohr, K.-H., Buhse, H. E. Jr. and Zeuthen, E. (1974) *J. Protozool.* **21**, 552–555.
Rastl, E. and Swetly, P. (1978) *J. biol. Chem.* **253**, 4333–4340.

Rattner, J. B. and Phillips, S. G. (1973) *J. Cell Biol.* **57**, 359–372.
Ravkin, I. A. and Pryanish, V. A. (1977) *Tsitologiya* **19**, 625–631.
Rebhun, L. I. (1977) *In* "International Review of Cytology" (G. H. Bourne and J. F. Danielli, Eds.) **49**, 1–54. Academic Press, New York and London.
Reddy, S. B., Linden, W. A., Zywietz, F., Baisch, H. and Struck, U. (1977) *Arznei-For* **27**, 1549–1552.
Reeve, J. N. and Clarke, D. J. (1972) *J. Bact.* **110**, 117–121.
Reeve, J. N., Groves, D. J. and Clarke, D. J. (1970) *J. Bact.* **104**, 1052–1064.
Reich, J. G. and Sel'kov, E. E. (1974) *FEBS Lett.* (Suppl.) **40**, S119–S127.
Reid, B. J. and Hartwell, L. H. (1977) *J. Cell Biol.* **75**, 355–365.
Rensing, L. and Goedeke, K. (1976) *Chronobiologia* **3**, 53–65.
Rentschler, R. E., Barlogie, B., Johnston, D. A. and Bodey, G. P. (1978) *Cancer Res.* **38**, 2209–2215.
Rentzepis, P. M. (1978) *Biophys. J.* **24**, 272–284.
Ribbons, D. W. and Brew, K. (1976) Eds. "Proteolysis and Physiological Regulation" p. 428. Academic Press, London and New York.
Ricard, M. and Hirota, J. (1973) *J. Bact.* **116**, 314–322.
Richard, H. and Broda, E. (1976) *Experientia* **32**, 1158–1159.
Richards, L. and Thurston, C. (1980) *J. gen. Microbiol.* **121**, 49–61.
Richardson, C. C. (1969) *Ann. Rev. Biochem.* **38**, 795–840.
Richarm, M. and Hirota, Y. (1973) *J. Bact.* **116**, 314–322.
Richmond, K. M. V. (1976) *Proc. Genet. Soc.* 182nd meeting, Univ. Coll. London.
Ricketts, T. R. (1977a) *J. exp. Bot.* **28**, 416–424.
Ricketts, T. R. (1977b) *J. exp. Bot.* **28**, 1278–1288.
Ricketts, T. R. (1979) *Br. Phycol. J.* **14**, 219–223.
Ridder, G. M. and Margerum, D. W. (1977) *Analyt. Chem.* **49**, 2098–2108.
Riddle, J. C. and Hsie, A. W. (1978) *Mutation Res.* **52**, 409–420.
Riggs, A. D., Reiness, G. and Zubay, G. (1971) *Proc. natn. Acad. Sci. USA* **68**, 1222–1225.
Riley, P. A. and Dean, R. T. (1978) *Expl. Cell Biol.* **46**, 367–373.
Rivin, C. J. and Fangman, W. L. (1980a) *J. Cell Biol.* **85**, 96–107.
Rivin, C. J. and Fangman, W. L. (1980b) *J. Cell Biol.* **85**, 108–115.
Rixon, R. H. and Whitfield, J. F. (1976) *J. Cell Physiol.* **87**, 147–156.
Robberson, D. L., Kasamatsu, H. and Vinograd, J. (1972) *Proc. natn. Acad. Sci. USA* **69**, 737–741.
Robberson, D. L., Crawford, L. V., Syrett, C. and James, W. (1975) *J. gen. Virol.* **26**, 59–69.
Robbins, E. and Marcus, P. (1964) *Science* **144**, 1152–1153.
Robbins, E. and Morrill, G. A. (1969) *J. Cell Biol.* **43**, 629–633.
Robbins, E. and Scharff, M. D. (1967) *J. Cell Biol.* **34**, 684–686.
Robbins, E., Jentzsch, G. and Micali, A. (1968) *J. Cell Biol.* **36**, 329–339.
Roberts, D. McL. (1980) *J. gen. Microbiol.* **120**, 211–218.
Roberts, J. H., Stark, P., Giri, C. P. and Smulson, M. (1975) *Arch. biochem. Biophys.* **171**, 305–315.
Robinson, J. E., Wizenburg, M. J. and McCready, W. A. (1974) *Nature Lond.* **251** 521–522.
Robinow, C. F. and Marak, J. (1966) *J. Cell Biol.* **29**, 129–151.
Romsdahl, M. M. (1968) *Expl. Cell Res.* **50**, 463–467.
Ron, E. Z., Rozenhak, S. and Grossman, N. (1975) *J. Bact.* **123**, 374–376.
Ron, E. Z., Grossman, N. and Helmstetter, E. (1977) *J. Bact.* **129**, 569–573.

Rønning, Ø. W., Pettersen, E. O. and Seglen, P. O. (1979) *Expl. Cell Res.* **123**, 63–72.
Rooney, D. W. and Costello, J. P. (1977) *J. theoret. Biol.* **69**, 597–611.
Rooney, D. W. and Eiler, J. J. (1967) *Expl. Cell Res.* **48**, 649–652.
Rooney, D. W. and Eiler, J. J. (1979) *Expl. Cell Res.* **54**, 49–52.
Rooney, D. W., Yen, B. C. and Mikita, D. J. (1971) *Expl. Cell Res.* **65**, 94–98.
Roscoe, D. H., Read, M., and Robinson, H. (1973a) *J. Cell Physiol.* **82**, 325–332.
Roscoe, D. H., Robinson, H. and Carbonell, A. W. (1973b) *J. Cell Physiol.* **82**, 333–338.
Rosenberg, B., van Camp, L., Trosko, J. E. and Mansour, V. H. (1969) *Nature Lond.* **222**, 385–386.
Rosenberg, B. H., Cavalieri, L. F. and Unders, G. (1969) *Proc. natn. Acad. Sci. USA* **63**, 1410–1417.
Rosenberger, R. F., Grover, N. B., Zaritsky, A. and Woldringh, C. L. (1978a) *J. theoret. Biol.* **73**, 711–721.
Rosenberger, R. F., Grover, N. B., Zaritsky, A. and Woldringh, C. L. (1978b) *Nature Lond.* **271**, 244–245.
Rosenfeld, A., Zackroff, R. V. and Weisenberg, R. C. (1976) *FEBS Lett.* **65**, 144–147.
Rosin, M. P. and Zimmerman, A. M. (1977) *Mutation Res.* **44**, 207–216.
Ross, D. W. (1976) *Cell Tiss. Kinet.* **9**, 379–387.
Ross, D. W. and Mel, H. C. (1972) *Biophys. J.* **12**, 1562–1572.
Rossini, M. and Baserga, R. (1978) *Biochemistry* **17**, 858–863.
Rossini, M., Weinmann, R. and Baserga, R. (1979) *Proc. natn. Acad. Sci. USA* **76**, 4441–4445.
Rossini, M., Baserga, S., Huang, C. H., Ingles, C. J. and Baserga, R. (1980) *J. Cell Physiol.* **103**, 97–103.
Rossow, W., Riddle, V. G. H. and Pardee, A. B. (1979) *Proc. natn. Acad. Sci. USA* **76**, 4446–4450.
Ross-Riveros, P. and Alpen, E. L. (1978) *Radiat. Res.* **74**, 480.
Rotenberg, M. (1977) *J. theoret. Biol.* **66**, 389–398.
Roth, R. (1975) *Genetics* **83**, 675–686.
Roufa, D. J. and Reed, S. J. (1975) *Genetics* **80**, 549–566.
Rowen, L. and Kornberg, A. (1978a) *J. biol. Chem.* **253**, 758–764.
Rowen, L. and Kornberg, A. (1978b) *J. biol. Chem.* **253**, 770–774.
Rowley, R., Leeper, D. B. and Schneider, M. H. (1980) *Radiat. Res.* **83**, p. 369.
Rubin, A. H., Terasaki, M. and Sanui, H. (1979) *Proc. natn. Acad. Sci. USA* **76**, 3917–3921.
Rubin, H. (1975a) *Proc. natn. Acad. Sci. USA* **72**, 3551–3555.
Rubin, H. (1975b) *Proc. natn. Acad. Sci. USA* **72**, 1676–1680.
Rubin, H. (1976) *J. Cell Physiol.* **89**, 613–626.
Rubin, H. and Koide, T. (1976) *Proc. natn. Acad. Sci. USA* **73**, 168–172.
Rubin, L. B. and Rubin, A. B. (1978) *Biophys. J.* **24**, 84–92.
Rubin, R. W. and Everhart, L. P. (1973) *J. Cell Biol.* **57**, 837–844.
Rudland, P. S. and Jimenez de Asua, L. (1979) *Biochim. biophys. Acta* **560**, 91–133.
Rudner, R., Prokop-Schneider, B. and Chargaff, E. (1964) *Nature Lond.* **203**, 479–483.
Rudner, R., Rejman, E. and Chargaff, E. (1965) *Proc. natn. Acad. Sci. USA* **54**, 904–911.

Rueckert, R. R. and Mueller, G. C. (1960) *Cancer Res.* **20**, 1584–1591.
Rusch, H. P., Sachsenmaier, W., Behrens, K. and Gruter, V. (1966) *J. Cell Biol.* **31**, 204–209.
Russell, D. H. and Snyder, S. H. (1969) *Mol. Pharmacol.* **5**, 253–262.
Russell, D. H. and Stambrook, P. J. (1975) *Proc. natn. Acad. Sci. USA* **72**, 1482–1486.
Rustad, R. C. (1959) *Exp. Cell Res.* **15**, 444–446.
Ryder, O. A., Kavenoff, R. and Smith, D. W. (1975) *In* "DNA Synthesis and its Regulation". (M. Goulian, P. Hanawalt and C. F. Fox, Eds.) 159–186. W. A. Benjamin Inc., Menlo Park.
Ryser, U., Fakan, S. and Braun, R. (1973) *Exp. Cell Res.* **78**, 89–97.
Ryter, A., Shuman, H. and Schwartz, H. (1975) *J. Bact.* **122**, 295–301.
Sachs, H. G., Stambrook, P. J. and Ebert, J. D. (1974) *Exp. Cell Res.* **83**, 362–366.
Sachsenmaier, W. (1976) *In* "The Molecular Basis of Circadian Rhythms". (J. W. Hastings and H.-G. Schweiger, Eds.) 410–420. Dahlem Konferenzen, Berlin.
Sachsenmaier, W., Immich, H. Grunst., J., Scholtz, R. and Bücher, Th. (1969) *Eur. J. Biochem.* **8**, 557–561.
Sachsenmaier, W., Donges, K. H., Rupff, H. and Czihak, G. (1970) *Z. Naturf.* **256**, 866–871.
Sachsenmaier, W., Remy, U. and Plattner-Schobel, R. (1972) *Exp. Cell Res.* **73**, 41–48.
Sachsenmaier, W., Blessing, J., Brauser, B. and Hansen, K. (1973) *Protoplasma* **77**, 381–396.
Sacristán-Gárate, A. M., Navarrete, M. H. and De La Torre, C. (1974a) *J. Cell Sci.* **16**, 333–347.
Sacristán-Gárate, A. M., Navarrete, M. H. and De La Torre, C. (1974b) *Cytobios.* **11**, 21–31.
Sacristán-Gárate, A. M., Navarete, M. H. and De La Tore, C. (1974c) *J. Microscopie* **21**, 63–74.
Sakakibara, Y. and Tomizawa, J.-I. (1974) *Proc. natn. Acad. Sci. USA* **71**, 4935–4939.
Sakamoto, K. and Elkind, M. M. (1969) *Biophys. J.* **9**, 1115–1130.
Salmon, I. (1980) Ph.D. Thesis, University of London.
Salmon, I. and Poole, R. K. (1980a) *Soc. Gen. Microbiol. Quart.* **8**, 30.
Salmon, I. and Poole, R. K. (1980b) *Soc. Gen. Microbiol. Quart.* **7**, 85.
Salmond, G. P. C., Lutkenhaus, J. F. and Donachie, W. D. (1980) *J. Bact.* **144**, 438–440.
Sander, G. and Pardee, A. B. (1972) *J. Cell Physiol.* **80**, 267–271.
Sanders, J. F. and Carter, S. K. (Eds.) (1977) *Natn. Cancer Inst.* Monograph No. 45. Bethesda, Md.
Sanderson, R. J. and Bird, K. E. (1977) *In* "Methods in Cell Biology". (D. M. Prescott, Ed.) Vol. 15, 1–14. Academic Press, New York and London.
Sanderson, R. J., Bird, K. E., Palmer, N. F. and Brenman, J. (1976) *Anal. Biochem.* **71**, 615–622.
Sando, N. (1963) *J. gen. appl. Microbiol.* **9**, 233–241.
Sanger, J. W. and Sanger, J. N. (1980) *Cell Tissue Rev.* **209**, 177–186.
Santos, D. and de Almeida, D. F. (1975) *J. Bact.* **124**, 1502–1507.
Sapareto, S. A., Hopwood, L. E., Dewey, W. C., Raju, M. and Gray, J. W. (1977) *Radiat. Res.* **70**, p. 631.
Sapozink, M. D., Drescher, E. E. and Hahn, E. W. (1973) *Nature Lond.* **244**, 299–300.

Sargent, M. G. (1973a) *J. Bact.* **116**, 397–409.
Sargent, M. G. (1973b) *J. Bact.* **116**, 736–740.
Sargent, M. G. (1975a) *Biochim. biophys. Acta* **406**, 564–574.
Sargent, M. G. (1975b) *J. Bact.* **123**, 1218–1234.
Sargent, M. G. (1975c) *J. Bact.* **123**, 7–19.
Sargent, M. G. (1979) *In* "Advances in Microbial Physiology". (A. H. Rose, and J. G. Morris, Eds.) Vol. 18, 105–176. Academic Press, London, New York and San Francisco.
Sargent, M. G. (1980) *Soc. gen. Microbiol. Quart.* **8**, 23–24.
Sato, Ch. (1976) *Exp. Cell Res.* **101**, 251–259.
Sato, K. and Hama-Inaba, H. (1978) *Expl. Cell Res.* **14**, 484–486.
Sato, T., Ohki, M., Yura, T. and Ito, K. (1979) *J. Bact.* **138**, 305–313.
Satta, G. and Pardee, A. B. (1978) *J. Bact.* **133**, 1492–1500.
Saunders, D. S. (1977) "An Introduction to Biological Rhythms". Blackie, Glasgow.
Saunders, E. F. (1972) *Blood* **39**, 575–580.
Sawicki, C. A. and Gibson, Q. H. (1978) *Biophys. J.* **24**, 21–28.
Sawicki, W., Rowinski, J. and Swenson, R. (1974) *J. Cell Physiol.* **84**, 423–428.
Scandella, C. J., Devaux, P. and McConnell, H. M. (1972) *Proc. natn. Acad. Sci. USA* **69**, 2056–2060.
Schaap, G. H., Van der Kamp, A. W., Ory, F. G. and Jongkind, J. F. (1979) *Expl. Cell Res.* **122**, 422–426.
Schaechter, M., Maaløe, O. and Kjeldgaard, N. O. (1958) *J. gen. Microbiol.* **19**, 592–606.
Schaechter, M., Williamson, J. P., Hood, J. R. and Koch, A. L. (1962) *J. gen. Microbiol.* **29**, 421–434.
Schaer, J. C. and Maurer, U. (1977) *Experientia* **33**, p. 829.
Schaller, H. (1978) *Cold Spring Harbor Symp. Quant. Biol.* **43**, 401–408.
Schaller, H., Uhlmann, A. and Geider, K. (1976) *Proc. natn. Acad. Sci. USA* **73**, 49–53.
Scharff, M. D. and Robbins, E. (1965) *Nature Lond.* **208**, 464–466.
Scharff, M. D. and Robbins, E. (1966) *Science* **151**, 992–995.
Scheffler, I. E. and Buttin, G. (1973) *J. Cell Physiol.* **81**, 199–216.
Scherbaum, O. H. (1963) *Expl. Cell Res.* **33**, 89–98.
Scherbaum, O. and Zeuthen, E. (1954) *Expl. Cell Res.* **6**, 221–227.
Scherbaum, O. and Zeuthen, E. (1955) *Expl. Cell Res. Suppl.* **3**, 312–325.
Scherbaum, O. H., Chou, S. C., Seraydarian, K. H. and Byfield, J. E. (1962) *Can. J. Microbiol.* **8**, 753–760.
Schild, D. and Byers, B. (1978) *Chromsome* **70**, 109–130.
Schimke, R. T. and Katunuma, N. (1975) (Eds.) "Intracellular Protein Turnover". pp. 348. Academic Press, London and New York.
Schimmer, O. and Loppes, R. (1975) *Mol. gen. Genet.* **138**, 25–31.
Schindler, R. (1960) *Helv. Physiol. Pharmacol. Acta* **18**, p. C93.
Schindler, R. and Schaer, J. C. (1973) *In* "Methods in Cell Biology". (D. M. Prescott, Ed.) Vol. 6. 43–65. Academic Press, New York.
Schindler, R., Ramseier, L., Schaer, J. C. and Grieder, A. (1970) *Expl. Cell Res.* **59**, 90–95.
Schlag, H. and Lück-Hule, T. (1976) *Eur. J. Cancer* **12**, 827–831.
Schlag, H., Weibezah, K. F. and Luckehuh, C. (1978) *Int. J. Radiat. Biol.* **33**, 1–10.
Schlösser, U. (1966) *Arch. Microbiol.* **54**, 129–159.
Schmid, A.-M. M. and Schulz, D. (1979) *Protoplasma* **100**, 267–288.

Schmidt, G. E., Martin, A. P. and Vorzeck, M. L. (1977) *J. Ultrastruct. Res.* **60**, 52–62.
Schmidt, R. R. (1974a) *In Vitro* **10**, 306–320.
Schmidt, R. R. (1974b) *In* "Cell Cycle Controls". (G. M. Padilla, I. L. Cameron and Z. Zimmerman, Eds.) 201–234. Academic Press, New York and London.
Schmidt, R. R. (1975) *In* "Intracellular Protein Turnover". (R. T. Schimke and N. Katunuma, Eds.) 77–100. Academic Press, London and New York.
Schnedl, W. (1974) *Cytobiologie* **8**, 403–411.
Schnedl, W. and Schnedl, M. (1972) *Z. Zellforsch.* **126**, 374–382.
Schneider, D. O. and Whitmore, G. F. (1963) *Radiat Res.* **18**, 286–306.
Schneiderman, M. H., Dewey, W. C. and Highfield, D. P. (1971) *Expl. Cell Res.* **67**, 147–155.
Schnös, M. and Inman, R. B. (1970) *J. mol. Biol.* **51**, 61–73.
Schnös, M, and Inman, R. B. (1971) *J. mol. Biol.* **55**, 31–38.
Schoenheimer, R. (1942) "Dynamic State of Body Constituents". Harvard University Press, Cambridge, Mass. USA.
Schor, S. (1971) Ph.D. Thesis, The Rockefeller University, New York.
Schor, S., Siekevitz, P. and Palade, G. E. (1970) *Proc. natn. Acad. Sci. USA* **66**, 174–180.
Schreiner, Ø., Lien, T. and Knutsen, G. (1975) *Biochim. biophys. Acta* **384**, 180–193.
Schulmeister, Th. and Selkov, E. E. (1978) *Studia biophys.* **72**, 111–112.
Schweiger, H.-G. and Schweiger, M. (1977) *In* "International Review of Cytology". (G. H. Bourne, J. F. Danielli, and K. W. Jeon, Eds.) Vol. 51, 315–342. Academic Press, New York and London.
Scopes, A. W. and Williamson, D. H. (1964) *Expl. Cell Res.* **35**, 361–371.
Scopes, A. W. and Williamson, D. H. (1962) *Nature* **193**, 256–257.
Scott, J. F. and Kornberg, A. (1978) *J. biol. Chem.* **253**, 3292–3297.
Scott, J. F., Eisenberg, G., Bertch, L. L. and Kornberg, A. (1977) *Proc. natn. Acad. Sci. USA* **74**, 193–197.
Scott, J. R. (1970) *Virology* **41**, 66–71.
Scott, R. E., Carter, R. L. and Kidwell, W. R. (1971) *Nature Lond. New Biol.* **233**, 219–220.
Scott, R. I. and Poole, R. K. (1982) *J. gen. Microbiol* (In press).
Scott, R. I., Gibson, J. F. and Poole, R. K. (1980) *J. gen. Microbiol.* **120**, 183–198.
Scott, R. I., Poole, R. K. and Chance, B. (1981) *J. gen. Microbiol.* **122**, 255–261.
Seale, R. L. (1975) *Proc. natn. Acad. Sci. USA* **73**, 2270–2274.
Sebastian, J., Carter, B. L. A. and Halvorson, H. O. (1971) *J. Bact.* **108**, 1045–1050.
Sebastian, J., Carter, B. L. A. and Halvorson, H. O. (1973) *Eur. J. Biochem.* **37**, 516–522.
Sebastian, J., Takano, I. and Halvorson, H. O. (1974) *Proc. natn. Acad. Sci. USA* **71**, 769–773.
Sedgley, N. N. and Stone, G. E. (1969) *Expl. Cell Res.* **56**, 174–177.
Sedory, M. J. and Mitchell, J. L. (1977) *Expl. Cell Res.* **107**, 105–110.
Seifert, W. E. and Rudland, P. S. (1974a) *Proc. natn. Acad. Sci. USA* **71**, 4920–4924.
Seifert, W. E. and Rudland, P. S. (1974b) *Nature Lond.* **248**, 138–140.
Seki, T. and Okazaki, T. (1979) *Nucleic Acids Res.* **7**, 1603–1619.

Selawry, O. S., Goldstein, M. N. and McCormick, T. (1957) *Cancer Res.* **17**, 785–791.
Sel'kov, E. E. (1968) *Eur. J. Biochem.* **4**, 79–86.
Sel'kov, E. E. (1970) *Biophysika* **15**, 1065–1073.
Sel'kov, E. E. (1971) *In* "Oscillatory Processes in Biological and Chemical Systems". (E. E. Sel'kov, Ed.) 7–12. USSR Academy of Sciences, Puschino.
Sel'kov, E. E. (1972) *8th FEBS Meeting* (H. C. Hemker and B. Hess, Eds.) **25**, 145–161. North Holland Press, Amsterdam.
Sel'kov, E. E. (1973) Abstrt. IVth *Int. Biophys. Cong.*, Moscow, Vol. **3**, 453–475. IUPAB and USSR Academy of Sciences, Puschino.
Sel'kov, E. E. (1980) *Ber. Bunsen, Gesell. Phys. Chem.* **84**, 399–402.
Sena, E. P., Radin, O. N. and Fogel, S. (1973) *Proc. natn. Acad. Sci. USA* **70**, 1373–1377.
Sena, E. P., Welch, J. W., Halvorson, H. O. and Fogel, S. (1975) *J. Bact.* **123**, 497–504.
Senger, H. (1965) *Arch. Microbiol.* **51**, 307–322.
Senger, H. (1970a) *Planta* **90**, 243–266.
Senger, H. (1970b) *Planta* **92**, 327–346.
Senger, H. (1975) *Colloq. Int. CNRS* **240**, 101–107.
Senger, H. and Bishop, N. I. (1969) *In* "The Cell Cycle: Gene-Enzyme Interaction". (G. M. Padilla, G. L. Whitson and I. L. Cameron, Eds.) 179–202. Academic Press, New York and London.
Senger, H. and Bishop, N. I. (1977) *Pl. Physiol.* **59**, p. 130.
Senger, H. and Bishop, N. I. (1979) *Planta* **145**, 53–62.
Sentein, P. (1979) *Expl. Cell Biol.* **47**, 368–391.
Setterfield, G., Sheinin, R., Dardick, I., Kiss, G. and Dubsky, M. (1978) *J. Cell Biol.* **77**, 246–263.
Severs, N. J. (1977) *Cytobios.* **18**, 51–67.
Severs, N. J. and Jordan, E. G. (1975) *J. Ultrastruct. Res.* **52**, 85–99.
Severs, N. J., Jordan, E. G. and Williamson, O. H. (1976) *J. Ultrastruct. Res.* **54**, 374–387.
Shalitin, C. and Weiser, I. (1977) *J. Bact.* **131**, 735–740.
Shall, S. (1973) *In* "Methods in Cell Biology". (D. M. Prescott, Ed.) Vol. **7**, 269–285. Academic Press, New York and London.
Shall, S. and McClelland, A. J. (1971) *Nature Lond. New Biol.* **229**, 59–60.
Shannon, K. P., Spratt, B. G. and Rowbury, R. J. (1972) *Mol. gen. Genet.* **118**, 185–197.
Shapiro, L. (1976) *Ann. Rev. Microbiol.* **30**, 377–407.
Shapiro, L. and Maizel, J. V. (1973) *J. Bact.* **113**, 478–485.
Sharpe, H. B. A., Field, E. O. and Hellman, K. (1970) *Nature Lond.* **226**, 524–526.
Sheath, R. G., Hellebus, J. A. and Sawa, T. (1977) *J. Phycology* **13**, p. 62.
Shehata, T. E. and Kempner, E. S. (1979) *J. Protozool.* **26**, 626–630.
Sheinin, R. (1976a) *J. Virol.* **17**, 692–704.
Sheinin, R. (1976b) *Cell* **7**, 49–57.
Sheinin, R. and Guttman, S. (1977) *Biochim. biophys. Acta* **479**, 105–118.
Sheinin, R., Darragh, P. and Dubsky, M. (1977) *Can. J. Biochem.* **55**, 543–547.
Sheinin, R., Darragh, P. and Dubsky, M. (1978a) *J. Biol. Chem.* **253**, 922–926.
Sheinin, R., Humbert, J. and Pearlman, R. E. (1978b) *Ann. Rev. Biochem.* **47**, 277–316.
Shen, B. H. P. and Boos, W. (1973) *Proc. natn. Acad. Sci. USA* **70**, 1481–1485.

Sherman, L. A. and Gefter, M. L. (1976) *J. mol. Biol.* **103**, 61–76.
Sherman, F. E. and Lawrence, C. W. (1974) *In* "Handbook of Genetics". (R. C. King, Ed.) Vol. 1, 395–393. Plenum Press, New York.
Shields, R. (1977) *Nature* **267**, 704–707.
Shields, R. (1978) *Nature* **273**, 755–758.
Shields, R. (1979) *Cell Biol. Int. Reports.* **3**, 659–662.
Shields, R. (1980) *In* "Control Mechanisms in Animal Cells—Specific Growth Factors". (L. Jiminez de Asua, R. Shields, R Levi-Montalicihi, and Iacobell, Eds.) 157–165. Raven Press, New York.
Shields, R., Brooks, R. F., Riddle, P. N., Capellar, D. F. and Delia, D. (1978) *Cell* **15**, 496–474.
Shilo, B., Shilo, V. and Simchen, G. (1976) *Nature Lond.* **264**, 767–770.
Shilo, B., Shilo, V. and Simchen, G. (1977) *Nature Lond.* **267**, 648–649.
Shilo, B., Simchen, G. and Pardee, A. B. (1978) *J. Cell Physiol.* **97**, 177–188.
Shilo, B., Riddle, V. G. H. and Pardee, A. B. (1979) *Expl. Cell Res.* **123**, 221–227.
Shilo, V., Simchen, G. and Shilo, B. (1978) *Expl. Cell Res.* **112**, 241–248.
Shiomi, T. and Sato, K. (1976) *Expl. Cell Res.* **100**, 297–302.
Shiomi, T. and Sato, K. (1978) *Cell Struct. Funct.* **3**, 95–102.
Shlømai, J. and Kornberg, A. (1978) *J. biol. Chem.* **253**, 3305–3312.
Shockman, G. D., Daneo-Moore, L. and Higgins, A. L. (1974) *Ann. N.Y. Acad. Sci.* **235**, 161–197.
Shonkwiler, R. (1977) *J. math. Biol.* **39**, 613–618.
Shuler, M. L., Leung, S. and Dick, C. C. (1979) *In* "Biochemical Engineering". (W. R. Vieth, K. Venkatasubramanian and A. Constantinides, Eds.) *Ann. N.Y. Acad. Sci.* **326**, 33–35.
Shulman, R. W. (1978) *In* "Methods in Cell Biology". (D. M. Prescott, Ed.) Vol. 20, 35–43. Academic Press, New York and London.
Shulman, R. W., Hartwell, L. H. and Warner, J. R. (1973) *J. mol. Biol.* **73** 513–525.
Shulman, R. W., Sripati, C. E. and Warner, J. R. (1977) *J. biol. Chem.* **252**, 1344–1349.
Siccardi, A. G., Galizzi, A., Mazza, G., Clivio, A. and Albertini, A. M. (1975) *J. Bact.* **121**, 13–19.
Sieber-Blum, M. and Burger, M. M. (1977) *Biochem. biophys. Res. Comm.* **74**, 1–8.
Sierra, J. M., Sentandreu, R. and Villaneuva, J. R. (1973) *FEBS Lett.* **34**, 285–290.
Sigdestad, C. P., Grdina, D. J., Peters, L. J. and Stutesman, J. (1979) *Proc. Am. Ass. Cancer Res.* **20**, 178.
Sigal, N., Delius, H., Kornberg, T., Gefter, M. L. and Alberts, B. (1972) *Proc. natn. Acad. Sci. USA* **69**, 3537–3541.
Silver, L. L., Chandler, M. G. and Card, L. G. (1977a) *Experientia* **33**, 831.
Silver, L. L., Chandler, M. G., Boy de la Tour, E. and Caro, L. G. (1977b) *J. Bact.* **131**, 929–942.
Silvestrini, R., DiMarco, A. and Dasdia, T. (1970) *Cancer Res.* **30**, 966–973.
Simchen, G. (1974) *Genetics* **76**, 745–753.
Simchen, G. (1978) *Ann. Rev. Genet.* **12**, 161–191.
Simchen, G. and Hinschberg, J. (1977) *Genetics* **86**, 57–72.
Siminovitch, L. (1976) *Cell* **7**, 1–11.
Siminovitch, L. and Thompson, L. H. (1978) *J. Cell Physiol.* **95**, 361–366.
Simmons, T., Heywood, P. and Hodge, L. O. (1973) *J. Cell Biol.* **59**, 150–164.
Simpson, L. (1972) *Int. Rev. Cytology,* **32**, 139–207.

Simpson, L. and Braly, P. (1970) *J. Protozool.* **17**, 511–517.
Sims, J. and Dressler, D. (1978) *Proc. natn. Acad. Sci. USA* **75**, 3094–3098.
Sims, J., Koths, K. and Dressler, D. (1978) *Cold Spring Harbor Symp. Quant. Biol.* **43**, 349–365.
Sinclair, R. and Bishop, D. H. L. (1965) *Nature Lond.* **205**, 1271–1274.
Sinclair, W. K. (1965) *Science* **150**, 1729–1731.
Sinclair, W. K. (1967) *Cancer Res.* **27**, 297–301.
Sinclair, W. K. (1968) *Radiat. Res.* **33**, 620–643.
Sinclair, W. K. and Morton, R. A. (1963) *Nature Lond.* **199**, 1158–1160.
Sinclair, W. K. and Morton, R. A. (1965) *Biophys. J.* **5**, 1–12.
Singer, R. A. and Johnston, G. C. (1979) *Mol. gen. Genet.* **176**, 37–39.
Singer, R. A., Johnston, G. C. and Bedard, D. (1978) *Proc. natn. Acad. Sci. USA* **75**, 6083–6087.
Sisken, J. E. and Morasca, L. (1965) *J. Cell Biol.* **25**, 179–189.
Sisken, J. E., Morasca, L. and Kibby, S. (1965) *Expl. Cell Res.* **39**, 103–116.
Sissons, C. H., Mitchison, J. M. and Creanor, J. (1973) *Expl. Cell Res.* **82**, 63–72.
Sitz, T. O., Kent, A. B., Hopkins, H. A. and Schmidt, R. R. (1970) *Science* **168**, 1231–1232.
Skehan, P. and Friedman, S. J. (1979) *Cell Biol.* **3**, 535–542.
Skoog, K. L., Nordenskjöld, B. A. and Bjursell, K. G. (1973) *Eur. J. Biochem.* **33**, 428–432.
Skutelsky, E. and Bayer, E. A. (1979) *Expl. Cell Res.* **121**, 331–336.
Slater, M. and Schaechter, M. (1974) *Bact. Rev.* **38**, 199–221.
Slater, M. L. (1973) *J. Bact.* **113**, 263–270.
Slater, M. L. (1976) *J. Bact.* **126**, 1339–1341.
Slater, M. L. and Ozer, H. L. (1976) *Cell* **7**, 289–295.
Slater, M. L., Sharrow, S. O. and Gart, J. J. (1977) *Proc. natn. Acad. Sci. USA* **74**, 3850–3854.
Sloat, B. F. and Pringle, J. R. (1977) *J. Cell Biol.* **75**, A/CC 1195.
Sloat, B. F. and Pringle, J. R. (1978) *Science* **200**, 1171–1173.
Sluder, G. (1979) *J. Cell Biol.* **80**, 674–691.
Smit, J. and Nikaido, H. (1978) *J. Bact.* **105**, 687–702.
Smith, B. J. and Wigglesworth, N. M. (1972) *J. Cell Physiol.* **80**, 253–260.
Smith, B. J. and Wigglesworth, N. M. (1973) *J. Cell Physiol.* **82**, 339–348.
Smith, B. J. and Wigglesworth, N. M. (1974) *J. Cell Physiol.* **84**, 127–133.
Smith, D., Tauro, P., Schweizer, E. and Halvorson, H. O. (1968) *Proc. natn. Acad. Sci. USA* **60**, 936–942.
Smith, H. and Pardee, A. B. (1970) *J. Bact.* **101**, 901–909.
Smith, H. H., Fusseli, P. C. and Kugelman, B. H. (1963) *Science* **142**, 595–596.
Smith, J. A. and Martin, L. (1973) *Proc. natn. Acad. Sci. USA* **70**, 1263–1267.
Smith, J. A. and Martin, L. (1974) *In* "Cell Cycle Controls". (G. A. Padilla, I. L. Cameron and A. Zimmerman, Eds.) 43–60. Academic Press, New York and London.
Snyder, J. A. and Liskay, R. N. (1978) *J. Cell Biol.* **79**, 13a.
Söderhäll, S. and Lindahl, T. (1976) *FEBS Lett.* **67**, 1–8.
Soeiro, R., Vaughan, M. H., Warner, J. R. and Darnell, J. E. Jr. (1968) *J. Cell Biol.* **39**, 112–118.
Sogin, S. J., Carter, B. L. A. and Halvorson, H. O. (1974) *Expl. Cell Res.* **89**, 127–138.
Soll, D. R. (1979) *Science* **203**, 841–849.

Soll, D. R., Stasi, M. and Bedell, G. (1978) *Expl. Cell Res.* **116**, 207–215.
Sompayrac, L. and Maaløe, O. (1973) *Nature Lond. New Biol.* **241**, 133–135.
Soprano, K. J. and Kuchler, R. J. (1978) *Fed. Proc.* **37**, p. 240.
Sorokin, C. and Krauss, R. W. (1961) *Biochim. biophys. Acta* **48**, 314–319.
Spadari, S. and Weissbach, A. (1974) *J. mol. Biol.* **86**, 11–20.
Spenser, R. D. and Weber, G. (1969) *Ann. N.Y. Acad. Sci.* **158**, 361–376.
Spratt, B. G. (1975) *Proc. natn. Acad. Sci. USA* **72**, 2999–3003.
Spratt, B. G. (1977) *J. Bact.* **131**, 293–305.
Spratt, B. G. and Rowbury, R. J. (1971) *Mol. gen. Genet.* **114**, 35–49.
Spudich, J. L. and Sager, R. (1980) *J. Cell Biol.* **85**, 136–146.
Ssymank, V., Kaushik, B. D. and Lorenzen, H. (1977) *Planta* **135**, 13–17.
Stacey, V. J. and Pienaar, R. N. (1979) *S. Afr. J. Sci.* **75**, p. 467.
Stadtman, E. R. (1970) *In* "The Enzymes, Structure and Control". (P. D. Boyer, Ed.) 389–444. Academic Press, New York and London.
Stahelin, H. (1970) *Eur. J. Cancer* **6**, 303–311.
Stambrook, P. J. and Sisken, J. E. (1972) *J. Cell Biol.* **52**, 514–525.
Stanners, C. P. (1978) *J. Cell Physiol.* **95**, 407–416.
Steen, H. B. and Lindmo, T. (1978) *Cell Tiss. Kinet.* **11**, 69–81.
Stein, G., Park, W., Thrall, C., Mans, R. and Stein, J. (1975) *Nature Lond.* **257**, 764–767.
Stein, S. M. and Berestecky, J. M. (1975) *J. Cell Physiol.* **85**, 243–250.
Steinert, M. (1979) *FEBS Lett.* **5**, 291–294.
Stetten, G. and Lederberg, S. (1973) *J. Cell Biol.* **56**, 259–262.
Stevens, B. J. (1977) *Biol. Cell* **28**, 37–56.
Stewart, D. L., Shaeffer, J. R. and Humphrey, R. M. (1968) *Science* **161**, 791–793.
Stocco, D. M. and Zimmerman, A. M. (1975) *Mol. Cell Biochem.* **7**, 187–194.
Stone, A. B. (1973) *J. Bact.* **116**, 741–750.
Stone, G. (1968) *In* "Methods in Cell Physiology". (D. M. Prescott, Ed.) Vol. 3, 161–170. Academic Press, New York and London.
Stonehill, E. H. and Hutchison, D. J. (1966) *J. Bact.* **92**, 136–143.
Storrie, B. and Attardi, G. (1973) *J. Cell Biol.* **56**, 833–838.
Straus, S. E., Sebring, E. D. and Rose, J. A. (1976) *Proc. natn. Acad. Sci. USA* **73**, 742–746.
Streiblová, E. (1977) *Microbiologiya* **46**, 589–595.
Streiblová, E. and Girbardt, M. (1980) *Can. J. Microbiol.* **26**, 250–254.
Streiblová, E. and Wolf, A. (1972) *Z. Allg. Mikrobiol.* **12**, 678–684.
Stubblefield, E. (1968) *In* "Methods in Cell Physiology". (D. M. Prescott, Ed.) Vol. 3, 25–43. Academic Press, New York and London.
Stubblefield, E., Klevecz, R. and Deaven, L. (1967) *J. Cell Physiol.* **69**, 345–354.
Sturgeon, J. A. and Ingram, L. O. (1978) *J. Bact.* **133**, 256–264.
Sturman, A. J. and Archibald, A. R. (1978) *FEMS Microbiol. Lett.* **4**, 255–259.
Sudbery, P. E. and Grant, W. D. (1975) *Expl. Cell Res.* **95**, 405–415.
Sudbery, P. E. and Grant, W. D. (1976) *J. Cell Sci.* **22**, 59–65.
Sudbery, P., Haugli, K. and Haugli, F. (1978) *Genet. Res.* **31**, 1–12.
Sudbery, P. E., Goodey, A. R. and Carter, B. L. A. (1980) *Nature Lond.* **288**, 401–404.
Suggs, S. V. and Ray, D. S. (1978) *Cold Spring Harbor Symp. Quant. Biol.* **43**, 379–388.
Sugimoto, K., Oka, A., Sugisaki, H., Takanami, M., Nishimura, A., Yasuda, Y. and Hirota, Y. (1979) *Proc. natn. Acad. Sci. USA* **76**, 575–579.

Sugino, A., Hirose, S. and Okazaki, R. (1972) *Proc. natn. Acad. Sci. USA* **69**, 1863–1867.
Sugino, A., Peebles, C., Kreuzer, K. and Cozzarelli, N. R. (1977) *Proc. natn. Acad. Sci. USA* **74**, 4767–4771.
Suhr-Jessen, P. B. (1978) *Carlsberg Res. Communs.* **43**, 255–263.
Suhr-Jessen, P. B., Stewart, J. M. and Rasmussen, L. (1977) *J. Protozool.* **24**, 299–303.
Sullivan, C. W. (1977) *J. Phycology* **13**, 86–91.
Sullivan, C. W. (1979) *J. Phycology* **15**, 210–216.
Sumner, H. C. B., Collins, R. C. L. S. and Pasternak, C. A. (1973) *Tissue Antigens* **3**, 477–484.
Sumrada, R. and Cooper, T. G. (1978) *J. Bact.* **136**, 234–246.
Sunkara, P. S., Paragac, M. B. and Rao, P. N. (1977) *J. Cell Biol.* **75**, A125.
Sussenbach, J. S., Ellens, D. J., van der Vliet, P. Ch., Kuijk, M. G., Steenberg, P. H., Vlak, J. M., Rozijn, Th. H. and Jansz, H. S. (1975) *Cold Spring Harbor Symp. Quant. Biol.* **39**, 539–545.
Surzycki, S. (1971) *In* "Methods in Enzymology". (S. P. Colowick and N. O. Kaplan, Eds.) 23A, 67–73. Academic Press, New York and London.
Svetina, S. (1977) *Cell Tiss. Kinet.* **10**, 575–581.
Svetina, S. and Žekš, B. (1978) *In* "Biomathematics and Cell Kinetics". (A.-J. Valleron and P. D. M. Macdonald, Eds.) 71–82. Elsevier North Holland Press, Amsterdam.
Swann, M. M. (1957) *Cancer Res.* **7**, 727–757.
Swann, M. M. (1962) *Nature Lond.* **193**, 1222–1227.
Sweeney, B. M. (1960) *Cold Spring Harbor Symp. Quant. Biol.* **25**, 145–148.
Sweeney, B. M. (1961) *Cold Spring Harbor Symp. Quant. Biol.* **25**, 145–153.
Sweeney, B. M. (1963) *Ann. Rev. Pl. Physiol.* **14**, 411–440.
Sweeney, B. M. (1976) *In* "The Molecular Basis of Circadian Rhythms". 77–84. Dahlem Konferenzen, Berlin.
Sweeney, B. M. and Hastings, J. W. (1958) *J. Protozool.* **5**, 217–244.
Swierenga, S. H. H., MacManus, J. P. and Whitfield, J. F. (1976) *In Vitro* **12**, 31–36.
Swoboda, U. and Dow, C. S. (1979) *J. gen. Microbiol.* **112**, 235–239.
Szyszko, A. H., Prazak, B. L., Ehret, C. F., Eisler, W. J. Jr. and Wille, J. J. Jr. (1968) *J. Protozool.* **15**, 781–785.
Tabak, H. F., Griffith, J., Geider, K., Schaller, H. and Kornberg, A. (1974) *J. biol. Chem.* **249**, 3049–3054.
Takabaya, A., Nishimur, T. and Iwamura, T. (1976) *J. gen. Appl. Microbiol.* **22**, 183–196.
Takeda, H. and Hirokawa, T. (1978) *Pl. Cell Physiol.* **19**, 591–598.
Takeda, H. and Hirokawa, T. (1979) *Pl. Cell Physiol.* **20**, 989–991.
Talavera, A. and Basilico, C. (1977) *J. Cell Physiol.* **92**, 425–436.
Talavera, A., Nishimoto, T. and Basilico, C. (1976) *J. Cell Biol.* **70**, 341A.
Talbert, D. M. and Sorokin, C. (1971) *Arch. Microbiol.* **78**, 281–294.
Talley, D. J., White, L. H. and Schmidt, R. R. (1972) *J. biol. Chem.* **247**, 7927–7935.
Tamiya, H. (1966) *Ann. Rev. Pl. Physiol.* **17**, 1–26.
Tamiya, H., Iwamura, T., Shibata, K., Hase, E. and Nihei, T. (1953) *Biochim. biophys. Acta.* **12**, 23–40.
Tamiya, H., Morimura, Y., Yokota, M. and Kunieda, R. (1961) *Pl. Cell Physiol. Tokyo* **2**, 383–403.

Tan, I., Hartmann, W., Guntermann, U., Hüttermann, A. and Kühlwein, H. (1974) *Arch. Microbiol.* **100**, 389–396.
Tandler, B. and Hoppel, C. L. (1972) *Anat. Rec.* **173**, 309–324.
Tang, M. S. and Helmstetter, C. E. (1980) *J. Bact.* **141**, 1148–1156.
Tanida, S., Higashide, E. and Yoneda, M. (1980) *J. gen. Microbiol.* **118**, 411–417.
Tanuma, S. I., Enomoto, T. and Yamada, M. A. (1978) *Expl. Cell Res.* **117**, 421–430.
Tarnowka, M. A. and Yuyama, S. (1978) *J. Cell Physiol.* **95**, 85–93.
Tashiro, F., Mita, T. and Higashinakagawa, T. (1976) *Eur. J. Biochem.* **65**, 123–130.
Tauro, P. and Halvorson, H. O. (1966) *J. Bact.* **92**, 652–661.
Taylor, J. H. (1960) *Proc. natn. Acad. Sci. USA* **90**, 409–421.
Taylor, J. H., Woods, P. S. and Hughes, W. L. (1957) *Proc. natn. Acad. Sci USA* **43**, 122–128.
Teather, R. M., Collins, J. F. and Donachie, W. D. (1974) *J. Bact.* **118**, 407–413.
Teng, M.-H., Bartholomew, J. C. and Bissell, M. J. (1977) *Nature Lond.* **268**, 739–741.
Tenner, A., Zieg, J. and Scheffler, I. E. (1977) *J. Cell Physiol.* **90**, 145–160.
Terasima, T. and Tolmach, L. J. (1961) *Nature Lond.* **190**, 1210–1211.
Terasima, T. and Tolmach, L. J. (1963a) *Expl. Cell Res.* **30**, 344–362.
Terasima, T. and Tolmach, L. J. (1963b) *Biophys. J.* **3**, 11–33.
Terasima, T. and Yasukawa, M. (1977) *Cryobiology* **14**, 379–381.
Terry, O. W. and Edmunds, L. N. (1970) *Planta* **93**, 106–127.
Theiss-Seuberling, H. B. (1975) *Arch. Microbiol.* **104**, 139–146.
Thilly, W. G. (1976) *In* "Methods in Cell Biology" (D. M. Prescott, Ed.) Vol. 14, 273–285. Academic Press, New York and London.
Thilly. W. G., Nowak, T. S. Jr. and Wogan, G. N. (1974a) *Biotech. Bioeng.* **16**, 149–156.
Thomas, D. B. and Lingwood, C. A. (1975) *Cell* **5**, 37–42.
Thomas, J. O. and Kornberg, R. D. (1975) *Proc. natn. Acad. Sci. USA* **72**, 2626–2630.
Thomas, K. C. and Dawson, P. S. S. (1977) *J. Bact.* **132**, 36–43.
Thomas, K. C., Dawson, P. S. S. and Gamborg, B. L. (1980a) *J. Bact.* **141**, 1–9.
Thomas, K. C., Dawson, P. S. S., Gamborg, B. L. and Steinhuer, L. (1980b) *J. Bact.* **141**, 10–19.
Thompson, L. H. and Humphrey, R. M. (1969) *Int. J. Radiat. Biol.* **15**, 181–184.
Thompson, L. H. and Lindl, P. A. (1976) *Somatic Cell Genet.* **2**, 387–400.
Thompson, L. H., Mankovitz, R., Baker, R. A., Till, J. E., Siminovitch, L., Whitmore, G. F. (1970) *Proc. natn. Acad. Sci. USA* **66**, 377–384.
Thompson, L. H., Mankovitz, R., Baker, R. M., Wright, J. A., Till, J. E., Siminovitch, L. and Whitmore, G. F. (1971) *J. Cell Physiol.* **78**, 431–440.
Thompson, L. H., Lofgren, D. J. and Adair, G. M. (1977) *Cell* **11**, 157–168.
Thorell, B., Chance, B. and Legallais, V. (1965) *J. Cell Biol.* **26**, 741–746.
Throm, E. and Duntze, W. (1970) *J. Bact.* **104**, 1388–1390.
Thuriaux, P., Nurse, P. and Carter, B. (1978) *Mol. gen. Genet.* **161**, 215–220.
Thurston, C. F. and Richards, L. (1980) *J. gen. Microbiol.* **121**, 63–68.
Tien, P. K. (1977) *Rev. Mod. Physiol.* **49**, 361–420.
Tippe-Schindler, R., Zahn, G. and Messer, W. (1979) *Mol. gen. Genet.* **168**, 185–195.
Tischner, R. (1976) *Planta* **132**, 285–290.
Tischner, R. and Lorenzen, H. (1979) *Planta* **146**, 287–292.

Tkacz, J. S. and Lampen, J. O. (1972) *J. gen. Microbiol.* **72**, 243–247.
Tkacz, J. S., Cybulska, E. B. and Lampen, J. O. (1971) *J. Bact.* **105**, 1–5.
Tobey, R. A. (1972) *Cancer Res.* **32**, 2720–2725.
Tobey, R. A. (1973) *In* "Methods in Cell Biology". (D. M. Prescott, Ed.) Vol. 6, 67–112. Academic Press, New York and London.
Tobey, R. A. and Crissman, H. A. (1972) *Cancer Res.* **32**, 2726–2732.
Tobey, R. A. and Crissman, H. A. (1975) *Cancer Res.* **35**, 460–470.
Tobey, R. A. and Ley, K. D. (1970) *J. Cell Biol.* **46**, 151–157.
Tobey, R. A. and Ley, K. D. (1971) *Cancer Res.* **31**, 46–15.
Tobey, R. A., Anderson, E. C. and Petersen, D. F. (1967) *J. Cell Physiol.* **70**, 63–68.
Tobey, R. A., Oka, M. S. and Crissman, H. A. (1975) *Eur. J. Cancer* **11**, 433–441.
Toma, Z. (1980) Ph.D. Thesis, University of Wales.
Tomasovic, S. P., Henle, K. J. and Dethlefsen, L. A. (1979) *Radiat. Res.* **80**, 378–388.
Tomizawa, J.-I. and Selzer, G. (1979) *Ann. Rev. Biochem.* **48**, 999–1034.
Tomizawa, J.-I., Sakakibara, Y. and Kakefuda, T. (1974) *Proc. natn. Acad. Sci. USA* **72**, 1050–1054.
Tomita, K. and Plager, J. E. (1979) *Cancer Res.* **39**, 4407–4411.
Toniolo, D., Meiss, H. K. and Basilico, C. (1973) *Proc. natn. Acad. Sci. USA* **70**, 1273–1277.
Tonnesen, T. and Andersen, H. A. (1977) *Expl. Cell Res.* **106**, 408–412.
Tormo, A., Martínez-Salas, E. and Vicente, M. (1980) *J. Bact.* **141**, 806–813.
Torpier, G., Montagnier, L., Biquard, J.-M. and Vigier, P. (1975) *Proc. natn. Acad. Sci. USA* **72**, 1695–1698.
Torres, J. (1979) *Anat. Rec.* **193**, p. 705.
Torti, S. V. and Park, J. T. (1976) *Nature Lond.* **263**, 323–326.
Torti, S. and Park, J. T. (1980) *J. Bact.* **143**, 1289–1294.
Traganos, F., Evenson, D. P., Staiano-Coico, L., Darzynkiewicz, Z. and Melamed, M. R. (1980a) *Cancer Res.* **40**, 671–681.
Traganos, F., Stainoc L., Darzynkiewicz, Z. and Melamed, M. R. (1980b) *Cancer Res.* **40**, 2390–2399.
Trakht, N. N., Grozdova, I. D., Severin, E. S. and Gnuchev, N. V. (1980) *Biochemistry (USSR)* **45**, 487–493.
Travis, S. L. and Mendelson, N. H. (1977) *Mol. gen. Genet.* **150**, 309–316.
Trenfield, K. and Masters, C. (1980) *Int. J. Biochem.* **11**, 55–67.
Trinci, A. P. J. (1972) *Trans. Br. mycol. Soc.* **58**, 467–473.
Trinci, A. P. J. (1978) *Sci. Prog. Oxf.* **65**, 75–79.
Trinci, A. P. J. (1979) *In* "Fungal Walls and Hyphal Growth". (J. H. Burnett and A. P. J. Trinci, Eds.) 319–338. Cambridge University Press/British Mycological Society.
Trueba, F. J. and Woldringh, E. L. (1980) *J. Bact.* **142**, 869–878.
Tsanev, R. and Russev, G. (1974) *Eur. J. Biochem.* **43**, 257–263.
Tsilimigras, C. W. A. and Gilbert, D. A. (1977) *S. Afr. J. Sci.* **73**, 123–125.
Tsukagoshi, N. and Fox, C. F. (1979) *Agric. biol. Chem.* **43**, 1911–1917.
Tsukagoshi, N., Fielding, P. and Fox, C. F. (1971) *Biochem. biophys. Res. Comm.* **44**, 497–502.
Tucker, R. W., Pardee, A. B. and Fujiwara, K. (1979) *Cell* **17**, 527–535.
Tulp, A., Van der Steen, J. and Barnhoorn, M. G. (1979) *In* "Separation of Cells and Subcellular Elements". (H. Peeters, Ed.) 45–50. Pergamon Press, Oxford.
Turing, A. M. (1952) *Phil. Trans. R. Soc.* **B 237**, 37–72.

Turner, K. J., Gronostajski, R. M. and Schmidt, R. R. (1978) *J. Bact.* **134**, 1013–1019.
Tyson, C. B., Lord, P. G. and Wheals, A. E. (1979) *J. Bact.* **138**, 92–98.
Tyson, J. J. (1976) *In* "The Molecular Basis of Circadian Rhythms". (J. W. Hastings and H.-C. Schweiger, Eds.) 85–108. Dahlem Konferenzen, Berlin.
Tyson, J. J. (1979) *J. theor. Biol.* **80**, 27–38.
Tyson, J. and Kauffman, S. A. (1975) *J. math. Biol.* **1**, 289–310.
Tyson, J. and Sachsenmaier, W. (1978) *J. theor. Biol.* **73**, 723–738.
Tyson, J., Garcia-Herdugo, G. and Sachsenmaier, W. (1979) *Expl. Cell Res.* **119**, 87–98.
Udden, M. M. and Finkelstein, D. B. (1978) *J. Bact.* **133**, 1501–1507.
Ueda, K. and Noguchi, T. (1976) *Protoplasma* **87**, 145–162.
Ueno, A. M., Goldin, E. M., Cox, A. B. and Lett, J. T. (1979) *Radiat. Res.* **79**, 377–389.
Ułaszewski, S., Mamouneas, T., Shen, W.-K., Rosenthal, P. J., Woodward, J. R., Cirillo, V. P. and Edmunds, J. R. Jr. (1979) *J. Bact.* **138**, 523–529.
Umbarger, H. E. (1961) *In* "Control Mechanisms in Cellular Processes". (D. M. Bonner, Ed.) Roland Press, New York.
Unger, M. W. (1977) *J. Bact.* **130**, 11–19.
Unger, M. W. and Hartwell, L. H. (1976) *Proc. natn Acad. Sci. USA* **73**, 1664–1668.
Unitt, M. D. (1980) Ph.D. Thesis, University of Wales.
Unrau, P. and Holliday, R. (1970) *Genet. Res.* **15**, 157–169.
Vaage, R. (1973) Cand.Real Thesis, University of Bergen, Norway.
Vadlamudi, S. and Goldin, A. (1971) *Cancer Chemother. Rep.* **55**, 547–555.
Valleron, A.-J. (Ed.) (1975) "Mathematical Models in Cell Kinetics". European Press, Medikon.
Van Alstyne, O. and Simon, M. I. (1971) *J. Bact.* **108**, 1366–1379.
Van Assel, S. and Steinert, M. (1971) *Expl. Cell Res.* **65**, 353–358.
Vanbuul, P. P. W., Ricordy, R., Spirito, F. and Tates, A. D. (1978) *Mutation Res.* **50**, 377–382.
Van de Putte, P., Van Dillewijn, J. and Rörsch, A. (1964) *Mutation Res.* **1**, 121–128.
Van Wijk, R., Wicks, W. D. and Clay, K. (1972) *Cancer Res.* **32**, 1905–1911.
Vanden Driessche, T. (1975) *Biosystems* **6**, 188–201.
Vanden Driessche, T. (1966) *Expl. Cell. Res.* **42**, 18–30.
Vandevoorde, J. P. and Hansen, H. S. (1970) *Proc. Am. Ass. Cancer Res.* **11**, p. 317.
Van Putten, L. M. and Lelieveld, P. (1976) *In* "Scientific Foundations of Oncology". (T. Symington and R. Carter, Eds.) 136–145. Heinemann, London.
Van Putten, L. M., Keizer, H. J. and Mulder, J. H. (1976) *Eur. J. Cancer* **12**, 79–85.
Van Dilla, M. A., Trujillo, T. T., Mullaney, P. F. and Coulter, J. R. (1969) *Science* **163**, 1213–1214.
Venkov, P. V., Staynov, D. Z. and Hadjiolov, A. A. (1977) *J. Bact.* **129**, 47–51.
Verbin, R. S. and Farber, E. (1975) *In* "Methods in Cell Biology". (D. M. Prescott, Ed.) Vol. 9, 51–69. Academic Press, New York and London.
Vicuna, R., Hurwitz, J., Wallace, S. and Girard, M. (1977a) *J. biol. Chem.* **252**, 2524–2533.
Vicuna, R., Ikeda, J. E. and Hurwitz, J. (1977b) *J. biol. Chem.* **252**, 2534–2544.
Villanueva, V. R., Adlakha, R. C. and Calvayrac, R. (1980) *Phytochem.* **19**, 962–964.
Vogler, W. R., Kremer, W. B., Knospe, W. H., Omura, G. A. and Tornyos, K. (1976) *Cancer Treat. Rep.* **60**, 1845–1859.
Volm, M. (1964) *Z. vergl. Physiol.* **48**, 157–180.

Volm, M., Mattern, J., Weber, N. and Wayss, K. (1980) *Cancer Chemot.* **42**, 89–93.
Volpe, P. and Eremenko, T. (1973) *In* "Methods in Cell Biology". (D. M. Prescott, Ed.) 113–126. Academic Press, New York and London.
Von Bertalanffy, L. (1952) "Problems of Life", Harper, New York.
Von Lenhossek, M. (1898) *Verh. Anat. Ges. Kiel* **12**, 106–128.
Von Meyenburg, H. K. (1968) *Path. Microbiol.* **31**, 117–127.
Von Meyenburg, H. K. (1969) *Arch. Microbiol.* **66**, 289–303.
Voorhees, J. J., Duell, E. A., Bass, L. J. and Harrell, E. R. (1973) *Natn. Cancer Inst. Monogr.* **38**, 47–53.
Vorbrodt, A. and Borun, T. W. (1979) *J. hist. Cytol.* **27**, 1596–1603.
Vos, O., Schenk, H. A. E. M. and Bootsma, D. (1966) *Int. J. Radiat. Biol.* **11**, 495–503.
Vršanská, M., Krátký, Z., Biely, P. and Machala, S. (1979) *Z. allg. Mikrobiol.* **19**, 357–362.
Wain, W. H. and Staatz, W. D. (1973) *Expl. Cell Res.* **81**, 269–278.
Wain, W. H., Price, M. F., Brayton, A. R. and Cawson, R. A. (1976) *J. gen. Microbiol.* **97**, 211–217.
Waldron, C., Jund, R. and Lacrovte, F. (1977) *Biochem. J.* **168**, 409–415.
Walker, E. and Wheatley, D. N. (1979) *J. Cell Physiol.* **99**, 1–13.
Walker, G. M. (1978) Ph.D. Thesis, Heriot-Watt University, Edinburgh.
Walker, G. M. and Duffus, J. H. (1979) *J. gen. Microbiol.* **114**, 391–400.
Walker, G. M. and Duffus, J. (1980) *J. Cell Science* **42**, 329–356.
Walker, G. M. and Zeuthen, E. (1980) *Expl. Cell Res.* **127**, 487–490.
Walker, G. M., Thompson, J. C., Slaughter, J. C. and Duffus, J. H. (1980) *J. gen. Microbiol.* **119**, 543–546.
Walker, J. R. and Pardee, A. B. (1967) *J. Bact.* **93**, 107–114.
Walker, J. R., Kovarik, A., Allen, J. J. and Gustafson, R. A. (1975) *J. Bact.* **123**, 693–703.
Wang, E.-C. and Furth, J. J. (1977) *J. biol. Chem.* **252**, 116–124.
Wang, R. J. (1974) *Nature Lond.* **248**, 76–78.
Wang, R. J. (1976) *Cell* **8**, 257–261.
Wang, R. J. and Yin, L. (1976) *Expl. Cell Res.* **101**, 331–336.
Wang, T., Sheppard, J. R. and Foker, J. E. (1978) *Science* **201**, 155–157.
Ward, C. B. and Glaser, D. A. (1969) *Proc. natn. Acad. Sci. USA* **64**, 905–912.
Warmsley, A. M. H. and Pasternak, C. A. (1970) *Biochem. J.* **119**, 493–499.
Warmsley, A. M. H., Phillips, B. and Pasternak, C. A. (1970) *Biochem. J.* **120**, 683–688.
Warr, J. R. (1968) *J. gen. Microbiol.* **52**, 243–251.
Warr, J. R. and Durber, S. (1971) *Expl. Cell Res.* **64**, 463–469.
Warr, J. R. and Gibbons, D. (1973) *Expl. Cell Res.* **78**, 454–456.
Warr, J. R. and Quinn, D. (1977) *Expl. Cell Res.* **104**, 442–445.
Wanka, F., Joosten, H. F. D. and De Grip, W. J. (1970) *Arch. Mikrobiol.* **75**, 25–36.
Watanabe, M. and Horikawa, M. (1980) *Mutation Res.* **71**, 219–231.
Watanabe, T. and Nakamura, T. (1980) *J. Biochem.* **88**, 815–817.
Watson, C. D. and Berry, D. R. (1977) *FEMS Microbiol. Lett.* **1**, 175–178.
Watson, J. V. and Chambers, S. H. (1979) *Br. J. Cancer* **40**, 315–316.
Watson, J. V., Chambers, S. H., Workman, P. and Horsnell, T. S. (1977) *FEBS Lett.* **81**, 179–182.
Wayss, K., Mattern, J. and Volm, M. (1980) *Strahlen.* **156**, 41–45.

Weber, G. (1976) *In* "Horizons in Biochemistry and Biophysics". Vol. 2, 163–198. Addison-Wesley Publishing Company.
Weeks, D. P. and Collis, P. S. (1979) *Devel. Biol.* **69**, 400–407.
Weeks, D. P., Collis, P. S. and Gealt, M. A. (1977) *J. Cell Biol.* **75**, Abst. CD 196.
Wegmann, K. and Metzner, H. (1971) *Arch. Microbiol.* **78**, 360–367.
Weiner, J. H., Bertsch, L. L. and Kornberg, A. (1975) *J. biol. Chem.* **250**, 1972–1980.
Weinstein, G., Newburger, A., Troner, M. and Colton, A. (1979) *Proc. Am. Ass. Cancer* **20**, 403.
Weis, D. S. (1977) *Trans. Am. microscop. Soc.* **96**, 82–86.
Weisenberg, R. C. (1972) *Science* **77**, 1104–1105.
Weiser, I. and Shalitin, C. (1978) *Expl. Mycol.* **2**, 326–336.
Weiss, B. G. and Tolmach, L. J. (1967) *Biophys. J.* **7**, 779–795.
Weissbach, A. (1977) *Ann. Rev. Biochem.* **46**, 25–47.
Wells, J. R. and James, T. W. (1972) *Expl. Cell Res.* **75**, 465–474.
Wells, R. D., Flügel, R. M., Larson, J. E., Schendel, P. F. and Sweet, R. W. (1972) *Biochemistry* **11**, 621–629.
Westergaard, O., Brutlag, D. and Kornberg, A. (1973) *J. biol. Chem.* **248**, 1361–1364.
Westmacott, D. and Primrose, S. B. (1976) *J. gen. Microbiol.* **94**, 117–125.
Westra, B. and Dewey, W. C. (1971) *Int. J. Radiat. Biol.* **19**, 467–477.
Wetherell, D. F. (1958) *Physiol. Pl.* **11**, 260–274.
Whatley, J. M. (1974) *New Phytol.* **73**, 139–142.
Wheals, A. E. (1970) *Genetics* **66**, 623–633.
Wheals, A. E. (1977) *Nature Lond.* **267**, 647–648.
Wheals, A. E., Grant, W. D. and Jockusch, B. M. (1976) *Mol. gen. Genet.* **149**, 111–114.
Wheatley, D. N. (1976) *In* "Methods in Cell Biology". (D. M. Prescott, Ed.) Vol. 14, 297–317. Academic Press, New York.
Wheeler, G. P., Bowden, B. J., Adamson, D. J. and Vail, M. H. (1970) *Cancer Res.* **30**, 100–111.
Wheeler, G. P., Bowden, B. J., Adamson, D. J. and Vail, M. H. (1972) *Cancer Res.* **32**, 2661–2669.
White, R. A. and Thames, H. D. (1979) *J. theor. Biol.* **77**, 141–160.
Whitfield, J. F., Boynton, A. L., MacManus, J. P., Sikorska, M. and Tsang, B. K. (1979) *Mol. Cell Biochem.* **27**, 155–179.
Whitmore, G. F. and Gulyas, S. (1966) *Science* **151**, 691–694.
Whitmore, G. F., Stanners, C. P., Till, J. E. and Gulyas, S. (1961) *Biochim. biophys. Acta.* **47**, 66–77.
Whitmore, G. F., Gulyas, S. and Botond, J. (1965) *In* "Cellular Radiation Biology". p. 423. Williams and Wilkins, Baltimore Md.
Whitney, R. B. and Sutherland, R. M. (1973) *J. Cell Physiol.* **82**, 9–20.
Whitson, G. L., Padilla, G. M. and Fisher, W. J. (1966) *Expl. Cell Res.* **42**, 438–446.
Whittaker, J. A., Alismail, S. A. and Khurshid, M. (1977) *Lancet* **2**, p. 557.
Whittenbury, R. and Dow, C. S. (1977) *Bact. Rev.* **41**, 754–808.
Wibe, E., Oftenbro, R., Laland, S. G., Pettersen, E. O. and Lindmo, T. (1979) *Br. J. Cancer* **39**, 391–397.
Wickner, R. B. (1976) *Bact. Rev.* **40**, 757–773.
Wickner, S. (1976) *Proc. natn. Acad. Sci. USA* **73**, 3511–3515.
Wickner, S. (1977) *Proc. natn. Acad. Sci. USA* **74**, 2815–2819.

Wickner, S. H. (1978a) *Ann. Rev. Biochem.* **47**, 1163–1191.
Wickner, S. H. (1978b) *Cold Spring Harbor Symp. Quant. Biol.* **43**, 303–310.
Wickner, S. and Hurwitz, J. (1974) *Proc. natn. Acad. Sci. USA* **71**, 4120–4124.
Wickner, S. and Hurwitz, J. (1975a) *Proc. natn. Acad. Sci. USA* **72**, 921–925.
Wickner, S. and Hurwitz, J. (1975b) *Proc. natn. Acad. Sci. USA* **72**, 3342–3346.
Wickner, S. and Hurwitz, J. (1976) *Proc. natn. Acad. Sci. USA* **73**, 1053–1057.
Wickner, W. and Kornberg, A. (1974a) *Proc. natn. Acad. Sci. USA* **71**, 4425–4428.
Wickner, W. and Kornberg, A. (1974b) *J. biol. Chem.* **249**, 6244–6249.
Wickner, W., Brutlag, D., Schekman, R. and Kornberg, A. (1972a) *Proc. natn. Acad. Sci. USA* **69**, 965–969.
Wickner, R. B., Wright, M., Wickner, S. and Hurwitz, J. (1972b) *Proc. natn. Acad. Sci. USA* **69**, 3233–3237.
Wickner, S., Wright, M. and Hurwitz, J. (1973) *Proc. natn. Acad. Sci. USA* **70**, 1613–1618.
Wickner, S., Wright, M. and Hurwitz, J. (1974) *Proc. natn. Acad. Sci. USA* **71**, 783–787.
Wiemken, A., Matile, P. and Moore, H. (1970a) *Arch. Microbiol.* **70**, 89–103.
Wiemken, A., von Meyenburg, H. K. and Matile, Ph. (1970b) *Acta Fac. med. Univ. brun.* **37**, 47–52.
Wijsman, H. J. W. (1972) *Genet. Res.* **20**, 65–74.
Wilkinson, L. E. and Pringle, J. R. (1974) *Expl. Cell Res.* **89**, 175–187.
Wille, J. J. Jr. (1974) *In* "Chronobiology". (L. E. Scheving, F. Halberg, and J. E. Pauley, Eds.) 72–77. Igaku Shoin, Tokyo.
Wille, J. J. (1979) *In* "Biochemistry and Physiology of Protozoa". (M. Levandowsky and S. H. Hutner, Eds.) 2nd end. Vol. 2, 67–149. Academic Press, New York.
Wille, J. J. and Ehret, C. F. (1968) *J. Protozool.* **15**, 785–788.
Wille, J. J., Scheffey, C. and Kauffman, S. A. (1977) *J. Cell Sci.* **27**, 91–104.
Williams, C. A. and Ockey, C. H. (1970) *Expl. Cell Res.* **63**, 365–372.
Williams, E. (1979) M.Sc. Thesis, University of Wales.
Williams, K. L. (1976) *Nature Lond.* **260**, 785–786.
Williams, N. E. and Zeuthen, E. (1966) *C. r. Trav. Lab. Carlsberg.* **35**, 101–130.
Williamson, D. H. (1964a) *In* "Synchrony in Cell Division and Growth". (E. Zeuthen, Ed.) 351–379 and 589–591, Wiley (Interscience), New York.
Williamson, D. H. (1964b) *Biochem. J.* **90**, 25–26P.
Williamson, D. H. (1965) *J. Cell Biol.* **25**, 517–528.
Williamson, D. H. (1966) *In* "Cell Synchrony. Studies in Biosynthetic Regulation". (I. L. Cameron and G. Padilla, Eds.) 81–101. Academic Press, New York.
Williamson, D. H. (1970) *Symp. Soc. exp. Biol.* **24**, 247–276.
Williamson, D. H. (1973) *Biochem. biophys. Res. Commun.* **52**, 731–740.
Williamson, D. H. (1974) *In* "Cell Cycle Controls". (G. M. Padilla, I. L. Cameron and A. Zimmerman, Eds.) 143–152. Academic Press, London and New York.
Williamson, D. H. and Scopes, A. W. (1960) *Expl. Cell Res.* **20**, 338–349.
Williamson, D. H. and Scopes, A. W. (1961) *Symp. Soc. J. gen. Microbiol.* **11**, 217–242.
Williamson, D. H. and Scopes, A. W. (1961b) *J. Inst. Brew.* **67**, 39–42.
Williamson, D. H. and Scopes, A. W. (1962) *Nature Lond.* **193**, 256–257.
Williamson, D. H. and Moustacchi, E. (1971) *Biochem. biophys. Res. Commun.* **42**, 195–201.
Willingham, M. C., Johnson, G. S. and Pastan, I. (1972) *Biochem. biophys. Res. Commun.* **48**, 743–748.

Willison, J. H. M. and Johnston, G. C. (1978) *J. Bact.* **136**, 318–323.
Wilson, A. T. and Calvin, M. (1955) *J. Am. Chem. Soc.* **77**, 5948.
Wilson, B. W. and James, T. W. (1966) *In* "Cell Synchrony". (I. L. Cameron and G. M. Padilla, Eds.) p. 236. Academic Press, New York and London.
Wilson, G. and Fox, C. F. (1971) *Biochem. biophys. Res. Commun.* **44**, 503–509.
Wilson, D. F., Owen, C. S. and Holian, A. (1977) *Arch Biochem. Biophys.* **182**, 749–762.
Wilson, R. and Chiang, K. S. (1977) *J. Cell Biol.* **72**, 470–481.
Winfree, A. T. (1967) *J. theor. Biol.* **16**, 15–42.
Winfree, A. T. (1972) *Arch. Biochem. Biophys.* **149**, 388–401.
Winfree, A. T. (1975) *Nature Lond.* **253**, 315–319.
Winfree, A. T. (1976) *In* "The Molecular Basis of Circadian Rhythms". (J. W. Hastings and H.-G. Schweiger, Eds.) 109–129. Dahlem Konferenzen, Berlin.
Winfree, A. T. (1980) "The Geometry of Biological Time". Springer-Verlag, Berlin.
Wintersberger, E. (1978) *Rev. Physiol. biochem. Pharmacol.* **84**, 93–142.
Wintersberger, U., Binder, M. and Fischer, P. (1975) *Mol. gen. Genet.* **142**, 13–17.
Wintersberger, U., Hirsch, J. and Fink, A. M. (1974) *Mol. gen. Genet.* **131**, 291–299.
Wintzerith, M., Wittendrop, E., Ittel, M. E., Rechenmann, R. V. and Mandel, P. (1975) *Expl. Cell Res.* **91**, 279–284.
Wise, G. E. and Prescott, D. M. (1973) *Proc. natn. Acad. Sci USA* **70**, 714–717.
Wissinger, W. and Wang, R. J. (1978) *Expl. Cell Res.* **112**, 82–94.
Witkin, E. M. (1975) *Mol. gen. Genet.* **142**, 87–103.
Wittes, R. E. and Ozer, H. L. (1973) *Exp. Cell Res.* **80**, 127–136.
Wolfe, J. (1973) *Expl. Cell Res.* **77**, 232–238.
Wolfe, J. (1976) *Dev. Biol.* **54**, 116–126.
Wolff, D. A. and Pertoft, H. (1972) *J. Cell Biol.* **55**, 579–585.
Wolff, S. (1968) *J. Cell comp. Physiol.* **58** Suppl., 151–162.
Woffendin, C. and Griffiths, A. J. (1981) *J. gen. Microbiol.* (Submitted in press).
Wolfner, M., Yep, D., Messenguy, F. and Fink, G. R. (1975) *J. mol. Biol.* **96**, 273–290.
Wolfson, J. and Dressler, D. (1972) *Proc. natn. Acad. Sci. USA* **69**, 2682–2686.
Wolosker, H. B. M. and de Almeida, D. F. (1975) *J. gen. Microbiol.* **88**, 381–383.
Wolosker, H. B. M. and de Almeida, D. F. (1979) *J. gen. Microbiol.* **110**, 225–227.
Wolpert, L. (1969) *J. theoret. Biol.* **25**, 1–47.
Wolpert, L. and O'Neill, C. H. (1962) *Nature Lond.* **196**, 1261–1266.
Wolpert-Defilippes, M. K., Adamson, R. H., Cysyk, R. L. and Johns, D. G. (1975) *Biochem. Pharmacol.* **24**, 751–754.
Wolstenholme, D. R., Koike, K. and Cochranfouts, P. (1973) *J. Cell Biol.* **56**, 230–245.
Woo, K. B., Brenkus, L. B. and Wiig, K. M. (1975) *Cancer Chemother. Rep.* **59**, 847–860.
Woodard, J., Gelber, B. and Swift, H. (1961) *Expl. Cell Res.* **23**, 258–264.
Woodcock, C. L. F. (1973) *J. Cell Biol.* **59**, 368a.
Woodward, J., Rasch, E. and Swift, H. (1961) *J. Biophys. biochem. Cytol.* **9**, 445–462.
Worthington, D. A. and Nachtwey, D. S. (1976) *Cell Tiss. Kinet.* **9**, 469–478.
Worthington, D. A., Salamone, M. and Nachtwey, D. S. (1976) *Cell Tiss. Kinet.* **9**, 119–130.
Wraight, C. A., Lucking, D. R., Fraley, R. T. and Kaplan, S. (1978) *J. biol. Chem.* **253**, 465–471.

Wright, M. and Tollon, Y. (1979) *Eur. J. Biochem.* **96**, 177–181.
Wright, M., Tollon, Y., Moisand, A., Oustrin, M. L. and DelCastillo, L. (1976) Abstr., 3rd European Physarum Workshop, Bern, Switzerland.
Wright, M., DelCastillo, L. and Oustrin, M. L. (1980) *Current Genetics* **1**, 203–209.
Wu, P. C. and Pardee, A. B. (1973) *J. Bact.* **114**, 603–611.
Wunderlich, F. and Peyk, D. (1969) *Expl. Cell Res.* **57**, 142–144.
Wunderlich, F., Berezney, R. and Kleinig, H. (1976) in Biological Membranes (Chapman, O. and Wallach, D. F. H., Eds). pp. 241–333, Academic Press, London.
Wurster, B. (1976) *Nature Lond.* **260**, 703–704.
Xeros, N. (1962) *Nature Lond.* **194**, 682.
Yamada, K. and Ito, M. (1979) *Pl. Cell Physiol.* **20**, 1471–1479.
Yamada, M. A. and Puck, T. T. (1961) *Proc. natn. Acad. Sci. USA* **47**, 1181–1191.
Yamaguchi, H., Ishiguro, S., Oka, Y. and Miyamoto, H. (1977) *Cell Struct.* **2**, 111–118.
Yamaguchi, K. and Yoshikawa, H. (1973) *Nature Lond. New Biol.* **244**, 204–206.
Yamaguchi, K. and Yoshikawa, H. (1977) *J. mol. Biol.* **110**, 219–233.
Yamamoto, I., Anraku, Y. and Ohki, M. (1979) *J. biol. Chem.* **264**, 8584–8589.
Yamamoto, M. C. (1980) *Mol. gen. Genet.* **180**, 231–284.
Yanagi, K., Talavera, A., Nishimoto, T. and Rush, M. G. (1978) *J. Virol.* **25**, 42–50.
Yarygin, K. N., Nechaeva, N. V., Fateeva, V. I., Trushina, E. D. and Brodsky, V. Ya. (1978) *Bull. Expl. Biol. Med.* **86**, 726–727.
Yarygin, K. N., Nechaeva, N. V., Fateeva, V. I., Novikova, T. E. and Brodsky, V. Ya. (1979) *Bull. exp. Biol. Med.* **88**, 711–712.
Yashphe, J. and Halvorson, H. O. (1976) *Science* **191**, 1283–1284.
Yates, I., Darley, W. and Kochert, G. (1975) *Cytobios.* **12**, 211–223.
Yen, A. and Pardee, A. B. (1977) *Science* **204**, 1315–1317.
Young, V. R., Steffel, W. P., Pencharz, P. B., Winterer, J. C. and Scrimshaw, N. S. (1975) *Nature Lond.* **253**, 192–193.
Yu, C. K. and Sinclair, W. K. (1967) *J. natn. Cancer Inst.* **39**, 619–632.
Yuyama, S. (1975) *Expl. Cell Res.* **90**, 381–391.
Zada-Hames, I. M. and Ashworth, J. M. (1978) *J. Cell Sci.* **32**, 1–20.
Zaikin, A. N. and Zhabotinsky, A. M. (1970) *Nature Lond.* **225**, 535–537.
Zajicek, G., Michaeli, Y. and Regev, J. (1979) *Cell Tiss. Kinet.* **12**, 229–237.
Zalkinder, V. (1979) *Biosystems* **11**, 295–307.
Zamb, T. J. and Roth, R. (1977) *Proc. natn. Acad. Sci. USA* **74**, 3951–3955.
Zaritsky, A. (1975) *J. theoret. Biol.* **54**, 243–248.
Zaritsky, A. and Pritchard, R. H. (1973) *J. Bact.* **114**, 824–837.
Zaritsky, A. and Woldringh, C. C. (1978) *J. Bact.* **135**, 581–587.
Zeilig, C. E. and Goldberg, N. D. (1977) *Proc. natn. Acad. Sci. USA* **74**, 1052–1056.
Zeilig, C. E., Johnson, R. A., Sitherland, E. W. and Friedman, D. L. (1976) *J. Cell Biol.* **71**, 515–534.
Zetterberg, A. (1966) *Expl. Cell Res.* **43**, 517–525.
Zetterberg, A. and Killander, D. (1965a) *Expl. Cell Res.* **39**, 22–32.
Zetterberg, A. and Killander, D. (1965b) *Expl. Cell Res.* **40**, 1–11.
Zeuthen, E. (1953) *J. embryol. exp. Morph.* **1**, 239–249.
Zeuthen, E. (1964) *In* "Synchrony in Cell Division and Growth". (E. Zeuthen, Ed.) 99–158. Wiley (Interscience), New York.
Zeuthen, E. (1971) *Expl. Cell Res.* **68**, 49–60.
Zeuthen, E. (1974) *In* "Cell Cycle Controls". (G. M. Padilla, I. L. Cameron and Z. Zimmerman, Eds.) 1–30. Academic Press, New York and London.

Zeuthen, E. (1978) *Expl. Cell Res.* **116**, 39–46.
Zeuthen, E. and Rasmussen, L. (1972) *In* "Research in Protozoology". (T. T. Chen, Ed.) Vol. 4, 9–145. Pergamon Press, New York, Oxford.
Zeuthen, E. and Williams, N. E. (1969) *In* "Nucleic Acid Metabolism, Cell Differentiation and Cancer Growth". (E. V. Cowdry and S. Seno, Eds.) 203–216. Pergamon Press, Oxford.
Zielke, H. R. and Littlefield J. W. (1974) *In* "Methods in Cell Biology". (D. M. Prescott, Ed.) Vol. 8, 107–121. Academic Press, New York.
Zieve, G. W., Turnbull, D., Mullins, J. M. and McIntosh, J. R. (1980) *Expl. Cell Res.* **126**, 397–405.
Zimmerman, A. M. and Laurence, H. L. (1975) *Expl. Cell Res.* **90**, 119–136.
Zingales, B. and Colli, W. (1977) *Biochim. biophys. Acta* **474**, 562–577.
Zuchowski, C. and Pierucci, O. (1978) *J. Bact.* **33**, 1533–1535.
Zucker, R. M., D'Alisa, R. M. and Gershey, E. L. (1979a) *Exp. Cell Res.* **122**, 1–8.
Zucker, R. M., Tershakovec, A., D'Alisa, R. M. and Gershey, E. L. (1979b) *Expl. Cell Res.* **122**, 15–22.
Zucker, R. M., Wu, N. C., Krishan, A. and Silverman, M. (1979c) *Expl. Cell Res.* **123**, 383–387.
Zurkowski, W. and Lorkiewicz, Z. (1977) *Mol. gen. Genet.* **156**, 215–217.
Zusman, D. and Rosenberg, E. (1971) *J. Bact.* **105**, 801–810.
Zusman, D. R., Inouye, M. and Pardee, A. B. (1972) *J. mol. Biol.* **69**, 119–136.
Zusman, D. R., Krotski, D. M. and Cumsky, M. (1978) *J. Bact.* **133**, 122–129.
Zylber, E. N. and Penman, S. (1971) *Science* **172**, 947–949.
Zywietz, F. and Jung, H. (1980) *Eur. J. Cancer* **16**, 1381–1388.

Index